Birkhäuser

Progress in Probability

Volume 78

Progress in Probability is designed for the publication of workshops, seminars and conference proceedings on all aspects of probability theory and stochastic processes, as well as their connections with and applications to other areas such as mathematical statistics and statistical physics.

All manuscripts are peer-reviewed to meet the highest standards of scientific literature.

- Series editors provide their final decision for inclusion in the series after manuscripts have been fully peer-reviewed;
- For contributed volumes, volume editors conduct the anonymous peer-review process before final submission;
- The peer-review process is anonymous and confidential in order to ensure reviewers provide impartial opinions

Interested authors may submit proposals by email to the series editors or to the relevant Birkhäuser editor listed under "Contacts."

More information about this series at http://www.springer.com/series/4839

Loïc Chaumont • Andreas E. Kyprianou
Editors

A Lifetime of Excursions Through Random Walks and Lévy Processes

A Volume in Honour of Ron Doney's 80th Birthday

Birkhäuser

Editors
Loïc Chaumont
LAREMA
Université d'Angers
Angers, France

Andreas E. Kyprianou
Department of Mathematical Sciences
University of Bath
Bath, UK

ISSN 1050-6977 ISSN 2297-0428 (electronic)
Progress in Probability
ISBN 978-3-030-83308-4 ISBN 978-3-030-83309-1 (eBook)
https://doi.org/10.1007/978-3-030-83309-1

Mathematics Subject Classification: 60J65, 60H05, 60F05, 60E07, 60G70, 62G32, 60E15, 60G51, 60G18, 33C20; Primary: 60G50; Secondary: 60G40, 60F17, 60G55, 60H25, 60J10, 60J80, 60E10, 60E10

This book is published under the imprint Birkhäuser, www.birkhauser-science.com, by the registered company Springer Nature Switzerland AG.
The registered company address is: Gewerbestrasse 11, 6330 Cham, Switzerland

Contents

A Lifetime of Excursions Through Random Walks and Lévy Processes

Loïc Chaumont and Andreas E. Kyprianou

Abstract We recall the many highlights of Professor Ron Doney's career summarising his main contributions to the theory of random walks and Lévy processes.

Keywords Ron Doney · Random walks · Lévy processes

Through this volume, it is with the greatest of admiration that we pay tribute to the mathematical achievements of Professor Ron Doney. His career has spanned generations of probabilists and his work continues to play a significant role in the community. In addition to the major contributions he has made in the theory of random walks and Lévy processes, Ron is equally appreciated for the support he has given to younger colleagues. The sentiment and desire to organise both a workshop and this volume to honour his lifetime achievements surfaced naturally a couple of years ago at the Lévy process meeting in Samos, as it became apparent that Ron was approaching his 80th birthday. A huge appreciation of his standing in the community meant it was very easy to find willing participation in both projects. As a prelude to this *hommage à Ron*, let us spend a little time reflecting on his career and his main achievements.

Ron grew up in working-class Salford in the North West of Greater Manchester at a time when few leaving school would attend university. Having a veracious appetite for reading, Ron spent long hours as a schoolboy in Manchester Central Library where he cultivated his intellect. Coupled with an obvious talent for mathematics, he found his way to the University of Durham. There he studied Mathematics as an undergraduate and continued all the way to a PhD under the supervision of Harry

L. Chaumont (✉)
LAREMA – UMR CNRS 6093, Université d'Angers, Angers cedex, France
e-mail: loic.chaumont@univ-angers.fr

A. E. Kyprianou
Department of Mathematical Sciences, University of Bath, Claverton Down, Bath, UK
e-mail: a.kyprianou@bath.ac.uk

L. Chaumont, A. E. Kyprianou (eds.), *A Lifetime of Excursions Through Random Walks and Lévy Processes*, Progress in Probability 78,
https://doi.org/10.1007/978-3-030-83309-1_1

Reuter, graduating in 1964 aged just 24. Ron's thesis entitled *'Some problems on random walks'* was no doubt inspired by the shift in interests of Harry Reuter from analysis to probability. At the time of his PhD, all doctoral activity in probability theory in the UK was essentially supervised either by Harry Reuter or David Kendall. Because of this, Reuter and Kendall formed StAG, the *Stochastic Analysis Group*, which would meet regularly under the auspices of the London Mathematical Society. From a very young age, Ron thus had the opportunity to engage with his contemporaries. Back in the 1960s they were precious few in number compared with the vast numbers that make up the probability PhD community today in the UK, and included the likes of Rollo Davidson, David Williams, Daryl Daley, David Vere-Jones, Nick Bingham, John Hawkes and John Kingman.

The time that Ron completed his PhD coincided with a period of expansion in the UK higher education system which proved to offer numerous opportunities during the first years of his academic career. From his PhD, Ron was successfully appointed directly to a lectureship at the University of East Anglia. He spent the academic year 1964–1965 there, but quickly moved on to lectureship at Imperial College London in 1965, coinciding with the appointment of Harry Reuter to a chair there. During this early phase, Ron had a slow start to his publication record. His first two papers [1, 2] concern random walks in three dimensions, followed by a paper concerning higher dimensional version of the renewal theory [3].

By 1970, Ron had moved back to his native Manchester. He joined the then world-famous Manchester-Sheffield School of Probability and Statistics, formally taking up a lectureship at Manchester's Statistical Laboratory. At this point in time he moved away from his initial work on random walks to the theory of Galton-Watson processes and, what were then called, general branching processes (today they would rather come under the heading of Crump-Mode-Jagers (CMJ) processes). Although seemingly a change in direction, this was a very natural move for anyone who harboured interests in random walks and renewal processes. Indeed, whilst the theory of branching processes has become significantly more exotic in recent years, the interplay of these two fields still remains highly pertinent today. Concurrently with the work of Peter Jagers, Ron produced a cluster of articles through to the mid 1970s looking at growth properties of CMJ processes in which, among other things, he demonstrated the central role that renewal processes play; [4–12]. As early as 1972, one finds computations in his work which echo what would later become known as the method of *spines*; [6]. Here, Ron also made contributions to underlying functional equations and the so-called $x \log x$ condition, which precede a number of similar results in the setting of more general spatial models such as branching random walks. The papers, [8, 9] with Nick Bingham, are also interesting to reflect upon in terms of how collaborations of the day were conducted. In a period of no internet or email, and with the probability community in the UK being very few in number and thinly spread, Ron maintained communication with Nick through hand-written and mailed letters during the epoch of their overlapping interests. In a process that has largely been replaced by googling, this involved sharing sample calculations, summaries of articles they had found and broader mathematical ideas. In the case of Ron and Nick, this led not only two these two papers, but calculations that lay dormant, surfacing over a decade later in [13].

Ron Doney in 1972, Manchester

By 1977, Ron was back to random walks, albeit now in one dimension. As a passive observer, it appears that there was an awakening in Ron's understanding of how many problems still remained open for general one-dimensional random walks, particularly when looking at them in terms of excursions. Aside from some works on Markov chains and Brownian motion, [14–16], Ron's work focused mostly on random walks through the 1980s. The prominence of his contributions lay with his use of Spitzer's condition as well as the problem of characterising the moments and tail behaviour of ladder variables in relation to assumptions on the underlying random walk; [17–24]. He began a growing interest in arcsine laws and random walks in the domain of attraction of stable laws as well as stable processes themselves; [13, 25–27]. Most notable of the latter is a remarkable paper on the Wiener-Hopf factorisation of stable processes [27], the significance of which would become apparent many years later after 2010 thanks to continued work of a number of authors, most notably Alexey Kuznetsov. It was also during the 1980s that Ron took two sabbaticals in Canada. The first stay was in Vancouver in 1980–1981, visiting Cindy Greenwood, which also allowed him the chance to connect with Sidney Port, Ed Perkins and John Walsh. For the second, he visited George O'Brien in York University, Toronto in 1988–1989.

Moving into the 1990s, many things changed for Ron's research, least of all, his rate of publication, [28–51]. The previous work he had done on ladder heights and Spitzer's condition culminated in one of Ron's most important and widely appreciated results: For random walks (and shortly after for Lévy processes), he proved that Spitzer's condition was equivalent to the convergence of the positivity probability, [40, 50]. Ron found himself catapulted into a rapidly growing and much better organised global community of probabilists with shared interests to his own. Although he largely continued publishing in the context of random walks, it was during this decade that Ron became increasingly exposed to the theory of Lévy processes.

The 1990s was also the decade that saw Ron begin to jointly co-author many more of his papers. There were a number of factors at play here. During the first twenty five years of Ron's career it was quite common for academics to

Ron Doney's departmental photo as it appeared on the entrance to the Mathematics Tower in the 1990s

author papers alone, least of all because, within the field of probability theory, probabilists were few and far between. The 1990s was the decade of globalisation and the prevalence of email and internet made international connectivity and communication much easier. But perhaps most importantly in this respect was Ron's collaboration with a young French mathematician by the name of Jean Bertoin, which dramatically opened up his relationship with the Parisian (and more generally the French) school of probability for the rest of his career.

In the early 1990s, Ron had made contact with Jean Bertoin, who was very interested in random walks, but also in capturing and expanding on the less-well explored theory of Lévy processes. It was Ron's first paper on Lévy processes, [30], which had caught the interest of Jean and they began sending each other preprints. They first met in person in Luminy during one of the *Journées de Probabilités* in 1992 organised by Azéma and Yor. At the time, Ron was interested in a paper by Keener on the simple RW conditioned to stay positive that had been published that year. Although others had written on the topic of conditioned random walks, it was the joint work of Jean and Ron [39] that ensued, which formalised a robust approach to the notion of such conditionings and led the way for a number of articles on conditioned processes, particularly in the setting of Lévy processes. Reflecting on earlier remarks about how the theory of random walks and branching processes are so closely intertwined, it is worth noting that the paper [39] ended up playing a hugely influential role in the theory of branching random walks, branching Brownian motion and (growth) fragmentation processes, where conditioning of the *spine* in the spirit of their work proved to be instrumental in understanding the so-called *derivative martingale*.

The collaboration between Jean and Ron was relatively intensive for a period of 3–4 years with six of their seven co-authored papers, [36–39, 41, 44, 50], appearing between 1994 and 1997. During this period, Jean would visit Ron regularly, often working at his home in the village of Whaley Bridge, South East of Manchester. Their daily working routine would be interlaced with regular hikes out into the Peak District which lies just beyond Ron's back door.

Through his collaboration with Jean Bertoin, Ron also started to visit Paris more often. There, he became acquainted with a new generation of young probabilists who were being guided towards the theory of Lévy processes. Loïc Chaumont was the first PhD student of Jean who also became a long-time collaborator of Ron. Their collaboration spans 8 papers to date, [51–58], in which they cover the study of perturbed Brownian motion, distributional decompositions of the general Wiener-Hopf factorisation and Lévy processes conditioned to stay positive. It is perhaps the latter, [55, 56], for which they are best known as co-authors, building on the PhD thesis of Loïc for conditioned Lévy processes that had, in turn, grown out of the formalisation for conditioned random walks that Ron had undertaken with Jean Bertoin. It was also during the mid 1990s that Larbi Alili, a contemporary of Loïc and PhD student of Marc Yor, became the postdoc of Ron on a competitively funded EPSRC project; cf. [49, 54, 59]. Philippe Marchal, another gifted young probabilist from the Paris school, was also a regular visitor to Manchester during this period. Another young French probabilist whose work greatly impressed Ron at the turn of the Millennium was Vincent Vigon. Aside from Ron's admiration of Vincent's unexpected emergence from Rouen rather than Paris, what impressed him the most was that Vigon had established a necessary and sufficient condition, in the form of an explicit integral test, for when a Lévy process of unbounded variation with no Gaussian component creeps; in particular, this result showed that Ron's previous conjecture on this matter, which had been assimilated from the long-term behaviour of random walks, was wrong.

The new Millennium brought about yet further change for Ron. His publication rate went up yet another gear, with almost as many articles published during this decade as in the previous two. This was all thanks to his increased exposure to collaborative partnership as well as the inevitable depth of understanding of random walks and Lévy processes he had acquired; [44, 52–56, 59–84]. It was also during this decade that Ron began an extremely fruitful collaboration with Ross Maller, publishing 10 papers together; [44, 63, 64, 68, 70, 75, 76, 79, 81, 83, 85]. On a visit to the UK, Ross was advised by Charles Goldie to go and spend time in Manchester visiting Ron. He did and they immediately started producing material. In a number of important papers [64, 68, 83], Ross and Ron first investigated the asymptotic behaviour of random walks and Lévy processes at deterministic times and at first passage times across a fixed level. In another series of remarkable works [44, 76, 79, 81] they considered first passage times across power law boundaries of random walks, Lévy processes and their reflected version at the infimum. In particular they obtained necessary and sufficient conditions for these first passage times to have finite moments.

By now, there was widespread renewed interest in the theory of fluctuations for Lévy processes and the study of their overshoots had become very popular. Two articles by Ron, written with Phil Griffin [65, 71] bear witness to this. It was also during this time that Ron collaborated with Andreas Kyprianou and wrote one of his most cited articles on the so-called *quintuple law*, [77]. The latter gives a distributional identity for a suite of five important and commonly used path functionals of a Lévy process at first passage over a fixed level in terms of its ladder

Taken on the 24th July 2009 at the 8th International Conference on Lévy processes, held in Angers, Ron Doney sits with Marc Yor to celebrate Marc's 60th birthday

potentials and Lévy measure. In essence, the result constitutes a 'disintegration' of the Wiener–Hopf factorisation. Ron also began a fruitful collaboration with Mladen Savov during this period; [83, 86, 87]. Mladen, who came to Manchester from Bulgaria, proved to be the most accomplished of Ron's several PhD students in the field of probability theory.

A very important event of this decade for Ron was his invitation to give a lecture at the famous Saint-Flour summer school in 2005. It was arguably the first major recognition of Ron's career by the mathematical community. For Ron, this also presented the opportunity to write a book on fluctuation theory for Lévy processes, [80], which remains an important reference in the domain to date. Ron found himself centre-stage as part of a huge community of researchers now working specifically in the field of random walks and Lévy processes. He spoke at many venues, including a rapid succession of workshops and congresses devoted to Lévy processes and was elected as a Fellow of the Institute of Mathematical Statistics in 2006. Moreover, in 2005, Manchester hosted the 4th international workshop on Lévy processes. This was a huge undertaking given the large number of attendees, but nonetheless an important moment that asserted the importance of Manchester's probability group and, in particular, Ron's identity as a highly accomplished researcher in this field.

The next decade, 2010–2020, saw an invitation by the Bernoulli Society for Ron to give a prestigious plenary lecture at the 2013 Stochastic Processes and their Applications congress in Boulder, Colorado. Together with his lectures at Saint Flour, this stands as quite an important recognition of his achievements from the international mathematical community. Technically speaking, Ron retired in 2006, however, as a special case given the quality of his research output, Manchester had gladly continued his appointment beyond his 67th birthday, right through until 2014, on a purely research basis. Now in his seventies, one might say Ron worked at a slower rate, however, although publishing in lower volume, one sees remarkable crowning quality in his work with an exceptional number of his papers appearing in the top two probability journals *Annals of Probability* and *Probability Theory and Related Fields*; [57, 58, 85–99].

These works are marked by two projects in which Ron revisited old obsessions. The first is a cluster of papers, co-authored with Víctor Rivero. Ron went back to the first passage times of Lévy processes and obtained sharp results for the local behaviour of their distribution, [93, 95, 96]. The second concerns improving the classical Gnedenko and Stone local limit theorems. In collaboration with Francesco Caravenna [98], Ron obtained necessary and sufficient conditions for random walks in the domain of attraction of a stable law to satisfy the strong renewal theorem. This work, as well as [99], solve a long-standing problem which dates back to the 1960s. This second achievement is the one that Ron himself quietly admits he is most proud of, and rightly so.

As Ron's career wound down, he had the pleasure of watching the probability group in Manchester dramatically grow in size. With 10 members around the time of his final retirement, this was something he had dreamed of for many years. Ron was appointed in Manchester at a time that it stood as a global stronghold for probability and statistics. The Manchester Statistical Laboratory was half of the very unique two-university Manchester-Sheffield School of Probability and Statistics, the brain child of Joe Gani. With changing times the Manchester-Sheffield School disbanded and, aside from Fredos Papangelou, who joined in 1973, Ron was the only other probabilist who remained in the Statistical Laboratory for a number of years to come. The 1980s were hard times for British academia, but the growth in the international community around Ron's research interests through the 1990s was mirrored by the growth of the probability group in Manchester. The turn of the Millennium saw the merger of the University of Manchester (more formally, the Victoria University of Manchester) with UMIST (University of Manchester Institute of Science and Technology), which opened the door to new opportunities. Many of the probability appointments in Manchester since then clearly reflect the strong association of Manchester with the theory of random walks and Lévy processes; something that is directly tethered to Ron's towering achievements as a researcher.

As alluded to at the start of this article, Ron is appreciated as much for his encouragement of young researchers as he is for the mathematics that he has produced. Just as the authors of this article see Ron as one of the major influencing characters in their own careers, both through academic mentorship and mathematical discourse, so do many others among our community, both in the UK and around the globe. There are simply too many to list here, moreover, an attempt to do so would carry the risk that we forget names. But it should be said that, when the idea of holding a workshop for Ron's 80th birthday surfaced, this was carried forward with gusto by an emotional surge of support from the many who belong to the aforementioned list.

As a humble researcher who cares little for the limelight, Ron did not always get the honours he deserved. Towards his retirement, the number of researchers in probability theory exploded exponentially and Ron's contribution to a classical field, which now lies in the DNA of many modern research endeavours, is often overlooked. One goal of this volume and the accompanying conference is to try to correct this.

We hope the contents of this volume will stimulate Ron to think about his next piece of work. He currently has 99 publications and we are all keenly awaiting his 100th!

References

1. Doney, R.A.: Recurrent and transient sets for 3-dimensional random walks. Z. Wahrscheinlichkeitstheorie und Verw. Gebiete **4**, 253–259 (1965)
2. Doney, R.A.: Hits on an axis by the simple random walk in three dimensions. Z. Wahrscheinlichkeitstheorie und Verw. Gebiete **5**, 146–155 (1966)
3. Doney, R.A.: An analogue of the renewal theorem in higher dimensions. Proc. Lond. Math. Soc. (3) **16**, 669–684 (1966)
4. Doney, R.A.: The progeny of a branching process. J. Appl. Probab. **8**, 407–412 (1971)
5. Doney, R.A.: Age-dependent birth and death processes. Z. Wahrscheinlichkeitstheorie und Verw. Gebiete **22**, 69–90 (1972)
6. Doney, R.A.: A limit theorem for a class of supercritical branching processes. J. Appl. Probab. **9**, 707–724 (1972)
7. Doney, R.A.: On a functional equation for general branching processes. In: J. Appl. Probab. **10**, 198–205 (1973)
8. Bingham, N.H., Doney, R.A.: Asymptotic properties of supercritical branching processes. I. The Galton-Watson process. Adv. Appl. Probab. **6**, 711–731 (1974)
9. Bingham, N.H., Doney, R.A.: Asymptotic properties of supercritical branching processes. II. Crump-Mode and Jirina processes. Adv. Appl. Probab. **7**, 66–82 (1975)
10. Doney, R.A.: On single- and multi-type general age-dependent branching processes. J. Appl. Probab. **13**(2), 239–246 (1976)
11. Doney, R.A.: A note on the subcritical generalized age-dependent branching process. J. Appl. Probab. **13**(4), 798–803 (1976)
12. Doney, R.A.: A note on a condition satisfied by certain random walks. J. Appl. Probab. **14**(4), 843–849 (1977)
13. Bingham, N.H., Doney, R.A.: On higher-dimensional analogues of the arc-sine law. J. Appl. Probab. **25**(1), 120–131 (1988)
14. Doney, R.A.: Letter to the editor: "On conditional passage time structure of birthdeath processes" [J. Appl. Probab. **21**(1), 10–21 (1984); MR0732667 (85k:60119)] by U. Sumita". J. Appl. Probab. **21**(3), 673–674 (1984)
15. Doney, R.A.: A note on some results of Schuh. J. Appl. Probab. **21**(1), 192–196 (1984)
16. Doney, R.A., Grey, D.R.: Some remarks on Brownian motion with drift. J. Appl. Probab. **26**(3), 659–663 (1989)
17. Doney, R.A.: Moments of ladder heights in random walks. J. Appl. Probab. **17**(1), 248–252 (1980)
18. Doney, R.A.: Spitzer's condition for asymptotically symmetric random walk. J. Appl. Probab. **17**(3), 856–859 (1980)
19. Doney, R.A.: On the exact asymptotic behaviour of the distribution of ladder epochs. Stoch. Process. Appl. **12**(2), 203–214 (1982)
20. Doney, R.A.: On the existence of the mean ladder height for random walk. Z. Wahrsch. Verw. Gebiete **59**(3), 373–382 (1982)

21. Doney, R.A.: A note on conditioned random walk. J. Appl. Probab. **20**(2), 409–412 (1983)
22. Doney, R.A.: On the asymptotic behaviour of first passage times for transient random walk. Probab. Theory Relat. Fields **81**(2), 239–246 (1989)
23. Doney, R.A.: A large deviation local limit theorem. Math. Proc. Camb. Philos. Soc. **105**(3), 575–577 (1989)
24. Doney, R.A.: Last exit times for random walks. Stoch. Process. Appl. **31**(2), 321–331 (1989)
25. Doney, R.A.: Conditional limit theorems for asymptotically stable random walks. Z. Wahrsch. Verw. Gebiete **70**(3), 351–360 (1985)
26. Doney, R.A.: On the maxima of random walks and stable processes and the arc-sine law. Bull. Lond. Math. Soc. **19**(2), 177–182 (1987)
27. Doney, R.A.: On Wiener-Hopf factorisation and the distribution of extrema for certain stable processes. Ann. Probab. **15**(4), 1352–1362 (1987)
28. Doney, R.A.: A bivariate local limit theorem. J. Multivariate Anal. **36**(1), 95–102 (1991)
29. Doney, R.A., O'Brien, G.L.: Loud shot noise. Ann. Appl. Probab. **1**(1), 88–103 (1991)
30. Doney, R.A.: Hitting probabilities for spectrally positive Lévy processes. J. Lond. Math. Soc. (2) **44**(3), 566–576 (1991)
31. Doney, R.A., Goldie, C.M.: Letter to the editor: "Autoregressive processes and first-hit probabilities for randomized random walks" [J. Appl. Probab. **28**(2), 347–359 (1991); MR1104571 (93a:60100)] by R. Epstein-Feldman". J. Appl. Probab. **29**(1), 244 (1992)
32. Doney, R.A., O'Brien, G.L.: Thickened renewal processes. Stoch. Process. Appl. **43**(1), 1–8 (1992)
33. Doney, R.A., Greenwood, P.E.: On the joint distribution of ladder variables of random walk. Probab. Theory Relat. Fields **94**(4), 457–472 (1993)
34. Doney, R.A.: A path decomposition for Lévy processes. Stoch. Process. Appl. **47**(2), 167–181 (1993)
35. Doney, R.A.: Some results involving the maximum of Brownian motion. J. Appl. Probab. **30**(4), 805–818 (1993)
36. Bertoin, J., Doney, R.A.: On the local behaviour of ladder height distributions. J. Appl. Probab. **31**(3), 816–821 (1994)
37. Bertoin, J., Doney, R.A.: Cramér's estimate for Lévy processes. Statist. Probab. Lett. **21**(5), 363–365 (1994)
38. Bertoin, J., Doney, R.A.: On conditioning random walks in an exponential family to stay nonnegative. In: Séminaire de Probabilités, XXVIII. Lecture Notes in Mathematics, vol. 1583, pp. 116–121. Springer, Berlin (1994)
39. Bertoin, J., Doney, R.A.: On conditioning a random walk to stay nonnegative. Ann. Probab. **22**(4), 2152–2167 (1994)
40. Doney, R.A.: Spitzer's condition and ladder variables in random walks. Probab. Theory Relat. Fields **101**(4), 577–580 (1995)
41. Bertoin, J., Doney, R.A.: Some asymptotic results for transient random walks. Adv. Appl. Probab. **28**(1), 207–226 (1996)
42. Doney, R.A.: Increase of Lévy processes. Ann. Probab. **24**(2), 961–970 (1996)
43. Doney, R.A.: One-sided local large deviation and renewal theorems in the case of infinite mean. Probab. Theory Relat. Fields **107**(4), 451–465 (1997)
44. Bertoin, J., Doney, R.A., Maller, R.A.: Passage of Lévy processes across power law boundaries at small times. Ann. Probab. **36**(1), 160–197 (2008)
45. Doney, R.A., Yor, M.: On a formula of Takács for Brownian motion with drift. J. Appl. Probab. **35**(2), 272–280 (1998)
46. Doney, R.A.: Some calculations for perturbed Brownian motion. In: Séminaire de Probabilités, XXXII. Lecture Notes in Mathematics, vol. 1686, pp. 231–236. Springer, Berlin (1998)
47. Doney, R.A., Warren, J., Yor, M.: Perturbed Bessel processes. In: Séminaire de Probabilités, XXXII. Lecture Notes in Mathematics, vol. 1686, pp. 237–249. Springer, Berlin (1998)
48. Doney, R.A.: The Martin boundary and ratio limit theorems for killed random walks. J. Lond. Math. Soc. (2) **58**(3), 761–768 (1998)

49. Alili, L., Doney, R.A.: Wiener-Hopf factorization revisited and some applications. Stoch. Stoch. Rep. **66**(1–2), 87–102 (1999)
50. Bertoin, J., Doney, R.A.: Spitzer's condition for random walks and Lévy processes. Ann. Inst. H. Poincaré Probab. Statist. **33**(2), 167–178 (1997)
51. Chaumont, L., Doney, R.A.: Pathwise uniqueness for perturbed versions of Brownian motion and reflected Brownian motion. Probab. Theory Relat. Fields **113**(4), 519–534 (1999)
52. Chaumont, L., Doney, R.A.: Some calculations for doubly perturbed Brownian motion. Stoch. Process. Appl. **85**(1), 61–74 (2000)
53. Chaumont, L., Doney, R.A., Hu, Y.: Upper and lower limits of doubly perturbed Brownian motion. Ann. Inst. H. Poincaré Probab. Statist. **36**(2), 219–249 (2000)
54. Alili, L., Chaumont, L., Doney, R.A.: On a fluctuation identity for random walks and Lévy processes. Bull. Lond. Math. Soc. **37**(1), 141–148 (2005)
55. Chaumont, L., Doney, R.A.: On Lévy processes conditioned to stay positive. Electron. J. Probab. **10**(28), 948–961 (2005)
56. Chaumont, L., Doney, R.A.: Corrections to: "On Lévy processes conditioned to stay positive" [Electron J. Probab. **10**(28), 948–961 (2005); MR2164035]. Electron. J. Probab. **13**(1), 1–4 (2008)
57. Chaumont, L., Doney, R.A.: Invariance principles for local times at the maximum of random walks and Lévy processes. Ann. Probab. **38**(4), 1368–1389 (2010)
58. Chaumont, L., Doney, R.A.: On distributions determined by their upward, space-time Wiener-Hopf factor. J. Theor. Probab. **33**(2), 1011–1033 (2020)
59. Alili, L., Doney, R.A.: Martin boundaries associated with a killed random walk. Ann. Inst. H. Poincaré Probab. Statist. **37**(3), 313–338 (2001)
60. Doney, R.A.: A local limit theorem for moderate deviations. Bull. Lond. Math. Soc. **33**(1), 100–108 (2001)
61. Doney, R.A.: Fluctuation theory for Lévy processes. Lévy processes, pp. 57–66. Birkhäuser, Boston (2001)
62. Doney, R.A., Nakhi, Y.B.: Perturbed and non-perturbed Brownian taboo processes. Ann. Inst. H. Poincaré Probab. Statist. **37**(6), 725–736 (2001)
63. Doney, R.A., Maller, R.A.: Random walks crossing curved boundaries: a functional limit theorem, stability and asymptotic distributions for exit times and positions. Adv. Appl. Probab. **32**(4), 1117–1149 (2000)
64. Doney, R.A., Maller, R.A.: Stability of the overshoot for Lévy processes. Ann. Probab. **30**(1), 188–212 (2002)
65. Doney, R.A., Griffin, P.S.: Overshoots over curved boundaries. II. Adv. Appl. Probab. **36**(4), 1148–1174 (2004)
66. Doney, R.A., Marchal, P.: A third arc-sine theorem. Bull. Lond. Math. Soc. **35**(4), 536–540 (2003)
67. Doney, R.A.: Small-time behaviour of Lévy processes. Electron. J. Probab. **9**(8), 209–229 (2004)
68. Doney, R.A., Maller, R.A.: Stability and attraction to normality for Lévy processes at zero and at infinity. J. Theor. Probab. **15**(3), 751–792 (2002)
69. Doney, R.A.: Stochastic bounds for Lévy processes. Ann. Probab. **32**(2), 1545–1552 (2004)
70. Doney, R.A., Maller, R.A.: Moments of passage times for Lévy processes. Ann. Inst. H. Poincaré Probab. Statist. **40**(3), 279–297 (2004)
71. Doney, R.A., Griffin, P.S.: Overshoots over curved boundaries. Adv. Appl. Probab. **35**(2), 417–448 (2003)
72. Doney, R.A., Zhang, T.: Perturbed Skorohod equations and perturbed reflected diffusion processes. Ann. Inst. H. Poincaré Probab. Statist. **41**(1), 107–121 (2005)
73. Doney, R.A.: Tanaka's construction for random walks and Lévy processes. In: Séminaire de Probabilités XXXVIII. Lecture Notes in Mathematics, vol. 1857, pp. 1–4. Springer, Berlin (2005)

74. Doney, R.A.: Some excursion calculations for spectrally one-sided Lévy processes. In: Séminaire de Probabilités XXXVIII. Lecture Notes in Mathematics, vol. 1857, pp. 5–15. Springer, Berlin (2005)
75. Doney, R.A., Maller, R.A.: Cramér's estimate for a reflected Lévy process. Ann. Appl. Probab. **15**(2), 1445–1450 (2005)
76. Doney, R.A., Maller, R.A.: Passage times of random walks and Lévy processes across power law boundaries. Probab. Theory Relat. Fields **133**(1), 57–70 (2005)
77. Doney, R.A., Kyprianou, A.E.: Overshoots and undershoots of Lévy processes. Ann. Appl. Probab. **16**(1), 91–106 (2006)
78. Bryn-Jones, A., Doney, R.A.: A functional limit theorem for random walk conditioned to stay non-negative. J. Lond. Math. Soc. (2) **74**(1), 244–258 (2006)
79. Doney, R.A., Maller, R.A.: Almost sure relative stability of the overshoot of power law boundaries. J. Theor. Probab. **20**(1), 47–63 (2007)
80. Doney, R.A.: Fluctuation Theory for Lévy Processes. Lecture Notes in Mathematics, vol. 1897. Lectures from the 35th Summer School on Probability Theory held in Saint-Flour, 6–23 July 2005, Edited and with a foreword by Jean Picard, pp. x+147. Springer, Berlin (2007)
81. Doney, R.A., Maller, R.A.: Curve crossing for random walks reflected at their maximum. Ann. Probab. **35**(4), 1351–1373 (2007)
82. Doney, R.A.: A note on the supremum of a stable process. In: Stochastics vol. **80**(2–3). IMS Lecture Notes–Monograph Series (2008), pp. 151–155
83. Doney, R.A., Maller, R.A., Savov, M.S.: Renewal theorems and stability for the reflected process. Stoch. Process. Appl. **119**(4), 1270–1297 (2009)
84. Doney, R.A.: Asymptotic analysis of random walks: heavy-tailed distributions [book review of MR2424161]. J. Am. Statist. Assoc. **104**(488), 1714 (2009)
85. Doney, R.A., Klüppelberg, C., Maller, R.A.: Passage time and fluctuation calculations for subexponential Lévy processes. Bernoulli **22**(3), 1491–1519 (2016)
86. Doney, R.A., Savov, M.S.: The asymptotic behavior of densities related to the supremum of a stable process. Ann. Probab. **38**(1), 316–326 (2010)
87. Doney, R.A., Savov, M.S.: Right inverses of Lévy processes. Ann. Probab. **38**(4), 1390–1400 (2010)
88. Doney, R.A., Korshunov, D.A.: Local asymptotics for the time of first return to the origin of transient random walk. Statist. Probab. Lett. **81**(9), 1419–1424 (2011)
89. Doney, R.A., Jones, E.M.: Conditioned random walks and Lévy processes. Bull. Lond. Math. Soc. **44**(1), 139–150 (2012)
90. Doney, R.A.: Local behaviour of first passage probabilities. Probab. Theory Relat. Fields **152**(3–4), 559–588 (2012)
91. Bertoin, J., et al.: Lévy processes at Saint-Flour. Probability at saint-Flour, pp. viii+476. Springer, Heidelberg (2012)
92. Doney, R.A., Jones, E.M.: Large deviation results for random walks conditioned to stay positive. Electron. Commun. Probab. **17**(38), 11 (2012)
93. Doney, R.A., Rivero, V.: Asymptotic behaviour of first passage time distributions for Lévy processes. Probab. Theory Relat. Fields **157**(1–2), 1–45 (2013)
94. Doney, R.A., Vakeroudis, S.: Windings of planar stable processes. In: Séminaire de Probabilités XLV. Lecture Notes in Mathematics, vol. 2078, pp. 277–300. Springer, Cham (2013)
95. Doney, R.A., Víctor M. Rivero. Asymptotic behaviour of first passage time distributions for subordinators. Electron. J. Probab. **20**(91), 28 (2015)
96. Doney, R.A., Rivero, V.: Erratum to: asymptotic behaviour of first passage time distributions for Lévy processes [MR3101839]. Probab. Theory Relat. Fields **164**(3–4), 1079–1083 (2016)
97. Doney, R.A., Griffin, P.S.: Cramér's estimate for the reflected process revisited. Ann. Appl. Probab. **28**(6), 3629–3651 (2018)
98. Caravenna, F., Doney, R.A.: Local large deviations and the strong renewal theorem. Electron. J. Probab. **24**, Paper No. 72, 48 (2019)
99. Doney, R.A.: The remainder in the renewal theorem. Electron. Commun. Probab. **25**, Paper No. 5, 8 (2020)

Path Decompositions of Perturbed Reflecting Brownian Motions

Elie Aïdékon, Yueyun Hu, and Zhan Shi

Abstract We are interested in path decompositions of a perturbed reflecting Brownian motion (PRBM) at the hitting times and at the minimum. Our study relies on the loop soups developed by Lawler and Werner (Probab Theory Relat Fields 4:197–217, 2004) and Le Jan (Ann Probab 38:1280–1319, 2010; Markov Paths, Loops and Fields. École d'été Saint-Flour XXXVIII 2008. Lecture Notes in Mathematics vol 2026. Springer, Berlin, 2011), in particular on a result discovered by Lupu (Mém Soc Math Fr (N.S.) 158, 2018) identifying the law of the excursions of the PRBM above its past minimum with the loop measure of Brownian bridges.

Keywords Perturbed reflecting Brownian motion · Path decomposition · Brownian loop soup · Poisson–Dirichlet distribution

1 Introduction

Let $(B_t,\ t \geq 0)$ be a standard one-dimensional Brownian motion. Let

$$\mathfrak{L}_t := \lim_{\varepsilon \to 0} \frac{1}{\varepsilon} \int_0^t 1_{\{0<B_s\leq \varepsilon\}}\, \mathrm{d}s\,, \qquad \text{a.s.},$$

E. Aïdékon
Fudan University, Shanghai, China

LPSM, Sorbonne Université Paris VI, Paris, France

Institut Universitaire de France, Paris, France
e-mail: elie.aidekon@upmc.fr

Y. Hu (✉)
LAGA, Université Sorbonne Paris Nord, Villetaneuse, France
e-mail: yueyun@math.univ-paris13.fr

Z. Shi
LPSM, Sorbonne Université Paris VI, Paris, France
e-mail: zhan.shi@upmc.fr

L. Chaumont, A. E. Kyprianou (eds.), *A Lifetime of Excursions Through Random Walks and Lévy Processes*, Progress in Probability 78,
https://doi.org/10.1007/978-3-030-83309-1_2

be the local time at time t and position 0. We take a continuous version of $(\mathfrak{L}_t,\ t \geq 0)$. Let $\mu \in \mathbb{R}\backslash\{0\}$ be a fixed parameter. Consider the perturbed reflecting Brownian motion (PRBM)

$$X_t := |B_t| - \mu \mathfrak{L}_t, \qquad t \geq 0. \tag{1.1}$$

The PRBM family contains two important special members: Brownian motion ($\mu = 1$; this is seen using Lévy's identity), and the three-dimensional Bessel process ($\mu = -1$; seen by means of Lévy's and Pitman's identities).

The PRBM, sometimes also referred to as the μ-process and appearing in the literature as the limiting process in the winding problem for three-dimensional Brownian motion around lines (Le Gall and Yor [12]), turns out to have remarkable properties such as the Ray–Knight theorems (Le Gall and Yor [11], Werner [24], Perman [18], Perman and Werner [19]), and Lévy's arc sine law (Petit [20], Carmona, Petit and Yor [4]). The process can also be viewed as (non-reflecting) Brownian motion perturbed by its one-sided maximum (Davis [6], Perman and Werner [19], Chaumont and Doney [5]). As explained on page 100 of Yor [25], these simply formulated and beautiful results were proved for the PRBM because of the scaling property and the strong Markov property of $(|B|, \mathfrak{L})$ via excursion theory.

We study in this paper path decompositions of the PRBM. Apart from their own interests, these decompositions can be used to understand the dual of general Jacobi stochastic flows which will be given in a forthcoming work, extending the work of Bertoin and Le Gall [2] who proved that the Jacobi flows of parameters (0, 0) and (2, 2) are dual with each other. These stochastic flows are connected to other important probabilistic objets in the study of population genetics such as flows of Fleming–Viot processes. Technically, our study of the PRBM often relies on the powerful tool of loop soups (Lawler and Werner [10], Le Jan [13, 14]), and in particular, on a result discovered by Lupu [15] identifying the law of the excursions of the PRBM above its past minimum with the loop measure of Brownian bridges. The point of view via loop soups has two main advantages: (i) its nice properties under rerooting allows to shed light on or extend some previously known results, see Proposition 2.6 or Theorem 5.3, and (ii) thanks to the independence structure in the Poisson point process representation of the loop measure, it helps to make arguments of conditioning rigorous: for instance, we show in Lemma 3.1 a path decomposition for PRBM, originated from Perman [18].

Our path decompositions focus on two families of random times of a recurrent PRBM: first hitting times (Sect. 3), and times at which the PRBM reaches its past minimum (Sect. 4). To illustrate the kind of results we have obtained, let us state two examples. The first one, Theorem 3.2, yields in the special case $\mu = 1$ the classical

Williams' Brownian path decomposition theorem (Revuz and Yor [21], Theorem VII.4.9). The second, Theorem 4.3, in the special case $\mu = 1$, states as follows:

Theorem 4.3 (special case $\mu = 1$***)*** *Consider B up to the first time that $\mathfrak{L}$ reaches 1, time-changed to remove its excursions above zero. We decompose this path at the minimum into the post- and time reversed pre- minimum processes. The two processes are independent and distributed as X^1 and X^2, time-changed to remove their excursions above level $H^{1,2}$, where X^1 and X^2 are independent three-dimensional Bessel processes and $H^{1,2}$ is the last level at which the sum of the total local times of X^1 and X^2 equals 1.*

The rest of this paper is organized as follows.

- Section 2: we recall Lupu [15]'s connection between the PRBM and the Brownian loop soup, and some known results on the PRBM. We also study the minimums of the PRBM considered up to its inverse local times (the process J defined in (2.4)). The main (new) result in this section is Proposition 2.6, a description of the jumping times of J;
- Section 3: we study the path decomposition at the hitting time of the PRBM. We prove that conditioned on its minimum, the PRBM can be split into four independent processes (see Fig. 1) and describe their laws in Theorems 3.2 and 3.3;
- Section 4: we study the path decomposition at the minimum of the PRBM considered up to its inverse local time. Theorem 4.3 deals with the recurrent case and describes the laws of the post- and time reversed pre- minimum processes. A similar decomposition is obtained in Proposition 4.4 for the transient case (see Fig. 2);

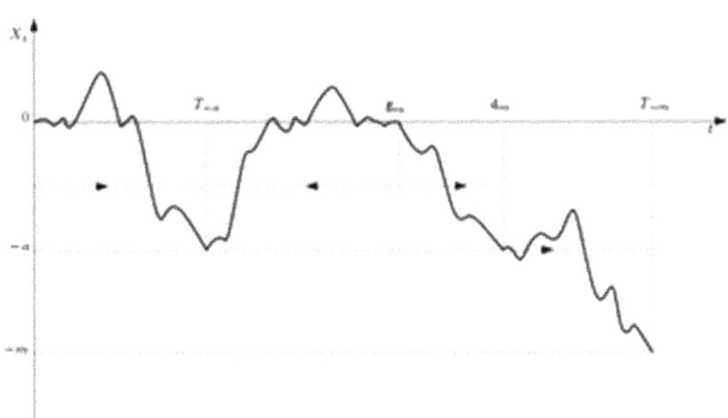

Fig. 1 The four processes in Theorem 3.2

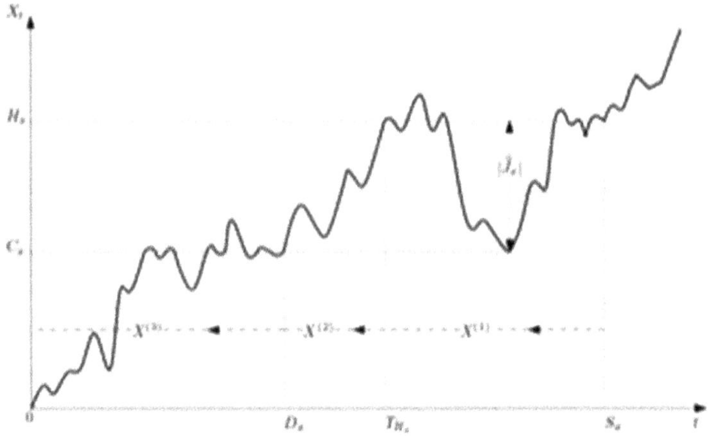

Fig. 2 Under $\mathbb{P}^{(-\delta)}$, $X_t \to \infty$ a.s.

- Section 5: we extend the perturbed Bessel process studied in Doney, Warren and Yor [7] to the perturbed Bessel process with a positive local time at 0. The main result in this section (Theorem 5.3) gives an extension of Theorem 2.2 of Doney, Warren and Yor [7].

2 Preliminaries

This section is divided into three subsections. We recall in Sect. 2.1 Lupu [15]'s description (Proposition 2.2) on the excursions of the PRBM above its past minimum in terms of Brownian loop soup, and we collect the intensities of various Poisson point processes in Lemma 2.3. In Sect. 2.2, we study the minimum process $J(x)$ defined in (2.4) and describe the jump times of $J(\cdot)$ by means of Poisson–Dirichlet distributions in Proposition 2.6. Finally in Sect. 2.3 we recall some known results on the PRBM. We also introduce some notations (in particular Notations 2.1 and 2.11) which are used throughout the paper.

2.1 *The Brownian Loop Soup*

Lupu [15] showed a connection between perturbed reflecting Brownian motions and the Brownian loop soup. We rely on [15] and review this connection in this subsection. Let $\mathcal{K}$ denote the set of continuous functions $\gamma : [0, T(\gamma)] \to \mathbb{R}$ with some $T(\gamma) \in (0, \infty)$, endowed with a metric $d_{\mathcal{K}}(\gamma, \widehat{\gamma}) := |\log T(\gamma) - \log T(\widehat{\gamma})| + \sup_{0 \le s \le 1} |\gamma(sT(\gamma)) - \widehat{\gamma}(sT(\widehat{\gamma}))|$ for any $\gamma, \widehat{\gamma} \in \mathcal{K}$. A rooted loop is an element γ

of $\mathcal{K}$ such that $\gamma(0) = \gamma(T(\gamma))$ (Section 3.1, p. 29). On the space of rooted loops, one defines the measure (Definition 3.8, p.37)

$$\mu_{\text{loop}}(\mathrm{d}\gamma) := \int_{t>0} \int_{x \in \mathbb{R}} P^t_{x,x}(\mathrm{d}\gamma) p_t(x,x) \mathrm{d}x \frac{\mathrm{d}t}{t},$$

where $P^t_{x,x}$ is the distribution of the Brownian bridge of length t from x to x, and $p_t(x,x)$ is the heat kernel $p_t(x,x) = \frac{1}{\sqrt{2\pi t}}$. An unrooted loop is the equivalence class of all loops obtained from one another by time-shift, and μ^*_{loop} denotes the projection of μ_{loop} on the space of unrooted loops. For any fixed $\beta > 0$, the Brownian loop soup of intensity measure β is the Poisson point process on the space of unrooted loops with intensity measure given by $\beta\mu^*_{\text{loop}}$ (Definition 4.2, p. 60). We denote it by $\mathcal{L}_\beta$.

For any real q, we let $\gamma - q$ denote the loop $(\gamma(t) - q,\ 0 \le t \le T(\gamma))$. We write $\min\gamma$, resp. $\max\gamma$ for the minimum, resp. maximum of γ. If γ denotes a loop, the loop γ rooted at its minimum is the rooted loop obtained by shifting the starting time of the loop to the hitting time of $\min\gamma$. Similarly for the loop γ rooted at its maximum. By an abuse of notation, we will often write γ for its range. For example $0 \in \gamma$ means that γ visits the point 0.

Similarly to Lupu [15], Section 5.2, define

$$\mathcal{Q}^{\uparrow}_{\beta} := \{\min\gamma,\ \gamma \in \mathcal{L}_\beta\}, \qquad \mathcal{Q}^{\downarrow}_{\beta} := \{\max\gamma,\ \gamma \in \mathcal{L}_\beta\}.$$

For any $q \in \mathcal{Q}^{\uparrow}_{\beta}$ and $\gamma \in \mathcal{L}_\beta$ such that $\min\gamma = q$, define $\mathfrak{e}^{\uparrow}_q$ as the loop $\gamma - q$ rooted at its minimum. It is an excursion above 0. Define similarly, for any $q \in \mathcal{Q}^{\downarrow}_{\beta}$, $\mathfrak{e}^{\downarrow}_q$ as the excursion below 0 given by $\gamma - q$ rooted at its maximum. The point measure $\{(q,\ \mathfrak{e}^{\downarrow}_q),\ q \in \mathcal{Q}^{\downarrow}_{\beta}\}$ has the same distribution as $\{-(q, \mathfrak{e}^{\uparrow}_q),\ q \in \mathcal{Q}^{\uparrow}_{\beta}\}$.

Notation 2.1 *For $\delta > 0$, let $\mathbb{P}^{\delta}$ (resp. $\mathbb{P}^{(-\delta)}$) be the probability measure under which $(X_t)_{t\ge 0}$ is distributed as the PRBM $(|B_t| - \mu\mathfrak{L}_t)_{t\ge 0}$ defined in (1.1) with $\mu = \frac{2}{\delta}$ (resp. $\mu = -\frac{2}{\delta}$).*

Note that under $\mathbb{P}^{\delta}$, $(X_t)_{t\ge 0}$ is recurrent whereas under $\mathbb{P}^{(-\delta)}$, $\lim_{t\to\infty} X_t = +\infty$ a.s. Define

$$I_t := \inf_{0\le s\le t} X_s,\ t \ge 0.$$

We will use the same notation $\{(q, \mathfrak{e}^{\uparrow}_{X,q}),\ q \in \mathcal{Q}^{\uparrow}_X\}$ to denote

- under $\mathbb{P}^{\delta}$: the excursions away from $\mathbb{R} \times \{0\}$ (q is seen as a real number) of the process $(I_t, X_t - I_t)$;
- under $\mathbb{P}^{(-\delta)}$: the excursions away from $\mathbb{R} \times \{0\}$ of the process $(\widehat{I}_t, X_t - \widehat{I}_t)$ where $\widehat{I}_t := \inf_{s\ge t} X_s$.

The following proposition is for example Proposition 5.2 of [15]. One can also see it from [11] or [1] (together with Proposition 3.18 of [15]).

Proposition 2.2 (Lupu [15]) *Let $\delta > 0$. The point measure $\{(q, \mathfrak{e}^{\uparrow}_{X,q}),\ q \in \mathcal{Q}^{\uparrow}_X\}$ is distributed under $\mathbb{P}^{\delta}$, respectively $\mathbb{P}^{(-\delta)}$, as $\{(q, \mathfrak{e}^{\uparrow}_q),\ q \in \mathcal{Q}^{\uparrow}_{\frac{\delta}{2}} \cap (-\infty, 0)\}$, resp. $\{(q,\ \mathfrak{e}^{\uparrow}_q),\ q \in \mathcal{Q}^{\uparrow}_{\frac{\delta}{2}} \cap (0, \infty)\}$.*

Actually, Proposition 5.2 [15] states the previous proposition in a slightly different way. In the same way that standard Brownian motion can be constructed from its excursions away from 0, Lupu shows that one can construct the perturbed Brownian motions from the Brownian loop soup by "gluing" the loops of the Brownian loop soup rooted at their minimum and ordered by decreasing minima.

We close this section by collecting the intensities of various Poisson point processes. It comes from computations of [15].

Denote by $\mathfrak{n}$ the Itô measure on Brownian excursions and $\mathfrak{n}^+$ (resp. $\mathfrak{n}^-$) the restriction of $\mathfrak{n}$ on positive excursions (resp. negative excursions). For any loop γ, let $\ell^0_\gamma := \lim_{\varepsilon\to 0} \int_0^{T(\gamma)} 1_{\{0<\gamma(t)<\varepsilon\}} \mathrm{d}t$ be its (total) local time at 0.

In the following lemma, we identify a Poisson point process with its atoms.

Lemma 2.3 *Let $\delta > 0$.*

(i) *The collection $\{(q, \mathfrak{e}^{\uparrow}_q),\ q \in \mathcal{Q}^{\uparrow}_{\frac{\delta}{2}}\}$ is a Poisson point process of intensity measure $\delta \mathrm{d}a \otimes \mathfrak{n}^+(\mathrm{d}\mathfrak{e})$.*

(ii) *The collection $\{(q, \mathfrak{e}^{\downarrow}_q),\ q \in \mathcal{Q}^{\downarrow}_{\frac{\delta}{2}}$ such that $q + \mathfrak{e}^{\downarrow}_q \subset (0, \infty)\}$ is a Poisson point process of intensity measure $\delta 1_{\{a>0\}} \mathrm{d}a \otimes 1_{\{\min\ \mathfrak{e}>-a\}} \mathfrak{n}^-(\mathrm{d}\mathfrak{e})$.*

(iii) *The collection $\{\min \gamma,\ \gamma \in \mathcal{L}_{\frac{\delta}{2}}$ such that $0 \in \gamma\}$ is a Poisson point process of intensity measure $\frac{\delta}{2|a|} 1_{\{a<0\}} \mathrm{d}a$.*

(iv) *Let $m > 0$. The collection $\{\ell^0_\gamma,\ \gamma \in \mathcal{L}_{\frac{\delta}{2}}$ such that $\min \gamma \in [-m, 0],\ 0 \in \gamma\}$ is a Poisson point process of intensity measure $1_{\{\ell>0\}} \frac{\delta}{2\ell} e^{-\ell/2m} \mathrm{d}\ell$.*

(v) *The collection $\{(\ell^0_\gamma, \gamma),\ \gamma \in \mathcal{L}_{\frac{\delta}{2}}$ such that $0 \in \gamma\}$ is a Poisson point process of intensity measure $\frac{\delta}{2} 1_{\{\ell>0\}} \frac{\mathrm{d}\ell}{\ell} \mathbb{P}^*((B_t,\ 0 \le t \le \tau^B_\ell) \in \mathrm{d}\gamma)$, where $\tau^B_\ell := \inf\{s > 0 : \mathfrak{L}_s > \ell\}$ denotes the inverse of the Brownian local time, and $\mathbb{P}^*((B_t,\ 0 \le t \le \tau^B_\ell) \in \bullet)$ is the projection of $\mathbb{P}((B_t,\ 0 \le t \le \tau^B_\ell) \in \bullet)$ on the space of unrooted loops.*

Proof Item (i) is Proposition 3.18, p. 44 of [15]. Item (ii) follows from the equality in distribution $\{(q,\ \mathfrak{e}^{\downarrow}_q),\ q \in \mathcal{Q}^{\downarrow}_{\frac{\delta}{2}}\} \overset{(\text{law})}{=} \{-(q, \mathfrak{e}^{\uparrow}_q),\ q \in \mathcal{Q}^{\uparrow}_{\frac{\delta}{2}}\}$ and (i). Item (iii) comes from (i) and the fact that $\mathfrak{n}^+(r \in \mathfrak{e}) = \frac{1}{2r}$ for any $r > 0$ where, here and in the sequel, $\mathfrak{e}$ denotes a Brownian excursion. We prove now (iv). The intensity measure is given by

$$\delta \int_{-m}^{0} \mathfrak{n}^+\big(\ell^{|a|}_{\mathfrak{e}} \in \mathrm{d}\ell,\ |a| \in \mathfrak{e}\big)\, \mathrm{d}a$$

where $\ell_{\mathfrak{e}}^r$ denotes the local time at r of the excursion $\mathfrak{e}$. Under $\mathfrak{n}^+$, conditionally on $|a| \in \mathfrak{e}$, the excursion after hitting $|a|$ is a Brownian motion killed at 0. Therefore

$$\mathfrak{n}^+\left(\ell_{\mathfrak{e}}^{|a|} \in \mathrm{d}\ell \,\middle|\, |a| \in \mathfrak{e}\right) = \mathbb{P}_{|a|}(\mathfrak{L}_{T_0^B}^{|a|} \in \mathrm{d}\ell) = \frac{1}{2|a|} e^{-\frac{\ell}{2|a|}} \,\mathrm{d}\ell,$$

where under $\mathbb{P}_{|a|}$, the Brownian motion B starts at $|a|$ and $\mathfrak{L}_{T_0^B}^{|a|}$ denotes its local time at position $|a|$ up to $T_0^B := \inf\{t > 0 : B_t = 0\}$, and the last equality follows from the standard Brownian excursion theory. Hence the intensity measure is given by

$$\delta \int_{-m}^{0} \frac{1}{4a^2} e^{-\frac{\ell}{2|a|}} \,\mathrm{d}\ell \,\mathrm{d}a = \frac{\delta}{2\ell} e^{-\ell/2m} \,\mathrm{d}\ell.$$

Finally, (v) comes from Corollary 3.12, equation (3.3.5) p. 39 of [15]. □

2.2 *The Poisson–Dirichlet Distribution*

For a vector $\mathcal{D} = (D_1, D_2, \ldots)$ and a real r, we denote by $r\mathcal{D}$ the vector $(rD_1, rD_2, \ldots)$. We recall that for $a, b > 0$, the density of the gamma (a, b) distribution is given by

$$\frac{1}{\Gamma(a)b^a} x^{a-1} e^{-\frac{x}{b}} 1_{\{x>0\}},$$

and the density of the beta (a, b) distribution is

$$\frac{\Gamma(a+b)}{\Gamma(a)\Gamma(b)} x^{a-1} (1-x)^{b-1} 1_{\{x\in(0,1)\}}.$$

We introduce the Poisson–Dirichlet distribution, relying on Perman, Pitman and Yor [17]. Let $\beta > 0$. Consider a Poisson point process of intensity measure $\frac{\beta}{x} e^{-x} 1_{\{x>0\}} \,\mathrm{d}x$ and denote by $\Delta_{(1)} \geq \Delta_{(2)} \geq \ldots$ its atoms. We can see them also as the jump sizes, ordered decreasingly, of a gamma subordinator of parameters $(\beta, 1)$ up to time 1. The sum $T := \sum_{i\geq 1} \Delta_{(i)}$ has a gamma $(\beta, 1)$ distribution. The random variable on the infinite simplex defined by

$$(P_{(1)}, P_{(2)}, \ldots) := \left(\frac{\Delta_{(1)}}{T}, \frac{\Delta_{(2)}}{T}, \ldots\right)$$

has the Poisson–Dirichlet distribution with parameter β [9], and is independent of T ([16], also Corollary 2.3 of [17]).

Consider a decreasingly ordered positive vector $(\xi_{(1)}, \xi_{(2)}, \ldots)$ of finite sum $\sum_{i\geq 1} \xi_{(i)} < \infty$. A size-biased random permutation, denoted by $(\xi_1, \xi_2, \ldots)$, is a permutation of $(\xi_{(1)}, \xi_{(2)}, \ldots)$ such that, conditionally on $\xi_1 = \xi_{(i_1)}, \ldots, \xi_j = \xi_{(i_j)}$, the term ξ_{j+1} is chosen to be $\xi_{(k)}$ for $k \notin \{i_1, \ldots, i_j\}$ with probability $\frac{\xi_{(k)}}{\sum_{i\geq 1} \xi_{(i)} - (\xi_1 + \ldots + \xi_j)}$. The indices $(i_j,\ j \geq 1)$ can be constructed by taking i.i.d. exponential random variables of parameter 1, denoted by $(\varepsilon_i,\ i \geq 1)$, and by ordering $\mathbb{N}$ increasingly with respect to the total order $k_1 \leq k_2$ if and only if $\xi_{(k_1)}/\varepsilon_{k_1} \geq \xi_{(k_2)}/\varepsilon_{k_2}$ (Lemma 4.4 of [17]). A result from McCloskey [16] says that the $(P_i,\ i \geq 1)$ obtained from the $(P_{(1)}, P_{(2)}, \ldots)$ by size-biased ordering can also be obtained via the stick-breaking construction:

$$P_i = (1 - U_i) \prod_{j=1}^{i-1} U_j$$

where $U_i,\ i \geq 1$ are i.i.d. with law beta $(\beta, 1)$. Let

$$\mathcal{D}_\beta := (D_1, D_2, \ldots) \tag{2.1}$$

be the point measure in $[0, 1]$ defined by $D_i := \prod_{j=1}^{i} U_j$, for $i \geq 1$.

Lemma 2.4 *Let $m > 0$, $\beta > 0$ and Ξ_β be a Poisson point process of intensity measure $\frac{\beta}{a} 1_{\{a>0\}}\mathrm{d}a$. We denote by $a_1^{(m)} > a_2^{(m)} > \ldots$ the points of Ξ_β belonging to $[0, m]$. Then $\left(a_i^{(m)},\ i \geq 1\right)$ is distributed as $m\mathcal{D}_\beta$.*

Proof For $0 \leq a \leq 1$,

$$\mathbb{P}\big(\frac{1}{m} a_1^{(m)} \leq a\big) = \exp\big(- \int_{am}^{m} \frac{\beta}{x} \mathrm{d}x\big) = a^\beta.$$

Therefore it is a beta $(\beta, 1)$ distribution. Conditionally on $\{a_1^{(m)} = a\}$, the law of Ξ_β restricted to $[0, a_1^{(m)})$ is the one of Ξ_β restricted to $[0, a)$. By iteration we get the Lemma. □

Denote by $L(t, r)$, $r \in \mathbb{R}$ and $t \geq 0$, the local time of X at time t and position r. Let

$$\tau_r(t) := \inf\{s \geq 0 : L(s, r) > t\}, \tag{2.2}$$

be the inverse local time of X. Denote by

$$T_r := \inf\{t \geq 0 : X_t = r\} \tag{2.3}$$

the hitting time of r. We are interested in the process $(J(x))_{x\geq 0}$ defined as follows:

$$J(x) := \inf\{X_s,\ s \leq \tau_0(x)\}, \qquad x \geq 0. \tag{2.4}$$

Observe that under $\mathbb{P}^\delta$, J is a Markov process. It has been studied in Section 4 of [3]. The perturbed reflecting Brownian motions are related to the Poisson–Dirichlet via the following proposition.

Proposition 2.5 *Let $\delta > 0$. Under $\mathbb{P}^\delta$, the range $\{J(x),\ x > 0\}$ is distributed as $-\Xi_\beta$ with $\beta = \frac{\delta}{2}$. Consequently, for any $m > 0$, the range of J in $[-m, 0]$, ordered increasingly, is distributed as $-m\mathcal{D}_\beta$.*

Proof Recall Proposition 2.2. Note that the range $\{J(x),\ x > 0\}$ is equal to $\{\min\gamma\ :\ 0 \in \gamma,\ \gamma \in \mathcal{L}_{\frac{\delta}{2}}\}$, hence the first statement is (iii) of Lemma 2.3. The second statement is Lemma 2.4. □

Under $\mathbb{P}^\delta$, for $m > 0$, let

$$T^J_{-m} := \inf\{x > 0 : J(x) \leq -m\} \tag{2.5}$$

be the first passage time of $-m$ by J. The main result in this subsection is the following description of the jump times of J before its first passage time of $-m$:

Proposition 2.6 *For $m > 0$, let $x^{(m)}_1 > x^{(m)}_2 > \ldots$ denote the jumping times of J before time T^J_{-m}. Under $\mathbb{P}^\delta$:*

(i) *[25] T^J_{-m} follows a gamma $(\frac{\delta}{2}, 2m)$ distribution. Consequently, for any $x > 0$, $\frac{-1}{J(x)}$ follows a gamma $(\frac{\delta}{2}, \frac{2}{x})$ distribution.*
(ii) *T^J_{-m} is independent of $\frac{1}{T^J_{-m}}(x^{(m)}_1, x^{(m)}_2, \ldots)$.*
(iii) *$\frac{1}{T^J_{-m}}(x^{(m)}_1, x^{(m)}_2, \ldots)$ is distributed as $\mathcal{D}_\beta$ with $\beta = \frac{\delta}{2}$.*

Statement (i) of Proposition 2.6 is not new. It is contained in Proposition 9.1, Chapter 9.2, p. 123, of Yor [25]. For the sake of completeness we give here another proof of (i) based on Lemma 2.3.

Proof (i) Lemma 2.3, (iv) says that $\{\ell^0_\gamma,\ \gamma \in \mathcal{L}_{\frac{\delta}{2}}$ such that $\min\gamma \in (-m, 0],\ 0 \in \gamma\}$ forms a Poisson point process of intensity measure $1_{\{\ell>0\}}\frac{\delta}{2\ell}e^{-\ell/2m}\mathrm{d}\ell$, whose atoms are exactly the (non-ordered) sequence $\{x^{(m)}_{i-1} - x^{(m)}_i,\ i \geq 1\}$ where $x^{(m)}_0 := T^J_{-m}$. Note that $T^J_{-m} = L(T_{-m}, 0) = \sum_{\min\gamma\in(-m,0],\, 0\in\gamma} \ell^0_\gamma$.

Let for $i \geq 1$, $d^{(m)}_i := (x^{(m)}_{i-1} - x^{(m)}_i)/T^J_{-m}$ and denote by $\{d^{(m)}_{(1)} > d^{(m)}_{(2)} > \ldots\}$ the sequence ordered decreasingly. Then the properties of the Poisson–Dirichlet distribution recalled at the beginning of the section imply that T^J_{-m} is independent of the point measure $\{d^{(m)}_{(1)}, d^{(m)}_{(2)}, \ldots\}$ and that $T^J_{-m}/2m$ follows a gamma $(\frac{\delta}{2}, 1)$ distribution. Also, the second statement of (i) comes from the observation that $\{J(x) > -m\} = \{T^J_{-m} > x\}$. This proves (i).

(ii) and (iii): It remains to show that the vector $(d_1^{(m)}, d_2^{(m)}, \dots)$ is a size-biased ordering of $\{d_{(1)}^{(m)}, d_{(2)}^{(m)}, \dots\}$, and that this size-biased ordering is still independent of T_{-m}^J.

To this end, denote by $\{(-m_i,\ i \in \mathcal{I})\}$ the range of J. By Proposition 2.5, the point measure $\{\ln(m_i),\ i \in \mathcal{I}\}$ is a Poisson point process on $\mathbb{R}$ of intensity measure $\frac{\delta}{2}\mathrm{d}t$.

When J jumps at some $-m_i$, the time to jump at $-m_{i+1}$ is exponentially distributed with parameter $\mathfrak{n}^-(\min \mathfrak{e} \le -m_i) = \frac{1}{2m_i}$ (it is the local time at 0 of a Brownian motion when it hits level $-m_i$, by Markov property of the process (X, I) under $\mathbb{P}^\delta$).

Denote by ε_i the exponential of parameter 1 obtained as the waiting time between jumps to $-m_i$ and to $-m_{i+1}$, divided by $2m_i$. Conditionally on $\{(m_i,\ i \in \mathcal{I})\}$, the random variables $(\varepsilon_i,\ i \in \mathcal{I})$ are i.i.d. and exponentially distributed with parameter 1. Then $\{(\ln(m_i), \varepsilon_i),\ i \in \mathcal{I}\}$ is a Poisson point process on $\mathbb{R} \times \mathbb{R}_+$ of intensity measure $\frac{\delta}{2}\mathrm{d}t \otimes e^{-x}\mathrm{d}x$. It is straightforward to check that $\{(\ln(2m_i\varepsilon_i), \varepsilon_i),\ i \in \mathcal{I}\}$ is still a Poisson point process with the same intensity measure.

Suppose that we enumerated the range of J with $\mathcal{I} = \mathbb{Z}$ so that $(-m_i,\ i \ge 1)$ are the atoms of the range in $(-m, 0)$ ranked increasingly. Then, $2m_i\varepsilon_i = T_{-m}^J d_i^{(m)} =: \xi_i$ for any $i \ge 1$. We deduce that, conditionally on $\{T_{-m}^J d_i^{(m)}, i \ge 1\}$, the vector $(\varepsilon_1, \varepsilon_2, \dots)$ consists of i.i.d. random variables exponentially distributed with parameter 1, and $i \le j$ if and only if $\xi_i/\varepsilon_i \ge \xi_j/\varepsilon_j$. From the description of size-biased ordering at the beginning of this section, we conclude that the $(\xi_i,\ i \ge 1)$ are indeed size-biased ordered, hence also $(d_1^{(m)}, d_2^{(m)}, \dots)$. □

Since $T_{-m}^J = L(T_{-m}, 0)$, the statement (i) of Proposition 2.6 says

$$L(T_{-m}, 0) \overset{\text{(law)}}{=} gamma(\frac{\delta}{2}, 2m). \tag{2.6}$$

Corollary 2.7 *Let $\delta > 0$. Under $\mathbb{P}^\delta$, the collection of jumping times of J is distributed as Ξ_β with $\beta = \frac{\delta}{2}$.*

Proof From Proposition 2.6 and Lemma 2.4, we can couple the jumping times of J which are strictly smaller than the passage time of $-m$ with Ξ_β restricted to $[0, Z_m]$ where Z_m is gamma $(\frac{\delta}{2}, 2m)$ distributed, independent of Ξ_β. Letting $m \to +\infty$ gives the Corollary. □

2.3 Some Known Results

At first we recall two Ray–Knight theorems:

Proposition 2.8 (Le Gall and Yor [11]) *Let $\delta > 0$. Under $\mathbb{P}^{(-\delta)}$, the process $\big(L(\infty, t),\ t \ge 0\big)$ is the square of a Bessel process of dimension δ starting from 0 reflected at 0.*

Proposition 2.9 (Carmona, Petit and Yor [4], Werner [24]) *Let $\delta > 0$ and $a \geq 0$. Under $\mathbb{P}^{\delta}$, the process $\big(L(\tau_0(a), -t),\ t \geq 0\big)$ is the square of a Bessel process of dimension $(2-\delta)$ starting from a absorbed at 0.*

We have the following independence result.

Proposition 2.10 (Yor [25], Proposition 9.1) *Let $\delta > 0$. Under $\mathbb{P}^{\delta}$, for any fixed $x > 0$, $L(T_{J(x)}, 0)/x$ is independent of $J(x)$ and follows a beta $(\frac{\delta}{2}, 1)$ distribution.*

We introduce some notations which will be used in Sect. 4.

Notation 2.11 *Let $h \in \mathbb{R}$. We define the process $X^{-,h}$ obtained by gluing the excursions of X below h as follows. Let for $t \geq 0$,*

$$A_t^{-,h} := \int_0^t 1_{\{X_s \leq h\}} \mathrm{d}s, \qquad \alpha_t^{-,h} := \inf\{u > 0,\ A_u^{-,h} > t\},$$

with the usual convention $\inf \emptyset := \infty$. *Define*

$$X_t^{-,h} := X_{\alpha_t^{-,h}}, \qquad t < A_\infty^{-,h} := \int_0^\infty 1_{\{X_s \leq h\}} \mathrm{d}s.$$

Similarly, we define $A_t^{+,h}$, $\alpha_t^{+,h}$ and $X^{+,h}$ by replacing $X_s \leq h$ by $X_s > h$. When the process is denoted by X with some superscript, the analogous quantities will hold the same superscript. For example for $r \in \mathbb{R}$, $\ell > 0$, $\tau_r^{+,h}(\ell) = \inf\{t > 0 : L^{+,h}(t,r) > \ell\}$, where $L^{+,h}(t,r)$ denotes the local time of $X^{+,h}$ at position r and time t.

Proposition 2.12 (Perman and Werner [19]) *Let $\delta > 0$. Under $\mathbb{P}^{\delta}$, the two processes $X^{+,0}$ and $X^{-,0}$ are independent. Moreover, $X^{+,0}$ is a reflecting Brownian motion, and the process $(X_t^{-,0}, \inf_{s \leq t} X_s^{-,0})_{t \geq 0}$ is strongly Markovian.*

Let $m > 0$. We look at these processes up to the first time the process X hits level $-m$. In this case, there is a dependence between $X^{-,0}$ and $X^{+,0}$ due to their duration. This dependence is taken care of by conditioning on the (common) local time at 0 of $X^{-,0}$ and $X^{+,0}$. It is the content of the following corollary.

Corollary 2.13 *Let $\delta > 0$. Fix $m > 0$. Under $\mathbb{P}^{\delta}$, conditionally on $(X_t^{-,0}, t \leq A_{T_{-m}}^{-,0})$, the process $(X_t^{+,0}, t \leq A_{T_{-m}}^{+,0})$ is a reflecting Brownian motion stopped at time $\tau_0^{+,0}(\ell)$ where $\ell = L^{-,0}(A_{T_{-m}}^{-,0}, 0) = L(T_{-m}, 0)$.*

Proof By Proposition 2.12, conditionally on $X^{-,0}$, the process $(X^{+,0}_{\tau_0^{+,0}(t)}, t \geq 0)$ is a reflecting Brownian motion indexed by its inverse local time. Observe that $A_{T_{-m}}^{+,0} = \tau_0^{+,0}(\ell)$ with $\ell = L^{-,0}(0, T_{-m}^{-,0})$. It proves the Corollary. □

As mentioned in Section 3 of Werner [24], we have the following duality between $\mathbb{P}^{(-\delta)}$ and $\mathbb{P}^{\delta}$.

Proposition 2.14 (Werner [24]) *Let $\delta > 0$. For any $m > 0$, the process $(X_{T_{-m}-t} + m,\ t \le T_{-m})$ under $\mathbb{P}^{\delta}$ has the distribution of $(X_t,\ t \le \mathscr{D}_m)$ under $\mathbb{P}^{(-\delta)}$, where $\mathscr{D}_m := \sup\{t > 0 : X_t = m\}$ denotes the last passage time at m.*

3 Decomposition at a Hitting Time

The following lemma is Lemma 2.3 in Perman [18], together with the duality stated in Proposition 2.14. Recall that under $\mathbb{P}^{(-2)}$, X is a Bessel process of dimension 3. We refer to (2.3) for the definition of the first hitting time T_r and to (2.4) for the process $J(x)$.

Lemma 3.1 (Perman [18]) *Let $\delta > 0$. Let $m, x > 0$ and $y \in (0, x)$. Define the processes*

$$Z^1 := (X_{T_{J(x)}-t} - J(x))_{t\in[0,T_{J(x)}]} \qquad Z^2 := (X_{T_{J(x)}+t} - J(x))_{t\in[0,\tau_0(x)-T_{J(x)}]}.$$

Under $\mathbb{P}^{\delta}(\cdot|J(x) = -m, L(T_{J(x)}, 0) = y)$:

(i) *Z^1 and Z^2 are independent,*
(ii) *Z^1 is distributed as $(X_t)_{t\in[0,\mathscr{D}_m]}$ under $\mathbb{P}^{(-\delta)}(\cdot|L(\infty, m) = y)$,*
(iii) *Z^2 is distributed as $(X_t)_{t\in[0,\mathscr{D}_m]}$ under $\mathbb{P}^{(-2)}(\cdot|L(\infty, m) = x - y)$,*

with $\mathscr{D}_m := \sup\{t > 0 : X_t = m\}$.

Proof For the sake of completeness, we give here a proof which is different from Perman [18]'s.

By Proposition 2.2, we can identify under $\mathbb{P}^{\delta}$ the point measure $\{(q, \mathfrak{e}^{\uparrow}_{X,q}),\ q \in \mathcal{Q}^{\uparrow}_X\}$ with $\{(q, \mathfrak{e}^{\uparrow}_q),\ q \in \mathcal{Q}^{\uparrow}_{\frac{\delta}{2}} \cap (-\infty, 0)\}$. Using the notations in Lemma 2.3, we have

$$L(T_{J(x)}, 0) = \sum_{q\in\mathcal{Q}^{\uparrow}_{\frac{\delta}{2}}\cap(J(x),0)} \ell^0_\gamma < x \le \sum_{q\in\mathcal{Q}^{\uparrow}_{\frac{\delta}{2}}\cap[J(x),0)} \ell^0_\gamma,$$

where in the above sum γ is the (unique) loop in $\mathcal{L}_{\frac{\delta}{2}}$ such that $\min\gamma = q$. Let $\ell^r(\mathfrak{e})$ be the local time of the excursion $\mathfrak{e}$ at level r. We claim that conditioning on $\{J(x) = -m, L(T_{J(x)}, 0) = y\}$, $\mathfrak{e}^{\uparrow}_{J(x)}$ and $\{(q, \mathfrak{e}^{\uparrow}_q),\ q \in \mathcal{Q}^{\uparrow}_{\frac{\delta}{2}} \cap (J(x), 0)\}$ are independent and distributed as a Brownian excursion $\mathfrak{e}$ under $\mathfrak{n}^+(\cdot \mid \ell^m(\mathfrak{e}) > x - y)$, and $\{(q,\ \mathfrak{e}^{\uparrow}_q),\ q \in \mathcal{Q}^{\uparrow}_{\frac{\delta}{2}} \cap (-m, 0)\}$ conditioned on $\{\xi_m = y\}$ respectively, where $\xi_m := \sum_{q\in\mathcal{Q}^{\uparrow}_{\frac{\delta}{2}}\cap(-m,0)} \ell^0_\gamma$.

In fact, let $F : \mathbb{R}_- \times \mathcal{K} \to \mathbb{R}_+$, $G : \mathcal{K} \to \mathbb{R}_+$ and $f : \mathbb{R}^2 \to \mathbb{R}_+$ be three measurable functions. Note that $\xi_m = \sum_{q\in\mathcal{Q}^{\uparrow}_{\frac{\delta}{2}}\cap(-m,0)} \ell^{|q|}(\mathfrak{e}^{\uparrow}_q)$. By Proposition 2.2, we deduce from the master formula that

$$\mathbb{E}^{\delta}\left[\mathrm{e}^{-\sum_{q\in\mathcal{Q}^{\uparrow}_{\frac{\delta}{2}}\cap(J(x),0)}F(q,\mathfrak{e}^{\uparrow}_q)} G(\mathfrak{e}^{\uparrow}_{J(x)})\, f(J(x),L(T_{J(x)},0))\right]$$

$$=\mathbb{E}^{\delta}\left[\sum_{m>0}\mathrm{e}^{-\sum_{q\in\mathcal{Q}^{\uparrow}_{\frac{\delta}{2}}\cap(-m,0)}F(q,\mathfrak{e}^{\uparrow}_q)} G(\mathfrak{e}^{\uparrow}_{-m})\, f(-m,\xi_m)\,1_{\{\xi_m<x,\ell^m(\mathfrak{e}^{\uparrow}_{-m})>x-\xi_m\}}\right]$$

$$=\delta\int_0^{\infty}\mathrm{d}m\,\mathbb{E}\left[\mathrm{e}^{-\sum_{q\in\mathcal{Q}^{\uparrow}_{\frac{\delta}{2}}\cap(-m,0)}F(q,\mathfrak{e}^{\uparrow}_q)} f(-m,\xi_m)\,1_{\{\xi_m<x\}}\int \mathfrak{n}^{+}(\mathrm{d}\mathfrak{e})G(\mathfrak{e})\,1_{\{\ell^m(\mathfrak{e})>x-\xi_m\}}\right],$$

by using Lemma 2.3(i). The claim follows.

Now we observe that Z^2 is measurable with respect to $\mathfrak{e}^{\uparrow}_{J(x)}$ whereas Z^1 is to $\{(q,\mathfrak{e}^{\uparrow}_q),\ q\in\mathcal{Q}^{\uparrow}_{\frac{\delta}{2}}\cap(J(x),0)\}$. It yields (i). Moreover, conditioning on $\{J(x)=-m, L(T_{J(x)},0)=y\}$, Z^2 is distributed as $(\mathfrak{e}_t)_{t\in[0,\sigma^m_{x-y}]}$, under $\mathfrak{n}^{+}(\cdot\,|\,\ell^m(\mathfrak{e})>x-y)$, where $\sigma^m_{x-y}:=\inf\{t>0:\ell^m_t(\mathfrak{e})=x-y\}$ with $\ell^m_t(\mathfrak{e})$ being the local time at level m at time t. The latter process has the same law as $(X_t)_{t\in[0,\mathscr{D}_m]}$ under $\mathbb{P}^{(-2)}(\cdot|L(\infty,m)=x-y)$.[1] We get (iii).

To prove (ii), we denote by $\widehat{e}$ the time-reversal of a loop e. By Proposition 2.14, $\{(m+q,\widehat{\mathfrak{e}}^{\uparrow}_{q,X}),\ q\in\mathcal{Q}^{\uparrow}_X\cap(-m,0)\}$ under $\mathbb{P}^{\delta}$ is distributed as $\{(q,\mathfrak{e}^{\uparrow}_{q,X}),\ q\in\mathcal{Q}^{\uparrow}_X\cap(0,m)\}$ under $\mathbb{P}^{(-\delta)}$. Note that under $\mathbb{P}^{\delta}(\cdot|J(x)=-m, L(T_{J(x)},0)=y)$, Z^1 can be constructed from $\{(m+q,\widehat{\mathfrak{e}}^{\uparrow}_{q,X}),\ q\in\mathcal{Q}^{\uparrow}_X\cap(-m,0)\}$. Then Z^1 is distributed as $(X_t)_{0\le t\le\mathscr{D}_m}$ under $\mathbb{P}^{(-\delta)}(\cdot|\sum_{q\in\mathcal{Q}^{\uparrow}_X\cap(0,m)}\ell^{m-q}(\mathfrak{e}^{\uparrow}_{q,X})=y)$. Finally remark that $\sum_{q\in\mathcal{Q}^{\uparrow}_X\cap(0,m)}\ell^{m-q}(\mathfrak{e}^{\uparrow}_{q,X})=L(\mathscr{D}_m,m)=L(\infty,m)$. We get (ii). This completes the proof of Lemma 3.1.

□

Fix $m>0$. The following Theorems 3.2 and 3.3 describe the path decomposition of (X_t) at T_{-m}. Let $\mathfrak{g}_m:=\sup\{t\in[0,T_{-m}]:X_t=0\}$. Recall that $I_{\mathfrak{g}_m}=\inf_{0\le s\le\mathfrak{g}_m}X_s$. Define

$$\mathfrak{d}_m:=\inf\{t>\mathfrak{g}_m:X_t=I_{\mathfrak{g}_m}\}.$$

[1]Under $\mathfrak{n}^{+}(\cdot\,|\,\ell^m(\mathfrak{e})>x-y)$, an excursion up to the inverse local time $x-y$ at position m is a three-dimensional Bessel process, up to the hitting time of m, followed by a Brownian motion starting at m stopped at local time at level m given by $x-y$, this Brownian motion being conditioned on not touching 0 during that time. By excursion theory, the time-reversed process is distributed as a Brownian motion starting at level m stopped at the hitting time of 0 conditioned on the local time at m being equal to $x-y$. We conclude by William's time reversal theorem (Corollary VII.4.6 of [21]).

Theorem 3.2 *Let $\delta > 0$. Fix $m > 0$. Under $\mathbb{P}^\delta$, the random variable $\frac{1}{m}|I_{\mathfrak{g}_m}|$ is beta $(\frac{\delta}{2}, 1)$ distributed. Moreover, for $0 < a < m$, conditionally on $\{I_{\mathfrak{g}_m} = -a\}$, the four processes*

$$
\begin{aligned}
&(X_t,\ t \in [0, T_{-a}]),\\
&(X_{\mathfrak{g}_m - t},\ t \in [0, \mathfrak{g}_m - T_{-a}],\\
&(-X_{\mathfrak{g}_m + t},\ t \in [0, \mathfrak{d}_m - \mathfrak{g}_m]),\\
&(X_{\mathfrak{d}_m + t} + a,\ t \in [0, T_{-m} - \mathfrak{d}_m]),
\end{aligned}
$$

are independent, with law respectively the one of:

(i) *X under $\mathbb{P}^\delta$ up to the hitting time of $-a$;*
(ii) *a Brownian motion up to the hitting time of $-a$;*
(iii) *a Bessel process of dimension 3 from 0 stopped when hitting a;*
(iv) *X under $\mathbb{P}^\delta$ conditionally on $\{T_{-(m-a)} < T_a\}$.*

Proof To get the distribution of $\frac{1}{m}I_{\mathfrak{g}_m}$ we proceed as follows: under $\mathbb{P}^\delta$, X is measurable with respect to its excursions above the infimum, that we denoted by $(\mathfrak{e}^{\uparrow}_{q,X},\ q \in \mathcal{Q}^{\uparrow}_X)$, that we identify with $(\mathfrak{e}^{\uparrow}_q,\ q \in \mathcal{Q}^{\uparrow}_{\frac{\delta}{2}})$ by Proposition 2.2. The variable $I_{\mathfrak{g}_m}$ is the global minimum of the loops γ such that $\min \gamma > -m$ and $0 \in \gamma$. By Lemma 2.3(iii), we get the law of $I_{\mathfrak{g}_m}$ (it is also a consequence of Lemma 2.4 together with Proposition 2.5).

Let $\tilde{\gamma}$ be the loop such that $\min \tilde{\gamma} = I_{\mathfrak{g}_m}$ and call $-a = I_{\mathfrak{g}_m}$ its minimum. Conditioning on $\tilde{\gamma}$ and loops hitting $(-\infty, -a)$, the loops γ such that $\min \gamma > -a$ are distributed as the usual Brownian loop soup $\mathcal{L}_{\frac{\delta}{2}}$ in $(-a, \infty)$. It gives (i) by Proposition 2.2. Conditioning on $\min \tilde{\gamma} = I_{\mathfrak{g}_m} = -a$ and on loops hitting $(-\infty, -a)$, the loop $\tilde{\gamma} - \min \tilde{\gamma}$ has the measure $\mathfrak{n}^+(d\mathfrak{e}|\max \mathfrak{e} > a)$. Therefore (ii) and (iii) come from the usual decomposition of the Itô measure. Finally conditioning on $\min \tilde{\gamma} = I_{\mathfrak{g}_m} = -a$, the collection of loops γ with $\min \gamma \in (-m, -a)$ is distributed as the Brownian loop soup $\mathcal{L}_{\frac{\delta}{2}}$ restricted to loops γ such that $\min \gamma \in (-m, -a)$ conditioned on the event that none of these loops hit 0. We deduce (iv). □

The following theorem gives the path decomposition when conditioning on $(L(T_{-m}, 0), I_{\mathfrak{g}_m})$. Recall (2.6) for the law of $L(T_{-m}, 0)$.

Theorem 3.3 *We keep the notations of Theorem 3.2. Under $\mathbb{P}^\delta$,*

(i) *the density of $(L(T_{-m}, 0), |I_{\mathfrak{g}_m}|)$ is given by $\frac{a^{-2}}{2\Gamma(\frac{\delta}{2})}(\frac{x}{2m})^{\frac{\delta}{2}} e^{-\frac{x}{2a}}$ for $x > 0$ and $0 < a < m$.*

(ii) *conditionally on* $\{L(T_{-m}, 0) = x, I_{\mathrm{g}_m} = -a\}$, *the three processes*

$$\begin{aligned}&(X_t,\ t \in [0, \mathrm{g}_m]),\\&(-X_{\mathrm{g}_m+t},\ t \in [0, \mathrm{d}_m - \mathrm{g}_m]),\\&(X_{\mathrm{d}_m+t} + a,\ t \in [0, T_{-m} - \mathrm{d}_m]),\end{aligned}$$

are independent and distributed respectively as

$(X_t, t \in [0, \tau_0(x)])$ *under* $\mathbb{P}^\delta(\cdot \mid J(x) = -a)$,

a Bessel process of dimension 3 starting from 0 *stopped when hitting* a,

X *under* $\mathbb{P}^\delta$ *conditionally on* $\{T_{-(m-a)} < T_a\}$;

Proof By Theorem 3.2, conditionally on $\{I_{\mathrm{g}_m} = -a\}$, the three processes

$$\begin{aligned}&(X_t,\ t \in [0, \mathrm{g}_m]),\\&(-X_{\mathrm{g}_m+t},\ t \in [0, \mathrm{d}_m - \mathrm{g}_m]),\\&(X_{\mathrm{d}_m+t} + a,\ t \in [0, T_{-m} - \mathrm{d}_m]),\end{aligned}$$

are independent. Since $L(T_{-m}, 0)$ is measurable with respect to $\sigma(X_t,\ t \in [0, \mathrm{g}_m])$, we obtain the independence of the three processes in (ii) and the claimed laws of the last two processes in (ii).

To complete the proof, it is enough to show that for any bounded continuous functional Φ on $\mathcal{K}$ and any bounded continuous function $f : \mathbb{R}^2 \to \mathbb{R}$,

$$\begin{aligned}&\mathbb{E}^\delta[\Phi(X_t,\ t \in [0, \mathrm{g}_m]) f(L(T_{-m}, 0), I_{\mathrm{g}_m})] \qquad (3.1)\\&= \int_0^\infty \int_0^m \mathbb{E}^\delta[\Phi(X_t, t \in [0, \tau_0(x)]) \mid J(x) = -a] f(x, -a) \frac{a^{-2}}{2\Gamma(\frac{\delta}{2})} (\frac{x}{2m})^{\frac{\delta}{2}} e^{-\frac{x}{2a}} \mathrm{d}a\mathrm{d}x.\end{aligned}$$

By Theorem 3.2,

$$\begin{aligned}&\mathbb{E}^\delta[\Phi(X_t,\ t \in [0, \mathrm{g}_m]) f(L(T_{-m}, 0), I_{\mathrm{g}_m})]\\&= \int_0^m \frac{\delta}{2} m^{-\frac{\delta}{2}} a^{\frac{\delta}{2}-1} \mathbb{E}^\delta[\Phi(X^{1,a} \oplus X^{2,a}) f(L^0(X^{1,a}) + L^0(X^{2,a}), -a)] \mathrm{d}a, \quad (3.2)\end{aligned}$$

where $X_s^{1,a} := X_s, s \le T_{-a}$, $X^{2,a}$ is the time-reversal of an independent Brownian motion up to its hitting time of $-a$ (so $X^{2,a}$ starts from $-a$ and ends at 0), $X^{1,a} \oplus X^{2,a}$ denotes the process obtained by gluing $X^{2,a}$ and $X^{1,a}$ at time T_{-a}, and $L^0(X^{1,a})$ (resp. $L^0(X^{2,a})$) is the local time at position 0 of $X^{1,a}$ (resp. $X^{2,a}$).

The standard excursion theory says that $\mathbb{P}^\delta(L^0(X^{2,a}) \in \mathrm{d}z) = \frac{1}{2a} e^{-\frac{z}{2a}} \mathrm{d}z$, $z > 0$. By (2.6), $L^0(X^{1,a}) \overset{\text{(law)}}{=} \mathrm{gamma}(\frac{\delta}{2}, 2a)$. Then for any bounded Borel function h, we have

$$
\begin{aligned}
&\mathbb{E}^\delta[\Phi(X^{1,a} \oplus X^{2,a})\, h(L^0(X^{1,a}) + L^0(X^{2,a}))] \\
&= \int_0^\infty \int_0^\infty h(y+z) \mathbb{E}^\delta[\Phi(X^{1,a} \oplus X^{2,a}) | L^0(X^{1,a}) = y, L^0(X^{2,a}) = z] \frac{(2a)^{-1-\frac{\delta}{2}}}{\Gamma(\frac{\delta}{2})} y^{\frac{\delta}{2}-1} e^{-\frac{y+z}{2a}} \mathrm{d}y\mathrm{d}z \\
&= \int_0^\infty h(x) \int_0^x \mathbb{E}^\delta[\Phi(X^{1,a} \oplus X^{2,a}) | L^0(X^{1,a}) = y, L^0(X^{2,a}) = x - y] \frac{(2a)^{-1-\frac{\delta}{2}}}{\Gamma(\frac{\delta}{2})} y^{\frac{\delta}{2}-1} e^{-\frac{x}{2a}} \mathrm{d}y\mathrm{d}x \\
&= \int_0^\infty h(x) \int_0^x \mathbb{E}^\delta[\Phi(X_t, t \le \tau_0(x)) | J(x) = -a, L(T_{J(x)}, 0) = y] \frac{(2a)^{-1-\frac{\delta}{2}}}{\Gamma(\frac{\delta}{2})} y^{\frac{\delta}{2}-1} e^{-\frac{x}{2a}} \mathrm{d}y\mathrm{d}x,
\end{aligned}
$$

where the last equality is due to Lemma 3.1. Since $\mathbb{P}^\delta(L(T_{J(x)}, 0) \in \mathrm{d}y) = \frac{\delta}{2} x^{-\frac{\delta}{2}} y^{\frac{\delta}{2}-1} 1_{\{0<y<x\}} \mathrm{d}y$ (see Proposition 2.10), we get that

$$
\begin{aligned}
&\mathbb{E}^\delta[\Phi(X^{1,a} \oplus X^{2,a})\, h(L^0(X^{1,a}) + L^0(X^{2,a}))] \\
&= \int_0^\infty h(x) \mathbb{E}^\delta[\Phi(X_t, t \le \tau_0(x)) | J(x) = -a] \frac{(2a)^{-1-\frac{\delta}{2}}}{\Gamma(1+\frac{\delta}{2})} x^{\frac{\delta}{2}} e^{-\frac{x}{2a}} \mathrm{d}x,
\end{aligned}
$$

which in view of (3.2) yields (3.1) and completes the proof of the Proposition. □

Remark 3.4 We may also directly prove (i) as follows: In view of (2.6), it is enough to show

$$
\mathbb{P}^\delta\big(|I_{\mathfrak{g}_m}| \in \mathrm{d}a \mid L(T_{-m}, 0) = x\big) = \frac{x}{2a^2} e^{-\frac{x}{2a} + \frac{x}{2m}} 1_{\{0<a<m\}} \mathrm{d}a. \tag{3.3}
$$

To this end, we shall prove that conditionally on $\{L(T_{-m}, 0) = x\}$, $I_{\mathfrak{g}_m}$ is distributed as $\inf_{0 \le t \le \tau_0^B(x)} B(t)$ conditioned on $\{\inf_{0 \le t \le \tau_0^B(x)} B(t) > -m\}$, where $\tau_0^B(x) := \inf\{t > 0 : \mathfrak{L}_t > x\}$ denotes the first time when the local time at 0 of B attains x.

Consider the Brownian loop soup $\mathcal{L}_{\frac{\delta}{2}}$. In this setting (recalling Proposition 2.2),

$$
I_{\mathfrak{g}_m} = \inf_{\gamma \in \mathcal{L}_{\frac{\delta}{2}}} \Big\{ q : q = \min \gamma > -m, 0 \in \gamma \Big\},
$$

and

$$
L(T_{-m}, 0) = \sum_{\gamma \in \mathcal{L}_{\frac{\delta}{2}}} \ell_\gamma^0 \, 1_{\{\min \gamma \in (-m,0), 0 \in \gamma\}}.
$$

From (v) of Lemma 2.3, conditionally on $\{\ell^0_\gamma : \gamma \in \mathcal{L}_{\frac{\delta}{2}}, 0 \in \gamma\}$, the loops γ such that $0 \in \gamma$ are (the projection on the space of unrooted loops of) independent Brownian motions stopped at τ_0^ℓ with $\ell = \ell^0_\gamma$. Then, the loops γ such that $0 \in \gamma$ and $\min \gamma > -m$ are merely (the projection of) independent Brownian motions stopped at local time given by ℓ^0_γ, conditioned on not hitting $-m$. The conditional density (3.3) of $I_{\mathfrak{g}_m}$ follows from standard Brownian excursion theory. □

Corollary 3.5 *Let us keep the notations of Theorem 3.2. Let $x > 0$ and $m > a > 0$. Under $\mathbb{P}^\delta$, the conditional law of the process $(X_t, t \in [0, \mathfrak{g}_m])$ given $\{L(T_{-m}, 0) = x\}$ is equal to the (unconditional) law of $(X_t, t \in [0, \tau_0(x)])$ biased by $c_{m,x,\delta} |J(x)|^{\frac{\delta}{2}-1} 1_{\{J(x) > -m\}}$, with*

$$c_{m,x,\delta} := \Gamma(\frac{\delta}{2}) \, (\frac{x}{2})^{1-\frac{\delta}{2}} e^{\frac{x}{2m}}.$$

Proof Let Φ be a bounded continuous functional on $\mathcal{K}$. Recall from (2.6) that the density function of $L(T_{-m}, 0)$ is $\frac{1}{\Gamma(\frac{\delta}{2})}(2m)^{-\frac{\delta}{2}} x^{\frac{\delta}{2}-1} e^{-\frac{x}{2m}}$, $x > 0$. Considering some f in (3.1) which only depends on the first coordinate, we see that for all $x > 0$,

$$\begin{aligned} &\mathbb{E}^\delta[\Phi(X_t, t \in [0, \mathfrak{g}_m]) \mid L(T_{-m}, 0) = x] \\ &= \int_0^m \mathbb{E}^\delta[\Phi(X_t, t \in [0, \tau_0(x)]) \mid J(x) = -a] \frac{x}{2} a^{-2} e^{-\frac{x}{2a} + \frac{x}{2m}} \mathrm{d}a \\ &= c_{m,x,\delta} \, \mathbb{E}^\delta[\Phi(X_t, t \in [0, \tau_0(x)]) \, |J(x)|^{\frac{\delta}{2}-1} \, 1_{\{J(x) > -m\}}], \end{aligned} \tag{3.4}$$

by using the fact that the density of $|J(x)|$ is $a \to \frac{1}{\Gamma(\frac{\delta}{2})} (\frac{x}{2})^{\frac{\delta}{2}} a^{-\frac{\delta}{2}-1} e^{-\frac{x}{2a}}$. This proves Corollary 3.5. □

Remark 3.6 Note that the conditional expectation term on the left-hand-side of (3.4) is a continuous function of (m, x), this fact will be used later on.

As an application of the above decomposition results, we give an another proof of Proposition 2.6(ii) and (iii).

Another proof of Proposition 2.6(ii) and (iii) Notice that $T^J_{-m} = L(T_{-m}, 0)$. Conditioning on $T^J_{-m} = x$: by Corollary 3.5, $x_1^{(m)}$ is distributed as $L(T_{J(x)}, 0)$ under P^δ biased by $J(x)^{\frac{\delta}{2}-1} 1_{\{J(x) > -m\}}$. By the independence of $L(T_{J(x)}, 0)$ and $J(x)$ of Proposition 2.10, the biased law of $L(T_{J(x)}, 0)$ is the same as under $\mathbb{P}^\delta$, hence is x times a beta $(\frac{\delta}{2}, 1)$ random variable. Moreover, conditionally on $x_1^{(m)} = y$ and $J(x) = -m_1$, the process before $T_{J(x)}$ is simply the process X under $\mathbb{P}^\delta$ before hitting T_{-m_1} conditioned on $L(T_{-m_1}, 0) = y$ (by Corollary 3.5 and Lemma 3.1). Therefore we can iterate and get Proposition 2.6. □

4 Decomposition at the Minimum

Let $\delta > 0$. Let X_1 be a copy of the process X under $\mathbb{P}^{(-\delta)}$ and X_2 be an independent Bessel process of dimension 3, both starting at 0. Recall Notation 2.11. From our notations, $L^1(\infty, r)$, resp. $L^2(\infty, r)$, denotes the total local time at height r of X_1, resp. X_2, while $X_1^{-,h}$, $X_2^{-,h}$ are obtained by gluing the excursions below h of X_1 and X_2 respectively. We set

$$H^{1,2} := \sup\{r \geq 0 \,:\, L^1(\infty, r) + L^2(\infty, r) = 1\}. \tag{4.1}$$

Proposition 2.8 yields that the process $L^1(\infty, r)+L^2(\infty, r), r \geq 0$ is distributed as the square of a Bessel process of dimension $\delta+2$, starting from 0. Then $H^{1,2} < \infty$ a.s.

Lemma 4.1 *Let $\delta > 0$. Let $m > 0$ and $x \in (0, 1)$. Conditionally on $\{H^{1,2} = m, L^1(\infty, H^{1,2}) = x\}$:*

(i) *$X_1^{-,H^{1,2}}$ and $X_2^{-,H^{1,2}}$ are independent;*
(ii) *$X_1^{-,H^{1,2}}$ is distributed as $(X_t^{-,m}, t < A_\infty^{-,m})$ under $\mathbb{P}^{(-\delta)}(\cdot|L^1(\infty, m) = x)$;*
(iii) *$X_2^{-,H^{1,2}}$ is distributed as $(X_t^{-,m}, t < A_\infty^{-,m})$ under $\mathbb{P}^{(-2)}(\cdot|L^2(\infty, m) = 1 - x)$,*

where $A_\infty^{-,m} = \int_0^\infty 1_{\{X_t \leq m\}} dt$ is the total lifetime of the process $X^{-,m}$.

Proof First we describe the law of $(X_t^{-,m}, t < A_\infty^{-,m})$ under $\mathbb{P}^{(-\delta)}(\cdot|L^1(\infty, m) = x)$. Let $\mathscr{D}_m := \sup\{t > 0 : X_t \leq m\}$ be the last passage time of X at m [note that under $\mathbb{P}^{(-\delta)}$, $X_t \to \infty$ as $t \to \infty$]. By the duality of Proposition 2.14, $\{X_{\mathscr{D}_m - t} - m, 0 \leq t \leq \mathscr{D}_m\}$, under $\mathbb{P}^{(-\delta)}$, has the same law as $\{X_t, 0 \leq t \leq T_{-m}\}$ under $\mathbb{P}^{\delta}$. Corollary 3.5 gives then the law of $(X_t, t \leq \mathscr{D}_m)$ under $\mathbb{P}^{(-\delta)}(\cdot|L(\infty, m) = x)$. The process $(X_t^{-,m}, t < A_\infty^{-,m})$ is a measurable function of $(X_t, t \leq \mathscr{D}_m)$. Note that $m \to (X_1^{-,m}, X_2^{-,m})$ is continuous.[2] From Corollary 3.5, we may find a regular

[2]For instance, we may show that for any $T > 0$, almost surely $\sup_{0 \leq t \leq T} |X_t^{-,m'} - X_t^{-,m}| \to 0$ as $m' \to m$. Let us give a proof by contradiction. Suppose there exists some $\varepsilon_0 > 0$, a sequence (t_k) in $[0, T]$ and $m_k \to m$ such that $|X_{t_k}^{-,m_k} - X_{t_k}^{-,m}| > \varepsilon_0$. Write for simplification $s_k := \alpha_{t_k}^{-,m}$ and $s'_k := \alpha_{t_k}^{-,m_k}$. Consider the case $m_k > m$ (the other direction can be treated in a similar way). Then $s_k \geq s'_k$ and $|X_{s'_k} - X_{s_k}| > \varepsilon_0$ for all k. Since $X_{s_k} \leq m$ and $X_{s'_k} \leq m_k$, either $X_{s_k} \leq m - \frac{\varepsilon_0}{2}$ or $X_{s'_k} \leq m - \frac{\varepsilon_0}{2}$ for all large k. Consider for example the case $X_{s'_k} \leq m - \frac{\varepsilon_0}{2}$. By the uniform continuity of X_t on every compact, there exists some $\delta_0 > 0$ such that $X_u \leq m$ for all $|u - s'_k| \leq \delta_0$ and $k \geq 1$. Then for any $s \geq s'_k$, $\int_{s'_k}^{s} 1_{\{X_u \leq m\}} du \geq \min(\delta_0, s - s'_k)$. Note that by definition, $\int_{s'_k}^{s_k} 1_{\{X_u \leq m\}} du = t_k - \int_0^{s'_k} 1_{\{X_u \leq m\}} du = \int_m^{m_k} L(s'_k, x) dx \leq \zeta\,(m_k - m)$, with $\zeta := \sup_{x \in \mathbb{R}} L(\alpha_T^{-,m}, x)$. It follows that for all sufficiently large k, $0 \leq s_k - s'_k \leq \zeta\,(m_k - m)$. Consequently $X_{s_k} - X_{s'_k} \to 0$ as $m_k \to m$, in contradiction with the assumption that $|X_{s'_k} - X_{s_k}| > \varepsilon_0$ for all k. This proves the continuity of $m \to X^{-,m}$.

version of the law of $(X_t^{-,m},\ t < A_\infty^{-,m})$ under $\mathbb{P}^{(-\delta)}(\cdot|L(\infty,m)=x)$ such that for any bounded continuous functional F on $\mathcal{K}$, the application

$$(m,x)\mapsto \mathbb{E}^{(-\delta)}\big[F(X_t^{-,m},\ t<A_\infty^{-,m})|L(\infty,m)=x\big]$$

is continuous.

Now let us write $H := H^{1,2}$ for concision. Let F_1 and F_2 be two bounded continuous functionals on $\mathcal{K}$ and $g : \mathbb{R}_+^2 \to \mathbb{R}$ be a bounded continuous function. Let $H_n := 2^{-n}\lfloor 2^n H\rfloor$ for any $n \geq 1$. By the continuity of $m \to (X_1^{-,m}, X_2^{-,m})$ and that of $(L^1(\infty,m), L^2(\infty,m))$, we have

$$\mathbb{E}\Big[F_1(X_1^{-,H})F_2(X_2^{-,H})g(H,L^1(\infty,H))\Big] = \lim_{n\to\infty}\mathbb{E}\Big[F_1(X^{1,H_n})F_2(X^{2,H_n})g(H_n,L^1(\infty,H_n))\Big].$$

Note that

$$\begin{aligned}&\mathbb{E}\Big[F_1(X_1^{-,H_n})F_2(X_2^{-,H_n})g(H_n,L^1(\infty,H_n))\Big]\\&=\sum_{j=0}^{\infty}\mathbb{E}\Big[F_1(X_1^{-,\frac{j}{2^n}})F_2(X_2^{-,\frac{j}{2^n}})g\big(\frac{j}{2^n},L^1(\infty,\frac{j}{2^n})\big)1_{\{\frac{j}{2^n}\leq H<\frac{j+1}{2^n}\}}\Big].\end{aligned}$$

By the independence property of Corollary 2.13 and the duality of Proposition 2.14, conditioning on $\{L^1(\infty,\frac{j}{2^n}), L^2(\infty,\frac{j}{2^n})\}$, the processes $(X_1^{-,\frac{j}{2^n}}, X_2^{-,\frac{j}{2^n}})$ are independent, and independent of $(X_1^{+,\frac{j}{2^n}}, X_2^{+,\frac{j}{2^n}})$. Since $\{\frac{j}{2^n}\leq H<\frac{j+1}{2^n}\}$ is measurable with respect to $\sigma(X_1^{+,\frac{j}{2^n}}, X_2^{+,\frac{j}{2^n}})$, we get that for each $j\geq 0$,

$$\begin{aligned}&\mathbb{E}\Big[F_1(X_1^{-,\frac{j}{2^n}})F_2(X_2^{-,\frac{j}{2^n}})g\big(\frac{j}{2^n},L^1(\infty,\frac{j}{2^n})\big)1_{\{\frac{j}{2^n}\leq H<\frac{j+1}{2^n}\}}\Big]\\&=\mathbb{E}\Big[\Phi_1\big(\frac{j}{2^n},L^1(\infty,\frac{j}{2^n})\big)\,\Phi_2\big(\frac{j}{2^n},L^2(\infty,\frac{j}{2^n})\big)\,g\big(\frac{j}{2^n},L^1(\infty,\frac{j}{2^n})\big)1_{\{\frac{j}{2^n}\leq H<\frac{j+1}{2^n}\}}\Big],\end{aligned}$$

where

$$\Phi_1(m,x):=\mathbb{E}[F_1(X_1^{-,m})\,|L^1(\infty,m)=x],\ \ \Phi_2(m,x):=\mathbb{E}[F_2(X_2^{-,m})\,|L^2(\infty,m)=x].$$

By Remark 3.6, Φ_1 and Φ_2 are continuous functions in (m,x). Taking the sum over j we get that

$$\begin{aligned}&\mathbb{E}\Big[F_1(X_1^{-,H_n})F_2(X_2^{-,H_n})g(H_n,L^1(\infty,H_n))\Big]\\&=\mathbb{E}\Big[\Phi_1(H_n,L^1(\infty,H_n))\Phi_2(H_n,L^2(\infty,H_n))g(H_n,L^1(\infty,H_n))\Big].\end{aligned}$$

Since Φ_1 and Φ_2 are bounded and continuous, the dominated convergence theorem yields that

$$
\begin{aligned}
&\mathbb{E}\Big[F_1(X_1^{-,H})F_2(X_2^{-,H})g(H, L^1(\infty, H))\Big] \\
&= \lim_{n\to\infty} \mathbb{E}\Big[F_1(X_1^{-,H_n})F_2(X_2^{-,H_n})g(H_n, L^1(\infty, H_n))\Big] \\
&= \mathbb{E}\Big[\Phi_1(H, L^1(\infty, H))\Phi_2(H, L^2(\infty, H))g(H, L^1(\infty, H))\Big], \qquad (4.2)
\end{aligned}
$$

proving Lemma 4.1 as $L^2(\infty, H) = 1 - L^1(\infty, H)$. □

Remark 4.2 Let $\delta > 2$. Consider the process X under $\mathbb{P}^{(-\delta)}$ and the total local time $L(\infty, r)$ of X at position $r \geq 0$. For $x > 0$, let

$$H_x := \sup\{r \geq 0 : L(\infty, r) = x\}.$$

By Proposition 2.8, $H_x < \infty$, $\mathbb{P}^{(-\delta)}$-a.s. Note that the same arguments leading to (4.2) shows that

$$\mathbb{E}\Big[F_1(X^{-,H_x})g(H_x, L(\infty, H_x))\Big] = \mathbb{E}\Big[\Phi_1(H_x, L(\infty, H_x))g(H_x, L(\infty, H_x))\Big],$$

where $\Phi_1(m, x) := \mathbb{E}[F_1(X^{-,m})\,|L(\infty, m) = x]$ for $m > 0, x > 0$. Since $L(\infty, H_x) = x$, we obtain that conditionally on $\{H_x = m\}$, the process X^{-,H_x} is distributed as $(X_t^{-,m},\ t < A_\infty^{-,m})$ under $\mathbb{P}^{(-\delta)}(\cdot|L(\infty, m) = x)$.

Recall (2.3), (2.4) and (4.1). The main result in this section is the following path decomposition of (X_t) at $T_{J(1)} = \inf\{t \in [0, \tau_0(1)] : X_t = J(1)\}$, the unique time before $\tau_0(1)$ at which X reaches its minimum $J(1)$.

Theorem 4.3 *Let $\delta > 0$. Define $Z_1 := (X_{T_{J(1)}-t} - J(1))_{t\in[0,T_{J(1)}]}$ and $Z_2 := (X_{T_{J(1)}+t} - J(1))_{t\in[0,\tau_0(1)-T_{J(1)}]}$. Under $\mathbb{P}^\delta$, the couple of processes*

$$\left(Z_1^{-,|J(1)|},\ Z_2^{-,|J(1)|}\right)$$

is distributed as $(X_1^{-,H^{1,2}}, X_2^{-,H^{1,2}})$.

Proof From Lemmas 3.1 and 4.1, it remains to prove that the joint law of $(|J(1)|, L(T_{J(1)}, 0))$ is the same as $(H^{1,2}, L^1(\infty, H^{1,2}))$. Recall the law of $(J(1), L(T_{J(1)}, 0))$ from Propositions 2.6(i) and 2.10. Define a process $(Y_t,\ -\infty < t < \infty)$ with values in $[0, 1]$ defined by time-change as

$$Y_{A_m} = \frac{L^1(\infty, m)}{L^1(\infty, m) + L^2(\infty, m)},$$

where $A_m := \int_1^m \frac{\mathrm{d}h}{L^1(\infty,h)+L^2(\infty,h)}$ for any $m > 0$ (as such $\lim_{m\to 0} A_m = -\infty$ a.s.). Following Warren and Yor [23], equation (3.1), we call Jacobi process of parameters $d, d' \geq 0$ the diffusion with generator $2y(1-y)\frac{\mathrm{d}^2}{\mathrm{d}y^2} + (d-(d+d')y)\frac{\mathrm{d}}{\mathrm{d}y}$. We claim that Y is a stationary Jacobi process of parameter $(\delta, 2)$, independent of $(L^1(\infty, m) + L^2(\infty, m),\ m \geq 0)$. It is a consequence of Proposition 8 of Warren and Yor [23]. Let us see why.

First, notice that $\frac{L^1(\infty,m)}{L^1(\infty,m)+L^2(\infty,m)}$ is a beta $(\frac{\delta}{2}, 1)$-random variable for any $m > 0$, because $L^1(\infty, m)$ and $L^2(\infty, m)$ are independent and distributed as gamma $(\frac{\delta}{2}, 2m)$ and gamma $(1, 2m)$ respectively, by Proposition 2.8 and the duality in Proposition 2.14. It is independent of $L^1(\infty, m) + L^2(\infty, m)$, hence by the Markov property, also of $(L^1(\infty, h) + L^2(\infty, h),\ h \geq m)$.

Let $t_0 \in \mathbb{R}$. By Proposition 8 of [23], for any $m \in (0, 1)$, conditioning on $(L^1(\infty, h) + L^2(\infty, h),\ h \geq m)$, the process $(Y_{h+A_m},\ h \geq 0)$ is distributed as a Jacobi process starting from a beta $(\frac{\delta}{2}, 1)$ random variable, hence stationary. Notice that A_m is measurable with respect to $\sigma(L^1(\infty, h) + L^2(\infty, h),\ h \geq m)$. We deduce that, conditioned on $(L^1(\infty, h) + L^2(\infty, h),\ h \geq m)$ and $A_m \leq t_0$, the process $(Y_h,\ h \geq t_0)$ is a Jacobi process starting from a beta $(\frac{\delta}{2}, 1)$ random variable. Letting $m \to 0$ we see that Y is a stationary Jacobi process of parameter $(\delta, 2)$, independent of $(L^1(\infty, m) + L^2(\infty, m),\ m \geq 0)$.

Since $L^1(\infty, H^{1,2}) = Y_{A_{H^{1,2}}}$ and $A_{H^{1,2}}, H^{1,2}$ are measurable with respect to $\sigma\{L^1(\infty, m) + L^2(\infty, m),\ m \geq 0\}$, we deduce that the random variable $L^1(\infty, H^{1,2})$ follows the beta $(\frac{\delta}{2}, 1)$ distribution and that $H^{1,2}$ and $L^1(\infty, H^{1,2})$ are independent. Finally, the random variable $H^{1,2}$ is the exit time at 1 of a square Bessel process of dimension $2 + \delta$ by Proposition 2.8, whose density is equal to $\frac{1}{\Gamma(\frac{\delta}{2})} 2^{-\frac{\delta}{2}} t^{-\frac{\delta}{2}-1} e^{-\frac{1}{2t}}$ for $t > 0$ (Exercise (1.18), Chapter XI of Revuz and Yor [21]). By Proposition 2.6(i), we see that $|J(1)|$ is distributed as $H^{1,2}$. This completes the proof. □

The rest of this section is devoted to a path decomposition of X under $\mathbb{P}^{(-\delta)}$ for $\delta > 2$. For $x > 0$, let as in Remark 4.2,

$$H_x := \sup\{r \geq 0 \ : \ L(\infty, r) = x\}.$$

Define

$$S_x := \sup\{t \geq 0 \ : \ X_t = H_x\}, \qquad \widehat{J}_x := \inf\{X_t,\ t \geq T_{H_x}\} - H_x,$$

where as before $T_{H_x} := \inf\{t \geq 0 : X_t = H_x\}$ is the hitting time of H_x by X.

Write $C_x := H_x + \widehat{J}_x$. We consider the following three processes:

$$X^{(1)} := (X_{S_x - t} - H_x,\ t \in [0, S_x - T_{H_x}]),$$
$$X^{(2)} := -(X_{T_{H_x} - t} - H_x,\ t \in [0, T_{H_x} - D_x]),$$
$$X^{(3)} := (X_{D_x - t} - C_x,\ t \in [0, D_x]),$$

where $D_x := \sup\{t < T_{H_x} : X_t = C_x\}$.

Furthermore, let $X^{(1),-}$ be the process $X^{(1)}$ obtained by removing all its positive excursions:

$$X_t^{(1),-} := X^{(1)}_{\alpha_t^{(1),-}},$$

with $\alpha_t^{(1),-} := \inf\{s > 0 : \int_0^s 1_{\{X_u^{(1)} \le 0\}} du$ and $t \le \int_0^{S_x - T_{H_x}} 1_{\{X_u^{(1)} \le 0\}} du$.

Proposition 4.4 *Let $\delta > 2$ and $x, a > 0$.*

(i) *Under $\mathbb{P}^{(-\delta)}$, $\frac{1}{|\widehat{J}_x|}$ is distributed as gamma$(\frac{\delta}{2}, \frac{2}{x})$.*

(ii) *Under $\mathbb{P}^{(-\delta)}(\cdot \mid \widehat{J}_x = -a)$, the three processes $X^{(1),-}$, $X^{(2)}$, $X^{(3)}$ are independent and distributed respectively as*

- *$(X_t, 0 \le t \le \tau_0(x))$, under $\mathbb{P}^\delta(\cdot | J(x) = -a)$, after removing all excursions above 0;*
- *a Bessel process $(R_t)_{0 \le t \le T_a}$ of dimension 3 starting from 0 killed at $T_a := \inf\{t > 0 : R_t = a\}$;*
- *$(X_t, 0 \le t \le T_{-a(1-u)/u})$ under $\mathbb{P}^\delta(\cdot | T_{-a(1-u)/u} < T_a)$, where $u \in [0, 1]$ is independently chosen according to the law beta $(\frac{\delta}{2} - 1, 1)$.*

Since $\widehat{J}_x$ under $\mathbb{P}^{(-\delta)}$ is distributed as $J(x)$ under $\mathbb{P}^\delta$, we observe that the (unconditional) law of $X^{(1),-}$ under $\mathbb{P}^{(-\delta)}$ is equal to that $(X_t, 0 \le t \le \tau_0(x))$, under $\mathbb{P}^\delta$, after removing all excursions above 0.

Proof Let $m > 0$. By Remark 4.2, conditionally on $\{H_x = m\}$, the process X^{-,H_x} is distributed as $(X_t^{-,m}, t < A_\infty^{-,m})$ under $\mathbb{P}^{(-\delta)}(\cdot | L(\infty, m) = x)$.

Note that conditionally on $\{H_x = m\}$, $S_x = \sup\{t > 0 : X_t \le m\}$ is the last passage time of X at m. By the duality of Proposition 2.14, under $\mathbb{P}^{(-\delta)}(\cdot | L(\infty, m) = x)$, the process $(X^{-,m}_{A_\infty^{-,m} - t} - m, t < A_\infty^{-,m})$ is distributed as $\{X_t^{-,0}, 0 \le t \le A^{-,0}_{T_{-m}}\}$ under $\mathbb{P}^\delta(\cdot | L(T_{-m}, 0) = x)$, the process $(X_t, 0 \le t \le T_{-m})$ obtained by removing all positive excursions. Furthermore, remark that $\widehat{J}(x)$ corresponds to $I_{\mathfrak{g}_m}$ which is defined for the process $(X_t, 0 \le t \le T_{-m})$ under $\mathbb{P}^\delta(\cdot | L(T_{-m}, 0) = x)$. Then $\mathbb{P}^{(-\delta)}(|\widehat{J}_x| \in \cdot \mid H_x = m) = \mathbb{P}^\delta(|I_{\mathfrak{g}_m}| \in \cdot \mid L(T_{-m}, 0) = x)$. We deduce that for any $0 < a < m$, the conditional law of the process $(X^{-,m}_{A_\infty^{-,m} - t} - m, t < A_\infty^{-,m})$ under $\mathbb{P}^{(-\delta)}(\cdot \mid H_x = m, \widehat{J}_x = -a)$

is the same as the conditional law of the process $\{X_t^{-,0}, 0 \le t \le A_{T_{-m}}^{-,0}\}$ under $\mathbb{P}^\delta(\cdot | L(T_{-m}, 0) = x, I_{\mathfrak{g}_m} = -a)$.

Then we may apply Theorem 3.3 (ii) to see that conditionally on $\{H_x = m, \widehat{J}_x = -a\}$, $X^{(1),-}$, $X^{(2)}$, and $X^{(3)}$ are independent, and

- $X^{(1),-}$ is distributed as $(X_t^{-,0}, t \le A_{\tau_0(x)}^{-,0})$ under $\mathbb{P}^\delta(\cdot | J(x) = -a)$, where $(X_t^{-,0}, t \le A_{\tau_0(x)}^{-,0})$ is the process obtained from $(X_t, 0 \le t \le \tau_0(x))$ by removing all positive excursions;
- $X^{(2)}$ is distributed as a three-dimensional Bessel process $(R_t)_{0 \le t \le T_a}$ killed at $T_a := \inf\{t > 0 : R_t = a\}$;
- $X^{(3)}$ is distributed as $(X_t, 0 \le t \le T_{-(m-a)})$ under $\mathbb{P}^\delta(\cdot | T_{-(m-a)} < T_a)$.

Moreover

$$\begin{aligned}\mathbb{P}^{(-\delta)}(|\widehat{J}_x| \in \mathrm{d}a \mid H_x = m) &= \mathbb{P}^\delta(I_{\mathfrak{g}_m} \in \mathrm{d}a \mid L(T_{-m}, 0) = x)\\ &= \frac{x}{2a^2} e^{-\frac{x}{2a} + \frac{x}{2m}} 1_{\{0 < a < m\}} \mathrm{d}a,\end{aligned}$$

where the last equality follows from (3.3).

Recall [8] the law of H_x under $\mathbb{P}^{(-\delta)}$: For $\delta > 2$,

$$\mathbb{P}(H_x \in \mathrm{d}m)/\mathrm{d}m = \frac{(\frac{x}{2})^{\frac{\delta}{2}-1}}{\Gamma(\frac{\delta}{2} - 1)} m^{-\frac{\delta}{2}} e^{-\frac{x}{2m}}, \qquad m > 0.$$

We get

$$\mathbb{P}^{(-\delta)}(|\widehat{J}_x| \in \mathrm{d}a) = \frac{1}{\Gamma(\frac{\delta}{2})} (\frac{x}{2})^{\frac{\delta}{2}} a^{-(\frac{\delta}{2}+1)} e^{-\frac{x}{2a}} \mathrm{d}a, \qquad a > 0,$$

which implies (i).

For any bounded continuous functionals F_1, F_2, F_3 on $\mathcal{K}$, we have

$$\begin{aligned}&\mathbb{E}^{(-\delta)}[F_1(X^{(1,\le 0)}) F_2(X^{(2)}) F_3(X^{(3)}) \mid \widehat{J}_x = -a]\\ &= \int_a^\infty \frac{\mathbb{P}^{(-\delta)}(|\widehat{J}_x| \in \mathrm{d}a, H_x \in \mathrm{d}m)}{\mathbb{P}(|\widehat{J}_x| \in \mathrm{d}a)} \mathbb{E}^{(-\delta)}[F_1(X^{(1),-} F_2(X^{(2)}) F_3(X^{(3)}) \mid \widehat{J}_x = -a, H_x = m]\\ &= (\frac{\delta}{2} - 1) \int_a^\infty \mathrm{d}m a^{\frac{\delta}{2}-1} m^{-\frac{\delta}{2}} \mathbb{E}^\delta[F_1(X_t^{-,0}, t \le A_{\tau_0(x)}^{-,0}) | J(x) = -a] \mathbb{E}[F_2(R_t, t \le T_a)]\\ &\qquad \times \mathbb{E}^\delta[F_3(X_t, t \le T_{-(m-a)}) \mid T_{-(m-a)} < T_a]\\ &= \mathbb{E}^\delta[F_1(X_t^{-,0}, t \le A_{\tau_0(x)}^{-,0}) | J(x) = -a] \mathbb{E}[F_2(R_t, t \le T_a)] \times\\ &\qquad (\frac{\delta}{2} - 1) \int_0^1 \mathrm{d}u u^{\frac{\delta}{2}-2} \mathbb{E}^\delta[F_3(X_t, t \le T_{-a(1-u)/u}) \mid T_{-a(1-u)/u} < T_a],\end{aligned}$$

which gives (ii) and completes the proof of the Proposition. □

5 The Perturbed Bessel Process and Its Rescaling at a Stopping Time

We rely on the paper of Doney, Warren and Yor [7], restricting our attention to the case of dimension $d = 3$. For $\kappa < 1$, the κ-perturbed Bessel process of dimension $d = 3$ starting from $a \geq 0$ is the process $(R_{3,\kappa},\ t \geq 0)$ solution of

$$R_{3,\kappa}(t) = a + W_t + \int_0^t \frac{\mathrm{d}s}{R_{3,\kappa}(s)} + \kappa(S_t^{R_{3,\kappa}} - a), \tag{5.1}$$

where $S_t^{R_{3,\kappa}} = \sup_{0\leq s\leq t} R_{3,\kappa}(s)$ and W is a standard Brownian motion. For $a > 0$, it can be constructed as the law of X under the measure $\mathbb{P}_a^{3,\kappa}$ defined by

$$\mathbb{P}_a^{3,\kappa}\big|_{\mathcal{F}_t} = \frac{1}{a^{1-\kappa}} \frac{X_{t\wedge T_0}}{(S_{t\wedge T_0})^{\kappa}} \tilde{\mathbb{P}}_a^{\delta}\big|_{\mathcal{F}_t} \tag{5.2}$$

where: $\delta := 2(1-\kappa)$, $S_t := \sup_{0\leq s\leq t} X_s$, and for any $a \geq 0$, X under $\tilde{\mathbb{P}}_a^{\delta}$ is distributed as $a - X$ under $\mathbb{P}^{\delta}$. Roughly speaking, the κ-perturbed Bessel process of dimension 3 can be thought of as the process $-X$ under $\mathbb{P}^{\delta}$ "conditioned to stay positive". The next proposition is very related to Lemma 5.1 of [7]. We let $\mathbb{P}^{3,\kappa} = \mathbb{P}_0^{3,\kappa}$ be a probability measure under which X is distributed as the process $R_{3,\kappa}$ starting from 0. Under $\mathbb{P}_a^{3,\kappa}$ and $\tilde{\mathbb{P}}_a^{\delta}$, we denote by $\{(q, \mathfrak{e}_{X,q}^{\downarrow}),\ q \in \mathcal{Q}_X^{\downarrow}\}$ the excursions of the process $(S_t, X_t - S_t)$ away from $\mathbb{R} \times \{0\}$. Notice that, under $\tilde{\mathbb{P}}_0^{\delta}$, by Proposition 2.2 and the invariance in distribution of the loop soup by the map $x \to -x$, these excursions are distributed as $\{(q, \mathfrak{e}_q^{\downarrow}),\ q \in \mathcal{Q}_{\frac{\delta}{2}}^{\downarrow} \cap (0, \infty)\}$.

Proposition 5.1 *Let $\kappa < 1$ and $\delta := 2(1-\kappa)$. The point process $\{(q,\ \mathfrak{e}_{X,q}^{\downarrow}),\ q \in \mathcal{Q}_X^{\downarrow}\}$ under $\mathbb{P}^{3,\kappa}$ is distributed as the Poisson point process*

$$\{(q, \mathfrak{e}_q^{\downarrow}),\ q \in \mathcal{Q}_{\frac{\delta}{2}}^{\downarrow} \textit{ such that } q + \mathfrak{e}_q^{\downarrow} \subset (0, \infty)\}.$$

In other words, the excursions of $R_{3,\kappa}$ below its supremum, seen as unrooted loops, are distributed as the loops of the Brownian loop soup $\mathcal{L}_{\frac{\delta}{2}}$ which entirely lie in the positive half-line.

Remark 5.2

(i) The intensity measure of this Poisson point process has been computed in (ii) of Lemma 2.3.
(ii) Similarly to Section 5.1 of Lupu [15], one can construct the process $R_{3,\kappa}$ from the loops of the Brownian loop soup $\mathcal{L}_{\frac{\delta}{2}}$ which entirely lie in $(0, \infty)$, by rooting them at their maxima and gluing them in the increasing order of their maxima.

(iii) The process $(R_{3,\kappa}, S^{R_{3,\kappa}})$ is a Markov process. Hence, applying the strong Markov property under $\mathbb{P}^{3,\kappa}$ to X at time T_a, we deduce that under $\mathbb{P}_a^{3,\kappa}$, the excursions below supremum of the process, seen as unrooted loops, are distributed as the loops of the Brownian loop soup $\mathcal{L}_{\frac{\delta}{2}}$ which entirely lie in the positive half-line and with maximum larger than a. It entails that for $a > 0$, the process X under $\mathbb{P}_a^{3,\kappa}$ is the limiting distribution as $m \to \infty$ of the process X under $\widetilde{\mathbb{P}}_a^{\delta}$ conditioned on hitting m before 0 (the process X is measurable with respect to its excursions below supremum, which are equally distributed before time T_m under $\mathbb{P}_a^{3,\kappa}$ and under $\widetilde{\mathbb{P}}_a^{\delta}(\cdot|T_m < T_0)$).

Proof of Proposition 5.1 Let $f : \mathbb{R}_+ \times \mathcal{K} \to \mathbb{R}_+$ be measurable. For any $0 < s < s'$, we compute

$$\mathbb{E}^{3,\kappa}\Big[\mathrm{e}^{-\sum_{q\in\mathcal{Q}_X^{\downarrow}\cap[s,s']} f(q,\mathfrak{e}_{X,q}^{\downarrow})}\Big].$$

Notice that the integrand is measurable with respect to the σ-algebra $\sigma(X_t,\ t \in [T_s, T_{s'}])$. By the strong Markov property at time T_s and the absolute continuity (5.2) with $a = s$ there, the previous expectation is equal to

$$s^{\kappa-1}\frac{s'}{(s')^{\kappa}}\,\widetilde{\mathbb{E}}_0^{\delta}\Big[\mathrm{e}^{-\sum_{q\in\mathcal{Q}_X^{\downarrow}\cap[s,s']} f(q,\mathfrak{e}_{X,q}^{\downarrow})}\,1_{\{T_0\circ\theta_{T_s} > T_{s'}\}}\Big]$$

where θ is the shift operator. Notice that

$$\mathrm{e}^{-\sum_{q\in\mathcal{Q}_X^{\downarrow}\cap[s,s']} f(q,\mathfrak{e}_{X,q}^{\downarrow})}\,1_{\{T_0\circ\theta_{T_s} > T_{s'}\}} = \mathrm{e}^{-\sum_{q\in\mathcal{Q}_X^{\downarrow}\cap[s,s'],\,q+\,\mathfrak{e}_{X,q}^{\downarrow}\subset(0,\infty)} f(q,\mathfrak{e}_{X,q}^{\downarrow})}\,1_{\mathcal{E}}$$

where $\mathcal{E}$ is the event that the set of $q \in \mathcal{Q}_X^{\downarrow} \cap [s, s']$ such that $q + \mathfrak{e}_{X,q}^{\downarrow} \not\subset (0,\infty)$ is empty.

We already mentioned that the collection of $(q, \mathfrak{e}_{q,X}^{\downarrow})$ for $q \in \mathcal{Q}_X^{\downarrow}$ is a Poisson point process under $\widetilde{\mathbb{P}}_0^{\delta}$, distributed as $\{(q, \mathfrak{e}_q^{\downarrow}),\ q \in \mathcal{Q}_{\frac{\delta}{2}}^{\downarrow} \cap (0,\infty)\}$. By the independence property of Poisson point processes, we deduce that

$$\mathbb{E}^{3,\kappa}\Big[\mathrm{e}^{-\sum_{q\in\mathcal{Q}_X^{\downarrow}\cap[s,s']} f(q,\mathfrak{e}_{X,q}^{\downarrow})}\Big] = c\widetilde{\mathbb{E}}^{\delta}\Big[\mathrm{e}^{-\sum_{q\in\mathcal{Q}_X^{\downarrow}\cap[s,s'],q+\mathfrak{e}_{X,q}^{\downarrow}\subset(0,\infty)} f(q,\mathfrak{e}_{X,q}^{\downarrow})}\Big]$$

for some constant c which is necessarily 1. The Proposition follows. □

In the rest of this section, we will extend the definition of perturbed Bessel processes to allow some positive local time at 0.

Let $x \geq 0$. We define a kind of *perturbed Bessel process* $R_{3,\kappa}^x$ *with local time* x *at position* 0. More precisely, for $\kappa < 1$ and $\delta = 2(1-\kappa)$, we denote by $R_{3,\kappa}^x$ the process obtained by concatenation in the following way: take $-X$ under $\mathbb{P}^{\delta}$ up to time $\tau_0(x)$, biased by $|J(x)|^{\frac{\delta}{2}-1}$ followed by a Bessel of dimension 3 killed when hitting $|J(x)|$, followed by the κ-pertubed process $R_{3,\kappa}$ starting from $|J(x)|$. Recall Theorem 3.3. Corollary 3.5 and Remark 5.2(iii) show that, when $x > 0$, $R^x_{3,\kappa}$ is the limit in distribution of the process $(-X_t,\, t \le T_{-m})$ under $\mathbb{P}^\delta(\cdot|L(T_{-m},0) = x)$ as $m \to \infty$. Clearly when $x = 0$, $R^0_{3,\kappa}$ coincides with $R_{3,\kappa}$ defined previously in (5.1) with $a = 0$. The following theorem shows that one can recover the process $R^x_{3,\kappa}$ by a suitable time-space scaling of a conditioned PRBM up to a hitting time. It is an extension of Theorem 2.2 equation (2.5) of Doney, Warren, Yor [7] [which corresponds to the case $x = 0$ and $m = 1$].

Theorem 5.3 *Suppose $\kappa < 1$ and let $\delta := 2(1-\kappa)$. Fix $m > 0$ and $x \ge 0$. Let the space-change*

$$\theta(z) := \begin{cases} -\frac{mz}{m+z} & \text{if } z \ge 0, \\ -z & \text{if } z < 0, \end{cases}$$

and the time-change

$$A_t := \int_0^t \left(\theta'\left(R^x_{3,\kappa}(s)\right)\right)^2 \mathrm{d}s, \qquad t \ge 0.$$

If $\widetilde{X}$ is defined via $\theta\left(R^x_{3,\kappa}(t)\right) := \widetilde{X}_{A_t}$, then $\widetilde{X}$ is distributed as $(X_t, 0 \le t \le T_{-m})$ under $\mathbb{P}^\delta(\cdot|L(T_{-m},0) = x)$.

Proof First we describe the excursions above infimum of X under $\mathbb{P}^\delta(\cdot|L(T_{-m},0) = x)$ in terms of the Brownian loop soup $\mathcal{L}_{\frac{\delta}{2}}$. For the loops which hit 0, we use again the same observation as in the proof of (3.3): conditionally on $\{\ell^0_\gamma : \gamma \in \mathcal{L}_{\frac{\delta}{2}}, 0 \in \gamma\}$, the loops γ such that $\min\gamma > -m$ are (the projection on the space of unrooted loops of) independent Brownian motions stopped at local time given by ℓ^0_γ, conditioned on not hitting $-m$.

Remark that the set $\{\ell^0_\gamma : \gamma \in \mathcal{L}_{\frac{\delta}{2}}, 0 \in \gamma, \min\gamma > -m\}$ is equal to the (non-ordered) set $\{x^{(m)}_{i-1} - x^{(m)}_i\}_{i\ge 1}$ in the notation of Proposition 2.6, and $L(T_{-m},0) = T^J_{-m}$. By Proposition 2.6, conditionally on $\{L(T_{-m},0) = x\}$, the ordered sequence of $(\ell^0_\gamma : \gamma \in \mathcal{L}_{\frac{\delta}{2}}, 0 \in \gamma, \min\gamma > -m)$ is distributed as $x(P_{(1)}, P_{(2)}, \ldots)$, where $(P_{(1)}, P_{(2)}, \ldots)$ has the Poisson–Dirichlet distribution of parameter $\frac{\delta}{2}$.

Note that the loops γ of $\mathcal{L}_{\frac{\delta}{2}}$ such that $\gamma \subset (-m,0)$ are independent of $L(T_{-m},0)$. Then the excursions above infimum of X (seen as unrooted loops) under $\mathbb{P}^\delta(\cdot|L(T_{-m},0) = x)$ consists in the superposition of:

- the loops γ of the Brownian loop soup $\mathcal{L}_{\frac{\delta}{2}}$ such that $\gamma \subset (-m,0)$;
- an independent collection of independent Brownian motions stopped at local time $x(P_{(1)}, P_{(2)}, \ldots)$, where $(P_{(1)}, P_{(2)}, \ldots)$ has the Poisson–Dirichlet distribution of parameter $\frac{\delta}{2}$.

Let $m \to \infty$, we deduce that the excursions below supremum of $R^x_{3,\kappa}$ (seen as unrooted loops) consists in the superposition of:

- the loops γ of the Brownian loop soup $\mathcal{L}_{\frac{\delta}{2}}$ such that $\min \gamma > 0$;
- an independent collection of independent Brownian motions stopped at local time $x(P_{(1)}, P_{(2)}, \dots)$, where $(P_{(1)}, P_{(2)}, \dots)$ has the Poisson–Dirichlet distribution of parameter $\frac{\delta}{2}$.

Note that via the transformation $\theta(R^x_{3,\kappa}(t)) = \widetilde{X}_{A_t}$, the excursions of $R^x_{3,\kappa}$ below their current supremum are transformed into excursions of $\widetilde{X}$ above their current infimum. Since the law of $R^x_{3,\kappa}$ is characterized by the law of its excursions below their current supremum (exactly as Remark 5.2(ii)), and the law of $\widetilde{X}$ is characterized by the law of its excursions above their current infimum in a similar way, we only have to focus on the law of those excursions and show that applying the transformation in space θ and in time A_t^{-1}, say Φ,[3]

(a) the loops γ of the Brownian loop soup $\mathcal{L}_{\frac{\delta}{2}}$ such that $\min \gamma > 0$ are transformed into loops $\widetilde{\gamma}$ of $\mathcal{L}_{\delta/2}$ such that $\max \widetilde{\gamma} < 0$ and $\min \widetilde{\gamma} > -m$;
(b) for any $\ell > 0$, a Brownian motion $(B_t, 0 \le t \le \tau^B_\ell)$ stopped at local time ℓ is transformed into $(B_t, 0 \le t \le \tau^B_\ell)$ conditioned on $\{\inf_{0 \le t \le \tau^B_\ell} B_t > -m\}$.

Let us prove (a). For a loop γ, we let $\gamma^{\uparrow}$ be the loop $\gamma - \min \gamma$ rooted at its minimum, and $\gamma^{\downarrow}$ be the loop $\gamma - \max \gamma$ rooted at its maximum. Notice that $\gamma^{\uparrow}$ is a positive excursion above 0, and $\gamma^{\downarrow}$ is a negative excursion below 0. We remark that for any loop γ with $\min \gamma > 0$, $\min \Phi(\gamma) = \theta(a)$ with $a := \max \gamma$, and $\Phi(\gamma)^{\uparrow} = \Phi(a + \gamma^{\downarrow}) - \theta(a)$. By Lemma 2.3 (ii), for any nonnegative measurable function f on $\mathbb{R}_- \times \mathcal{K}$, we have

$$\mathbb{E}\Big[e^{-\sum_{\gamma \in \mathcal{L}_{\delta/2}, \min \gamma > 0} f(\min \Phi(\gamma), \Phi(\gamma)^{\uparrow})}\Big]$$

$$= \exp\Big(-\delta \int_0^\infty \mathrm{d}a \int \mathfrak{n}^-(\mathrm{d}\mathfrak{e})(1 - e^{-f(\theta(a), \Phi(a+\mathfrak{e})-\theta(a))}) 1_{\{\min \mathfrak{e} > -a\}}\Big)$$

$$= \exp\Big(-\delta \int_0^\infty \mathrm{d}a \int \mathfrak{n}^+(\mathrm{d}\mathfrak{e})(1 - e^{-f(\theta(a), \Phi(a-\mathfrak{e})-\theta(a))}) 1_{\{\max \mathfrak{e} < a\}}\Big).$$

Let $h > 0$. Williams' description of the Itô measure says that under $\mathfrak{n}^+(\cdot \mid \max \mathfrak{e} = h)$, the excursion e can be split into two independent three-dimensional Bessel processes run until they hit h. For $a \ge h$, and a three-dimensional Bessel process R starting from 0 stopped when hitting h, the Itô formula together with the Dubins-Schwarz representation yield that $\Phi(a-R)-\theta(a)$ is still a three dimensional Bessel process run until it hits $\theta(a-h) - \theta(a)$ (this can also be seen as a special case of Theorem 2.2 equation (2.5) of Doney, Warren, Yor

[3] More precisely for any process $(\gamma_t, t \ge 0)$, $\Phi(\gamma)$ is the process defined by $\theta(\gamma_t) = \Phi(\gamma)\big(\int_0^t (\theta'(\gamma_s))^2 \mathrm{d}s\big)$.

[7] by taking $\alpha = 0$ there). It follows that under $\mathfrak{n}^+(\cdot \mid \max \mathfrak{e} = h)$, $\Phi(a - \mathfrak{e}) - \theta(a)$ is distributed as $\mathfrak{e}$ under $\mathfrak{n}^+(\cdot \mid \max \mathfrak{e} = \theta(a-h) - \theta(a))$. Consequently, for any $a > 0$,

$$
\begin{aligned}
&\int \mathfrak{n}^+(\mathrm{d}\mathfrak{e})(1 - e^{-f(\theta(a), \Phi(a-\,\mathfrak{e}) - \theta(a))}) 1_{\{\max \mathfrak{e} < a\}} \\
&= \int_0^a \frac{\mathrm{d}h}{2h^2} \int (1 - e^{-f(\theta(a), \mathfrak{e})}) \mathfrak{n}^+\big(\mathrm{d}\mathfrak{e} \mid \max \mathfrak{e} = \theta(a-h) - \theta(a)\big) \\
&= \frac{m^2}{(m+a)^2} \int_0^{|\theta(a)|} \frac{\mathrm{d}s}{2s^2} \int (1 - e^{-f(\theta(a), \mathfrak{e})}) \mathfrak{n}^+(\mathrm{d}\mathfrak{e} \mid \max \mathfrak{e} = s) \\
&= \frac{m^2}{(m+a)^2} \int \mathfrak{n}^+(\mathrm{d}\mathfrak{e})(1 - e^{-f(\theta(a), \mathfrak{e})}) 1_{\{\max \mathfrak{e} < |\theta(a)|\}},
\end{aligned}
$$

where the second equality follows from a change of variables $s = \theta(a-h) - \theta(a)$. It follows that

$$
\begin{aligned}
&\mathbb{E}\Big[e^{-\sum_{\gamma \in \mathcal{L}_{\frac{\delta}{2}}, \min \gamma > 0} f(\min \Phi(\gamma), \Phi(\gamma)^{\uparrow})}\Big] \\
&= \exp\Big(- \delta \int_0^\infty \mathrm{d}a \frac{m^2}{(m+a)^2} \int \mathfrak{n}^+(\mathrm{d}\mathfrak{e})(1 - e^{-f(\theta(a), \mathfrak{e})}) 1_{\{\max \mathfrak{e} < |\theta(a)|\}} \Big) \\
&= \exp\Big(- \delta \int_0^m \mathrm{d}y \int \mathfrak{n}^+(\mathrm{d}\,\mathfrak{e})(1 - e^{-f(-y, \mathfrak{e})}) 1_{\{\max \mathfrak{e} < y\}} \Big),
\end{aligned}
$$

after a change of variables $y = |\theta(a)|$. This proves (a).

It remains to show (b). Let $(\mathfrak{e}_s, s > 0)$ be the standard Brownian excursion process. It is well known that $(B_s, 0 \le s \le \tau_\ell^B)$ can be constructed from $(\mathfrak{e}_s, s \le \ell)$ (see Revuz and Yor [21] Chapter XII, Proposition 2.5). Observe that the process $\Phi(B_s, 0 \le s \le \tau_\ell^B)$ can be constructed from $(\Phi(\mathfrak{e}_s), s \le \ell)$ in the same way. To prove (b), it is enough to show that $(\Phi(\mathfrak{e}_s), s \le \ell)$ under the Itô measure $\mathfrak{n}$, is distributed as $(\mathfrak{e}_s, s \le \ell)$ under $\mathfrak{n}(\cdot \mid \inf_{s \le \ell} \min \mathfrak{e}_s > -m)$. To this end, we use the same observation as in the proof of (a): for any $h > 0$, under $\mathfrak{n}^+(\cdot \mid \max \mathfrak{e} = h)$, $\Phi(\mathfrak{e})$ is distributed as $-\mathfrak{e}$ under $\mathfrak{n}^+(\cdot \mid \max \mathfrak{e} = |\theta(h)|)$. Consequently, for any nonnegative measurable function f on $\mathcal{K}$,

$$
\begin{aligned}
\int \mathfrak{n}^+(\mathrm{d}\mathfrak{e})(1 - e^{-f(\Phi(\mathfrak{e}))}) &= \int_0^\infty \frac{\mathrm{d}h}{2h^2} \int (1 - e^{-f(-\mathfrak{e})})\, \mathfrak{n}^+(\mathrm{d}\mathfrak{e} \mid \max \mathfrak{e} = |\theta(h)|) \\
&= \int_0^m \frac{\mathrm{d}s}{2s^2} \int (1 - e^{-f(-\mathfrak{e})}) \mathfrak{n}^+(\mathrm{d}\mathfrak{e} \mid \max \mathfrak{e} = s) \\
&= \int \mathfrak{n}^+(\mathrm{d}\mathfrak{e})(1 - e^{-f(-\mathfrak{e})}) 1_{\{\max \mathfrak{e} < m\}},
\end{aligned}
$$

where the second equality follows from a change of variables $s = |\theta(h)|$. It follows that

$$\int \mathfrak{n}(d\mathfrak{e})(1 - e^{-f(\Phi(\mathfrak{e}))}) = \int \mathfrak{n}^-(d\mathfrak{e})(1 - e^{-f(-\mathfrak{e})}) + \int \mathfrak{n}^+(d\mathfrak{e})(1 - e^{-f(-\mathfrak{e})})1_{\{\max \mathfrak{e} < m\}}$$

$$= \int \mathfrak{n}(d\mathfrak{e})(1 - e^{-f(\mathfrak{e})})1_{\{\min \mathfrak{e} > -m\}},$$

which together with the exponential formula for the excursion process, yield that $(\Phi(\mathfrak{e}_s), s \le \ell)$ under $\mathfrak{n}$ is distributed as $(\mathfrak{e}_s, s \le \ell)$ under $\mathfrak{n}(\cdot \mid \inf_{s \le \ell} \min \mathfrak{e}_s > -m)$. This completes the proof of Theorem 5.3. □

Acknowledgments We are grateful to Titus Lupu for stimulating discussions on the link between the PRBM and the Brownian loop soup. We also thank an anonymous referee for useful suggestions on the paper. The project was partly supported by ANR MALIN.

References

1. Abraham, R., Mazliak, L.: Branching properties of Brownian paths and trees. Expo. Math. **16**(1), 59–73 (1998)
2. Bertoin, J., Le Gall, J.-F.: Stochastic flows associated to coalescent processes. II. Stochastic differential equations. Ann. Inst. H. Poincaré Probab. Stat. **41**, 307–333 (2005)
3. Carmona, P., Petit, F., Yor, M.: Sur les fonctionnelles exponentielles de certains processus de Lévy. Stoch. Stoch. Rep. **42**, 71–101 (1994)
4. Carmona, P., Petit, F., Yor, M.: Some extensions of the arc sine law as partial consequences of the scaling property of Brownian motion. Probab. Theory Relat. Fields **100**, 1–29 (1994)
5. Chaumont, L., Doney, R.A.: Pathwise uniqueness for perturbed versions of Brownian motion and reflected Brownian motion. Probab. Theory Relat. Fields **113**, 519–534 (1999)
6. Davis, B.: Brownian motion and random walk perturbed at extrema. Probab. Theory Relat. Fields **113**, 501–518 (1999)
7. Doney, R.A., Warren, J., Yor, M.: Perturbed Bessel processes. Séminaire de probabilités **32**, 237–249 (1998)
8. Gôing-Jaeschke, A., Yor, M.: A survey and some generalizations of Bessel processes. Bernoulli **9**, 313–349 (2003)
9. Kingman, J.F.C.: Random discrete distributions. J. R. Statist. Soc. **37**, 1–22 (1975)
10. Lawler, G.F., Werner, W.: The Brownian loop soup. Probab. Theory Relat. Fields **44**, 197–217 (2004)
11. Le Gall, J.-F., Yor, M.: Excursions Browniennes et carrés de processus de Bessel. C. R. Acad. Sci. Paris **302**, 641–643 (1986)
12. Le Gall, J.-F., Yor, M.: Étude asymptotique des enlacements du mouvement brownien autour des droites de l'espace. Probab. Theory Relat. Fields **74**, 617–635 (1987)
13. Le Jan, Y.: Markov loops and renormalization. Ann. Probab. **38**, 1280–1319 (2010)
14. Le Jan, Y.: Markov Paths, Loops and Fields. École d'été Saint-Flour XXXVIII 2008. Lecture Notes in Mathematics, vol. 2026. Springer, Berlin (2011)
15. Lupu, T.: Poisson Ensembles of Loops of One-Dimensional Diffusions. Mém. Soc. Math. Fr. (N.S.), vol. 158. Sociéte Mathématique de France, Paris (2018)
16. McCloskey, J.W.: A model for the distribution of individuals by species in an environment. Unpublished Ph.D. thesis, Michigan State University (1965)

17. Perman, M., Pitman, J., Yor, M.: Size-biased sampling of Poisson point processes and excursions. Probab. Theory Relat. Fields **92**, 21–39 (1992)
18. Perman, M.: An excursion approach to Ray–Knight theorems for perturbed Brownian motion. Stoch. Proc. Appl. **63**, 67–74 (1996)
19. Perman, M., Werner, W.: Perturbed Brownian motions. Probab. Theory Relat. Fields **108**, 357–383 (1997)
20. Petit, F.: Quelques extensions de la loi de l'arcsinus. C. R. Acad. Sci. Paris **315**, 855–858 (1992)
21. Revuz, D., Yor, M.: Continuous martingales and Brownian motion. Grundlehren der Mathematischen Wissenschaften [Fundamental Principles of Mathematical Sciences], 3rd edn. Springer, Berlin (1999)
22. Warren, J.: A stochastic flow arising in the study of local times. Probab. Theory Relat. Fields **133**, 559–572 (2005)
23. Warren, J., Yor, M.: The Brownian burglar: conditioning Brownian motion by its local time process. In: Séminaire de Probabilités XXXII. Lecture Notes in Mathematics, vol. 1686, pp. 328–342. Springer, Berlin (1998)
24. Werner, W.: Some remarks on perturbed reflecting brownian motion. Sém. Probab. **29**, 37–43 (1995)
25. Yor, M.: Some Aspects of Brownian Motion. Part I. Lectures in Mathematics ETH Zürich. Birkhäuser, Basel (1992)

On Doney's Striking Factorization of the Arc-Sine Law

Larbi Alili, Carine Bartholmé, Loïc Chaumont, Pierre Patie, Mladen Savov, and Stavros Vakeroudis

Abstract In Doney (Bull Lond Math Soc 19(2):177–182, 1987), R. Doney identifies a striking factorization of the arc-sine law in terms of the suprema of two independent stable processes of the same index by an elegant random walks approximation. In this paper, we provide an alternative proof and a generalization of this factorization based on the theory recently developed for the exponential functional of Lévy processes. As a by-product, we provide some interesting distributional properties for these variables and also some new examples of the factorization of the arc-sine law.

The authors are grateful to an anonymous referee for constructive comments.

L. Alili
Department of Statistics, The University of Warwick, Coventry, UK
e-mail: L.Alili@warwick.ac.uk

C. Bartholmé
Nordstad Lycée, Lycée classique de Diekirch et Unité de Recherche en Mathématiques, Université du Luxembourg, Luxembourg City, Luxembourg
e-mail: carine.bartholme@ext.uni.lu

L. Chaumont (✉)
LAREMA UMR CNRS 6093, Université d'Angers, Angers Cedex, France
e-mail: loic.chaumont@univ-angers.fr

P. Patie
School of Operations Research and Information Engineering, Cornell University, Ithaca, NY, USA
e-mail: pp396@cornell.edu

M. Savov
Institute of Mathematics and Informatics, Bulgarian Academy of Sciences, Sofia, Bulgaria
e-mail: mladensavov@math.bas.bg

S. Vakeroudis
Department of Mathematics, Statistics and Actuarial-Financial Mathematics, University of the Aegean, Karlovasi, Samos, Greece

L. Chaumont, A. E. Kyprianou (eds.), *A Lifetime of Excursions Through Random Walks and Lévy Processes*, Progress in Probability 78,
https://doi.org/10.1007/978-3-030-83309-1_3

Keywords Lévy processes · arc-sine law · mellin transform · recurrence equation

1 Introduction

Let $\mathrm{M}_\rho = \sup_{0\le t\le 1} X_t$ and $\widehat{\mathrm{M}}_\rho = \sup_{0\le t\le 1} -\widetilde{X}_t$ where $X = (X_t)_{t\ge 0}$ and $\widetilde{X} = (\widetilde{X}_t)_{t\ge 0}$ are two independent copies of a stable process of index $\alpha \in (0, 2)$ and positivity parameter $\rho \in (0, 1)$. Doney [9, Theorem 3] proved the following factorization of the arc-sine random variable $\mathcal{A}_\rho$ of parameter ρ

$$\frac{\mathrm{M}_\rho^\alpha}{\mathrm{M}_\rho^\alpha + \widehat{\mathrm{M}}_\rho^\alpha} \stackrel{(d)}{=} \mathcal{A}_\rho \tag{1.1}$$

where $\stackrel{(d)}{=}$ stands for the identity in distribution, and the law of $\mathcal{A}_\rho$ is absolutely continuous with a density given by

$$\frac{\sin(\pi\rho)}{\pi} x^{\rho-1}(1-x)^{-\rho},\ x \in (0, 1)\,.$$

The distributional identity (1.1) is remarkable because the law of the supremum of a stable process is usually a very complicated object whereas the arc-sine law has a simple distribution. In recent years, the law of M_ρ has been the interest of many researchers, see e.g. [11, 14, 15, 23, 25] where we can find series or Mellin-Barnes integral representations for the density of the supremum of a stable process valid for some set of parameters (α, ρ). We mention that Doney resorts to a limiting procedure to derive the factorization (1.1) of the arc-sine law. More specifically, his proof stems on a combination of an identity for each path of a random walk in the domain of attraction of a stable law with the arc-sine theorem which can be found in Spitzer [30]. We also mention that the arc-sine law appears surprisingly in different contexts in probability theory and in particular in the study of functionals of Brownian motion, see e.g. [8, 19, 22, 31].

The aim of this work is to provide an alternative proof and offer a generalization of Doney's factorization of the arc-sine law. The first key step relies on the well-known fact by now that, through the so-called Lamperti mapping, one can relate the law of the supremum of a stable process to the one of the exponential functional of a specific Lévy process, namely the Lamperti-stable process. It is then natural to wonder whether there are other factorizations of the arc-sine law given in terms of exponential functionals of more general Lévy processes. This will be achieved by resorting to the thorough study on the functional equation satisfied by the Mellin transform of the exponential functional of Lévy processes carried out in Patie and Savov [27].

Besides proving these identities in a more general framework, the problem of identifying a factorization of the exponential functionals as a simple distribution is interesting on its own since we shall show, on the way, that the law of the ratio of independent exponential functionals of some Lévy processes is the Beta prime's one which is known to belong to some remarkable sets of probability laws. This new fact is also relevant as the exponential functional of Lévy processes has attracted the attention of many researchers over the last two decades. The law of this random variable plays an important role in the study of self-similar processes, fragmentation and branching processes and is related to other theoretical problems as for example the moment problem and spectral theory of some non-self-adjoint semigroups, see [28]. Moreover it also plays an important role in more applied domains as for example in mathematical finance for the evaluation of Asian options, in actuarial sciences for random annuities, as well as in domains like astrophysics and biology. We refer to the survey paper [5] for a more detailed account on some of the mentioned fields. The remaining part of the paper is organized as follows. We state our main factorization of the arc-sine law along with some consequences and examples in the next section. The last section is devoted to the proofs.

2 The Arc-Sine Law and Exponential Functional of Lévy Processes

Throughout this paper we denote by $\xi = (\xi_t)_{t\geq 0}$ a possibly killed Lévy process issued from 0 and defined on the probability space $(\Omega, \mathcal{F}, \mathbb{P})$. It means that ξ is a real-valued stochastic process having independent and stationary increments and possibly killed at the random time $\mathbf{e}_q$, which is independent of ξ and exponentially distributed with parameter $q \geq 0$, where we understand that $\mathbf{e}_0 = +\infty$. We denote by Ψ its Lévy-Khintchine exponent, which, for any $z \in i\mathbb{R}$, takes the form

$$\log \mathbb{E}\left[e^{z\xi_1}\right] = \Psi(z) = az + \frac{1}{2}\sigma^2 z^2 + \int_{\mathbb{R}} \left(e^{zy} - 1 - zy\mathbb{I}_{\{|y|<1\}}\right) \Pi(dy) - q \tag{2.1}$$

where $a \in \mathbb{R}$, $\sigma \geq 0$ and Π is a Radon measure on $\mathbb{R}$ satisfying the conditions $\int_{\mathbb{R}}(1 \wedge y^2)\, \Pi(dy) < +\infty$ and $\Pi(\{0\}) = 0$. The law of ξ_1 is infinitely divisible and the one of ξ is uniquely characterized by the quadruplet (q, a, σ, Π). An excellent account on Lévy processes can be found in the monographs [4, 10, 16, 29]. Next, we define the exponential functional associated to the Lévy process ξ by

$$\mathrm{I}_\Psi = \int_0^\infty e^{\xi_t} dt = \int_0^{\mathbf{e}_q} e^{\xi_t} dt. \tag{2.2}$$

The variable I_Ψ is well defined if either $\Psi(0) = -q < 0$ or $\lim_{t\to+\infty} \xi_t = -\infty$ a.s. This last condition is equivalent to Erickson's integral tests involving the Lévy measure Π and the drift a, see Bertoin and Yor [5, Theorem 1]. With this remark in mind, we denote by $\mathcal{N}$ the set of Lévy Khintchine exponents of the form (2.1) for which the exponential functional I_Ψ is well defined, i.e.

$$\mathcal{N} = \left\{ \Psi \text{ of the form (2.1); } \Psi(0) < 0 \text{ or } \lim_{t \to +\infty} \xi_t = -\infty \text{ a.s.} \right\}. \tag{2.3}$$

Note that $\mathcal{N}$ is a subspace of the negative of continuous negative-definite functions, as defined in [12]. Next, it is well-known that $\Psi \in \mathcal{N}$ admits an analytical extension (still denoted by Ψ) to the strip $\mathbb{C}_{(0,\beta)} = \{z \in \mathbb{C};\ 0 < \Re(z) < \beta\}$ with $0 < \beta$ if and only if $\left|\mathbb{E}\left[e^{z\xi_1}\right]\right| < \infty$ for all $z \in \mathbb{C}_{(0,\beta)}$. Note that the existence of exponential moments for all $z \in \mathbb{C}_{(0,\beta)}$ is equivalent to

$$\int_{y>1} e^{uy}\Pi(dy) < \infty \text{ for all } u \in (0, \beta). \tag{2.4}$$

Under this condition, the restriction of Ψ on the real interval $(0, \beta)$ is convex and the condition $\lim_{t\to+\infty} \xi_t = -\infty$ a.s. is equivalent to $\Psi'(0^+) < 0$, see e.g. [5, Theorem 1 and Remark p.193]. We then define for any $\beta > 0$,

$$\mathcal{N}_\beta = \{\Psi \in \mathcal{N};\ (2.4) \text{ holds}\}.$$

Next, for any $\Psi \in \mathcal{N}_\beta$, let us denote

$$\rho = \sup\{u \in (0, \beta);\ \Psi(u) = 0\}$$

with the usual convention that $\sup\{\emptyset\} = +\infty$ and, introducing the notation

$$\overline{\Pi}_+(y) = \int_y^\infty \Pi(dr)\mathbb{I}_{\{y>0\}},$$

we define

$$\mathcal{N}_\beta(\rho) = \left\{\Psi \in \mathcal{N}_\beta;\, \rho < \infty,\, y \mapsto e^{\beta y}\overline{\Pi}_+(y) \text{ is non-increasing, } \infty < \lim_{u\uparrow 0} u\Psi(u+\beta) \leq 0\right\}.$$

We point out that if $\Psi \in \mathcal{N}_\beta(\rho)$ and $\lim_{u\uparrow 0} u\Psi(u+\beta)$ exists then necessarily $\lim_{u\uparrow 0} u\Psi(u+\beta) \leq 0$ as, by definition $\rho < \beta$, and Ψ is convex increasing on (ρ, β). Note also that for any $\Psi \in \mathcal{N}$ with $\overline{\Pi}_+ \equiv 0$, we always have $0 < \rho < \infty$ and thus $\Psi \in \mathcal{N}_\beta(\rho)$ for all $\beta > \rho$. We also point out if $|\Psi(\beta)| < \infty$, that is Ψ extends continuously to the line $\beta + i\mathbb{R}$, then plainly $\lim_{u\uparrow 0} u\Psi(u+\beta) = 0$. We are now ready to state our main result.

Theorem 2.1 *Assume that* $\Psi \in \mathcal{N}_1(\rho)$ *with* $0 < \rho < 1$, *then* $\widehat{\Psi}_1(z) = \Psi_1(-z) \in \mathcal{N}_1$ *with* $\widehat{\Psi}_1(1-\rho) = 0$, *where*

$$\Psi_1(z) = \frac{z}{z+1}\Psi(z+1), \ z \in i\mathbb{R},$$

and,

$$\frac{\mathrm{I}_{\widehat{\Psi}_1}}{\mathrm{I}_{\widehat{\Psi}_1} + \mathrm{I}_{\Psi}} \stackrel{(d)}{=} \mathcal{A}_\rho \text{ and } \frac{\mathrm{I}_{\Psi}}{\mathrm{I}_{\widehat{\Psi}_1} + \mathrm{I}_{\Psi}} \stackrel{(d)}{=} \mathcal{A}_{1-\rho} \tag{2.5}$$

where the variables I_Ψ *and* $\mathrm{I}_{\widehat{\Psi}_1}$ *are taken independent.*

We proceed by providing some consequences of this main result. We first derive some interesting distributional properties for the ratio of independent exponential functionals. To this end, we recall that a positive random variable is hyperbolically completely monotone if its law is absolutely continuous with a probability density f on $(0, \infty)$ which is such that the function h defined on $(0, \infty)$ by

$$h(w) = f(uv)\, f(u/v), \quad \text{with } w = v + v^{-1}, \tag{2.6}$$

is, for each fixed $u > 0$, completely monotone, i.e. $(-1)^n \frac{d^n}{dw^n} h(w) \geq 0$ on $(0, \infty)$ for all integers $n \geq 0$. This remarkable set of random variables was introduced by Bondesson and in [6, Theorem 2], he shows that it is a subset of the class of generalized gamma convolution. We recall that a positive random variable belongs to this latter class if it is self-decomposable, and hence infinitely divisible, such that its Lévy measure Π, concentrated on $\mathbb{R}^+$, is such that $\int_0^\infty (1 \wedge y)\Pi(dy) < \infty$ and $\Pi(dy) = \frac{k(y)}{y}dy$ where k is completely monotone. We also say that a positive random variable I is multiplicative infinitely divisible if $\log \mathrm{I}$ is infinitely divisible. It turns out that under some conditions the random variables I_Ψ is multiplicative infinitely divisible, see [1, Theorem 1.5] when Ψ is a Bernstein function and [27, Theorem 4.7] in the general case.

Corollary 2.2 *With the notation and assumptions of Theorem* 2.1, *the random variables* $\frac{\mathrm{I}_\Psi}{\mathrm{I}_{\widehat{\Psi}_1}}$ *and* $\frac{\mathrm{I}_{\widehat{\Psi}_1}}{\mathrm{I}_\Psi}$ *are hyperbolically completely monotone and multiplicative infinitely divisible. Moreover, when* $\rho = \frac{1}{2}$, *then* $\frac{\mathrm{I}_\Psi}{\mathrm{I}_{\widehat{\Psi}_1}}$ *is self-reciprocal, i.e. it has the same law than* $\frac{\mathrm{I}_{\widehat{\Psi}_1}}{\mathrm{I}_\Psi}$, *and it has the law of* C^2 *where* C *is a standard Cauchy variable.*

Another consequence of Theorem 2.1 is the following.

Corollary 2.3 *Doney's identity* (1.1) *holds.*

We close this section by describing another example illustrating our main factorization of the arc-sine law with some classical variables and refer the interested reader to the thesis [2] for the description of additional examples. Let us consider first $S(\alpha)$ a positive α-stable variable, with $0 < \alpha < 1$, and denote by $S_\gamma^{-\alpha}(\alpha)$ its γ-length-biased random variable, $\gamma > 0$, that is for any bounded measurable function g on $\mathbb{R}^+$, one has

$$\mathbb{E}\left[g(S_\gamma^{-\alpha}(\alpha))\right] = \frac{\mathbb{E}\left[S^{-\alpha\gamma}(\alpha)g(S^{-\alpha}(\alpha))\right]}{\mathrm{e}\left[S^{-\alpha\gamma}(\alpha)\right]}$$

where we recall that $\mathbb{E}\left[S^{-\alpha\gamma}(\alpha)\right] < \infty$, see e.g. [24, Section 3(3)]. We also denote by $\mathcal{G}_a$ a gamma variable of parameter $a > 0$.

Corollary 2.4 *Let $0 < \alpha, \rho < 1$, then we have the following factorization of the arc-sine law*

$$\frac{\mathcal{G}_{\alpha(1-\rho)}^{-\alpha}}{\mathcal{G}_{1-\rho}^{-1}S_\rho^{-\alpha}(\alpha) + \mathcal{G}_{\alpha(1-\rho)}^{-\alpha}} \stackrel{(d)}{=} \mathcal{A}_\rho$$

where the three variables $\mathcal{G}_{\alpha(1-\rho)}$, $S_\rho(\alpha)$ and $\mathcal{G}_{1-\rho}$ are taken independent.

3 Proofs

The proof of Theorem 2.1 is split into several intermediate results which might be of independent interests. First, let $(\mathcal{T}_\beta)_{\beta\in\mathbb{R}}$ be the group of transformations defined, for a function f on the complex plane, by

$$\mathcal{T}_\beta f(z) = \frac{z}{z+\beta} f\left(z+\beta\right). \tag{3.1}$$

In what follows, which is a slight extension of [26, Proposition 2.1], we show that under mild conditions, this family of transformations enables to identify an invariance property of the subset of Lévy-Khintchine exponents. Note that this lemma contains the first claim of Theorem 2.1.

Lemma 3.1 *Let $\beta_+ > 0$ and Ψ be of the form* (2.1) *such that for any $\beta \in (0, \beta_+)$, $|\Psi(\beta)| < \infty$. Then, for any $\beta \in (0, \beta_+]$ such that*

$$y \mapsto e^{\beta y}\overline{\Pi}_+(y) \text{ is non-increasing on } \mathbb{R}^+ \text{ and } -\infty < q_\beta = \lim_{u\uparrow 0} \mathcal{T}_\beta\Psi(u) \leq 0,$$

we have that $\mathcal{T}_\beta\Psi$ is also of the form (2.1)*. More specifically, its killing rate is $-q_\beta$, its Gaussian coefficient is σ and its Lévy measure takes the form*

$$\Pi_\beta(dy) = e^{\beta y}\left(\Pi(dy) + \beta dy\left((\Pi(-\infty, y) + q)\mathbb{I}_{\{y<0\}} - \overline{\Pi}_+(y)\right)\right), \; y \in \mathbb{R}. \tag{3.2}$$

Finally, if, in addition, $\Psi \in \mathcal{N}_\beta(\rho)$ *with* $\beta \in (\rho, \beta_+]$*, then* $\widehat{\Psi}_\beta = \widehat{\mathcal{T}_\beta \Psi} \in \mathcal{N}_\beta$ *with* $\widehat{\Psi}_\beta(\beta - \rho) = 0$.

Remark 3.2 Note that when $|\Psi(\beta)| < \infty$ then immediately $q_\beta = 0$. Moreover, the situation $|\Psi(\beta_+)| = \infty$ is allowed if 0 is a removable singularity for $\mathcal{T}_{\beta_+}\Psi$ with $q_{\beta_+} = \mathcal{T}_{\beta_+}\Psi(0) \leq 0$.

Proof For any $\beta \in (0, \beta_+)$, since in this case plainly $q_\beta = 0$, the claim is given in [26, Proposition 2.1] and thus it remains to prove it only for $\beta = \beta_+$. Note also that the expression of the characteristics of $\mathcal{T}_{\beta_+}\Psi$ follows from this aforementioned result and we now show that it is indeed a characteristic exponent of a Lévy process. To this end, we recall a few properties of the set of all negative definite functions $N(\mathbb{R})$ and the set of all continuous negative definite functions denoted by $CN(\mathbb{R})$ and refer to the monograph [12] for an excellent account on these sets of functions. A function $f : \mathbb{R} \longrightarrow \mathbb{C}$ is an element in $N(\mathbb{R})$ if and only if the following conditions are fulfilled $f(0) \geq 0$, $f(z) = \overline{f(-z)}$, and for any $k \in \mathbb{N}$ and any choice of values $z^1, \cdots, z^k \in \mathbb{R}$ and complex numbers $c_1, \cdots, c_k$

$$\sum_{j=1}^{k} c_j = 0 \text{ implies that } \sum_{j,l=1}^{k} f(z^j - z^l) c_j \overline{c_l} \leq 0. \tag{3.3}$$

It is easy to verify that $-\Psi(-z) \in CN(\mathbb{R})$, $z \in i\mathbb{R}$ and it is also well-known that any element of $CN(\mathbb{R})$ can be written as the negative of a characteristic function of a Lévy process. Now, we have, for any $\beta \in (0, \beta_+)$, $\mathcal{T}_\beta\Psi(z)$, $z \in i\mathbb{R}$, is the characteristic exponent of a conservative ($q_\beta = 0$) Lévy process. Denote, for $u \in (-\beta^+, 0)$

$$\mathcal{T}_{\beta^+}\Psi(u) = \lim_{\beta \to \beta^+} \mathcal{T}_\beta \Psi(u), \tag{3.4}$$

and set $\mathcal{T}_{\beta^+}\Psi(0) := \lim_{u \uparrow 0} \mathcal{T}_{\beta^+}\Psi(u) = q_\beta$ which is a non-negative constant by assumption. Then, let

$$\Phi(z) = \begin{cases} -\mathcal{T}_{\beta^+}\Psi(z) & \text{if } z \neq 0 \\ 0 & \text{if } z = 0. \end{cases}$$

Then Φ is an element of $N(\mathbb{R})$ since we know from [12, Lemma 3.6.7, p.123] that the set $N(\mathbb{R})$ is a convex cone which is closed under pointwise convergence. Let further

$$\widetilde{\Phi}(z) = \begin{cases} -q_\beta + \Phi(z) & \text{if } z \neq 0 \\ -q_\beta & \text{if } z = 0, \end{cases}$$

where we recall that $q_\beta \leq 0$. For any $k \in \mathbb{N}$ and any choice of values $z^1, \cdots, z^k \in \mathbb{R}$ and complex numbers $c_1, \cdots, c_k$ with $\sum_{j=1}^k c_j = 0$, since $\Phi \in N(\mathbb{R})$ and $-q_\beta \in N(\mathbb{R})$, we have

$$\sum_{j,l=1}^{k} \widetilde{\Phi}(z^j - z^l)c_j\overline{c_l} = \sum_{j,l=1;z^j \neq z^l}^{k} \Phi(z^j - z^l)c_j\overline{c_l} - q_\beta \sum_{j,l=1;z^j \neq z^l}^{k} c_j\overline{c_l} - q_\beta \sum_{j,l=1;z^j = z^l}^{k} c_j\overline{c_l}$$

$$= \sum_{j,l=1}^{k} \Phi(z^j - z^l)c_j\overline{c_l} - q_\beta \sum_{j,l=1}^{k} c_j\overline{c_l} \leq \sum_{j,l=1}^{k} \Phi(z^j - z^l)c_j\overline{c_l} \leq 0.$$

Hence $\widetilde{\Phi} \in N(\mathbb{R})$ and by continuity, $\mathcal{T}_{\beta^+}\Psi$ is the characteristic exponent of a possibly killed Lévy process. Next, let us assume that, in addition, $\Psi \in \mathcal{N}_\beta(\rho)$ with $\beta \in (\rho, \beta_+]$, then writing $\widehat{\Psi}_\beta = \widehat{\mathcal{T}_\beta \Psi}$, $\widehat{\Psi}_\beta$ is the characteristic exponent of the dual Lévy process associated to $\mathcal{T}_\beta\Psi$ and, as $\widehat{\Psi}_\beta(z) = \frac{z}{z-\beta}\Psi(-z+\beta)$, we have that $\widehat{\Psi}_\beta(z)$ is analytical on the strip $\mathbb{C}_{(0,\beta)}$ and $\widehat{\Psi}_\beta(\beta - \rho) = 0$. Moreover, since for $u \in (0, \beta)$, $\widehat{\Psi}'_\beta(u) = -\frac{u}{u-\beta}\Psi'(-u+\beta) - \frac{\beta}{(u-\beta)^2}\Psi(-u+\beta)$, we get, if $\beta \in (\rho, \beta_+)$, that $\widehat{\Psi}'_\beta(0) = -\frac{\Psi(\beta)}{\beta} < 0$ as $\beta > \rho$, either $\Psi(0) < 0$ or $\Psi'(0^+) < 0$ and Ψ is convex on $(0, \beta)$. Hence, in this case, $\widehat{\Psi}_\beta \in \mathcal{N}_\beta(\beta - \rho)$. Finally, if $q_{\beta_+} < 0$ then clearly $\widehat{\Psi}_{\beta_+} \in \mathcal{N}_{\beta_+}(\beta - \rho)$ whereas if $q_{\beta_+} = 0$ we complete the proof by recalling that $\widehat{\Psi}_{\beta_+}(\beta_+ - \rho) = 0$ with $\beta_+ - \rho > 0$ and the convexity of $\widehat{\Psi}_\beta$. □

Next, we denote by $\mathcal{M}_{\mathrm{I}}$ the Mellin transform of a random variable I, that is, for $z \in \mathbb{C}$,

$$\mathcal{M}_{\mathrm{I}}(z) = \mathbb{E}[\mathrm{I}^{z-1}]$$

and, mention that the mapping $t \mapsto \mathcal{M}_{\mathrm{I}}(it+1)$ for t real is a positive-definite function. We proceed by recalling a few basic facts about the Beta prime random variable, which we denote by $\mathcal{P}_{a,b}$, $a, b > 0$, that will be useful in the sequel of the proof. It can be defined via the identity

$$\mathcal{P}_{a,b} \stackrel{(d)}{=} \frac{\mathcal{G}_b}{\mathcal{G}_a} \tag{3.5}$$

where the two variables are gamma variables of parameter b and a respectively, are considered independent. It is well-known that the law of $\mathcal{G}_a$ is absolutely continuous with the following density

$$\frac{1}{\Gamma(a)} x^{a-1} e^{-x}, \ x > 0.$$

$\mathcal{P}_{a,b}$ is also a positive variable whose law is absolutely continuous with a density given by

$$\frac{1}{B(a,b)} x^{b-1} (1+x)^{-a-b}, \ x > 0,$$

where $B(a,b) = \frac{\Gamma(a)\Gamma(b)}{\Gamma(a+b)}$ is the Beta function. The Mellin transform of $\mathcal{P}_{a,b}$ is given by

$$\mathbb{E}\left[\mathcal{P}_{a,b}^{z}\right] = \frac{\Gamma(a-z)\Gamma(b+z)}{\Gamma(a)\Gamma(b)}, \ -b < \Re(z) < a. \tag{3.6}$$

From (3.6) it is easy to see that $\mathcal{P}_{a,b}$ admits moments of order u for any $u \in (-b, a)$. In particular $\mathcal{P}_{a,b}$ has infinite mean whenever $a \leq 1$. We refer to [13] for a nice exposition on these variables. When $\rho = a = 1 - b$, we write simply $\mathcal{P}_{\rho} = \mathcal{P}_{\rho,1-\rho}$ which is linked to the generalized arc-sine law $\mathcal{A}_{\rho}$ of order ρ in the following way

$$(1 + \mathcal{P}_{\rho})^{-1} \overset{(d)}{=} \mathcal{A}_{\rho}. \tag{3.7}$$

Simple algebra yields, from the identity (1.1), the following factorization

$$\frac{\widehat{M}_{\rho}^{\alpha}}{\mathrm{M}_{\rho}^{\alpha}} \overset{(d)}{=} \mathcal{P}_{\rho}. \tag{3.8}$$

Inspired by this reasoning, we shall prove the factorization of the variable $\mathcal{P}_{\rho}$ in terms of exponential functionals of Lévy processes. To this end, we shall need the following characterizations of the Mellin transform of the Beta prime variable $\mathcal{P}_{\rho}$.

Lemma 3.3 *For any* $0 < \rho < 1$, *we have*

$$\mathcal{M}_{\mathcal{P}_{\rho}}(z+1) = \frac{\Gamma(z+1-\rho)}{\Gamma(1-\rho)} \frac{\Gamma(-z+\rho)}{\Gamma(\rho)} \tag{3.9}$$

which defines an analytical function on the strip $\mathbb{C}_{(\rho-1,\rho)}$ *with simple poles at the edges of its domain of analyticity, that is at the points* ρ *and* $\rho - 1$. *Moreover, it is the unique positive-definite function solution to the recurrence equation, for* $z \in \mathbb{C}$,

$$\mathcal{M}_{\mathcal{P}_{\rho}}(z+1) = -\mathcal{M}_{\mathcal{P}_{\rho}}(z), \quad \mathcal{M}_{\mathcal{P}_{\rho}}(1) = 1. \tag{3.10}$$

Proof First, from the definition (3.5) of $\mathcal{P}_{\rho}$, one has, for any $0 < \rho < 1$ and $b > 0$,

$$\mathcal{M}_{\mathcal{P}_{\rho,b}}(z+1) = \mathcal{M}_{\mathcal{G}_b}(z+1)\mathcal{M}_{\mathcal{G}_{\rho}}(-z+1) = \frac{\Gamma(z+b)}{\Gamma(b)} \frac{\Gamma(-z+\rho)}{\Gamma(\rho)}, \tag{3.11}$$

and thus, as $\mathcal{P}_\rho = \mathcal{P}_{\rho,1-\rho}$,

$$\mathcal{M}_{\mathcal{P}_\rho}(z+1) = \frac{\Gamma(z+1-\rho)}{\Gamma(1-\rho)} \frac{\Gamma(-z+\rho)}{\Gamma(\rho)}. \tag{3.12}$$

As a by-product of classical properties of the gamma function, one gets that $z \mapsto \mathcal{M}_{\mathcal{P}_\rho}(z+1)$ defines an analytical function on the strip $\mathbb{C}_{(\rho-1,\rho)}$ and which extends as a meromorphic function on $\mathbb{C}$ with simple poles at the points $\rho+n$ and $\rho-1-n$, $n \in \mathbb{N}$. Then, using the recurrence relation of the gamma function $\Gamma(z+1) = z\Gamma(z)$, $z \in \mathbb{C}$, we deduce that, for any $z \in \mathbb{C}_{(\rho-1,\rho)}$,

$$\mathcal{M}_{\mathcal{P}_\rho}(z+1) = \frac{z-\rho}{-z+\rho} \frac{\Gamma(z-\rho)}{\Gamma(1-\rho)} \frac{\Gamma(-z+1+\rho)}{\Gamma(\rho)} = -\mathcal{M}_{\mathcal{P}_\rho}(z), \tag{3.13}$$

which is easily seen to be valid, in fact, for all $z \in \mathbb{C}$. To prove the uniqueness, one notes that any solution of (3.10) can be written as the product $\mathcal{M}_{\mathcal{P}_\rho} f$ where f is a periodic function with period 1 and $f(1) = 1$. However, since the Stirling's formula yields that for any $a \in \mathbb{R}$ fixed,

$$\lim_{|b|\to\infty} |\Gamma(a+ib)||b|^{-a+\frac{1}{2}} e^{|b|\frac{\pi}{2}} = C_a \tag{3.14}$$

where $C_a > 0$, see e.g. [18], one gets, from (3.13), that for large $|b|$,

$$|\mathcal{M}_{\mathcal{P}_\rho}(1+i|b|)| \sim \overline{C}_\rho |b|^{-\frac{1}{2}} e^{-|b|\pi}.$$

As the Mellin transform of a random variable is bounded on the line $1+i\mathbb{R}$, one has necessarily that $|f(z)| \leq e^{(|b|+\epsilon)\pi}$ for any $\epsilon > 0$ and some $C > 0$. An application of Carlson's theorem on the growth of periodic functions, see [20, p.96, (36)], gives that f is a constant which completes the proof. □

We state the following result which is proved in [27] regarding the recurrence equation solved by the Mellin transform of the exponential functional I_Ψ for a general Lévy process. Note that the exponential functional is defined in the aforementioned paper with $\widehat{\xi} = -\xi$, that is for $\widehat{\Psi}(-z) = \Psi(z)$.

Lemma 3.4 ([27], Theorem 2.4.) *For any $\Psi \in \mathcal{N}$, $\mathcal{M}_{\mathrm{I}_\Psi}$ is the unique positive-definite function solution to the functional equation*

$$\mathcal{M}_{\mathrm{I}_\Psi}(z+1) = \frac{-z}{\Psi(z)} \mathcal{M}_{\mathrm{I}_\Psi}(z), \quad \mathcal{M}_{\mathrm{I}_\Psi}(1) = 1, \tag{3.15}$$

which is valid (at least) on the dashed line $\mathcal{Z}_0^c(\Psi) \setminus \{0\}$, where we set $\mathcal{Z}_0(\Psi) = \{z \in i\mathbb{R};\ \Psi(z) = 0\}$. If $\Psi \in \mathcal{N}_1(\rho)$, $0 < \rho < 1$, the validity of the recurrence equation (3.15) extends to $\mathbb{C}_{(0,2)} \cup \mathcal{Z}_0^c(\Psi) \cup \mathcal{Z}_0^c(\Psi(.+1))$ and $\mathcal{M}_{\mathrm{I}_\Psi}(z+1)$ is analytical on the strip $\mathbb{C}_{(-1,\rho)}$ and meromorphic on $\mathbb{C}_{(-1,1)}$ with ρ as unique simple pole.

We mention that in [27, Theorem 2.1], an explicit representation of the solution on a strip of the functional equation (3.15) is provided in terms of the co-called Bernstein-gamma functions. This representation turns out to be very useful to provide substantial distributional properties, such as a Wiener-Hopf factorization, smoothness, small and large asymptotic behaviors, of the exponential functional of any Lévy processes, see Section 2 of the aforementioned paper.

We are now ready to complete the proof of Theorem 2.1. Let $\Psi \in \mathcal{N}_1(\rho)$ and $\widetilde{\Psi} \in \mathcal{N}$ and define the random variable $\bar{\mathrm{I}} = \frac{\mathrm{I}_\Psi}{\mathrm{I}_{\widetilde{\Psi}}}$, where we assumed that the exponential functionals I_Ψ and $\mathrm{I}_{\widetilde{\Psi}}$ are independent variables. Then, plainly

$$\mathcal{M}_{\bar{\mathrm{I}}}(z+1) = \mathcal{M}_{\mathrm{I}_\Psi}(z+1)\mathcal{M}_{\mathrm{I}_{\widetilde{\Psi}}}(-z+1) \tag{3.16}$$

and, Lemma 3.4 yields, after a shift by 1, and with $\mathcal{M}_{\bar{\mathrm{I}}}(1) = 1$, that

$$\mathcal{M}_{\bar{\mathrm{I}}}(z+2) = \frac{-z-1}{\Psi(z+1)} \frac{\widetilde{\Psi}(-z)}{z} \mathcal{M}_{\bar{\mathrm{I}}}(z+1),$$

for (at least) any z on the dashed line $\mathcal{Z}_0^c(\Psi(.+1) \cap \mathcal{Z}_0^c(\widetilde{\Psi}) \setminus \{0, 1\}$. Therefore, if one chooses $\widetilde{\Psi}$ of the form

$$\widetilde{\Psi}(-z) = \frac{z+1}{z}\Psi(z+1),$$

that is $\widetilde{\Psi}(-z) = \mathcal{T}_1\Psi(z)$ or $\widetilde{\Psi}(z) = \widehat{\mathcal{T}_1\Psi}(z) = \widehat{\Psi}_1(z)$, one gets that $\widehat{\Psi}_1 \in \mathcal{N}_1(1-\rho)$ and thus according to Lemma 3.4, $\mathcal{M}_{\mathrm{I}_{\widehat{\Psi}_1}}(-z+1)$ is analytical on the strip $\mathbb{C}_{(\rho-1,1)}$ with $1-\rho$ as a simple pole. Then we obtain, from (3.16), that $\mathcal{M}_{\bar{\mathrm{I}}}(z+1)$ is analytical on the strip $\mathbb{C}_{(\rho-1,\rho)}$ with $1-\rho$ and ρ as simple poles and it is solution to the recurrence equation

$$\mathcal{M}_{\bar{\mathrm{I}}}(z+1) = -\mathcal{M}_{\bar{\mathrm{I}}}(z).$$

Since plainly $\mathcal{M}_{\bar{\mathrm{I}}}(it+1)$ is a positive-definite function we conclude by the uniqueness argument given in Lemma 3.3 that

$$\frac{\mathrm{I}_\Psi}{\mathrm{I}_{\widehat{\Psi}_1}} \stackrel{(d)}{=} \mathcal{P}_\rho. \tag{3.17}$$

Invoking the identity (3.7), one obtains the first identity in (2.5). To get the second one, one deduces easily, from (3.17) and (3.5), that $\frac{\mathrm{I}_{\widehat{\Psi}_1}}{\mathrm{I}_\Psi}$ has the same law as $\mathcal{P}_{1-\rho}$, and, by means of (3.7) again completes the proof of the Theorem.

3.1 Proof of Corollary 2.2

First, from (3.17), we deduce easily that $\frac{\mathrm{I}_{\widehat{\Psi}_1}}{\mathrm{I}_\Psi}$ has the same law as $\mathcal{P}_{1-\rho}$. The fact that the density of the variable $\mathcal{P}_{1-\rho}$ is hyperbolic completely monotone was proved in [6]. Moreover, Berg [3] showed that $\log \mathcal{G}_a$ is infinitely divisible for any $a > 0$, we conclude the proof by recalling that the set of infinitely divisible variables is closed by linear combination of independent variables. Finally, the last claim follows readily from the definition of $\mathcal{P}_\rho$ and the connection with the standard Cauchy variable, which was observed by Pitman and Yor in [21].

3.2 Proof of Corollary 2.3

In order to prove the identity (1.1), we first recall the connection between the law of the maximum of a stable process and the exponential functional of a specific Lévy process, usually referred to as the Lamperti-stable process. This link has been established through the so-called Lamperti transform and we refer to [7, 15] and [17, Section 2.2] for more details. We proceed by providing the Lévy-Khintchine exponent $\Psi_{\alpha,\rho}$ of the Lamperti-stable process of parameters (α, ρ), $\alpha \in (0, 2)$ and $\rho \in (0, 1)$, which is given by

$$\Psi_{\alpha,\rho}(z) = -\frac{\Gamma(1+\alpha z)}{\Gamma(1-\alpha\rho+\alpha z)}\frac{\Gamma(\alpha-\alpha z)}{\Gamma(\alpha\rho-\alpha z)}, \; z \in \mathbb{C}_{(-\frac{1}{\alpha},1)}, \tag{3.18}$$

see [17, Theorem 2.3] where we consider here the exponent of $\alpha\xi^*$ in the notation of that paper. The following identity in law between the suprema of stable processes and the exponential functional of Lévy processes can be found, for example, in [15, p. 133],

$$\mathrm{M}_\rho^{-\alpha} \stackrel{(d)}{=} \mathrm{I}_{\Psi_{\alpha,\rho}} \tag{3.19}$$

where we recall that $\mathrm{M}_\rho = \sup_{0\le t\le 1} X_t$ and $X = (X_t)_{t\ge 0}$ is an α stable process with positivity parameter ρ. Next, observe that $\Psi_{\alpha,\rho}(\rho) = 0$ and using the recurrence relation of the gamma function, easy algebra yields

$$\begin{aligned}\frac{z}{z+1}\Psi_{\alpha,\rho}(z+1) &= -\frac{z}{z+1}\frac{\Gamma(1+\alpha+\alpha z)}{\Gamma(1+\alpha(1-\rho)+\alpha z)}\frac{\Gamma(-\alpha z)}{\Gamma(-\alpha(1-\rho)-\alpha z)}\\ &= -\frac{\Gamma(\alpha+\alpha z)}{\Gamma(\alpha(1-\rho)+\alpha z)}\frac{\Gamma(1-\alpha z)}{\Gamma(1-\alpha(1-\rho)-\alpha z)}.\end{aligned}$$

Then, we get that $\lim_{u\to 0}\frac{u}{u+1}\Psi_{\alpha,\rho}(u+1) = -\frac{\Gamma(\alpha)}{\Gamma(\alpha(1-\rho))}\frac{1}{\Gamma(1-\alpha(1-\rho))} \le 0$ as always $\alpha(1-\rho) \le 1$. We could easily check from the form (3.18) of $\Psi_{\alpha,\rho}$ and the expression of its Lévy measure given in [15] that $y \mapsto e^y\overline{\Pi}_+(y)$ is non-decreasing on $\mathbb{R}^+$. Instead, we simply observe from the computation above that

$$\widehat{\Psi}_1(z) = \mathcal{T}_1\Psi_{\alpha,\rho}(-z) = -\frac{\Gamma(\alpha-\alpha z)}{\Gamma(\alpha(1-\rho)-\alpha z)}\frac{\Gamma(1+\alpha z)}{\Gamma(1-\alpha(1-\rho)+\alpha z)} = \Psi_{\alpha,1-\rho}(z),$$

which is the characteristic exponent of the Lamperti-stable process with parameter $(\alpha, 1-\rho)$. Hence, similarly to (3.19), we have the following identity in law

$$\widehat{\mathrm{M}}_1^{-\alpha} \stackrel{(d)}{=} \mathrm{I}_{\widehat{\Psi}_1},$$

which by an application of Theorem 2.1 completes the proof.

3.3 *Proof of Corollary* 2.4

Let us now consider, for any $\alpha \in (0, 1)$,

$$\widehat{\Psi}_\alpha(z) = \frac{\Gamma(1+\alpha-\alpha z)}{\alpha\Gamma(-\alpha z)},\ z \in \mathbb{C}_{(-\infty,1+\frac{1}{\alpha})}.$$

In [24, Section 3.1], it is shown that $\widehat{\Psi}_\alpha$ is the Lévy-Khintchine exponent of a spectrally positive Lévy process with a negative mean and that $\mathrm{I}_{\widehat{\Psi}_\alpha}$ is a positive self-decomposable variable with

$$\mathrm{I}_{\widehat{\Psi}_\alpha} \stackrel{(d)}{=} \mathbf{e}^{-\alpha}.$$

In other words, $\bar{\mathrm{I}}_{\Psi_\alpha}$ has the Fréchet distribution of parameter $\rho = \frac{1}{\alpha} > 1$. Observe that $\widehat{\Psi}_\alpha(\rho) = 0$ and thus $\widehat{\Psi}_\alpha$ does not satisfy the hypothesis of Theorem 2.1. However, in [24, Section 3.1] it is also shown that, up to a positive multiplicative constant, the tail of the Lévy measure of $\widehat{\Psi}_\alpha$ is given by $\overline{\Pi}_\alpha(y) = e^{-(\alpha+1)y/\alpha}(1-e^{-y/\alpha})^{-\alpha-1}$, $y > 0$, and thus plainly the mapping $e^{\beta y}\overline{\Pi}_\alpha(y)$ is non-decreasing on $\mathbb{R}^+$ for any $\beta \le \frac{1}{\alpha}+1$. Then, Lemma 3.3 gives, for any $\rho = \beta - \frac{1}{\alpha} \le 1$, that

$$\widehat{\Psi}_{\alpha,\rho}(z) = \mathcal{T}_{\rho+\frac{1}{\alpha}}\widehat{\Psi}_\alpha(z) = \frac{z}{z+\rho+\frac{1}{\alpha}}\frac{\Gamma(\alpha-\alpha\rho-\alpha z)}{\alpha\Gamma(-1-\alpha\rho-\alpha z)} = -z\frac{\Gamma(\alpha-\alpha\rho-\alpha z)}{\Gamma(-\alpha\rho-\alpha z)}$$

is, since $\lim_{u\to 0}\widehat{\Psi}_{\alpha,1}(u) = \lim_{u\to 0}\frac{\Gamma(1-\alpha u)}{\alpha\Gamma(-\alpha-\alpha u)} = -\frac{1}{\Gamma(1-\alpha)} < 0$, the characteristic exponent of a Lévy process, as well as, by duality,

$$\Psi_{\alpha,\rho}(z) = z\frac{\Gamma(\alpha-\alpha\rho+\alpha z)}{\Gamma(-\alpha\rho+\alpha z)}.$$

Note that, if $0 < \rho < 1$ then $\Psi_{\alpha,\rho}(\rho) = 0$ and we deduce by convexity and since $\Psi_{\alpha,\rho}(0) = 0$ that $\Psi'_{\alpha,\rho}(0^+) < 0$. Hence $\Psi_{\alpha,\rho} \in \mathcal{N}_1(\rho)$. Using Lemma 3.4, one gets that

$$\mathcal{M}_{\mathrm{I}_{\Psi_{\alpha,\rho}}}(z+1) = -\frac{\Gamma(\alpha - \alpha\rho + \alpha z)}{\Gamma(-\alpha\rho + \alpha z)}\mathcal{M}_{\mathrm{I}_{\Psi_{\alpha,\rho}}}(z), \quad \mathcal{M}_{\mathrm{I}_{\Psi_{\alpha,\rho}}}(1) = 1. \tag{3.20}$$

As mentioned earlier, [27, Theorem 2.1] provides the solution of this functional equation which is derived as follows. First we recall that the analytical Wiener-Hopf factorization of $\Psi_{\alpha,\rho}$ is given by

$$\Psi_{\alpha,\rho}(z) = -(-z+\rho)\phi_\rho(z)$$

where $\phi_\rho(z) = \alpha z \frac{\Gamma(\alpha-\alpha\rho+\alpha z)}{\Gamma(1-\alpha\rho+\alpha z)}$ is a Bernstein function, see [10]. Then, the solution of (3.20) takes the form

$$\mathcal{M}_{\mathrm{I}_{\Psi_{\alpha,\rho}}}(z+1) = \frac{\Gamma(z+1)}{W_{\phi_\rho}(z+1)}\Gamma(\rho - z),$$

where $W_{\phi_\rho}(z+1) = \phi_\rho(z)W_{\phi_\rho}(z)$, $W_{\phi_\rho}(1) = 1$. To solve this latter recurrence equation, we note that $\phi_\rho(z) = \alpha\frac{z}{z-\rho}\frac{\Gamma(\alpha(z+1-\rho))}{\Gamma(\alpha(z-\rho))}$, then easy algebra and the uniqueness argument used in the proof of Lemma 3.3 yield that $W_{\phi_\rho}(z+1) = \alpha^z\frac{\Gamma(1-\rho)\Gamma(z+1)\Gamma(\alpha(z+1-\rho))}{\Gamma(z+1-\rho)\Gamma(\alpha(1-\rho))}$ and thus

$$\begin{aligned}\mathcal{M}_{\mathrm{I}_{\Psi_{\alpha,\rho}}}(z+1) &= \alpha^{-z}\frac{\Gamma(z+1-\rho)\Gamma(\alpha(1-\rho))}{\Gamma(1-\rho)\Gamma(z+1)\Gamma(\alpha(z+1-\rho))}\Gamma(z+1)\Gamma(\rho-z) \\ &= \alpha^{-z}\frac{\Gamma(z+1-\rho))}{\Gamma(\alpha(z+1-\rho))}\frac{\Gamma(\alpha(1-\rho))}{\Gamma(1-\rho)}\Gamma(\rho-z).\end{aligned}$$

Next, recalling that $S_\gamma(\alpha)$ is the γ-length biased variable of a positive α-stable random variable, we observe, from [24, Section 3(3)] that

$$\mathbb{E}\left[S_{1-\rho}^{-\alpha z}(\alpha)\right] = \frac{\mathbb{E}\left[S^{-\alpha(1-\rho+z)}(\alpha)\right]}{\mathbb{E}\left[S^{-\alpha(1-\rho)}(\alpha)\right]} = \frac{\Gamma(z+1-\rho))}{\Gamma(\alpha(z+1-\rho))}\frac{\Gamma(\alpha(1-\rho))}{\Gamma(1-\rho)}.$$

Thus, by Mellin transform identification, we get that

$$\mathrm{I}_{\Psi_{\alpha,\rho}} \stackrel{(d)}{=} \alpha^{-1}S_{1-\rho}^{-\alpha}(\alpha) \times \mathcal{G}_{1-\rho}^{-1}$$

where the variables on the right-hand side are taken independent. Next, we have, writing simply $\Psi_1 = \mathcal{T}_1 \Psi_{\alpha,\rho}$,

$$\mathcal{T}_1 \Psi_{\alpha,\rho}(z) = z \frac{\Gamma(\alpha(2-\rho+z))}{\Gamma(\alpha(1-\rho+z))}$$

which yields

$$\mathcal{M}_{\mathrm{I}_{\widehat{\Psi}_1}}(z+1) = \frac{\Gamma(\alpha(1-\rho-z))}{\Gamma(\alpha(2-\rho-z))} \mathcal{M}_{\mathrm{I}_{\widehat{\Psi}_1}}(z), \quad \mathcal{M}_{\widehat{\Psi}_1}(1) = 1.$$

It is not difficult to check that $\mathcal{M}_{\mathrm{I}_{\widehat{\Psi}_1}}(z+1) = \Gamma(\alpha(1-\rho-z))$ is the unique positive-definite solution of this equation. Invoking Theorem 2.1 completes the proof.

References

1. Alili, L., Jedidi, W., Rivero, V.: On exponential functionals, harmonic potential measures and undershoots of subordinators. ALEA Lat. Am. J. Probab. Math. Stat. **11**(1), 711–735 (2014)
2. Bartholmé, C.: Self-similarity and exponential functionals of Lévy processes. PhD thesis, Université Libre de Bruxelles (2014)
3. Berg, C.: On powers of stieltjes moment sequences, II. J. Comput. Appl. Math. **199**(1), 23–38 (2007)
4. Bertoin, J.: Lévy Processes. Cambridge Tracts in Mathematics, vol. 121. Cambridge University Press, Cambridge (1996)
5. Bertoin, J., Yor, M.: Exponential functionals of Lévy processes. Probab. Surv. **2**, 191–212 (2005)
6. Bondesson, L.: On univariate and bivariate generalized gamma convolutions. J. Stat. Plan. Inference **139**(11), 3759–3765 (2009)
7. Caballero, M.E., Chaumont, L.: Conditioned stable Lévy processes and the Lamperti representation. J. Appl. Probab. **43**(4), 967–983 (2006)
8. Chaumont, L., Yor, M.: Exercises in Probability. A Guided Tour from Measure Theory to Random Processes, Via Conditioning, 2nd edn. Cambridge University Press, Cambridge/New York (2012)
9. Doney, R.A.: On the maxima of random walks and stable processes and the arc-sine law. Bull. Lond. Math. Soc. **19**(2), 177–182 (1987)
10. Doney, R.A.: Fluctuation Theory for Lévy Processes. Ecole d'Eté de Probabilités de Saint-Flour XXXV-2005. Springer, Berlin (2007)
11. Doney, R.A., Savov, M.: The asymptotic behavior of densities related to the supremum of a stable process. Ann. Probab. **38**(1), 316–326 (2010)
12. Jacob, N.: Pseudo Differential Operators and Markov Processes. Vol. I, Fourier Analysis and Semigroups. Imperial College Press, London (2001)
13. Johnson, N.L., Kotz, S., Balakrishnan, N.: Continuous Univariate Distributions. Wiley Series in Probability and Mathematical Statistics: Applied Probability and Statistics. A Wiley-Interscience Publication, vol. 2, 2nd edn. Wiley, New York (1995)
14. Kuznetsov, A.: On extrema of stable processes. Ann. Probab. **39**(3), 1027–1060 (2011)
15. Kuznetsov, A., Pardo, J.C.: Fluctuations of stable processes and exponential functionals of hypergeometric Lévy processes. Acta Appl. Math. **123**, 113–139 (2013)

16. Kyprianou, A.E.: Introductory Lectures on Fluctuations of Lévy Processes with Applications. Universitext. Springer, Berlin (2006) KSK
17. Kyprianou, A.E.: Stable Lévy processes, self-similarity and the unit ball. ALEA Lat. Am. J. Probab. Math. Stat. **15**(1), 617–690 (2018)
18. Lebedev, N.N.: Special Functions and Their Applications. Dover Publications Inc., New York (1972). Revised edition, translated from the Russian and edited by Richard A. Silverman, Unabridged and corrected republication
19. Lévy, P.: Sur certains processus stochastiques homogènes. Compositio Math. **7**, 283–339 (1939)
20. Markushevich, A.I.: Entire functions. Translated from the Russian by Scripta Technica, Inc. Translation editor: Leon Ehrenpreis. American Elsevier Publishing Co., Inc., New York (1966)
21. Pitman, J., Yor, M.: Asymptotic laws of planar Brownian motion. Ann. Probab. **14**, 733–759 (1986)
22. Pitman, J., Yor, M.: Arcsine laws and interval partitions derived from a stable subordinator. Proc. Lond. Math. Soc. **65**(3), 326–356 (1992)
23. Patie, P.: A few remarks on the supremum of stable processes. Statist. Probab. Lett. **79**(8), 1125–1128 (2009)
24. Patie, P.: A refined factorization of the exponential law. Bernoulli **17**(2), 814–826 (2011)
25. Patie, P.: Law of the absorption time of some positive self-similar Markov processes. Ann. Probab. **40**(2), 765–787 (2012)
26. Patie, P., Savov, M.: Extended factorizations of exponential functionals of Lévy processes. Electron. J. Probab. **17**(38), 22 (2012)
27. Patie, P., Savov, M.: Bernstein-gamma functions and exponential functionals of Lévy processes. Electron. J. Probab. **23**(75), 1–101 (2018)
28. Patie, P., Savov, M.: Spectral expansion of non-self-adjoint generalized Laguerre semigroups. Mem. Am. Math. Soc. to appear, 1–179 (2019)
29. Sato, K.-I.: Lévy processes and infinitely divisible distributions. Cambridge Studies in Advanced Mathematics, vol. 68. Cambridge University Press, Cambridge (1999). Translated from the 1990 Japanese original, Revised by the author
30. Spitzer, F.: A combinatorial lemma and its application to probability theory. Trans. Am. Math. Soc. **82**, 323–339 (1956)
31. Vakeroudis, S., Yor, M.: A scaling proof for Walsh's Brownian motion extended arc-sine law. Electron. Commun. Probab. **17**(63), 1–9 (2012)
32. Yor, M.: Exponential functionals of Brownian motion and related processes. Springer Finance, Berlin (2001)

On a Two-Parameter Yule-Simon Distribution

Erich Baur and Jean Bertoin

Abstract We extend the classical one-parameter Yule-Simon law to a version depending on two parameters, which in part appeared in Bertoin (J Stat Phys 176(3):679–691, 2019) in the context of a preferential attachment algorithm with fading memory. By making the link to a general branching process with age-dependent reproduction rate, we study the tail-asymptotic behavior of the two-parameter Yule-Simon law, as it was already initiated in Bertoin (J Stat Phys 176(3):679–691, 2019). Finally, by superposing mutations to the branching process, we propose a model which leads to the two-parameter range of the Yule-Simon law, generalizing thereby the work of Simon (Biometrika 42(3/4):425–440, 1955) on limiting word frequencies.

Keywords Yule-Simon model · Crump-Mode-Jagers branching process · Population model with neutral mutations · Heavy tail distribution · Preferential attachment with fading memory

1 Introduction

The standard Yule process $Y = (Y(t))_{t\geq 0}$ is a basic population model in continuous time and with values in $\mathbb{N} := \{1, 2, \ldots\}$. It describes the evolution of the size of a population started from a single ancestor, where individuals are immortal and give birth to children at unit rate, independently one from the other. It is well-known that for every $t \geq 0$, $Y(t)$ has the geometric distribution with parameter e^{-t}. As a consequence, if T_ρ denotes an exponentially distributed random time with parameter

E. Baur
Bern University of Applied Sciences, Bern, Switzerland
e-mail: erich.baur@bfh.ch

J. Bertoin (✉)
Institut für Mathematik, Universität Zürich, Zürich, Switzerland
e-mail: jean.bertoin@math.uzh.ch

L. Chaumont, A. E. Kyprianou (eds.), *A Lifetime of Excursions Through Random Walks and Lévy Processes*, Progress in Probability 78,
https://doi.org/10.1007/978-3-030-83309-1_4

$\rho > 0$ which is independent of the Yule process, then for every $k \in \mathbb{N}$, there is the identity

$$\mathbb{P}\left(Y(T_\rho) = k\right) = \rho \int_0^\infty \mathrm{e}^{-\rho t}(1 - \mathrm{e}^{-t})^{k-1}\mathrm{e}^{-t}\mathrm{d}t = \rho B(k, \rho + 1), \tag{1}$$

where B is the beta function.

The discrete distribution in (1) has been introduced by H.A. Simon [20] in 1955 and is nowadays referred to as the Yule-Simon distribution with parameter ρ. It arises naturally in preferential attachment models and often explains the occurrence of heavy tail variables in stochastic modeling. Indeed, the basic estimate

$$B(k, \rho + 1) \sim \Gamma(\rho + 1)k^{-(\rho+1)} \qquad \text{as } k \to \infty,$$

implies that the Yule-Simon distribution has a fat tail with exponent ρ.

The present work is devoted to a two-parameter generalization of the Yule-Simon distribution, which results from letting the fertility (i.e. the reproduction rate) of individuals in the population model depend on their age. Specifically, imagine that now the rate at which an individual of age $a \geq 0$ begets children is $\mathrm{e}^{-\theta a}$ for some fixed $\theta \in \mathbb{R}$. So for $\theta > 0$ the fertility decays with constant rate θ as individuals get older, whereas for $\theta < 0$, the fertility increases with constant rate $-\theta$. Denote the size of the population at time t by $Y_\theta(t)$. In other words, $Y_\theta = (Y_\theta(t))_{t\geq 0}$ is a general (or Crump-Mode-Jagers) branching process, such that the point process on $[0, \infty)$ that describes the ages at which a typical individual begets a child is Poisson with intensity measure $\mathrm{e}^{-\theta t}\mathrm{d}t$. For $\theta = 0$, $Y_0 = Y$ is the usual Yule process.

Definition 1.1 Let $\theta \in \mathbb{R}$ and $\rho > 0$. Consider Y_θ as above and let T_ρ be an exponential random time with parameter $\rho > 0$, independent of Y_θ. We call the law of the discrete random variable

$$X_{\theta,\rho} := Y_\theta(T_\rho)$$

the Yule-Simon distribution with parameters (θ, ρ).

A key difference with the original Yule-Simon distribution, which corresponds to $\theta = 0$, is that no close expression for the two-parameter distribution is known.[1] Actually, the general branching process Y_θ is not even Markovian for $\theta \neq 0$, and its one-dimensional distributions are not explicit. This generalization of the Yule-Simon distribution has recently appeared in [1] for $\theta > 0$ and $\rho > (1 - \theta)^+$, in connection with a preferential attachment model with fading memory in the vein of Simon's original model. We shall point out in Sect. 5 that the range of parameters $\theta \leq 0$ and $\rho > 0$ arises similarly for a family of related models.

[1] Although the probability $\mathbb{P}(X_{\theta,\rho} = 1)$ can easily be computed in terms of an incomplete Gamma function, the calculations needed to determine $\mathbb{P}(X_{\theta,\rho} = k)$ for $k \geq 2$ become soon intractable.

One of the purposes of the present contribution is to describe some features of the two-parameter Yule-Simon law, notably by completing [1] and determining the tail-asymptotic behavior of $X_{\theta,\rho}$. It was observed in [1] that the parameter $\theta = 1$ is critical, in the sense that when $\theta < 1$, $X_{\theta,\rho}$ has a fat tail with exponent $\rho/(1-\theta)$, whereas when $\theta > 1$, some exponential moments of positive order of $X_{\theta,\rho}$ are finite. We show here in Sect. 4 that when $\theta > 1$, the tail of $X_{\theta,\rho}$ is actually decaying exponentially fast with exponent $\ln\theta - 1 + 1/\theta$. Further, in the critical case $\theta = 1$, we show that $X_{1,\rho}$ has a stretched exponential tail with stretching exponent $1/3$.

By superposing independent neutral mutations at each birth with fixed probability $1 - p \in (0, 1)$ to the classical Yule process, the original Yule-Simon law with parameter $\rho = 1/p$ captures the limit number of species of a genetic type chosen uniformly at random among all types, upon letting time tend to infinity. This fact is essentially a rephrasing of Simon's results in [20]. We give some (historical) background in Sect. 5 and extend Simon's observations to more general branching processes, for which the two-parameter distribution from Definition 1.1 is observed.

In a similar vein, the number of species belonging to a genus chosen uniformly at random has been studied for generalized Yule models in several works by Lansky, Polito, Sacerdote and further co-authors, both at fixed times t and upon letting $t \to \infty$. For instance, in [12], the linear birth process governing the growth of species is replaced by a birth-and-death process, whereas in [13], a fractional nonlinear birth process is considered instead. Both works are formulated in the framework of World Wide Web modeling. Recently, Polito [17] changed also the dynamics of how different genera appear, leading to a considerably different limit behavior.

The rest of this article is organized as follows. In the following Sect. 2, we analyze the branching process Y_θ introduced above and study its large-time behavior. In Sect. 3, we develop an integral representation for the tail distribution of the two-parameter Yule-Simon law, which lies at the heart of our study of the tail asymptotics of $X_{\theta,\rho}$ in the subsequent Sect. 4. This part complements the work [1] and contains our main results. In the last Sect. 5, we relate the generalized Yule-Simon distribution to a population model with mutations, in the spirit of Simon's original work [20].

2 Preliminaries on the General Branching Process Y_θ

The purpose of this section is to gather some basic features about the general branching process Y_θ that has been described in the introduction. We start with a construction of Y_θ in terms of a certain branching random walk.

Specifically, we consider a sequence $\mathbf{Z} = (\mathbf{Z}_n)_{n\geq 0}$ of point processes on $[0, \infty)$ which is constructed recursively as follows. First, $\mathbf{Z}_0 = \delta_0$ is the Dirac point mass at 0, and for any $n \geq 0$, $\mathbf{Z}_{n+1}$ is obtained from $\mathbf{Z}_n$ by replacing each and every atom of $\mathbf{Z}_n$, say located at $z \geq 0$, by a random cloud of atoms $\{z + \omega_i^z\}_{i=1}^{\mathrm{N}^z}$, where $\{\omega_i^z\}_{i=1}^{\mathrm{N}^z}$ is the family of atoms of a Poisson point measure on $[0, \infty)$ with intensity $\mathrm{e}^{-\theta t}\mathrm{d}t$ and to different atoms z correspond independent such Poisson point measures. In particular, each N^z has the Poisson distribution with parameter $1/\theta$ when $\theta > 0$, whereas $\mathrm{N}^z = \infty$ a.s. when $\theta \leq 0$. If we now interpret $[0, \infty)$ as a set of times, the locations of atoms as birth-times of individuals, and consider the number of individuals born on the time-interval $[0, t]$,

$$Y_\theta(t) := \sum_{n=0}^{\infty} \mathbf{Z}_n([0, t]), \qquad t \geq 0,$$

then $Y_\theta = (Y_\theta(t))_{t\geq 0}$ is a version of the general branching process generalizing the standard Yule process that was discussed in the introduction.

We readily observe the following formula for the first moments:

Proposition 2.1 *One has for every $t \geq 0$:*

$$\mathbb{E}(Y_\theta(t)) = \begin{cases} (\mathrm{e}^{(1-\theta)t} - \theta)/(1-\theta) & \text{if } \theta \neq 1, \\ 1+t & \text{if } \theta = 1. \end{cases}$$

Proof By definition, the intensity of the point process $\mathbf{Z}_1$ is $\mathrm{e}^{-\theta t}\mathrm{d}t$, and by the branching property, the intensity of $\mathbf{Z}_n$ is the nth convolution product of the latter. Considering Laplace transforms, we see that for any $q > 1 - \theta$:

$$\begin{aligned} q\int_0^\infty \mathbb{E}\left(Y_\theta(t)\right)\mathrm{e}^{-qt}\mathrm{d}t &= q\int_0^\infty \mathrm{e}^{-qt}\left(\sum_{n=0}^{\infty}\mathbb{E}\left(\mathbf{Z}_n([0,t])\right)\right)\mathrm{d}t \\ &= \sum_{n=0}^{\infty}\mathbb{E}\left(\int_0^\infty \mathrm{e}^{-qt}\mathbf{Z}_n(\mathrm{d}t)\right) \\ &= \sum_{n=0}^{\infty}(\theta+q)^{-n} \\ &= \frac{\theta+q}{\theta+q-1}. \end{aligned}$$

Inverting this Laplace transform yields our claim. □

Remark 2.2 The calculation above shows that a two-parameter Yule-Simon variable $X_{\theta,\rho}$, as in Definition 1.1, is integrable if and only if $\theta + \rho > 1$, and in that case we have

$$\mathbb{E}(X_{\theta,\rho}) = \frac{\theta+\rho}{\theta+\rho-1}.$$

Proposition 2.1 ensures the finiteness of the branching process Y_θ observed at any time. Further, it should be plain that the atoms of the branching random walk $\mathbf{Z}$ (at all generations) occupy different locations. Thus Y_θ is a counting process, in the sense that its sample paths take values in $\mathbb{N}$, are non-decreasing and all its jumps have unit size. We next discuss its large time asymptotic behavior, and in this direction, we write

$$Y_\theta(\infty) = \lim_{t\to\infty} \uparrow Y_\theta(t) \in \bar{\mathbb{N}} := \mathbb{N} \cup \{\infty\}$$

for its terminal value.

Proposition 2.3

(i) *If* $\theta > 0$, *then* $Y_\theta(\infty)$ *has the Borel distribution with parameter* $1/\theta$, *viz.*

$$\mathbb{P}(Y_\theta(\infty) = n) = \frac{\mathrm{e}^{-n/\theta}(n/\theta)^{n-1}}{n!} \qquad \text{for every } n \in \mathbb{N}.$$

In particular $\mathbb{P}(Y_\theta(\infty) < \infty) = 1$ *if and only if* $\theta \geq 1$.

(ii) *If* $\theta < 1$, *then*

$$\lim_{t\to\infty} \mathrm{e}^{(\theta-1)t} Y_\theta(t) = W_\theta \qquad \text{in probability,}$$

where $W_\theta \geq 0$ *is a random variable in* $L^k(\mathbb{P})$ *for any* $k \geq 1$. *Moreover, the events* $\{W_\theta = 0\}$ *and* $\{Y_\theta(\infty) < \infty\}$ *coincide a.s., and are both negligible (i.e. have probability* 0*) if* $\theta \leq 0$.

Proof

(i) When $\theta > 0$, $(\mathbf{Z}_n([0,\infty)))_{n\geq 0}$ is a Galton-Watson process with reproduction law given by the Poisson distribution with parameter $1/\theta$. In particular, it is critical for $\theta = 1$, sub-critical for $\theta > 1$, and super-critical for $\theta < 1$. In this setting, $Y_\theta(\infty)$ is the total population generated by a single ancestor in this Galton-Watson process; since the reproduction law is Poisson, it is well-known that $Y_\theta(\infty)$ is distributed according to the Borel distribution with parameter $1/\theta$.

(ii) The claims follow by specializing to our setting well-known results on general branching processes. More precisely, the fact that $\int_0^\infty \mathrm{e}^{-(1-\theta)t}\mathrm{e}^{-\theta t}\mathrm{d}t = 1$ shows that the so-called Malthus exponent of the general branching process Y_θ equals $1-\theta$. Then we just combine Theorem A of Doney [4], Theorem 1 of Bingham and Doney [2], and Theorem 3.1 in Nerman [15].

□

Finally, it will be convenient to also introduce

$$F_\theta(t) := \sum_{n=0}^{\infty} \int_0^t \mathrm{e}^{-\theta(t-s)} \mathbf{Z}_n(\mathrm{d}s), \qquad t \geq 0.$$

We call $F_\theta = (F_\theta(t))_{t\geq 0}$ the fertility process; it can be interpreted as follows. Recall that an atom, say at $s \geq 0$, of the branching random walk (at any generation n) is viewed as the birth-time of an individual, and $t - s$ is thus its age at time $t \geq s$. The times at which this individual begets children form a Poisson point measure on $[s, \infty)$ with intensity $\mathrm{e}^{-\theta(t-s)}\mathrm{d}t$. Hence, $F_\theta(t)$ should be viewed as the total rate of birth (therefore the name fertility) at time t for the population model described by Y_θ.

Proposition 2.4 *The fertility process F_θ is a Markov process on $(0, \infty)$ with infinitesimal generator*

$$\mathcal{G}_\theta f(x) = -\theta x f'(x) + x(f(x+1) - f(x)), \tag{2}$$

say for $f : (0, \infty) \to \mathbb{R}$ a bounded C^1 function with bounded derivative f'.

Remark 2.5 Specialists will have recognized from (2) that the fertility F_θ is a so-called continuous state branching process; see [10] and Chapter 12 in [11] for background.

Proof The fertility process starts from $F_\theta(0) = 1$, takes values in $(0, \infty)$, decays exponentially with constant rate θ (by convention, exponential decay with rate $\theta < 0$ means exponential increase with rate $-\theta > 0$), and makes jumps of unit size corresponding to birth events at time t. That is, there is the identity

$$F_\theta(t) = Y_\theta(t) - \theta \int_0^t F_\theta(s)\mathrm{d}s. \tag{3}$$

The claim should now be intuitively obvious since $F_\theta(t)$ is also the rate at time t at which the counting process Y_θ has a jump of unit size.

To give a rigorous proof, we introduce the filtration $\mathcal{F}_t = \sigma(\mathbf{1}_{[0,t]}\mathbf{Z}_n : n \in \mathbb{N})$ for $t \geq 0$. Since the point measure $\mathbf{Z}_1$ is Poisson with intensity $\mathrm{e}^{-\theta s}\mathrm{d}s$, the process

$$\mathbf{Z}_1([0, t]) - \int_0^t \mathrm{e}^{-\theta s}\mathrm{d}s, \qquad t \geq 0$$

is an $(\mathcal{F}_t)$-martingale. By the branching property, we have more generally that for any $n \geq 0$,

$$\mathbf{Z}_{n+1}([0, t]) - \int_0^t \int_0^s \mathrm{e}^{-\theta(s-r)}\mathbf{Z}_n(\mathrm{d}r)\mathrm{d}s, \qquad t \geq 0$$

is also an $(\mathcal{F}_t)$-martingale, and summing over all generations, we conclude that

$$Y_\theta(t) - \int_0^t F_\theta(s)\mathrm{d}s \qquad \text{is an } (\mathcal{F}_t)\text{-martingale.} \tag{4}$$

As Y_θ is a counting process, we deduce from (3) that for any bounded C^1 function $f : (0, \infty) \to \mathbb{R}$ with bounded derivative, there is the identity

$$f(F_\theta(t)) - f(1) = -\theta \int_0^t F_\theta(s) f'(F_\theta(s))\mathrm{d}s + \int_0^t (f(F_\theta(s-)+1) - f(F_\theta(s-))\mathrm{d}Y_\theta(s).$$

We now see from (4) that

$$f(F_\theta(t)) - \int_0^t \mathcal{G}_\theta(F_\theta(s))\mathrm{d}s \qquad \text{is an } (\mathcal{F}_t)\text{-martingale.}$$

It is readily checked that the martingale problem above is well-posed, and the statement follows; see Section 4.4 in [9] for background. □

We point out that for $f(x) = x$, we get $\mathcal{G}_\theta f = (1-\theta)f$, and it follows that $\mathbb{E}(F_\theta(t)) = \mathrm{e}^{(1-\theta)t}$ for all $t \geq 0$. We then see from (3) that for $\theta \neq 1$,

$$\mathbb{E}(Y_\theta(t)) = \mathrm{e}^{(1-\theta)t} + \theta \int_0^t \mathrm{e}^{(1-\theta)s}\mathrm{d}s = \frac{1}{1-\theta}(\mathrm{e}^{(1-\theta)t} - \theta),$$

and that $\mathbb{E}(Y_1(t)) = 1 + t$ for $\theta = 1$, hence recovering Proposition 2.1.

3 Poissonian Representation for the Tail Distribution

The purpose of this section is to point at the following representation of the tail distribution of the two-parameter Yule-Simon distribution. We first introduce a standard Poisson process $N = (N(t))_{t\geq 0}$. We write

$$\gamma(n) := \inf\{t > 0 : N(t) = n\}$$

for every $n \in \mathbb{N}$ (so that $\gamma(n)$ has the Gamma distribution with parameters $(n, 1)$), and

$$\zeta_\theta := \inf\{t > 0 : N(t) + 1 - \theta t = 0\} \tag{5}$$

for $\theta \in \mathbb{R}$ (in particular $\zeta_\theta = \infty$ a.s. when $\theta \leq 0$).

Proposition 3.1 *Let $\theta \in \mathbb{R}$ and $\rho > 0$. For every $n \in \mathbb{N}$, one has*

$$\mathbb{P}(X_{\theta,\rho} > n) = \mathbb{E}\left(\exp\left(-\rho \int_0^{\gamma(n)} (N(t) + 1 - \theta t)^{-1}\mathrm{d}t\right)\mathbf{1}_{\gamma(n)<\zeta_\theta}\right).$$

This identity could be inferred from [1]; for the sake of completeness, we shall provide here an independent proof based on Proposition 2.4 and the identity (3).

Proof of Proposition 3.1 Observe from Proposition 2.4 that the infinitesimal generator $\mathcal{G}_\theta$ of the fertility process fulfills

$$x^{-1}\mathcal{G}_\theta f(x) = -\theta f'(x) + (f(x+1) - f(x)), \qquad x > 0,$$

and that the right-hand side is the infinitesimal generator of a standard Poisson process with drift $-\theta$ absorbed at 0. If we write

$$\xi_\theta(t) := N(t \wedge \zeta_\theta) + 1 - \theta(t \wedge \zeta_\theta) \qquad \text{for } t \geq 0,$$

so that the process ξ_θ is that described above and started from $\xi_\theta(0) = 1$, then by Volkonskii's formula (see e.g. Formula (21.6) of Section III.21 in [19]), the fertility can be expressed as a time-change of ξ_θ. Specifically, the map $t \mapsto \int_0^t \xi_\theta(s)^{-1}\mathrm{d}s$ is bijective from $[0, \zeta_\theta)$ to $\mathbb{R}_+$, and if we denote its inverse by σ_θ, then the processes F_θ and $\xi_\theta \circ \sigma_\theta$ have the same distribution; we can henceforth assume that they are actually identical.

In this setting, we can further identify $\sigma_\theta(t) = \int_0^t F_\theta(s)\mathrm{d}s$ and then deduce from (3) that $Y_\theta(t) = 1 + N(\sigma_\theta(t))$. As a consequence, if we write

$$\tau_\theta(n) := \inf\{t > 0 : Y_\theta(t) > n\},$$

then we have also

$$\tau_\theta(n) = \inf\{t > 0 : N \circ \sigma_\theta(t) = n\} = \begin{cases} \int_0^{\gamma(n)} \xi_\theta(s)^{-1}\mathrm{d}s & \text{if } \gamma(n) < \zeta_\theta, \\ \infty & \text{otherwise.} \end{cases}$$

Finally, recall from Definition 1.1 that T_ρ has the exponential distribution with parameter $\rho > 0$ and is independent of Y_θ, so

$$\mathbb{P}(X_{\theta,\rho} > n) = \mathbb{P}(Y_\theta(T_\rho) > n) = \mathbb{E}\left(\exp(-\rho\tau_\theta(n))\mathbf{1}_{\tau_\theta(n)<\infty}\right).$$

This completes the proof. □

Remark 3.2 Following up Remark 2.5, the application of Volkonskii's formula in the proof above amounts to the well-known Lamperti's transformation that relates continuous state branching processes and Lévy processes without negative jumps via a time-change; see [3] for a complete account.

We conclude this section by pointing at a simple inequality between the tail distributions of Yule-Simon processes with different parameters.

Corollary 3.3

(i) *The random variable* $X_{\theta,\rho}$ *decreases stochastically in the parameters* θ *and* ρ. *That is, for every* $\theta' \geq \theta$ *and* $\rho' \geq \rho > 0$, *one has*

$$\mathbb{P}(X_{\theta',\rho'} > n) \leq \mathbb{P}(X_{\theta,\rho} > n) \qquad \textit{for all } n \in \mathbb{N}.$$

(ii) *For every* $\theta \in \mathbb{R}$, $\rho > 0$ *and* $a > 1$, *one has*

$$\mathbb{P}(X_{\theta,\rho} > n)^a \leq \mathbb{P}(X_{\theta,a\rho} > n) \qquad \text{for all } n \in \mathbb{N}.$$

Proof

(i) It should be plain from the construction of the general branching process Y_θ in the preceding section, that for any $\theta \leq \theta'$, one can obtain $Y_{\theta'}$ from Y_θ by thinning (i.e. random killing of individuals and their descent). In particular Y_θ and $Y_{\theta'}$ can be coupled in such a way that $Y_\theta(t) \geq Y_{\theta'}(t)$ for all $t \geq 0$. Obviously, we may also couple T_ρ and $T_{\rho'}$ such that $T_\rho \geq T_{\rho'}$ (for instance by defining $T_{\rho'} = \frac{\rho}{\rho'} T_\rho$), and our claim follows from the fact that individuals are eternal in the population model. Alternatively, we can also deduce the claim by inspecting Proposition 3.1.

(ii) This follows immediately from Hölder's inequality and Proposition 3.1.

□

4 Tail Asymptotic Behaviors

We now state the main results of this work which completes that of [1]. The asymptotic behavior of the tail distribution of a two parameter Yule-Simon distribution exhibits a phase transition between exponential and power decay for the critical parameter $\theta = 1$; here is the precise statement.

Theorem 4.1 *Let* $\rho > 0$.

(i) *If* $\theta < 1$, *then there exists a constant* $C = C(\theta, \rho) \in (0, \infty)$ *such that, as* $n \to \infty$:

$$\mathbb{P}(X_{\theta,\rho} > n) \sim C n^{-\rho/(1-\theta)}.$$

(ii) *If* $\theta > 1$, *then as* $n \to \infty$:

$$\ln \mathbb{P}(X_{\theta,\rho} > n) \sim -(\ln \theta - 1 + 1/\theta)n.$$

This phase transition can be explained as follows. We rewrite Proposition 3.1 in the form

$$\mathbb{P}(X_{\theta,\rho} > n) = \mathbb{E}\left(\exp\left(-\rho \int_0^{\gamma(n)} (N(t) + 1 - \theta t)^{-1} \mathrm{d}t\right) \mid \gamma(n) < \zeta_\theta\right) \times \mathbb{P}(\gamma(n) < \zeta_\theta).$$

On the one hand, the probability that $\gamma(n) < \zeta_\theta$ remains bounded away from 0 when $\theta < 1$ and decays exponentially fast when $\theta > 1$. On the other hand, for $\theta < 1$, the integral $\int_0^{\gamma(n)} (N(t) + 1 - \theta t)^{-1} \mathrm{d}t$ is of order $\ln n$ on the event $\{\gamma(n) < \zeta_\theta\}$, and therefore the first term in the product decays as a power of n when n tends to infinity. Last, when $\theta > 1$, the first term in the product decays sub-exponentially fast.

In the critical case $\theta = 1$, we observe from the combination of Theorem 4.1 and Corollary 3.3 that the tail of $X_{1,\rho}$ is neither fat nor light, in the sense that

$$\exp(-\alpha n) \ll \mathbb{P}(X_{1,\rho} > n) \ll n^{-\beta}$$

for all $\alpha, \beta > 0$. We obtain a more precise estimate of stretched exponential type. In the following statement, $f \lesssim g$ means $\limsup_{n\to\infty} f(n)/g(n) \le 1$.

Theorem 4.2 *Consider the critical case $\theta = 1$, and let $\rho > 0$. Then we have as $n \to \infty$:*

$$-20(\rho^2 n)^{1/3} \lesssim \ln \mathbb{P}(X_{1,\rho} > n) \lesssim -(1/2)^{1/3}(\rho^2 n)^{1/3}.$$

Remark 4.3 Note that Theorems 4.1 and 4.2 entail that the series $\sum_{n\ge 0} \mathbb{P}(X_{\theta,\rho} > n)$ converges if and only if $\theta + \rho > 1$, in agreement with Remark 2.2.

The methods we use to prove Theorem 4.2 seem not to be fine enough to obtain the exact asymptotics of $n^{-1/3} \ln \mathbb{P}(X_{1,\rho} > n)$. More specifically, for the lower bound we employ estimates for first exits through moving boundaries proved by Portnoy [18] first for Brownian motion and then transferred via the KMT-embedding to general sums of independent random variables. The constant $c_1 = 20$ is an (rough) outcome of our proof and clearly not optimal.

For obtaining the upper bound, we consider an appropriate exponential martingale and apply optional stopping. The constant $c_2 = (1/2)^{1/3}$ provides the best value given our method, but is very likely not the optimal value neither.

Theorem 4.1(i) has been established in Theorem 1(ii) of [1] in the case $\theta \in (0, 1)$ and $\rho > 1-\theta$. Specifically, the parameters α and $\bar{p}$ there are such that, in the present notation, $\theta = \alpha/\bar{p}(\alpha + 1)$ and $\bar{p}(\alpha + 1) = 1/\rho$. Taking this into account, we see that the claim here extends Theorem 1(ii) in [1] to a larger set of parameters. The argument is essentially the same, relying now on Proposition 2.3(ii) here rather than on the less general Corollary 2 in [1], and we won't repeat it.

We next turn our attention to the proof of Theorem 4.1(ii), which partly relies on the following elementary result on first-passage times of Poisson processes with drift (we refer to [5–7] for related estimates in the setting of general random walks and Lévy processes).

Lemma 4.4 *Let $b > 1$, $x > 0$, and define $\nu(x) := \inf\{t > 0 : bt - N(t) > x\}$. The distribution of the integer-valued variable $b\nu(x) - x$ fulfills*

$$\mathbb{P}(b\nu(x) - x = n) = \frac{1}{n!}\mathrm{e}^{-(x+n)/b} x(x+n)^{n-1} b^{-n} \sim \frac{x\mathrm{e}^{x(1-1/b)}}{\sqrt{2\pi n^3}} \mathrm{e}^{n(1-1/b-\ln b)} \qquad \text{as } n \to \infty.$$

As a consequence,

$$\lim_{t\to\infty} t^{-1}\ln \mathbb{P}(\nu(x)\geq t) = -(b\ln b - b + 1).$$

Proof The event $b\nu(x) = x$ holds if and only if the Poisson process N stays at 0 up to time x/b at least, which occurs with probability $e^{-x/b}$. The first identity in the statement is thus plain for $n = 0$. Next, note that, since the variable $b\nu(x) - x$ must take integer values whenever it is finite, there is the identity

$$b\nu(x) - x = \inf\{j \geq 0 : N((j + x)/b) = j\}.$$

On the event $N(x/b) = k \in \mathbb{N}$, set $N'(t) = N(t + x/b) - k$, and write

$$b\nu(x) - x = \inf\{j \in \mathbb{N} : N'(j/b) = j - k\}.$$

Since N' is again a standard Poisson process, Kemperman's formula (see, e.g. Equation (6.3) in [16]) applied to the random walk $N'(\cdot/b)$ gives for any $n \geq k$

$$\mathbb{P}(b\nu(x) - x = n \mid N(x/b) = k) = \frac{k}{n}\cdot\frac{e^{-n/b}(n/b)^{n-k}}{(n-k)!}.$$

Since $N(x/b)$ has the Poisson distribution with parameter x/b, this yields for any $n \geq 1$

$$\begin{aligned}\mathbb{P}(b\nu(x) - x = n) &= e^{-(x+n)/b}\sum_{k=1}^{n}\frac{(x/b)^k}{k!}\cdot\frac{k}{n}\cdot\frac{(n/b)^{n-k}}{(n-k)!}\\ &= \frac{1}{n!}e^{-(x+n)/b}(x/b)\sum_{k=1}^{n}(x/b)^{k-1}(n/b)^{n-k}\cdot\frac{(n-1)!}{(k-1)!(n-k)!}\\ &= \frac{1}{n!}e^{-(x+n)/b}x(x+n)^{n-1}b^{-n},\end{aligned}$$

where we used Newton's binomial formula at the third line. The second assertion in the claim follows from Stirling's formula, and the third one is a much weaker version. □

We can now proceed to the proof of Theorem 4.1(ii).

Proof of Theorem 4.1(ii) The upper-bound is easy. Indeed on the one hand, Proposition 3.1 yields

$$\mathbb{P}(X_{\theta,\rho} > n) \leq \mathbb{P}(\gamma(n) < \zeta_\theta),$$

and on the other hand, since $N(\zeta_\theta) + 1 = \theta\zeta_\theta$, on the event $\{\gamma(n) < \zeta_\theta\}$, one has obviously $N(\zeta_\theta) \geq n$, and *a fortiori* $\theta\zeta_\theta > n$. Thus $\mathbb{P}(X_{\theta,\rho} > n)$ is bounded from above by $\mathbb{P}(\zeta_\theta > n/\theta)$, and we conclude from Lemma 4.4 specialized for $x = 1$ and $b = \theta$ that

$$\limsup_{n\to\infty} n^{-1} \ln \mathbb{P}(X_{\theta,\rho} > n) \leq -(\ln\theta - 1 + 1/\theta).$$

In order to establish a converse lower bound, let $\varepsilon \in (0, 1)$ be arbitrarily small, and consider the event

$$\Lambda(n, \theta, \varepsilon) := \{N(t) + 1 - \varepsilon - (\theta + \varepsilon)t \geq 0 \text{ for all } 0 \leq t \leq \gamma(n)\}.$$

On that event, one has $N(t) + 1 - \theta t \geq \varepsilon(1 + t)$ for all $0 \leq t \leq \gamma(n)$, and hence

$$\exp\left(-\rho \int_0^{\gamma(n)} (N(t) + 1 - \theta t)^{-1} \mathrm{d}t\right) \geq (\gamma(n) + 1)^{-\rho/\varepsilon} \geq (n/(\theta + \varepsilon))^{-\rho/\varepsilon},$$

where for the second inequality, we used that $N(\gamma(n)) + 1 - \varepsilon \geq (\theta + \varepsilon)\gamma(n)$.

We are left with estimating $\mathbb{P}(\Lambda(n, \theta, \varepsilon))$. Set $b = \theta + \varepsilon$ and use the notation of Lemma 4.4, so that

$$\Lambda(n, \theta, \varepsilon) = \{\gamma(n) < \nu(1 - \varepsilon)\}.$$

On the event $\{b\nu(1 - \varepsilon) \geq n\}$, one has

$$N(\nu(1 - \varepsilon)) = b\nu(1 - \varepsilon) - 1 + \varepsilon \geq n + \varepsilon - 1,$$

so actually $\gamma(n) < \nu(1-\varepsilon)$. Hence $\{b\nu(1-\varepsilon) \geq n\} \subset \Lambda(n, \theta, \varepsilon)$, and we conclude from Lemma 4.4 that

$$\liminf_{n\to\infty} n^{-1} \ln \mathbb{P}(\Lambda(n, \theta, \varepsilon)) \geq 1 - \ln b - 1/b.$$

Putting the pieces together, we have shown that for any $b > \theta$

$$\liminf_{n\to\infty} n^{-1} \ln \mathbb{E}\left(\exp\left(-\rho \int_0^{\gamma(n)} (N(t) + 1 - \theta t)^{-1} \mathrm{d}t\right) \mathbf{1}_{\gamma(n) < \zeta_\theta}\right) \geq 1 - \ln b - 1/b.$$

Thanks to Proposition 3.1, this completes the proof. □

We next establish Theorem 4.2.

Proof of Theorem 4.2 We use the abbreviations $\xi(t) := N(t) + 1 - t$ and $\zeta := \zeta_1$, and start with the lower bound. We let $0 < \varepsilon < 1$. First note that there are the inclusions of events

$$\{\gamma(n) < \zeta\} \supset \{\gamma(n) < \min\{\zeta,\ (1+\varepsilon)n\}\} \supset \{\xi(t) > 0 \text{ for all } 0 \le t \le (1+\varepsilon)n\,,\ \gamma(n) < (1+\varepsilon)n\}$$
$$\supset \{\xi(t) > \rho^{1/3}t^{2/3} \text{ for all } 0 < t \le (1+\varepsilon)n\,,\ \gamma(n) < (1+\varepsilon)n\}.$$

In particular, with Proposition 3.1 at hand, we obtain for small ε and large n

$$\mathbb{P}(X_{1,\rho} > n) = \mathbb{E}\left(\exp\left(-\rho\int_0^{\gamma(n)} (\xi(t))^{-1}\mathrm{d}t\right)\mathbf{1}_{\gamma(n)<\zeta}\right)$$
$$\ge \exp\left(-\rho\int_0^{(1+\varepsilon)n} \frac{1}{\rho^{1/3}t^{2/3}}\mathrm{d}t\right)\mathbb{P}(\xi(t) > \rho^{1/3}t^{2/3} \text{ for all } 0 < t \le (1+\varepsilon)n\,,\ \gamma(n) < (1+\varepsilon)n)$$
$$\ge \exp\left(-4(\rho^2 n)^{1/3}\right)\mathbb{P}(\xi(t) > \rho^{1/3}t^{2/3} \text{ for all } 0 < t \le (1+\varepsilon)n) - \mathbb{P}\left(\gamma(n) \ge (1+\varepsilon)n\right),$$

where for the last line, we used that $\exp(-3(\rho^2(1+\varepsilon)n)^{1/3}) \ge \exp(-4(\rho^2 n)^{1/3})$ for small ε, and that $\mathbb{P}(A \cap B) \ge \mathbb{P}(A) - \mathbb{P}(B^c)$ for arbitrary events A, B. From an elementary large deviation estimate for a sum of n independent standard exponentials, we know that for some $\lambda > 0$

$$\mathbb{P}\left(\gamma(n) \ge (1+\varepsilon)n\right) = O(\exp(-\lambda\,\varepsilon^2 n)). \tag{6}$$

Therefore, our claim follows if we show a bound of the form

$$\mathbb{P}\left(\xi(t) > \rho^{1/3}t^{2/3} \text{ for all } 0 < t \le (1+\varepsilon)n\right) \ge \exp\left(-16(\rho^2 n)^{1/3}\right) \tag{7}$$

for large n. Essentially, this can be deduced from [18, Theorem 4.1]: In the notation from there, we may consider the random walk $S_j = N(j) - j$, $j \in \mathbb{N}$, and the function

$$g(t) := \frac{3}{2}\rho^{1/3}(t + \max\{\rho, 1\})^{2/3} - 2\max\{\rho, \rho^{1/3}\}, \quad t \ge 0.$$

The function g is monotone increasing with $g(0) < 0$ and regularly varying with index $2/3$. Moreover, it is readily checked that

$$\sup_{t\ge 1}\left(g((2/3)t) - g((2/3)(t-1))\right) \le 2/3\,.$$

Therefore, the assumptions of [18, Theorem 4.1] are fulfilled, which ensures after a small calculation that for ε sufficiently small and n large enough,

$$\mathbb{P}\left(S_j > g(j) \text{ for all } j = 1, \ldots, \lfloor (1+\varepsilon)n \rfloor\right) \geq \exp\left(-16(\rho^2 n)^{1/3}\right). \tag{8}$$

Now let us define for $0 \leq t_0 < t_1$ the event

$$\mathcal{E}(t_0, t_1) := \left\{\xi(t) > \rho^{1/3} t^{2/3} \text{ for all } t_0 < t \leq t_1\right\}.$$

For $j \in \mathbb{N}$ and $t \in \mathbb{R}$ with $j \leq t \leq j+1$, we have $\xi(t) \geq S_j$ and, provided $j \geq \rho_0 := 8\lceil \rho \rceil$, also

$$g(j) \geq \rho^{1/3}(j+1)^{2/3} \geq \rho^{1/3} t^{2/3}.$$

Therefore, by (8),

$$\mathbb{P}\left(\mathcal{E}(\rho_0, (1+\varepsilon)n)\right) \geq \mathbb{P}\left(S_j > g(j) \text{ for all } j = \rho_0, \ldots, \lfloor (1+\varepsilon)n \rfloor\right) \geq \exp\left(-16(\rho^2 n)^{1/3}\right). \tag{9}$$

Writing

$$\mathbb{P}\left(\mathcal{E}(0, (1+\varepsilon)n)\right) = \mathbb{P}\left(\mathcal{E}(\rho_0, (1+\varepsilon)n) \mid \mathcal{E}(0, \rho_0)\right) \cdot \mathbb{P}\left(\mathcal{E}(0, \rho_0)\right),$$

we note that $\mathbb{P}\left(\mathcal{E}(0, \rho_0)\right)$ is bounded from below by a strictly positive constant (depending on ρ). Moreover, since ξ is a spatially homogeneous Markov process, we clearly have

$$\mathbb{P}\left(\mathcal{E}(\rho_0, (1+\varepsilon)n) \mid \mathcal{E}(0, \rho_0)\right) \geq \mathbb{P}\left(\mathcal{E}(\rho_0, (1+\varepsilon)n)\right),$$

so that our claim (7) follows from (9). En passant, let us mention that $n^{1/3}$ is the correct stretch for the exponential in (7). Indeed, this can be seen from Theorem 4.2 in [18], where an analogous upper bound on the probability in (7) is given.

We now turn our attention to the upper bound. We fix a small $0 < \varepsilon < 1$. On the event

$$\left\{\gamma(n) \geq (1-\varepsilon)n \text{ and } \sup_{t \leq (1-\varepsilon)n} \xi(t) \leq (2\rho)^{1/3} n^{2/3}\right\},$$

we have

$$\exp\left(-\rho \int_0^{\gamma(n)} \xi(t)^{-1} \mathrm{d}t\right) \leq \exp\left(-(1-\varepsilon)(\rho^2/2)^{1/3} n^{1/3}\right),$$

and from Proposition 3.1, $\mathbb{P}(X_{1,\rho} > n)$ can be bounded from above by

$$\exp\left(-(1-\varepsilon)(\rho^2/2)^{1/3}n^{1/3}\right) + \mathbb{P}(\gamma(n) < (1-\varepsilon)n) + \mathbb{P}(\sup_{t\leq(1-\varepsilon)n} \xi(t) > (2\rho)^{1/3}n^{2/3}).$$

On the one hand, from an elementary large deviation estimate similar to (6), we get that for some $\lambda > 0$:

$$\mathbb{P}(\gamma(n) < (1-\varepsilon)n) = \mathbb{P}(N((1-\varepsilon)n) \geq n) = O(\exp(-\lambda\varepsilon^2 n)).$$

On the other hand, ξ is a Lévy process with no negative jumps started from 1 such that

$$\mathbb{E}(\exp(q(\xi(t)-1))) = \exp\left(t(\mathrm{e}^q - 1 - q)\right), \qquad t \geq 0.$$

It follows classically that the process

$$\exp\left(q\xi(t) - t(\mathrm{e}^q - 1 - q)\right), \qquad t \geq 0$$

is a martingale started from e^q. An application of the optional sampling theorem at the first passage time of ξ above $(2\rho)^{1/3}n^{2/3}$ yields the upper-bound

$$\exp\left(q(2\rho)^{1/3}n^{2/3} - (1-\varepsilon)n(\mathrm{e}^q - 1 - q)\right)\mathbb{P}(\sup_{t\leq(1-\varepsilon)n} \xi(t) > (2\rho)^{1/3}n^{2/3}) \leq \mathrm{e}^q.$$

Specializing this for $q = (2\rho)^{1/3}n^{-1/3}$, we deduce that for n large enough

$$\mathbb{P}(\sup_{t\leq(1-\varepsilon)n} \xi(t) > (2\rho)^{1/3}n^{2/3}) \leq \exp\left(-(1+(\varepsilon/2))(\rho^2/2)^{1/3}n^{1/3}\right).$$

Since $\varepsilon > 0$ can be taken arbitrarily small, this completes the proof. □

5 Connection with a Population Model with Neutral Mutations

The Yule-Simon distribution originates from [20], where Simon introduced a simple random algorithm to exemplify the appearance of (1) in various statistical models. More specifically, he proposed a probabilistic model for describing observed linguistic (but also economic and biological) data leading to (1).

We shall now give some details of Simon's model, which he described in terms of word frequencies. Imagine a book that is being written has reached a length of

n words. Write $f(k, n)$ for the number of different words which occurred exactly k times in the first n words. Simon works under the following two assumptions:

A1 The probability that the $(n+1)$-st word is a word that has already appeared exactly k times is proportional to $kf(k, n)$;
A2 The probability that the $(n+1)$-st word is a new word is a constant α independent of n.

Setting

$$\rho = \frac{1}{1-\alpha}, \tag{10}$$

he argues that under these two assumptions, the relative frequencies of words that have occurred exactly k times are in the limit $n \to \infty$ described by a Yule-Simon law (1) with parameter ρ.

Simon's paper initiated a lively dispute between Simon and Mandelbrot (known as the Simon-Mandelbrot debate) on the validity and practical relevance of Simon's model. We mention only Mandelbrot's reply [14] and Simon's response [21], but the discussions includes further (final) notes and post scripta. While Simon's derivation presupposes $\rho > 1$, see (10), the discussion between the two gentlemen evolved in particular around the adequacy and meaning of Simon's model when $0 < \rho < 1$; see pp. 95–96 in [14].

It is one of the purposes of this section to specify a probabilistic population model for which the Yule-Simon law is observed in all cases $\rho > 0$. To that aim, it is convenient to first recast Simon's model in terms of random recursive forests, and then interpret the latter as a population model with neutral mutations. By letting the rate of mutation asymptotically decrease to zero in an appropriate way, we will then obtain the whole family of one-parameter Yule-Simon laws.

However, we will go further: A natural generalization of Simon's algorithm, which we formulate in terms of a more general population model with age-dependent reproduction rate, will finally result in the two-parameter Yule-Simon laws (Proposition 5.1).

5.1 *Simon's Model in Terms of Yule Processes with Mutations*

Fix $p \in (0, 1)$, take $n \gg 1$ and view $[n] := \{1, \dots, n\}$ as a set of vertices. We equip every vertex $2 \leq \ell \leq n$ with a pair of variables $(\varepsilon(\ell), u(\ell))$, independently of the other vertices. Specifically, each $\varepsilon(\ell)$ is a Bernoulli variable with parameter p, i.e. $\mathbb{P}(\varepsilon(\ell) = 1) = 1 - \mathbb{P}(\varepsilon(\ell) = 0) = p$, and $u(\ell)$ is independent of $\varepsilon(\ell)$ and has the uniform distribution on $[\ell - 1]$. Simon's algorithm amounts to creating an edge between ℓ and $u(\ell)$ if and only if $\varepsilon(\ell) = 1$. The resulting random graph is a random forest and yields a partition of $[n]$ into random sub-trees. In this setting, Simon showed that for every $k \geq 1$, the proportion of trees of size k, i.e. the ratio between the number of sub-trees of size k and the total number of sub-trees in the random forest, converges on average as $n \to \infty$ to $\rho B(k, \rho + 1)$, where $\rho = 1/p$.

Let us next enlighten the connection with a standard Yule process $Y = Y_0$. We start by enumerating the individuals of the population model described by the Yule process in the increasing order of their birth dates (so the ancestor is the vertex 1, its first child the vertex 2, …), and stop the process at time

$$T(n) := \inf\{t \geq 0 : Y(t) = n\}$$

when the population has reached size n. Clearly, the parent $u(\ell)$ of an individual $2 \leq \ell \leq n$ has the uniform distribution on $[\ell - 1]$, independently of the other individuals. The genealogical tree obtained by creating edges between parents and their children is known as a random recursive tree of size n; see e.g. [8]. Next imagine that neutral mutations are superposed to the genealogical structure, so that each child is either a clone of its parent or a mutant with a new genetic type, and more precisely, the individual ℓ is a mutant if and only if $\varepsilon(\ell) = 0$, where $(\varepsilon(\ell))_{\ell \geq 2}$ is a sequence of i.i.d. Bernoulli variables with parameter p, independent of the sequence $(u(\ell))_{\ell \geq 2}$. The partition of the population into sub-populations of the same genetic type, often referred to as the allelic partition, corresponds to an independent Bernoulli bond percolation with parameter p on the genealogical tree, that is, it amounts to deleting each edge with probability $1 - p$, independently of the other edges. The resulting forest has the same distribution as that obtained from Simon's algorithm.

Simon's result can then be re-interpreted by stating that the distribution of the size of a typical sub-tree after percolation (i.e. the number of individuals having the same genetic type as a mutant picked uniformly at random amongst all mutants) converges as $n \to \infty$ to the Yule-Simon distribution with parameter $\rho = 1/p$. This can be established as follows. Observe first that a typical mutant is born at time $T(\lfloor Un \rfloor)$, where U is an independent uniform variable on $[0, 1]$. By the branching property, a typical sub-tree can thus be viewed as the genealogical tree of a Yule process with birth rate p per individual (recall that p is the probability for a child to be a clone of its parent), stopped at time $T(n) - T(\lfloor Un \rfloor)$. Then recall that

$$\lim_{t \to \infty} \mathrm{e}^{-t} Y(t) = W \qquad \text{a.s.},$$

where $W > 0$ is some random variable, and hence

$$T(n) - T(\lfloor Un \rfloor) \sim \ln(n/W) - \ln(Un/W) = -\ln U \qquad \text{as } n \to \infty.$$

Since a Yule process with birth rate p per individual and taken at time $t \geq 0$ has the geometric distribution with parameter e^{-pt}, and $-p \ln U$ has the exponential distribution with parameter $\rho = 1/p$, we conclude that the distribution of the size of a typical sub-tree after percolation converges as $n \to \infty$ to (1).

In the following section, we shall generalize Simon's algorithm in two different directions.

5.2 A Generalization of Simon's Model

The random algorithm described above only yields Yule-Simon distributions with parameter $\rho > 1$. A modification dealing with the case $\rho \leq 1$ has already been suggested in Simon's article, see Case II on page 431 in [20]; let us now elaborate on this more specifically.

The Full Range of the One-Parameter Yule-Simon Law

Let us now assume that the $\varepsilon(\ell)$ are independent Bernoulli variables with parameter $p = p(\ell)$ depending possibly on ℓ, again everything independent of the $u(\ell)$'s. As previously, the individuals ℓ such that $\varepsilon(\ell) = 0$ are viewed as mutants, and those with $\varepsilon(\ell) = 1$ as clones.

We shall consider two mutually exclusive asymptotic regimes:

(a) $\lim_{n\to\infty} p(n) = 1/\rho$ for some $\rho > 1$,
(b) $\lim_{n\to\infty} p(n) = 1$, and $\sum_{\ell=1}^{n}(1 - p(\ell))$ is regularly varying with index $\rho \in (0, 1]$.

Plainly, case (a) holds in particular when the $\varepsilon(\ell)$'s form an i.i.d. sequence of Bernoulli variables with parameter $p = 1/\rho$ as in the preceding section. Regime (b) is a situation where mutations are asymptotically rare. In terms of the number of mutants $m(n) = n - \sum_{\ell=2}^{n} \varepsilon(\ell)$, (b) is implying that, in probability, $m(n) = o(n)$, and $m(n)$ is regularly varying with index ρ.

Just as before, we consider the allelic partition at time $T(n)$, i.e. the partition of the population into sub-population bearing the same genetic type. As we shall see in the following Proposition 5.1, this population model leads under the two different regimes to the full range of the one-parameter Yule-Simon law when studying the limit size of a typical sub-population.

A Two-Parameter Generalization

It remains to appropriately extend the model in order to encompass the two-parameter Yule-Simon distributions. To that aim, we replace the underlying standard Yule process $Y = Y_0$ by a general branching process Y_θ as considered in the introduction. Again, we consider independently a sequence $(\varepsilon(\ell))_{\ell\geq 2}$ of $\{0, 1\}$-valued random variables indicating which individuals are clones or mutants, respectively, and exactly as before, we may study the allelic partition at the time

$$T_\theta(n) = \inf\{t \geq 0 : Y_\theta(t) = n\} \tag{11}$$

when the total population size n is reached. We stress that the case $\theta = 0$ corresponds to the one-parameter model described just above: We have $Y_0 = Y$, and consequently $T_0(n) = T(n)$.

We are now in position to formulate a limit result for the proportion of sub-populations of size k in our extended model, generalizing Simon's result to the two-parameter Yule-Simon distributions. For the sake of simplicity, we focus on the case $\theta \le 0$ when the total population in the general branching process Y_θ is infinite almost surely.

Proposition 5.1 *Let $\theta \le 0$ and $\rho > 0$, consider a general branching process Y_θ as in Sect. 2, and define $T_\theta(n)$ as in (11). Let further $(\varepsilon(\ell))_{\ell \ge 2}$ be a sequence of variables in $\{0, 1\}$ which is independent of the branching process and fulfills one of the regimes* (a) *or* (b)*. Regard every individual ℓ with $\varepsilon(\ell) = 0$ as a mutant, and consider at time $T_\theta(n)$ the (allelic) partition of the whole population into sub-populations of individuals with the same genetic type.*

For every $k \in \mathbb{N}$, write $Q_n(k)$ for the proportion of sub-populations of size k (i.e. the number of such sub-populations divided by the total number of mutants) in the allelic partition at time $T_\theta(n)$. Then

$$\lim_{n\to\infty} Q_n(k) = \mathbb{P}(X_{\vartheta,\varrho} = k) \qquad \textit{in probability},$$

where

$$(\vartheta, \varrho) = \begin{cases} (\theta\rho, (1-\theta)\rho) & \textit{in regime } \text{(a)}, \\ (\theta, (1-\theta)\rho) & \textit{in regime } \text{(b)}. \end{cases}$$

Remark 5.2 We stress that our model leads to the complete range of parameters (ϑ, ϱ) of the Yule-Simon distribution satisfying $\vartheta \le 0$ and $\varrho > 0$. Indeed, if $\vartheta + \varrho > 1$, then the size of a typical sub-tree converges in law with the choices $\theta := \vartheta/(\vartheta + \varrho)$ and $\rho := \vartheta + \varrho$ under regime (a) to $X_{\vartheta,\varrho}$. If $\vartheta + \varrho \le 1$, then $\theta := \vartheta$ and $\rho := \varrho/(1 - \vartheta)$ under regime (b) yield the law $X_{\vartheta,\varrho}$.

Remark 5.3 The conditional expectation of the size of a typical sub-tree given that there are $m(n)$ mutants in the population of total size n is clearly $n/m(n)$. Note that $m(n) \sim (1 - 1/\rho)n$ in regime (a), whereas $m(n) = o(n)$ in regime (b). We may thus expect from Proposition 5.1 that

$$\mathbb{E}(X_{\theta\rho,(1-\theta)\rho}) = \rho/(\rho - 1) \qquad \text{when } \theta \le 0 \text{ and } \rho > 1,$$

and that

$$\mathbb{E}(X_{\theta,(1-\theta)\rho}) = \infty \qquad \text{when } \theta \le 0 \text{and } \rho \le 1.$$

That these identities indeed hold has already been observed in Remark 2.2.

We now present the main steps of the proof of Proposition 5.1, leaving some of the technical details to the interested readers. We start with an elementary observation in the case constant mutation rates.

Lemma 5.4 *Suppose that $p(n) \equiv 1/\rho$ for some fixed $\rho > 1$. The process $\tilde{Y}_{\theta,\rho} = (\tilde{Y}_{\theta,\rho}(t))_{t\geq 0}$ of the size of the sub-population bearing the same type as the ancestor has then the same distribution as $(Y_{\theta\rho}(t/\rho))_{t\geq 0}$.*

Proof Since $\tilde{Y}_{\theta,\rho}$ is obtained from Y_θ by killing each of its child (of course, together with its descent) with probability $1 - 1/\rho$ and independently of the others, $\tilde{Y}_{\theta,\rho}$ is also a general branching process. More precisely, thanks to the thinning property of Poisson random measures, typical individuals in $\tilde{Y}_{\theta,\rho}$ reproduce at ages according to a Poisson point measure on $[0, \infty)$ with intensity $\rho^{-1}\mathrm{e}^{-\theta t}\mathrm{d}t$, and the statement follows from the change of variables $s = t/\rho$. □

We next return to the general situation where mutation rates $p(n)$ may depend on n, and state two technical lemmas involving convergence in distribution in $D([0,\infty), \mathbb{R}) \times D([0,\infty), \mathbb{R})$, where $D([0,\infty), \mathbb{R})$ is the Skorokhod space of càdlàg functions from $[0,\infty)$ to $\mathbb{R}$.

Lemma 5.5 *Suppose $\sum_{n\geq 1} p(n) = \infty$, so there are infinitely many mutants a.s. Let $i, j \geq 1$, and let $Y^i_\theta = (Y^i_\theta(t))_{t\geq 0}$ denote the process of the size of the sub-population (both clones and mutants descents) generated by the ith mutant as a function of its age. Then $Y^i_\theta \overset{(\mathrm{d})}{=} Y^j_\theta \overset{(\mathrm{d})}{=} Y_\theta$, and for $i, j \to \infty$, $i \neq j$,*

$$\left(Y^i_\theta, Y^j_\theta\right) \overset{(\mathrm{d})}{\Longrightarrow} \left(Y'_\theta, Y''_\theta\right),$$

where Y'_θ and Y''_θ are independent copies of Y_θ.

Proof This is an immediate consequence of the branching property of Y_θ, noting that the probability that the individuals labelled i and j are in the same sub-population tends to zero for $i \neq j \to \infty$. □

Lemma 5.6 *Let $i, j \geq 1$, and let $\tilde{Y}^i_\theta$ denote the process of the size of the clonal sub-population generated by the ith mutant (i.e. $\tilde{Y}^i_\theta(t)$ is the size of the sub-population bearing the same genetic type as the ith mutant when the latter has age t). Then $\tilde{Y}^i_\theta \overset{(\mathrm{d})}{=} \tilde{Y}^j_\theta$, and we have the following convergence for $i, j \to \infty$, $i \neq j$:*

$$\left(\tilde{Y}^i_\theta, \tilde{Y}^j_\theta\right) \overset{(\mathrm{d})}{\Longrightarrow} \begin{cases} \left(\tilde{Y}'_{\theta,\rho}, \tilde{Y}''_{\theta,\rho}\right) & \text{in regime (a)}, \\ \left(Y'_\theta, Y''_\theta\right) & \text{in regime (b)}, \end{cases}$$

where $\tilde{Y}'_{\theta,\rho}$, $\tilde{Y}''_{\theta,\rho}$ are independent copies of the process $\tilde{Y}_{\theta,\rho}$ in Lemma 5.4, and Y'_θ, Y''_θ are independent copies of Y_θ.

Proof Recall that $\tilde{Y}_\theta^i$ is obtained from Y_θ^i by superposing neutral mutations to the latter and keeping only the clones of i. Since the nth individual is a mutant with probability $1 - p(n)$, the claim now follows readily from Lemma 5.5 and the fact that $p(n) \to 1/\rho$ in regime (a), whereas $p(n) \to 1$ in regime (b). □

We now have all the ingredients needed for the proof of Proposition 5.1.

Proof of Proposition 5.1 We will argue that

$$\lim_{n\to\infty} \mathbb{E}\left(Q_n^i(k)\right) = \mathbb{P}(X_{\vartheta,\varrho} = k)^i \quad \text{for } i = 1, 2\,.$$

The claim then follows via the second moment method.

As before, let us write $m(n)$ for the number of mutants among the first n individuals, i.e. at time $T_\theta(n)$. Writing b_i for the birth-time of the ith mutant (with $b_1 := 0$, interpreting the first individual as a mutant), we have

$$\mathbb{E}\left(Q_n(k)\right) = \mathbb{E}\left(\frac{1}{m(n)} \sum_{i=1}^{m(n)} 1\!\!1_{\left\{\tilde{Y}_\theta^i(T_\theta(n)-b_i)=k\right\}}\right).$$

We will first prove convergence of the first moment of $Q_n(k)$. From Proposition 2.3 we deduce that the time $T_\theta(n)$ at which the population reaches size n satisfies

$$T_\theta(n) = (1-\theta)^{-1}\ln(n/W_\theta) + o(1) \tag{12}$$

in probability, where $W_\theta > 0$ denotes the limit in probability of $\mathrm{e}^{(\theta-1)t}Y_\theta(t)$ as $t \to \infty$.

Let us first consider regime (a), where $p(n) \to 1/\rho$ as $n \to \infty$. For i tending to infinity, it follows that

$$b_i = (1-\theta)^{-1}\ln\left(i/((1-1/\rho)W_\theta)\right) + o(1)$$

in probability. By the law of large numbers, we have for the number of mutants that $m(n) \sim (1-1/\rho)n$ almost surely. Combining the last display with (12) and Lemma 5.6, we deduce by a Riemann-type approximation that

$$\lim_{n\to\infty} \mathbb{E}\left(Q_n(k)\right) = \mathbb{P}\left(\tilde{Y}_{\theta,\rho}\left(-(1-\theta)^{-1}\ln U\right) = k\right)$$

for $n \to \infty$, where U is an independent uniform variable on $(0, 1)$.

Note that the random variable $-(1-\theta)^{-1}\ln U$ inside the last probability has the law of an exponentially distributed random variable $T_{1-\theta}$ with parameter $1-\theta$. Now, thanks to Lemma 5.4,

$$\mathbb{P}\left(\tilde{Y}_{\theta,\rho}\left(T_{1-\theta}\right) = k\right) = \mathbb{P}\left(Y_{\theta\rho}\left(T_{1-\theta}/\rho\right) = k\right) = \mathbb{P}(X_{\theta\rho,(1-\theta)\rho} = k),$$

where for the last equality, we used that $T_{1-\theta}/\rho$ is exponentially distributed with parameter $(1-\theta)\rho$.
In regime (b), mutations become asymptotically rare, i.e. $p(n) \to 1$ for $n \to \infty$, and the number of mutants $m(n)$ is regularly varying with index $\rho \in (0, 1]$. Now, if $i = i(n) = \lfloor r^{\rho} m(n) \rfloor$ for some $r \in (0, 1)$, it follows that the ith mutant is born when there are approximately rn individuals in the population system. Therefore, in probability as $n \to \infty$,

$$b_i \sim (1-\theta)^{-1} \ln (rn/W_\theta) + o(1).$$

A approximation similar to that in regime (a) then allows us to deduce that

$$\lim_{n\to\infty} \mathbb{E}\left(Q_n(k)\right) = \mathbb{P}\Big(Y_\theta\left(-(1-\theta)^{-1}(1/\rho)\ln U\right) = k\Big).$$

Since $-(1-\theta)^{-1}(1/\rho)\ln U$ is exponentially distributed with parameter $(1-\theta)\rho$, the last expressions is equal to $\mathbb{P}(X_{\rho,(1-\theta)\rho} = k)$.

As far as the second moment is concerned, it is a consequence of the asymptotic independence derived in Lemma 5.6 that

$$\mathbb{E}\left(Q_n^2(k)\right) \sim \mathbb{E}\left(Q_n(k)\right)^2$$

as $n \to \infty$, proving our claim in both regimes (a) and (b). □

We shall now conclude this work by presenting an alternative proof of Proposition 5.1 in the special (and important) case when $p(n) \equiv 1/\rho$ for some fixed $\rho > 1$, by making the connection with results of Nerman [15]. We thus consider the general branching process Y_θ, where each child is a clone of its parent with fixed probability $1/\rho$, or a mutant bearing a new genetic type with complementary probability $1 - 1/\rho$, independently of the other individuals. We regard this population model as a branching particle system, in which the particles represent the clonal sub-populations. This means that the birth of a mutant child in the population is viewed as the birth of a new particle in the system; the size of the latter then grows as time passes and is given by the size of the sub-population having the same genetic type as this mutant.

The process of the size of a typical particle as a function of the age of the mutant ancestor has the same distribution as the process $\tilde{Y}_{\theta,\rho}$ in Lemma 5.4. Moreover, each particle also gives birth as time passes to daughter particles (i.e. to new mutants) which in turn evolve independently one from the others and according to the same dynamics. In words, the particle system is another general branching process.

Observe that the rate at which a particle with size ℓ gives birth to a daughter particle equals $(1-1/\rho)\ell$; hence the reproduction intensity measure $\tilde{\mu}$ of branching particle system is given by

$$\tilde{\mu}(\mathrm{d}t) = (1 - 1/\rho)\mathbb{E}(\tilde{Y}_{\theta,\rho}(t))\mathrm{d}t.$$

So Proposition 2.1 and Lemma 5.4 yield

$$\tilde{\mu}(\mathrm{d}t) = \frac{1-1/\rho}{1-\theta\rho}\left(\mathrm{e}^{(1-\theta\rho)t/\rho} - \theta\rho\right)\mathrm{d}t.$$

It is now readily checked that the so-called Malthusian parameter (see Equation (1.4) in [15]) of the branching particle system equals $1-\theta$, namely we have

$$\int_0^\infty \mathrm{e}^{-(1-\theta)t}\tilde{\mu}(\mathrm{d}t) = 1.$$

Finally, let us write Z_t for the number of particles (i.e. of genetic types) at time t. For any $k \geq 1$, write Z_t^k for the number of particles with size k at time t, i.e. the number of sub-populations with size exactly k in the allelic partition at time t. According to Theorem 6.3 and Equation (2.7) in [15] (the reader will easily check that the assumptions there are fulfilled in our framework), one has

$$\lim_{t\to\infty}\frac{Z_t^k}{Z_t} = (1-\theta)\int_0^\infty \mathrm{e}^{-(1-\theta)t}\mathbb{P}(\tilde{Y}_{\theta,\rho}(t) = k)\mathrm{d}t, \qquad \text{almost surely.}$$

Recalling Lemma 5.4, we have thus

$$\lim_{t\to\infty}\frac{Z_t^k}{Z_t} = \mathbb{P}(Y_{\theta\rho}(T_{(1-\theta)\rho}) = k) = \mathbb{P}(X_{\theta\rho,(1-\theta)\rho} = k), \qquad \text{almost surely.}$$

It only remains to observe that for $t = T_\theta(n)$, one has $Z_t^k/Z_t = Q_n(k)$, and we thus have arrived at the claim made in Proposition 5.1.

Finally, we point out that the argument above also applies for $\theta \in (0, 1)$, provided that one works conditionally on the event that the total number of mutants is infinite, which has then a positive probability.

References

1. Bertoin, J.: A version of Herbert A. Simon's model with slowly fading memory and its connections to branching processes. J. Stat. Phys. **176**(3), 679–691 (2019)
2. Bingham, N.H., Doney, R.A.: Asymptotic properties of supercritical branching processes. II. Crump-Mode and Jirina processes. Adv. Appl. Probab. **7**, 66–82 (1975)
3. Caballero, M.E., Lambert, A., Uribe Bravo, G.: Proof(s) of the Lamperti representation of continuous-state branching processes. Probab. Surv. **6**, 62–89 (2009)
4. Doney, R.A.: A limit theorem for a class of supercritical branching processes. J. Appl. Probab. **9**, 707–724 (1972)
5. Doney, R.A.: On the asymptotic behaviour of first passage times for transient random walk. Probab. Theory Relat. Fields **81**(2), 239–246 (1989)
6. Doney, R.A.: Local behaviour of first passage probabilities. Probab. Theory Relat. Fields **152**(3–4), 559–588 (2012)

7. Doney, R.A., Rivero, V.: Asymptotic behaviour of first passage time distributions for Lévy processes. Probab. Theory Relat. Fields **157**(1–2), 1–45 (2013)
8. Drmota, M.: Random Trees: An Interplay Between Combinatorics and Probability, 1st edn. Springer Publishing Company, Incorporated, Wien/London (2009)
9. Ethier, S.N., Kurtz, T.G.: Markov Processes. Characterization and Convergence. Wiley Series in Probability and Mathematical Statistics: Probability and Mathematical Statistics. Wiley, New York (1986)
10. Jiřina, M.: Stochastic branching processes with continuous state space. Czechoslovak Math. J. **8**(83), 292–313 (1958)
11. Kyprianou, A.E.: Fluctuations of Lévy Processes with Applications. Introductory Lectures. Universitext, 2nd edn. Springer, Heidelberg (2014)
12. Lansky, P., Polito, F., Sacerdote, L.: The role of detachment of in-links in scale-free networks. J. Phys. A: Math. Theor. **47**(34), 345002 (2014)
13. Lansky, P., Polito, F., Sacerdote, L.: Generalized nonlinear Yule models. J. Stat. Phys. **165**(3), 661–679 (2016)
14. Mandelbrot, B.: A note on a class of skew distribution functions: Analysis and critique of a paper by H. A. Simon. Inf. Control **2**(1), 90–99 (1959)
15. Nerman, O.: On the convergence of supercritical general (C-M-J) branching processes. Z. Wahrsch. Verw. Gebiete **57**(3), 365–395 (1981)
16. Pitman, J.: Combinatorial Stochastic Processes. Lecture Notes in Mathematics, vol. 1875. Springer, Berlin (2006). Lectures from the 32nd Summer School on Probability Theory held in Saint-Flour, 7–24 July 2002, With a foreword by Jean Picard
17. Polito, F.: Studies on generalized Yule models. Modern Stoch.: Theory Appl. **6**(1), 41–55 (2018)
18. Portnoy, S.: Probability bounds for first exits through moving boundaries. Ann. Probab. **6**(1), 106–117 (1978)
19. Rogers, L.C.G., Williams, D.: Diffusions, Markov Processes, and Martingales. Vol. 1. Foundations. Wiley Series in Probability and Mathematical Statistics: Probability and Mathematical Statistics, 2nd edn. Wiley, Chichester (1994)
20. Simon, H.A.: On a class of skew distribution functions. Biometrika **42**(3/4), 425–440 (1955)
21. Simon, H.A.: Some further notes on a class of skew distribution functions. Inf. Control **3**(1), 80–88 (1960)

The Limit Distribution of a Singular Sequence of Itô Integrals

Denis Bell

Abstract We give an alternative, elementary proof of a result of Peccati and Yor concerning the limit law of a sequence of Itô integrals with integrands having singular asymptotic behavior.

Keywords Brownian motion · Itô integrals · Limit law

1 Introduction and Statement of the Theorem

The purpose of this note is to provide a short, elementary proof of the following result.

Theorem 1.1 *Let $w = \{w_t : t \geq 0\}$ be a standard Brownian motion. Consider the sequence of Itô integrals*

$$X_n = \sqrt{n} \int_0^1 t^n w_t dw_t, \quad n \geq 1.$$

Then X_n converges in distribution as $n \to \infty$ to the law of $\frac{1}{\sqrt{2}} w_1 \eta$, where η is a standard normal random variable independent of w.

Theorem 1.1 was originally proved by Peccati and Yor [4] in 2004. The subsequent introduction into this field of significant modern areas of stochastic analysis, namely Malliavin calculus, Skorohod integration and fractional Brownian motion, has resulted in a surge of ever more general limit theorems of this type. Recent examples are Nourdin, Nualart and Peccati [2], Pratelli and Rigo [5], and Bell, Bolaños and Nualart [1]. (See Peccati [3] for a survey of results in this area.)

D. Bell (✉)
Department of Mathematics, University of North Florida, Jacksonville, FL, USA
e-mail: dbell@unf.edu

L. Chaumont, A. E. Kyprianou (eds.), *A Lifetime of Excursions Through Random Walks and Lévy Processes*, Progress in Probability 78,
https://doi.org/10.1007/978-3-030-83309-1_5

Our proof of Theorem 1.1 was, in fact, the point of departure for the approach in [1]. However, the argument is especially transparent in the simpler context treated here. For this reason, we thought it worthwhile to present separately in this context.

2 Proof of the Theorem

The idea behind the proof of Theorem 1.1 is as follows. As $n \to \infty$, the integrands $t^n w_t$ defining the X_n, when scaled by the factor $\sqrt{n}$, converge (or, more accurately *diverge*) to a delta function based at the right-hand endpoint 1 of the range of integration. Thus the limit distribution is determined by the behavior of the integrals over infinitesimal neighborhoods of $t = 1$. But in such infinitesimal neighborhoods, the integrals behave asymptotically as the product of w_1 and an independent Gaussian increment. This observation both motivates the result and suggests a method of proof.

We define below a sequence

$$\alpha_n \uparrow 1 \tag{1}$$

and a sequence of deterministic functions g_n such that

$$\int_{\alpha_n}^{1} g_n^2(t)dt \to \frac{1}{2} \tag{2}$$

and

$$E[X_n - Y_n]^2 \to 0, \tag{3}$$

where

$$Y_n \equiv w(\alpha_n) \int_{\alpha_n}^{1} g_n(t)dw_t, \tag{4}$$

This will suffice to prove the theorem since L^2-convergence implies convergence in law, and the limit law of Y_n is easily seen to have the desired form.

The details of this construction are as follows. First, note that

$$E[X_n^2] = n \int_0^1 t^{2n+1}dt = \frac{n}{2n+2} \to 1/2$$

and, assuming we have chosen α_n and g_n so that (1) and (2) hold

$$E[Y_n^2] = \alpha_n \int_{\alpha_n}^{1} g_n^2(t)dt \to 1/2.$$

Hence (3) will follow provided also

$$E[X_n Y_n] \to \frac{1}{2}.$$

Now

$$E[X_n Y_n] = \sqrt{n} E\Big[w(\alpha_n) \int_{\alpha_n}^{1} t^n dw_t \int_{\alpha_n}^{1} g_n(t) dw_t\Big]$$

$$= \sqrt{n}\alpha_n \int_{\alpha_n}^{1} t^n g_n(t) dt.$$

So we require that

$$\sqrt{n} \int_{\alpha_n}^{1} t^n g_n(t) dt \to \frac{1}{2}.$$

Choose $g_n(t) = \sqrt{n} t^n$. We then need to define the α_n such that

$$\int_{\alpha_n}^{1} n t^{2n} dt = \frac{n}{2n+1}[1 - \alpha_n^{2n+1}] \to 1/2,$$

i.e.

$$\alpha_n^{2n+1} \to 0 \tag{5}$$

This is achieved by setting

$$\alpha_n = 1 - \frac{\log n}{n}.$$

Since $\log n/n \to 0$ and $\alpha_n^n \sim 1/n$, as $n \to \infty$, both (1) and (5) hold and we are done. ∎

3 Concluding Remarks

1. The scaling factor $\sqrt{n}$ contained in the sequence X_n, is an intrinsic feature of the Itô integrals considered here. In [1, 2, 5], where analogous results are obtained for Skorohod integrals with respect to *fractional* Brownian motion, the appropriate scaling for the integrals proves to be n^H, where H is the Hurst parameter of the driving fBm.
2. The choice of the weighting sequence $\alpha_n = 1 - \log n/n$ introduced above is a little delicate and is a crucial element in the proof of the theorem. Such a sequence also plays a role in the proofs of the more general results in [1]. It turns out that this same weighting works in the more general setting. In view of the dependence of the form of the result on the parameter H noted in Remark 1, this strikes the author as somewhat surprising.

Acknowledgments The author is grateful to David Nualart for bringing this problem to his attention.

References

1. Bell, D., Bolaños, R., Nualart, D.: Limit theorems for singular Skorohod integrals. Theory Probab. Math. Stat. In Publication (2021)
2. Nourdin, I., Nualart, D., Peccati, G.: Quantitative stable limit theorems on the Wiener space. Ann. Probab. **44**(1), 1–41 (2016)
3. Peccati, G.: Quantitative CLTs on a Gaussian space: a survey of recent developments. ESAIM: Proc. **44**, 61–78 (2014)
4. Peccati, G., Yor, M.: Four limit theorems for quadratic functionals of Brownian motion and Brownian bridge. In: Asymptotic Methods in Stochastics. Fields Institute Communication Series, pp. 75–87. AMS, Providence (2004)
5. Pratelli, L., Rigo, P.: Total variation bounds for Gaussian functionals. Stoch. Proc. Appl. **129**(7), 2231–2248 (2019)

On Multivariate Quasi-infinitely Divisible Distributions

David Berger, Merve Kutlu, and Alexander Lindner

Abstract A quasi-infinitely divisible distribution on $\mathbb{R}^d$ is a probability distribution μ on $\mathbb{R}^d$ whose characteristic function can be written as the quotient of the characteristic functions of two infinitely divisible distributions on $\mathbb{R}^d$. Equivalently, it can be characterised as a probability distribution whose characteristic function has a Lévy–Khintchine type representation with a "signed Lévy measure", a so called quasi–Lévy measure, rather than a Lévy measure. A systematic study of such distributions in the univariate case has been carried out in Lindner, Pan and Sato (Trans Am Math Soc 370:8483–8520, 2018). The goal of the present paper is to collect some known results on multivariate quasi-infinitely divisible distributions and to extend some of the univariate results to the multivariate setting. In particular, conditions for weak convergence, moment and support properties are considered. A special emphasis is put on examples of such distributions and in particular on $\mathbb{Z}^d$-valued quasi-infinitely divisible distributions.

Keywords Infinitely divisible distribution · quasi-infinitely divisible distribution · signed Lévy measure

---

In honour of Ron Doney on the occasion of his 80th birthday. This research was supported by DFG grant LI-1026/6-1. Financial support is gratefully acknowledged.

D. Berger
Institut für Mathematische Stochastik, Technische Universität Dresden, Dresden, Germany
e-mail: david.berger2@tu-dresden.de

M. Kutlu · A. Lindner (✉)
Institute of Mathematical Finance, Ulm University, Ulm, Germany
e-mail: merve.kutlu@uni-ulm.de; alexander.lindner@uni-ulm.de

L. Chaumont, A. E. Kyprianou (eds.), *A Lifetime of Excursions Through Random Walks and Lévy Processes*, Progress in Probability 78,
https://doi.org/10.1007/978-3-030-83309-1_6

1 Introduction

The class of Lévy processes may be considered to be one of the most important classes of stochastic processes. On the one hand, Lévy processes generalise Brownian motion, stable Lévy processes or compound Poisson processes, on the other hand they are the natural continuous time analogue of random walks and as those are interesting both from a purely theoretical point of view, but also from a practical point of view as drivers of stochastic differential equations, similar to time series being driven by i.i.d. noise. Excellent books on Lévy processes include the expositions by Applebaum [2], Bertoin [6], Doney [13], Kyprianou [20] or Sato [32]. We do not attempt to summarise Ron Doney's numerous, important and deep contributions to the understanding of Lévy processes, but confine ourselves to mentioning some of his works on small and large time behaviour of Lévy processes, such as Bertoin et al. [7], Doney [12] or Doney and Maller [14], which have been of particular importance for the research of one of the authors of this article.

The class of $\mathbb{R}^d$-valued Lévy processes corresponds naturally to the class of infinitely divisible distributions on $\mathbb{R}^d$, which is an important and well-studied class of distributions, see e.g. [32] for various of its properties. Infinitely divisible distributions (and hence Lévy processes) are completely characterised by the Lévy-Khintchine formula, according to which μ is infinitely divisible if and only if its characteristic function $\mathbb{R}^d \ni z \mapsto \widehat{\mu}(z) = \int_{\mathbb{R}^d} \mathrm{e}^{\mathrm{i}\langle z,x\rangle}\,\mu(\mathrm{d}x)$ can be expressed for all $z \in \mathbb{R}^d$ as

$$\widehat{\mu}(z) = \exp\left(\mathrm{i}\langle\gamma, z\rangle - \frac{1}{2}\langle z, Az\rangle + \int_{\mathbb{R}^d}\left(e^{\mathrm{i}\langle z,x\rangle} - 1 - \mathrm{i}\langle z, x\rangle \mathbb{1}_{[0,1]}(|x|)\right)\nu(\mathrm{d}x)\right), \tag{1.1}$$

with a symmetric non-negative definite matrix $A \in \mathbb{R}^{d\times d}$, a constant $\gamma \in \mathbb{R}^d$ and a Lévy measure ν on $\mathbb{R}^d$, that is, a Borel measure on $\mathbb{R}^d$ satisfying $\nu(\{0\}) = 0$ and $\int_{\mathbb{R}}(1 \wedge |x|^2)\nu(\mathrm{d}x) < \infty$. The triplet (A, ν, γ) is unique and called the *(standard) characteristic triplet* of the infinitely divisible distribution μ. By (1.1), the characteristic function of an infinitely divisible distribution must obviously be zero-free.

The class of quasi-infinitely divisible distributions on $\mathbb{R}^d$ is much less known. Their definition is as follows (see Lindner et al. [21, Remark 2.4]):

Definition 1.1 A probability distribution μ on $\mathbb{R}^d$ is called *quasi-infinitely divisible*, if its characteristic function $\widehat{\mu}$ admits the representation $\widehat{\mu}(z) = \widehat{\mu}_1(z)/\widehat{\mu}_2(z)$ for all $z \in \mathbb{R}^d$ with infinitely divisible distributions μ_1 and μ_2.

Rewriting this as $\widehat{\mu}(z)\widehat{\mu}_2(z) = \widehat{\mu}_1(z)$, we see that a probability distribution μ on $\mathbb{R}^d$ is quasi-infinitely divisible if and only if there are two infinitely divisible distributions μ_1 and μ_2 on $\mathbb{R}^d$ such that

$$\mu * \mu_2 = \mu_1.$$

So quasi-infinitely divisible distributions arise naturally in the factorisation problem of infinitely divisible distributions, where one of the factors is infinitely divisible, and the other is then necessarily quasi-infinitely divisible. If μ, μ_1, μ_2 are related as in Definition 1.1, and if (A_1, ν_1, γ_1) and (A_2, ν_2, γ_2) denote the characteristic triplets of μ_1 and μ_2, respectively, then it is easy to see that the characteristic function $\widehat{\mu}$ of μ has a Lévy–Khintchine type representation as in (1.1), with $\gamma = \gamma_1 - \gamma_2$, $A = A_1 - A_2$ and $\nu = \nu_1 - \nu_2$, where in the definition of ν one has to be a bit careful since $\nu_1(B) - \nu_2(B)$ will not be defined for Borel sets $B \subset \mathbb{R}^d$ with $\nu_1(B) = \nu_2(B) = \infty$. It is however defined if $\min\{\nu_1(B), \nu_2(B)\} < \infty$, in particular if B is bounded away from zero. We will formalise this in Definition 2.1 and call $\nu = \nu_1 - \nu_2$ a *quasi-Lévy (type) measure*, so basically a "signed" Lévy measure with some extra care taken for sets that are not bounded away from zero. In Theorem 2.2 we shall then give a Lévy–Khintchine type formula for quasi-infinitely divisible distributions with these quasi-Lévy (type) measures.

Quasi-infinitely divisible distributions have already appeared (although not under this name) in early works of Linnik [23], Linnik and Ostrovskiǐ [24], Gnedenko and Kolmogorov [15] or Cuppens [9, 10], to name just a few, but a systematic study of them in one dimension was only initiated in [21]. The name "quasi-infinitely divisible" for such distributions seems to have been used the first time in Lindner and Sato [22].

The class of quasi-infinitely divisible distributions is larger than it might appear on first sight. For example, Cuppens [9, Prop. 1], [10, Thm. 4.3.7] showed that a probability distribution on $\mathbb{R}^d$ that has an atom of mass greater than $1/2$ is quasi-infinitely divisible, and in [21, Theorem 8.1] it was shown that a distribution supported in $\mathbb{Z}$ is quasi-infinitely divisible if and only if its characteristic function has no zeroes, which in [5, Theorem 3.2] was extended to $\mathbb{Z}^d$-valued distributions. For example, a binomial distribution $b(n, p)$ on $\mathbb{Z}$ is quasi-infinitely divisible if and only if $p \neq 1/2$. Quasi-infinite divisibility of one-dimensional distributions of the form $\mu = p\delta_{x_0} + (1-p)\mu_{ac}$ with $p \in (0, 1]$ and an absolutely continuous μ_{ac} on $\mathbb{R}$ has been characterised in Berger [4, Thm. 4.6]. Further, as shown in [21, Theorem 4.1], in dimension 1 the class of quasi-infinitely divisible distributions on $\mathbb{R}$ is dense in the class of all probability distributions with respect to weak convergence. Since there are probability distributions that are not quasi-infinitely divisible, the class of quasi-infinitely divisible distributions obviously cannot be closed.

Recently, applications of quasi-infinitely divisible distributions have been found in physics (Demni and Mouayn [11], Chhaiba et al. [8]) and insurance mathematics (Zhang et al. [34]). Quasi-infinitely divisible processes and quasi-infinitely divisible random measures and integration theory with respect to them have been considered in Passeggieri [28]. Following Passeggeri, a stochastic process is called quasi-infinitely divisible if its finite-dimensional distributions are quasi-infinitely divisible. Quasi-infinitely divisible distributions have also found applications in number theory, see e.g. Nakamura [25, 26] or Aoyama and Nakamura [1]. We also mention the recent work of Khartov [18], where compactness criteria for quasi-infinitely divisible distributions on $\mathbb{Z}$ have been derived, and the paper by Kadankova et al. [16], where an example of a quasi-infinitely divisible distribution on the real line whose quasi-Lévy measure has a strictly negative density on $(a_*, 0)$ for some $a_* < 0$ has been constructed.

As mentioned, a systematic study of one-dimensional quasi-infinitely divisible distributions has only been initiated in [21]. The goal of the present paper is to give a systematic account of quasi-infinitely divisible distributions on $\mathbb{R}^d$. We will collect some known results and also extend some of the one-dimensional results in [4] and [21] to the multivariate setting. To get a flavour of the methods, we have decided to include occasionally also proofs of already known results, such as the previously mentioned result of Cuppens [10], according to which a probability distribution on $\mathbb{R}^d$ with an atom of mass greater than $1/2$ is quasi-infinitely divisible.

The paper is structured as follows. In Sect. 2 we will formalise the concept of quasi-Lévy measures and the Lévy–Khintchine type representation of quasi-infinitely divisible distributions. We show in particular that the matrix A appearing in the characteristic triplet must be non-negative definite (Lemma 2.5) and that the introduction of complex quasi-Lévy type measures and complex symmetric matrices does not lead to new distributions, in the sense that if a probability distribution has a Lévy–Khintchine type representation with a complex symmetric matrix $A \in \mathbb{C}^{d\times d}$ and a complex valued quasi-Lévy measure, then $A \in \mathbb{R}^{d\times d}$ and ν is real valued, i.e. a quasi-Lévy measure (Theorem 2.7). In Sect. 3 we give examples of quasi-infinitely divisible distributions, in particular we reprove Cuppens' result (Theorem 3.2) and state some of the examples mentioned previously. We also show how to construct multivariate quasi-infinitely divisible distributions from independent one-dimensional quasi-infinitely divisible distributions. Section 4 is concerned with sufficient conditions for absolute continuity of quasi-infinitely divisible distributions, by extending a classical condition of Kallenberg [17, pp.794–795] for one-dimensional infinitely divisible distributions to multivariate quasi-infinitely divisible distributions (Theorem 4.1). This condition seems to be new even in the case of multivariate infinitely divisible distributions. Section 5 is concerned with topological properties of the class of quasi-infinitely divisible distributions, like it being dense with respect to weak convergence in dimension 1. In Sect. 6 we give a sufficient condition for weak convergence of quasi-infinitely divisible distributions in terms of their characteristic triplets, and in Sect. 7 we consider some support properties. Section 8 is concerned with moment conditions for quasi-infinitely divisible distributions and formulae for the moments in terms of the characteristic triplet. We end this section by setting some notation.

Throughout, we denote by $\mathbb{N} = \{1, 2, 3, \dots\}$ the natural numbers, by $\mathbb{N}_0 = \mathbb{N} \cup \{0\}$ the natural numbers including 0, and by $\mathbb{Z}$, $\mathbb{Q}$, $\mathbb{R}$ and $\mathbb{C}$ the integers, rational numbers, real numbers and complex numbers, respectively. The real part a of a complex number $z = a + bi$ with $a, b \in \mathbb{R}$ is denoted by $a = \Re(z)$, the imaginary part by $b = \Im(z)$, and the complex conjugate by $\overline{z} = a - bi$. Vectors in $\mathbb{R}^d$ will be column vectors, we denote by $\mathbb{R}^{n\times d}$ the set of all $n \times d$ matrices with real entries, and the transpose of a vector or matrix A is denoted by A^T. The Euclidian inner product in the d-dimensional space $\mathbb{R}^d$ is denoted by $\langle \cdot, \cdot \rangle$, and the absolute value of $z = (z_1, \dots, z_d)^T$ by $|z| = (z_1^2 + \dots + z_d^2)^{1/2}$. For two real numbers a, b we denote the minimum of a and b by $a \wedge b$, and for a set A the indicator function $\omega \mapsto \mathbb{1}_A(\omega)$

takes the value 1 for $\omega \in A$ and 0 for $\omega \notin A$. For two sets $B_1, B_2 \subset \mathbb{R}^d$ and a vector $b \in \mathbb{R}^d$ we write $b + B_1 = \{b + c : c \in B_1\}$ and $B_1 + B_2 := \{c + d : c \in B_1, d \in B_2\}$. By a (probability) distribution on $\mathbb{R}^d$ we will always mean a probability measure on $(\mathbb{R}^d, \mathcal{B}(\mathbb{R}^d))$, where $\mathcal{B}(\mathbb{R}^d)$ denotes the Borel-σ-algebra on $\mathbb{R}^d$. The Dirac measure at a point $x \in \mathbb{R}^d$ is denoted by δ_x, the convolution of two distributions μ_1 and μ_2 by $\mu_1 * \mu_2$, and the product measure of them by $\mu_1 \otimes \mu_2$, with μ_1^{*n} and $\mu_1^{\otimes n}$ denoting the n-fold convolution and n-fold product measure of μ_1 with itself. The characteristic function $\mathbb{R}^d \ni z \mapsto \int_{\mathbb{R}^d} e^{i\langle z,x\rangle} \mu(dx)$ of a probability distribution μ on $\mathbb{R}^d$ is denoted by $\widehat{\mu}$, the support of a non-negative measure ν on $\mathbb{R}^d$ by $\operatorname{supp}(\nu)$, and weak convergence of probability measures is denoted by $\xrightarrow{w}$. The law (or distribution) of an $\mathbb{R}^d$-valued random vector will be denoted by $\mathcal{L}(X)$, its expectation by $\mathbb{E}(X)$. By a signed measure on a σ-algebra we mean a σ-additive $[-\infty, \infty]$-valued set function that assigns the value 0 to the empty set, and we say that it is finite if it is $\mathbb{R}$-valued. The restriction of a (signed) measure ν on a measurable space $(\Omega, \mathcal{F})$ to $\mathcal{A} \subset \mathcal{F}$ is denoted by $\nu_{|\mathcal{A}}$, and if $\mathcal{A}$ is of the form $\mathcal{A} = \{F \cap A : F \in \mathcal{F}\}$ with some $A \in \mathcal{F}$ we occasionally also write $\nu_{|A}$ rather than $\nu_{\mathcal{A}}$. The support of a signed measure is the support of its total variation measure.

2 The Lévy-Khintchine Type Representation

As already mentioned, quasi-infinitely divisible distributions admit a Lévy-Khintchine representation, with a quasi-Lévy type measure instead of a Lévy measure. A quasi-Lévy type measure is, in a sense, the difference between two Lévy measures ν_1 and ν_2 and can be seen as a "signed Lévy measure". However, this difference is not a signed measure if both, ν_1 and ν_2 are infinite. On the other hand, for any neighborhood U of 0, the restrictions of ν_1 and ν_2 to $\mathbb{R}^d \setminus U$ are finite, so that the difference is a finite signed measure. The following concept of [21] formalizes this statement.

Definition 2.1 For $r > 0$ let $\mathcal{B}_r^d := \{B \in \mathcal{B}(\mathbb{R}^d) : B \subset \{x \in \mathbb{R}^d : |x| \geq r\}\}$ and let $\mathcal{B}_0^d := \cup_{r>0}\mathcal{B}_r^d$. Let $\nu : \mathcal{B}_0^d \to \mathbb{R}$ be a function such that $\nu_{|\mathcal{B}_r^d}$ is a finite signed measure for every $r > 0$ and denote the total variation, the positive and the negative part of $\nu_{|\mathcal{B}_r^d}$ by $|\nu_{|\mathcal{B}_r^d}|$, $\nu_{|\mathcal{B}_r^d}^+$ and $\nu_{|\mathcal{B}_r^d}^-$, respectively.

(a) Let $E_0 := \{x \in \mathbb{R}^d : |x| > 1\}$ and for $n \in \mathbb{N}$ let $E_n := \{x \in \mathbb{R}^d : \frac{1}{n+1} < |x| \leq \frac{1}{n}\}$. The *positive part* $\nu^+ : \mathcal{B}(\mathbb{R}^d) \to [0, \infty)$ of ν is defined by

$$\nu^+(B) := \lim_{r\to 0} \nu_{|\mathcal{B}_r^d}^+(B \cap \{x \in \mathbb{R}^d : |x| > r\}) = \sum_{n\in\mathbb{N}_0} \nu_{|\mathcal{B}_{1/(n+1)}^d}^+(B \cap E_n), \quad B \in \mathcal{B}(\mathbb{R}^d).$$

The well-definedness of the series can easily be seen by a monotonicity argument, since $\nu^+_{|\mathcal{B}^d_r}(B) = \nu^+_{|\mathcal{B}^d_s}(B)$ for all $B \in \mathcal{B}^d_r$ and $0 < s \le r$. Moreover, if $(B_m)_{m\in\mathbb{N}}$ is a series of pairwise disjoint elements in $\mathcal{B}(\mathbb{R}^d)$, then

$$\sum_{m\in\mathbb{N}} \nu^+(B_m) = \sum_{m\in\mathbb{N}}\sum_{n\in\mathbb{N}} \nu^+(B_m \cap E_n) = \sum_{n\in\mathbb{N}} \nu^+(\cup_{m\in\mathbb{N}} B_m \cap E_n) = \nu^+(\cup_{m\in\mathbb{N}} B_m),$$

so that ν^+ is a measure on $(\mathbb{R}^d, \mathcal{B}(\mathbb{R}^d))$. Analogously, we define the *negative part* ν^- and *total variation* $|\nu|$ of ν by

$$\nu^-(B) := \lim_{r\to 0} \nu^+_{|\mathcal{B}^d_r}(B \cap \{x \in \mathbb{R}^d : |x| > r\})$$

and

$$|\nu|(B) := \lim_{r\to 0} \nu^+_{|\mathcal{B}^d_r}(B \cap \{x \in \mathbb{R}^d : |x| > r\})$$

for $B \in \mathcal{B}(\mathbb{R}^d)$.

(b) A function $f : \mathbb{R}^d \to \mathbb{C}$ is said to be *integrable* with respect to ν, if f is integrable with respect to $|\nu|$, i.e. if f is integrable with respect to ν^+ and ν^-. In this case, we define

$$\int_{\mathbb{R}^d} f(x)\nu(\mathrm{d}x) := \int_{\mathbb{R}^d} f(x)\nu^+(\mathrm{d}x) - \int_{\mathbb{R}^d} f(x)\nu^-(\mathrm{d}x).$$

(c) ν is called a *quasi-Lévy type measure* on $\mathbb{R}^d$, if the mapping $\mathbb{R}^d \to \mathbb{R}$, $x \mapsto 1 \wedge |x|^2$ is integrable with respect to ν.

Observe that $|\nu|(\{0\}) = \nu^+(\{0\}) = \nu^-(\{0\}) = 0$ by construction. Note that the mapping ν itself is defined on $\mathcal{B}^d_0$, which is not a σ-algebra, and hence ν is no signed measure. But whenever ν has an extension on $\mathcal{B}(\mathbb{R}^d)$ which is a signed measure, we will identify ν with this extension and speak of ν as a signed measure. Moreover, for two Lévy measures ν_1 and ν_2 on $\mathbb{R}^d$, the mapping $\nu := (\nu_1)_{|\mathcal{B}^d_0} - (\nu_2)_{|\mathcal{B}^d_0}$ is obviously a quasi-Lévy type measure on $\mathbb{R}^d$.

Next, we give the Lévy-Khintchine type representation for quasi-infinitely divisible distributions, which we immediately state for general representation functions: by a *representation function* on $\mathbb{R}^d$ we mean a bounded, Borel measurable function $c : \mathbb{R}^d \to \mathbb{R}^d$ that satisfies $\lim_{x\to 0} |x|^{-2}|c(x) - x| = 0$. The representation function $\mathbb{R}^d \ni x \mapsto x\mathbb{1}_{[0,1]}(|x|)$ is called the *standard representation function*. It is well known that for every fixed representation function c, a probability distribution μ is infinitely divisible if and only if its characteristic function has a representation as in (2.1) below with $A \in \mathbb{R}^{d\times d}$ being non-negative definite, $\gamma \in \mathbb{R}^d$ and ν being a Lévy measure on $\mathbb{R}^d$; Equation (1.1) then corresponds to the use of the standard representation function. The triplet (A, ν, γ) is then unique and called the characteristic triplet of μ with respect to the representation function c, also denoted by $(A, \nu, \gamma)_c$, cf. [32, Sect. 56]. Observe that only the location parameter γ depends on the specific choice of the representation function. For the standard representation function we get the standard characteristic triplet. Let us now come to the Lévy–Khintchine type representation of quasi-infinitely divisible distributions. This has already been observed in [21, Rem. 2.4], but we have decided to give the proof in detail.

Theorem 2.2 *Let c be a representation function on $\mathbb{R}^d$. A probability distribution μ on $\mathbb{R}^d$ is quasi-infinitely divisible if and only if its characteristic function $\widehat{\mu}$ admits the representation*

$$\widehat{\mu}(z) = \exp\left(\mathrm{i}\langle \gamma, z\rangle - \frac{1}{2}\langle z, Az\rangle + \int_{\mathbb{R}^d} \left(e^{\mathrm{i}\langle z,x\rangle} - 1 - \mathrm{i}\langle z, c(x)\rangle \right) \nu(dx) \right) \tag{2.1}$$

for all $z \in \mathbb{R}^d$ with a symmetric matrix $A \in \mathbb{R}^{d\times d}$, a constant $\gamma \in \mathbb{R}^d$ and a quasi-Lévy type measure ν on $\mathbb{R}^d$. In this case, the triplet (A, ν, γ) in the representation (2.1) of $\widehat{\mu}$ is unique.

Proof It is clear that if μ is quasi-infinitely divisible with $\widehat{\mu} = \widehat{\mu}_1/\widehat{\mu}_2$ and μ_1, μ_2 being infinitely divisible with characteristic triplets $(A_i, \nu_i, \gamma_i)_c$, $i = 1, 2$, then $\widehat{\mu}$ has the representation (2.1) with $A = A_1 - A_2$, $\nu = \nu_1 - \nu_2$ and $\gamma = \gamma_1 - \gamma_2$. Conversely, let μ be a probability distribution whose characteristic function $\widehat{\mu}$ admits the representation (2.1) with a symmetric matrix $A \in \mathbb{R}^{d\times d}$, a constant $\gamma \in \mathbb{R}^d$ and a quasi-Lévy measure ν on $\mathbb{R}^d$. Since $A \in \mathbb{R}^{d\times d}$ is symmetric, we can write $A = A^+ - A^-$ with non-negative symmetric matrices $A^+, A^- \in \mathbb{R}^{d\times d}$, which can be seen by diagonalising A, splitting the obtained diagonal matrix into a difference of two diagonal matrices with non-negative entries, and then transforming these matrices back. Let μ_1 and μ_2 be infinitely divisible distributions with characteristic triplets (A^+, ν^+, γ) and $(A^-, \nu^-, 0)$, respectively. Then $\mu_2 * \mu = \mu_1$, so that μ is quasi-infinitely divisible.

The uniqueness of the triplet is proved in Cuppens [10, Thm. 4.3.3] or also Sato [32, Exercise 12.2], but for clarity in the exposition we repeat the argument. So let (A_1, ν_1, γ_1) and (A_2, ν_2, γ_2) be two triplets satisfying (2.1). Defining

$$\Psi_j : \mathbb{R}^d \to \mathbb{R}, \quad z \mapsto \mathrm{i}\langle \gamma_j, z\rangle - \frac{1}{2}\langle z, A_j z\rangle + \int_{\mathbb{R}^d} \left(e^{\mathrm{i}\langle z,x\rangle} - 1 - \mathrm{i}\langle z, c(x)\rangle \right) \nu_j(\mathrm{d}x)$$

for $j \in \{1, 2\}$, it is easily seen that both Ψ_1 and Ψ_2 are continuous with $\Psi_j(0) = 0$, implying $\Psi_1 = \Psi_2$ by the uniqueness of the distinguished logarithm, cf. [32, Lem. 7.6]. As before we can find symmetric non-negative definite matrices A_1^+, A_1^-, A_2^+

and A_2^- such that $A_1 = A_1^+ - A_1^-$ and $A_2 = A_2^+ - A_2^-$. Therefore, the equation $\Psi_1 = \Psi_2$ can be rewritten to

$$\mathrm{i}\langle\gamma_1, z\rangle - \frac{1}{2}\langle z, (A_1^+ + A_2^-)z\rangle + \int_{\mathbb{R}^d} \left(e^{\mathrm{i}\langle z,x\rangle} - 1 - \mathrm{i}\langle z, c(x)\rangle\right)(\nu_1^+ + \nu_2^-)(\mathrm{d}x)$$
$$= \mathrm{i}\langle\gamma_2, z\rangle - \frac{1}{2}\langle z, (A_2^+ + A_1^-)z\rangle + \int_{\mathbb{R}^d} \left(e^{\mathrm{i}\langle z,x\rangle} - 1 - \mathrm{i}\langle z, c(x)\rangle\right)(\nu_2^+ + \nu_1^-)(\mathrm{d}x)$$

for all $z \in \mathbb{R}^d$. By the uniqueness of the Lévy-Khintchine representation of infinitely divisible distributions (e.g. [32, Thm. 8.1]), it follows that $\gamma_1 = \gamma_2$, $A_1^+ + A_2^- = A_2^+ + A_1^-$ and $\nu_1^+ + \nu_2^- = \nu_2^+ + \nu_1^-$, which implies that $A_1 = A_2$ and $\nu_1 = \nu_2$ (observe that ν_1^+, ν_1^-, ν_2^+ and ν_2^- are all finite on $\mathcal{B}_0^d$). □

Definition 2.3 Let c be a representation function on $\mathbb{R}^d$. For a quasi-infinitely divisible distribution μ on $\mathbb{R}^d$, the representation of $\widehat{\mu}$ in (2.1) is called the *Lévy-Khintchine representation* of μ and the function $\Psi_\mu : \mathbb{R}^d \to \mathbb{C}$ given by

$$\Psi_\mu(z) := \mathrm{i}\langle\gamma, z\rangle - \frac{1}{2}\langle z, Az\rangle + \int_{\mathbb{R}^d} \left(e^{\mathrm{i}\langle z,x\rangle} - 1 - \mathrm{i}\langle z, c(x)\rangle\right)\nu(\mathrm{d}x)$$

for all $z \in \mathbb{R}^d$ is called the *characteristic exponent* of μ. The triplet (A, ν, γ) is called the *generating triplet* or *characteristic triplet* of μ *with respect to* c and denoted by $(A, \nu, \gamma)_c$. The matrix A is called the *Gaussian covariance matrix* of μ, the mapping ν the *quasi-Lévy measure* of μ and the constant $\gamma \in \mathbb{R}^d$ the *location parameter* of μ with respect to c. When $d = 1$ we also speak of A as the *Gaussian variance* of μ. We write $\mu \sim \text{q.i.d.}(A, \nu, \gamma)_c$ to state that μ is a quasi-infinitely divisible distribution with characteristic triplet (A, ν, γ) with respect to c. If c is the standard representation function, then $(A, \nu, \gamma)_c$ is also called the *(standard) characteristic triplet* of μ and is denoted by (A, ν, γ), omitting the index c.

Remark 2.4

(a) It is easily seen that the Gaussian covariance matrix and the quasi-Lévy measure of a quasi-infinitely divisible distribution do not depend on the specific representation function, but the location parameter does.
(b) It is well known that the right-hand side of (2.1) defines the characteristic triplet of a probability distribution μ for all $\gamma \in \mathbb{R}^d$, all Lévy measures ν on $\mathbb{R}^d$ and all non-negative definite symmetric $A \in \mathbb{R}^{d\times d}$, in which case μ is necessarily infinitely divisible. It is however not true that the right-hand side of (2.1) defines the characteristic function of a probability distribution for all $\gamma \in \mathbb{R}^d$, symmetric matrices $A \in \mathbb{R}^{d\times d}$ and quasi-Lévy type measures ν. To see this, let $(A, \nu, \gamma)_c$ be the characteristic triplet of a quasi-infinitely divisible distribution such that A is not non-negative definite or such that ν is not non-negative. If all such triplets were to give rise to characteristic functions of a probability distribution, then in particular $(n^{-1}A, n^{-1}, n^{-1}\gamma)_c$ must be the characteristic triplet of some probability distribution μ_n, say, for all $n \in \mathbb{N}$. It is then easy to

see that the characteristic function of the n-fold convolution of μ_n with itself has Lévy-Khintchine type representation (2.1), so that $\mu_n^{*n} = \mu$ for each $n \in \mathbb{N}$. Hence μ is infinitely divisible, and the uniqueness of the characteristic triplet implies that A is non-negative definite and ν is non-negative, a contradiction. In Lemma 2.5 below we will actually see that only non-negative definite matrices A are possible. The quasi-Lévy measure does not need to be a Lévy measure, examples of which will be given in Sect. 3. However, not every quasi-Lévy type measure can occur as the quasi-Lévy measure of a quasi-infinitely divisible distribution, e.g. a quasi-Lévy type measure ν in $\mathbb{R}$ with $\nu^- \neq 0$ and ν^+ being the zero measure or a one-point measure can never be the quasi-Lévy measure of a quasi-infinitely divisible distribution, as mentioned in [21, Ex. 2.9]. This is the reason why we distinguish between quasi-Lévy type measures and quasi-Lévy measures. A quasi-Lévy type measure is any function $\nu : \mathcal{B}_0^d \to \mathbb{R}$ as in Definition 2.1, while a quasi-Lévy measure is a quasi-Lévy type measure that is linked to a (necessarily quasi-infinitely divisible) probability distribution.

In [21, Lem. 2.7] it was shown that if (a, ν, γ) is the characteristic triplet of a quasi-infinitely divisible distribution on $\mathbb{R}$, then necessarily $a \geq 0$. We now extend this to higher dimensions, by showing that the Gaussian covariance matrix must necessarily be non-negative definite.

Lemma 2.5 *If μ is a quasi-infinitely divisible distribution on $\mathbb{R}^d$ with characteristic triplet (A, ν, γ), then A is non-negative definite.*

Proof For $z \in \mathbb{R}^d$ and $t \in \mathbb{R}$ it holds

$$\Psi_\mu(tz) = \mathrm{i}t\langle \gamma, z\rangle - t^2 \frac{1}{2}\langle z, Az\rangle + \int_{\mathbb{R}^d} \left(e^{\mathrm{i}t\langle z,x\rangle} - 1 - \mathrm{i}t\langle z, x\rangle \mathbb{1}_{[0,1]}(|x|)\right) \nu(\mathrm{d}x).$$

Due to [32, Lem. 43.11 (i)] we have

$$\lim_{t\to\infty} t^{-2} \int_{\mathbb{R}^d} \left(e^{\mathrm{i}t\langle z,x\rangle} - 1 - \mathrm{i}t\langle z, x\rangle \mathbb{1}_{[0,1]}(|x|)\right) \nu^{\pm}(\mathrm{d}x) = 0,$$

hence $\lim_{t\to\infty} t^{-2}\Psi_\mu(tz) = -\frac{1}{2}\langle z, Az\rangle$. If $\langle z_0, Az_0\rangle < 0$ would hold for some $z_0 \in \mathbb{R}^d$, then we would obtain $|\widehat{\mu}(tz_0)| = |\exp(\Psi_\mu(tz_0))| \to \infty$ as $t \to \infty$, which is a contradiction. □

It is natural to ask why one should restrict to symmetric matrices $A \in \mathbb{R}^{d\times d}$ in the Lévy-Khintchine type representation and not allow arbitrary matrices $A \in \mathbb{R}^{d\times d}$. The next remark clarifies that this does not lead to new distributions, but that one would loose uniqueness of the characteristic triplet when allowing more generally non-symmetric matrices.

Remark 2.6 Given an arbitrary matrix $A \in \mathbb{R}^{d\times d}$ we can write $A = A_1 + A_2$, where $A_1 = \frac{1}{2}(A + A^T)$ and $A_2 = \frac{1}{2}(A - A^T)$. The matrix A_1 is symmetric and A_2 satisfies $A_2^T = -A_2$, which implies that $\langle z, A_2 z\rangle = z^T A_2 z = (z^T A_2 z)^T =$

$z^T A_2^T z = -\langle z, A_2 z\rangle$, and therefore $\langle z, A_2 z\rangle = 0$ for all $z \in \mathbb{R}^d$. It follows that $\langle z, Az\rangle = \langle z, A_1 z\rangle$ for all $z \in \mathbb{R}^d$. Hence, if we do not require that the matrix A in Theorem 2.2 is symmetric, then the representation of $\widehat{\mu}$ in (2.1) is not unique. Further, the class of distributions μ on $\mathbb{R}^d$ whose characteristic function $\widehat{\mu}$ allows the representation (2.1) with an arbitrary matrix $A \in \mathbb{R}^{d\times d}$ is exactly the class of quasi-infinitely divisible distributions.

Having seen the reason why we restrict to symmetric matrices, we would now like to know if we get new distributions (or non-unique triplets) if we also allow for complex $\gamma \in \mathbb{R}^d$, complex symmetric $A \in \mathbb{C}^{d\times d}$ and complex quasi-Lévy measures in the Lévy–Khintchine representation. Berger showed in [4, Thm. 3.2] that for $d = 1$ this does not lead to a greater class of distributions, and that then necessarily $\gamma \in \mathbb{R}$, $A \in [0, \infty)$ and that ν is real-valued, i.e. a quasi-Lévy measure. We now generalise this result to distributions on $\mathbb{R}^d$. To state this theorem, a *complex quasi-Lévy type measure* on $\mathbb{R}^d$ is a mapping $\nu : \mathcal{B}_0^d \to \mathbb{C}$ such that $\Re\nu$ and $\Im\nu$ are quasi-Lévy type measures on $\mathbb{R}^d$. A function $f : \mathbb{R}^d \to \mathbb{C}$ is said to be *integrable* with respect to ν, if it is integrable with respect to $\Re\nu$ and $\Im\nu$. In this case, we define

$$\int_{\mathbb{R}^d} f(x)\nu(\mathrm{d}x) := \int_{\mathbb{R}^d} f(x)(\Re\nu)(\mathrm{d}x) + \mathrm{i}\int_{\mathbb{R}^d} f(x)(\Im\nu)(\mathrm{d}x).$$

Theorem 2.7 *Let μ be a distribution on $\mathbb{R}^d$ such that its characteristic function admits the representation*

$$\widehat{\mu}(z) = \exp\left(\mathrm{i}\langle\gamma, z\rangle - \frac{1}{2}\langle z, Az\rangle + \int_{\mathbb{R}^d}\left(e^{\mathrm{i}\langle z,x\rangle} - 1 - \mathrm{i}\langle z, x\rangle \mathbb{1}_{[0,1]}(|x|)\right)\nu(dx)\right)$$

for every $z \in \mathbb{R}^d$ with a symmetric matrix $A \in \mathbb{C}^{d\times d}$, a complex quasi-Lévy type measure ν on $\mathbb{R}^d$ and $\gamma \in \mathbb{C}^d$. Then $A \in \mathbb{R}^{d\times d}$, $\gamma \in \mathbb{R}^d$ and $\Im\nu = 0$, that is, ν is a quasi-Lévy type measure and μ is quasi-infinitely divisible.

Proof The proof is very much the same as that of [4, Thm. 3.2] in dimension 1, but we give the full proof for convenience.

For $z \in \mathbb{R}^d$ we have

$$|\widehat{\mu}(z)|^2 = \widehat{\mu}(z)\widehat{\mu}(-z) = \exp\left(-\langle z, Az\rangle + 2\int_{\mathbb{R}^d}(\cos\langle z, x\rangle - 1)\,\nu(\mathrm{d}x)\right).$$

The function $g : \mathbb{R}^d \to \mathbb{C}$, $z \mapsto -\langle z, Az\rangle + 2\int_{\mathbb{R}^d}(\cos\langle z, x\rangle - 1)\,\nu(\mathrm{d}x)$ is continuous and satisfies $g(0) = 0$, implying that g is the distinguished logarithm of $|\widehat{\mu}|^2$, see [32, Lem. 7.6]. The uniqueness of the distinguished logarithm implies that g also has to be the natural logarithm of $|\widehat{\mu}|^2$, so that $g(z) \in \mathbb{R}$ for all $z \in \mathbb{R}^d$. Hence,

$$-\frac{1}{2}\langle z, (\Im A)z\rangle + \int_{\mathbb{R}^d}(\cos\langle z, x\rangle - 1)\,(\Im\nu)(\mathrm{d}x) = 0 \quad \text{for all } z \in \mathbb{R}^d. \tag{2.2}$$

Further, for $z \in \mathbb{R}^d$ it holds

$$\frac{\widehat{\mu}(z)}{\widehat{\mu}(-z)} = \exp\left(2\mathrm{i}\left(\langle \gamma, z\rangle + \int_{\mathbb{R}^d} \left(\sin\langle z, x\rangle - \langle z, x\rangle \mathbb{1}_{[0,1]}(|x|)\right) \nu(\mathrm{d}x)\right)\right)$$

and $|\widehat{\mu}(z)| = |\overline{\widehat{\mu}(z)}| = |\widehat{\mu}(-z)|$, so $\left|\frac{\widehat{\mu}(z)}{\widehat{\mu}(-z)}\right| = 1$ and thus

$$\langle \Im\gamma, z\rangle + \int_{\mathbb{R}^d} \left(\sin\langle z, x\rangle - \langle z, x\rangle \mathbb{1}_{[0,1]}(|x|)\right) (\Im\nu)(\mathrm{d}x) = 0 \quad \text{for all } z \in \mathbb{R}^d.$$

Adding this identity multiplied by i to (2.2) we obtain

$$\Psi_{\delta_0}(z) = 0 = \mathrm{i}\langle \Im\gamma, z\rangle - \frac{1}{2}\langle z, (\Im A)z\rangle + \int_{\mathbb{R}^d} \left(e^{\mathrm{i}\langle z,x\rangle} - 1 - \mathrm{i}\langle z, x\rangle \mathbb{1}_{[0,1]}(|x|)\right) (\Im\nu)(\mathrm{d}x)$$

for all $z \in \mathbb{R}^d$. By the uniqueness of the Lévy-Khintchine type representation for the Dirac measure δ_0, it follows that $\Im A = 0$, $\Im\nu = 0$ and $\Im\gamma = 0$. □

In the proof of Lemma 2.5, through $\Psi_\mu(tz)$ we implicitly were concerned with the projections of quasi-infinitely divisible distributions onto the lines $\{tz : t \in \mathbb{R}\}$ for given $z \in \mathbb{R}^d$. These projections are again quasi-infinitely divisible, and this holds more generally for affine linear images of random vectors with quasi-infinitely divisible distribution:

Lemma 2.8 *Let X be a random vector in $\mathbb{R}^d$ with $\mu = \mathcal{L}(X)$ being quasi-infinitely divisible with characteristic triplet (A, ν, γ). Let $b \in \mathbb{R}^m$ and $M \in \mathbb{R}^{m\times d}$. Then the distribution of the $\mathbb{R}^m$-valued random vector $U := MX + b$ is quasi-infinitely divisible with characteristic triplet (A_U, ν_U, γ_U), where*

$$A_U = MAM^T,$$

$$\gamma_U = b + M\gamma + \int_{\mathbb{R}^d} Mx \left(\mathbb{1}_{[0,1]}(|Mx|) - \mathbb{1}_{[0,1]}(|x|)\right) \nu(dx) \quad \textit{and}$$

$$\nu_U(B) = \nu(\{x \in \mathbb{R}^d : Mx \in B\}) \quad \textit{for } B \in \mathcal{B}_0^m.$$

Proof We see that

$$\widehat{\mathcal{L}(U)}(z) = \int_{\mathbb{R}^d} \mathrm{e}^{\mathrm{i}\langle Mx+b,z\rangle} \mu(\mathrm{d}x) = \mathrm{e}^{\mathrm{i}\langle b,z\rangle} \int_{\mathbb{R}^d} \mathrm{e}^{\mathrm{i}\langle x,M^T z\rangle} \mu(\mathrm{d}x) = \mathrm{e}^{\mathrm{i}\langle b,z\rangle}\widehat{\mu}(M^T z)$$

for $z \in \mathbb{R}^m$. The rest follows similar to [32, Prop. 11.10]. □

We conclude this section with a remark that it is also possible to define the drift or center of a quasi-infinitely divisible distribution, provided the quasi-Lévy measure satisfies a certain integrability condition.

Remark 2.9 Let $\mu \sim$ q.i.d.(A, ν, γ) and suppose that $\int_{|x|\le 1} |x|\,|\nu|(\mathrm{d}x) < \infty$. Then the characteristic function $\widehat{\mu}$ of μ can be rewritten to

$$\widehat{\mu}(z) = \exp\left(\mathrm{i}\langle \gamma_0, z\rangle - \frac{1}{2}\langle z, Az\rangle + \int_{\mathbb{R}^d} \left(e^{\mathrm{i}\langle z,x\rangle} - 1\right) \nu(\mathrm{d}x)\right) \tag{2.3}$$

for all $z \in \mathbb{R}^d$, where $\gamma_0 = \gamma - \int_{|x|\le 1} x\nu(\mathrm{d}x)$. This representation is unique and γ_0 is called the *drift* of μ. Conversely, is μ is a distribution on $\mathbb{R}^d$ such that its characteristic function admits the representation (2.3) for a symmetric matrix $A \in \mathbb{R}^{d\times d}$, a quasi-Lévy measure ν on $\mathbb{R}^d$ and $\gamma_0 \in \mathbb{R}^d$, then one can easily verify that μ is quasi-infinitely divisible with characteristic triplet (A, ν, γ), where $\gamma = \gamma_0 + \int_{|x|\le 1} x\nu(\mathrm{d}x)$. Then (A, ν, γ_0) is also called the *characteristic triplet* of μ with respect to $c(x) = 0$ and denoted by $(A, \nu, \gamma_0)_0$.
Similarly, if $\int_{|x|>1} |x|\,|\nu|(\mathrm{d}x) < \infty$, then $\widehat{\mu}$ admits the representation

$$\widehat{\mu}(z) = \exp\left(\mathrm{i}\langle \gamma_m, z\rangle - \frac{1}{2}\langle z, Az\rangle + \int_{\mathbb{R}^d} \left(e^{\mathrm{i}\langle z,x\rangle} - 1 - \mathrm{i}\langle z, x\rangle\right) \nu(\mathrm{d}x)\right) \tag{2.4}$$

for all $z \in \mathbb{R}^d$ with $\gamma_m = \gamma + \int_{|x|>1} x\,\nu(\mathrm{d}x)$. This representation is again unique and γ_m is called the *center* of μ.

3 Examples

A helpful tool to find examples of quasi-infinitely divisible distributions is the fact that the convolution of quasi-infinitely divisible distributions is again quasi-infinitely divisible.

Remark 3.1 Let $c : \mathbb{R}^d \to \mathbb{R}^d$ be a representation function. If $\mu_1 \sim$ q.i.d.$(A_1, \nu_1, \gamma_1)_c$ and $\mu_2 \sim$ q.i.d.$(A_2, \nu_2, \gamma_2)_c$, then $\mu_1 * \mu_2 \sim$ q.i.d.$(A_1 + A_2, \nu_1 + \nu_2, \gamma_1 + \gamma_2)_c$.

An important class of quasi-infinitely divisible distributions was established by Cuppens [9, Prop. 1], [10, Thm. 4.3.7]. He showed that every distribution which has an atom of mass $\lambda > \frac{1}{2}$ is quasi-infinitely divisible. We state his result and also prove it, in order to get an idea of what is behind the theorem.

Theorem 3.2 *Let $\mu = \lambda\delta_a + (1-\lambda)\sigma$ some $\lambda \in (\frac{1}{2}, 1]$, $a \in \mathbb{R}^d$ and a distribution σ on $\mathbb{R}^d$ that satisfies $\sigma(\{a\}) = 0$. Then μ is quasi-infinitely divisible with finite quasi-Lévy measure $\nu = \left(\sum_{k=1}^{\infty} \frac{(-1)^{k+1}}{k} \left(\frac{1-\lambda}{\lambda}\right)^k (\delta_{-a} * \sigma)^{*k}\right)_{|\mathcal{B}_0^d}$, Gaussian covariance matrix 0 and drift a.*

Proof Shifting μ by a, we can and do assume without loss of generality by Remark 3.1 that $a = 0$. Define $\rho := \left(\sum_{k=1}^{\infty} \frac{(-1)^{k+1}}{k} \left(\frac{1-\lambda}{\lambda}\right)^k \sigma^{*k}\right)$, where the sum converges absolutely to the finite signed measure ρ since $0 \le \frac{1-\lambda}{\lambda} < 1$. Denote $\nu := \rho_{|\mathcal{B}_0^d}$ as in the statement of the theorem, which then is a finite quasi-Lévy type measure (observe that $\rho(\{0\}) \neq 0$ is possible although $\sigma(\{0\}) = 0$; hence it is important to subtract any mass of ρ at 0, which is in particular achieved by restricting ρ to $\mathcal{B}_0^d$).

Next, observe that $\widehat{\mu}(z) = \lambda + (1-\lambda)\widehat{\sigma}(z) = \lambda(1 + \frac{1-\lambda}{\lambda}\widehat{\sigma}(z))$ for $z \in \mathbb{R}^d$. Again, since $0 \le \frac{1-\lambda}{\lambda} < 1$ and $|\widehat{\sigma}(z)| \le 1$ for all $z \in \mathbb{R}^d$, the series expansion of the principal branch of the complex logarithm of $\log(1+w)$ for $w \in \mathbb{C}$ such that $|w| < 1$ gives

$$
\begin{aligned}
\log \widehat{\mu}(z) &= \log \lambda + \sum_{k=1}^{\infty} \frac{(-1)^{k+1}}{k} \left(\frac{1-\lambda}{\lambda}\right)^k \widehat{\sigma}(z)^k \\
&= \log \lambda + \sum_{k=1}^{\infty} \frac{(-1)^{k+1}}{k} \left(\frac{1-\lambda}{\lambda}\right)^k \int_{\mathbb{R}^d} e^{i\langle z,x\rangle}\, \sigma^{*k}(dx) \\
&= \log \lambda + \int_{\mathbb{R}^d} e^{i\langle z,x\rangle}\, \rho(dx) \\
&= \log \lambda + \rho(\mathbb{R}^d) + \int_{\mathbb{R}^d} (e^{i\langle z,x\rangle} - 1)\, \nu(dx)
\end{aligned}
$$

for every $z \in \mathbb{R}^d$. Using the fact that $\widehat{\mu}(0) = 1$, we find $0 = \log \widehat{\mu}(0) = \log \lambda + \rho(\mathbb{R}^d)$, finishing the proof. □

Lindner et al. [21, Thm. 8.1] showed that a probability distribution μ on $\mathbb{Z}$ is quasi-infinitely divisible if and only if its characteristic function has no zeroes. This has been extended recently to distributions on $\mathbb{Z}^d$ by Berger and Lindner [5, Thm. 3.2]. The precise result is as follows:

Theorem 3.3 *Let μ be a distribution that is supported in $\mathbb{Z}^d$. Then μ is quasi-infinitely divisible if and only if $\widehat{\mu}(z) \neq 0$ for all $z \in [0, 2\pi]^d$. In that case, the Gaussian covariance matrix of μ is zero, the quasi-Lévy measure is finite and supported in $\mathbb{Z}^d \setminus \{0\}$ and the drift of μ is in $\mathbb{Z}^d$.*

The proof given in [5, 21] relies on the Lévy–Wiener theorem, according to which for a continuous function $f : \mathbb{R}^d \to \mathbb{C}$ that is 2π-periodic in all coordinates and is such that it has absolutely summable Fourier coefficients, and a holomorphic function $h : D \to \mathbb{C}$ on an open subset $D \subset \mathbb{C}$ such that $f([0, 2\pi]^d) \subset \mathbb{C}$, also the composition $h \circ f$ has absolutely summable Fourier coefficients (i.e. it is an element of the so called Wiener algebra). The given zero-free characteristic function $\widehat{\mu}$ then has to be modified appropriately in order to apply this Wiener–Lévy theorem to the distinguished logarithm, and then an argument is needed in order to show that the Fourier coefficients are indeed real-valued and not complex. This is carried out in detail in [5, 21] and we refer to these articles for the detailed proof. We only mention here that it is also possible to replace the proofs given there for the fact that the Fourier coefficients are real and not complex by Theorem 2.7 given in the present article.

Recall that a stochastic process is quasi-infinitely divisible if all of its finite-dimensional distributions are quasi-infinitely divisible. By Theorem 3.3, for a $\mathbb{Z}^d$-valued stochastic process this is the case, if and only if the characteristic functions of its finite-dimensional distributions have no zeros.

There is nothing special about the lattice $\mathbb{Z}^d$ and Theorem 3.3 continues to hold for more general lattices, which is the contents of the next result that generalises [21, Cor. 3.10] to higher dimensions.

Corollary 3.4 *Let $M \in \mathbb{R}^{d\times d}$ be invertible, $b \in \mathbb{R}^d$ and μ be a probability distribution supported in the lattice $M\mathbb{Z}^d + b = \{Mz + b : z \in \mathbb{Z}^d\}$. Then μ is quasi-infinitely divisible if and only if the characteristic function of μ has no zeroes on $(M^T)^{-1}([0, 2\pi]^d) = \{(M^T)^{-1}x : x \in [0, 2\pi]^d\}$. In that case, the Gaussian covariance matrix of μ is zero, the quasi-Lévy measure is finite and supported in $M\mathbb{Z}^d \setminus \{0\}$ and the drift of μ is in $M\mathbb{Z}^d + b$.*

Proof Let U be a random vector with distribution μ and define $X = M^{-1}(U - b)$. Then $\mathcal{L}(X)$ is supported in $\mathbb{Z}^d$ and $\widehat{\mu}(z) = \widehat{\mathcal{L}(U)}(z) = e^{i\langle b,z\rangle}\widehat{\mathcal{L}(X)}(M^T z)$ for $z \in \mathbb{R}^d$. The result is then an immediate consequence of Theorem 3.3 together with Lemma 2.8; here, an easy extension of Lemma 2.8 shows that the drift γ_U of U is $M\gamma_X + b$, where γ_X is the drift of $\mathcal{L}(X)$. □

An interesting application of the previous theorem on $\mathbb{Z}^d$ has been given in [5, Thm. 4.1], where a Cramér–Wold device for infinite divisibility of $\mathbb{Z}^d$-distributions was established. The precise statement is as follows:

Corollary 3.5 *Let X be a $\mathbb{Z}^d$-valued random vector with distribution μ. Then the following are equivalent:*

(i) *μ is infinitely divisible.*
(ii) *$\mathcal{L}(a^T X)$ is infinitely divisible for all $a \in \mathbb{R}^d$.*
(iii) *$\mathcal{L}(a^T X)$ is infinitely divisible for all $a \in \mathbb{N}_0^d$.*
(iv) *The characteristic function of μ has no zeroes on $\mathbb{R}^d$ and there exists some $a = (a_1, \ldots, a_d)^T \in \mathbb{R}^d$ such that $a_1, \ldots, a_d$ are linearly independent over $\mathbb{Q}$ and such that $\mathcal{L}(a^T X)$ is infinitely divisible.*

This result is striking in the sense that a Cramér-Wold device does not hold in full generality for infinite divisibility of $\mathbb{R}^d$-valued distributions. Indeed, it is even known that for every $\alpha \in (0, 1)$ there exists a d-dimensional random vector X such that $\mathcal{L}(a^T X)$ is α-stable for all $a \in \mathbb{R}^d$, but that $\mathcal{L}(X)$ is not infinitely divisible (see [31, Sect. 2.2]). The proof of Corollary 3.5 heavily relies on Theorem 3.3, and we refer to [5, Thm. 4.1] for the details of the proof.

In view of Theorem 3.3 it is natural to ask if every distribution μ whose characteristic function is zero-free must be quasi-infinitely divisible. That this is not the case was shown in [21, Ex. 3.3] by giving a counter example. Let us give another counter example to this fact:

Example 3.6 Consider the function $\varphi : \mathbb{R} \to (0, \infty)$ given by $\varphi(x) = \exp(1 - \mathrm{e}^{|x|})$. Then φ is continuous on $\mathbb{R}$, $\varphi(0) = 1$ and $\varphi''(x) = (\mathrm{e}^x - 1)\exp(1 + x - \mathrm{e}^x) > 0$ for all $x > 0$. Hence φ is strictly convex on $(0, \infty)$. Since $\varphi(x)$ tends to 0 as $|x| \to \infty$, Pólya's theorem implies that φ is the characteristic function of an absolutely continuous distribution μ on $\mathbb{R}$. The distinguished logarithm of μ is given by $\Psi_\mu(x) = 1 - \mathrm{e}^{|x|}$. Hence $\lim_{t\to\infty} t^{-2}\Psi_\mu(t) = -\infty$. It follows that μ cannot be quasi-infinitely divisible, for if it were, then $\lim_{t\to\infty} t^{-2}\Psi_\mu(t) = -A/2$ as shown in the proof of Lemma 2.5, where $A \in \mathbb{R}$ denotes the Gaussian variance of μ. Hence we have a one-dimensional distribution μ that is not quasi-infinitely divisible but whose characteristic function is zero-free. Using Lemma 2.8 it then is easily seen that $\delta_0^{\otimes(d-1)} \otimes \mu$ is a distribution in $\mathbb{R}^d$ that is not quasi-infinitely divisible but whose characteristic function is zero-free. Further examples of such distributions can be constructed using Proposition 3.10 below.

Let us now give some concrete examples of quasi-infinitely divisible distributions on $\mathbb{Z}^d$:

Example 3.7

(a) Consider the distribution $\mu := a\delta_{(0,0)} + b\delta_{(1,0)} + c\delta_{(0,1)}$ on $\mathbb{R}^2$ with $a, b, c \in (0, 1)$. If $\max\{a, b, c\} > 1/2$, then μ is quasi-infinitely divisible by Cuppens' result (cf. Theorem 3.2). If $\max\{a, b, c\} \le 1/2$, then μ cannot be quasi-infinitely divisible, since $\widehat{\mu}$ is not zero-free. Indeed, for $(x, y) \in \mathbb{R}^2$, $\widehat{\mu}(x, y) = a + b\mathrm{e}^{\mathrm{i}x} + c\mathrm{e}^{\mathrm{i}y} = 0$ if and only if $b\mathrm{e}^{\mathrm{i}x} = -a - c\mathrm{e}^{\mathrm{i}y}$. The set $\{b\mathrm{e}^{\mathrm{i}x} : x \in \mathbb{R}\}$ describes a circle in the complex plane with center 0 and radius b, intersecting the real axis at the points $-b$ and b, and $\{-a - c\mathrm{e}^{\mathrm{i}y} : y \in \mathbb{R}\}$ describes a circle in the complex plane with center $-a$ and radius c, intersecting the real axis at the points $-a - c$ and $-a + c$. Now, since $\max\{a, b, c\} \le 1/2$, this implies that $a \le b + c$, $b \le a + c$ and $c \le a + b$, and hence $-a - c \le -b \le -a + c \le b$. Therefore, the two circles intersect or touch each other, so they share at least one common point, which corresponds to a zero of the characteristic function of μ.

(b) Let $p, q \in (0, 1) \setminus \{1/2\}$ and consider the distributions $\mu_1 := (1 - p)\delta_{(0,0)} + p\delta_{(1,0)}$ and $\mu_2 := (1 - q)\delta_{(0,0)} + q\delta_{(0,1)}$. Due to Theorem 3.2, the distributions μ_1 and μ_2 are quasi-infinitely divisible. Hence, also the distribution $\mu := \mu_1 * \mu_2 = (1 - p)(1 - q)\delta_{(0,0)} + p(1 - q)\delta_{(1,0)} + (1 - p)q\delta_{(0,1)} + pq\delta_{(1,1)}$ is quasi-infinitely divisible. Observe that it is possible to choose p and q such that $\max\{pq, (1 - p)(1 - q), p(1 - q), q(1 - p)\} < 1/2$.

(c) Let $\mu_1, \ldots, \mu_d$ be distributions on $\mathbb{R}$ supported in $\mathbb{Z}$ such that $\Re\,\widehat{\mu_k}(z) > 0$ for all $k \in \{1, \ldots, d\}$ and $z \in \mathbb{R}^d$. Examples of such distributions can be obtained as symmetrisations of distributions supported in $\mathbb{Z}$, whose characteristic functions have no zeros. Let $0 \leq p_1, \ldots, p_d \leq 1$ be such that $\sum_{k=1}^d p_k = 1$ and define the distribution

$$\mu := \sum_{k=1}^{d} p_k \sum_{l \in \mathbb{Z}} \mu_k(\{l\}) \delta_{le_k}$$

on $\mathbb{R}^d$, where e_k is the k-th unit vector. Then for $z = (z_1, \ldots, z_d)^T \in \mathbb{R}^d$ we have

$$\begin{aligned}\widehat{\mu}(z) &= \sum_{l \in \mathbb{Z}^d} \mu(\{l\}) e^{\mathrm{i}\langle z, l\rangle} = \sum_{k=1}^{d} \sum_{l \in \mathbb{Z}\setminus\{0\}} p_k \mu(\{le_k\}) e^{\mathrm{i}\langle z, le_k\rangle} + \mu(\{0\}) \\ &= \sum_{k=1}^{d} p_k \sum_{l \in \mathbb{Z}\setminus\{0\}} \mu_k(\{l\}) e^{\mathrm{i}lz_k} + \sum_{k=1}^{d} p_k \mu_k(\{0\}) = \sum_{k=1}^{d} p_k \widehat{\mu_k}(z_k) \neq 0\end{aligned}$$

since $\Re\, \widehat{\mu_k}(z_k) > 0$. In particular, $\widehat{\mu}$ has no zeros, so μ is quasi-infinitely divisible by Theorem 3.3.

(d) Let $p \in [0, 1/4)$ and consider the symmetric distribution $\sigma := p\delta_{-1} + (1-2p)\delta_0 + p\delta_1$. We have $\widehat{\sigma}(z) = pe^{-\mathrm{i}z} + 1 - 2p + pe^{\mathrm{i}z} = 1 - 2p + 2p\cos(z) > 0$ for $z \in \mathbb{R}$ since $p < 1/4$. With the construction in part (c), choosing $\mu_1 = \mu_2 = \sigma$ it follows that the distribution

$$\mu := rp\delta_{(-1,0)} + rp\delta_{(1,0)} + (1-r)p\delta_{(0,-1)} + (1-r)p\delta_{(0,1)} + (1-2p)\delta_{(0,0)}$$

is quasi-infinitely divisible for any $r \in [0, 1]$.

Similar as in the one-dimensional case in [21, Cor. 8.3], one can show that a consequence of Theorem 3.3 is that every factor of a quasi-infinitely divisible distribution that is supported in $\mathbb{Z}^d$ is also quasi-infinitely divisible.

Corollary 3.8 *Let μ be a quasi-infinitely divisible distribution supported in $\mathbb{Z}^d$. If μ_1 and μ_2 are distributions on $\mathbb{R}^d$ such that $\mu = \mu_1 * \mu_2$, then also μ_1 and μ_2 are quasi-infinitely divisible.*

Proof This follows in complete analogy to the proof of [21, Cor. 8.3], and is an easy consequence of Corollary 3.4 and the fact that if μ is supported in $\mathbb{Z}^d$, then there must be $k \in \mathbb{R}^d$ such that μ_1 is supported in $\mathbb{Z}^d + k$ and μ_2 is supported in $\mathbb{Z}^d - k$. □

Theorem 3.3 is nice since it gives a complete characterisation of quasi-infinite divisibility in terms of the characteristic function. In the univariate setting Berger [4, Thm. 4.12] extended this characterisation of quasi-infinitely divisibility to a greater class of distributions, which we now state without proof:

Theorem 3.9 *Let μ be a distribution on $\mathbb{R}$ of the form $\mu = \mu_d + \mu_{ac}$ with an absolutely continuous measure μ_{ac} and a non-zero discrete measure μ_d which is supported in the lattice $h\mathbb{Z} + d$ for some $r \in \mathbb{R}$, $h > 0$, such that $\widehat{\mu_d}(z) \neq 0$ for all $z \in \mathbb{R}$. Then μ is quasi-infinitely divisible if and only if $\widehat{\mu}(z) \neq 0$ for all $z \in \mathbb{R}$. In that case, the Gaussian variance of μ is zero and the quasi-Lévy measure ν satisfies $\int_{[-1,1]} |x|\,|\nu|(\mathrm{d}x) < \infty$.*

It is remarkable that the quasi-Lévy measure ν in Theorem 3.9 can indeed be infinite although the distribution μ there has atoms; a concrete example for this phenomenon will be given in Remark 4.5 below. A special case of Theorem 3.9 is when $\mu_d = p\delta_x$ for some $x \in \mathbb{R}$ and $p > 0$. Then $\widehat{\mu_d}(z) \neq 0$ for all $z \in \mathbb{R}$ and a distribution of the form $\mu(\mathrm{d}x) = p\delta_x(\mathrm{d}x) + (1-p)f(x)\,\mathrm{d}x$, where $f : \mathbb{R} \to [0, \infty)$ is integrable with integral 1, is quasi-infinitely divisible if and only if $\widehat{\mu}(z) \neq 0$ for all $z \in \mathbb{R}$, cf. [4, Thm. 4.6]. Further applications of Theorem 3.9 include convex combinations of normal distributions: let $\mu = \sum_{i=1}^n p_i N(b_i, a_i)$, where $0 < p_1, \ldots, p_n < 1$, $\sum_{i=1}^n p_i = 1$, $0 < a_1 < a_2 < \ldots < a_n$ and $b_1, \ldots, b_n \in \mathbb{R}$, where $N(b_i, a_i)$ denotes the normal distribution with mean b_i and variance a_i. Then μ is quasi-infinitely divisible if and only if $\widehat{\mu}(z) \neq 0$ for all $z \in \mathbb{R}$, as shown in [4, Rem. 4.12]. Observe that the latter condition is in particular satisfied when additionally $b_1 = \ldots = b_n = 0$, and one can even show that $\sum_{i=1}^n p_i N(0, a_i)$ is quasi-infinitely divisible, even when some of the variances a_i coincide, cf. [4, Ex. 4.16].

It seems likely that Theorem 3.9 continues to hold in the multivariate setting, but so far we have not proved that. We intend to invest this case in future work. For the moment, we content ourselves with a recipe for constructing multivariate quasi-infinitely divisible distributions from independent one-dimensional quasi-infinitely divisible distributions.

Proposition 3.10 *Let $X_1, \ldots, X_d$ be independent real-valued random variables and let $X = (X_1, \ldots, X_d)^T$. Then the law $\mathcal{L}(X)$ of X is quasi-infinitely divisible if and only if $\mathcal{L}(X_k)$ is quasi-infinitely divisible for all $k \in \{1, \ldots, d\}$. In this case, if (A, ν, γ) denotes the (standard) characteristic triplet of $\mathcal{L}(X)$ and (a_k, ν_k, γ_k) the (standard) characteristic triplet of $\mathcal{L}(X_k)$, then*

$$A = \begin{pmatrix} a_1 & 0 & \ldots & 0 \\ 0 & a_2 & \ddots & \vdots \\ \vdots & \ddots & \ddots & 0 \\ 0 & \ldots & 0 & a_d \end{pmatrix}, \quad \gamma = \begin{pmatrix} \gamma_1 \\ \vdots \\ \gamma_d \end{pmatrix} \quad \textit{and} \quad \nu = \sum_{k=1}^d \delta_0^{\otimes(k-1)} \otimes \nu_k \otimes \delta_0^{\otimes(d-k)}.$$

Proof That quasi-infinite divisibility of $\mathcal{L}(X)$ implies quasi-infinite divisibility of $\mathcal{L}(X_k)$ for $k = 1, \ldots, d$ is clear from Lemma 2.8. Conversely, let $\mathcal{L}(X_k) \sim$ q.i.d.(a_k, ν_k, γ_k) for $k \in \{1, \ldots, d\}$ with independent $X_1, \ldots, X_d$. Using $\widehat{\mathcal{L}(X)}(z) = \prod_{k=1}^d \widehat{\mathcal{L}(X_k)}(z_k)$ for all $z = (z_1, \ldots, z_d)^T \in \mathbb{R}^d$ it is easy to see that μ has a Lévy–Khintchine type representation as in (2.1) with A, ν and γ as given in the theorem, and hence that $\mathcal{L}(X)$ is quasi-infinitely divisible with standard characteristic triplet (A, ν, γ). □

Using Lemma 2.8, the following is now immediate:

Corollary 3.11 *Let $X = (X_1, \ldots, X_d)^T$ with independent real-valued random variables $X_1, \ldots, X_d$ such that $\mathcal{L}(X_k)$ is quasi-infinitely divisible for each $k \in \{1, \ldots, d\}$. Let further $n \in \mathbb{N}$, $M \in \mathbb{R}^{n\times d}$ and $b \in \mathbb{R}^n$. Then also the distribution $\mathcal{L}(MX + b)$ is quasi-infinitely divisible.*

4 Conditions for Absolute Continuity

In this section we study absolute continuity of quasi-infinitely divisible distributions and give some sufficient conditions in terms of the characteristic triplet. Considering an infinitely divisible distribution on $\mathbb{R}$, Kallenberg [17, p. 794 f.] gave a sufficient condition on the Lévy measure for the distribution to have a smooth Lebesgue density. The following theorem generalizes this result for quasi-infinitely divisible distributions on $\mathbb{R}^d$. In its statement, we denote by $S^{d-1} := \{x \in \mathbb{R}^d : |x| = 1\}$ the unit sphere in $\mathbb{R}^d$.

Theorem 4.1 *Let μ be a quasi-infinitely divisible distribution on $\mathbb{R}^d$ with characteristic triplet (A, ν, γ). Define*

$$G^-(r) := \sup_{\xi\in S^{d-1}} \xi^T \left(\int_{|x|\le r} xx^T \nu^-(dx)\right)\xi \quad \text{and} \quad G^+(r) := \inf_{\xi\in S^{d-1}} \xi^T \left(\int_{|x|\le r} xx^T \nu^+(dx)\right)\xi$$

for $r > 0$. Suppose that A is strictly positive definite, or that

$$\lim_{r\to 0} r^{-2}|\log r|^{-1}G^+(r)\left(\frac{1}{3} - 2\,\frac{r^2\nu^-(\{x \in \mathbb{R}^d : |x| > r\})}{G^+(r)} - \frac{2}{3}\frac{G^-(r)}{G^+(r)}\right) = \infty \tag{4.1}$$

(when $G^+(r) = 0$ for small $r > 0$ we interpret the left-hand side of (4.1) as 0 and hence (4.1) to be violated). Then μ has an infinitely often differentiable Lebesgue density whose derivatives tend to 0 as $|x| \to \infty$.

Proof Suppose first that A is strictly positive definite and let $\lambda_0 > 0$ be the smallest eigenvalue of A. By [32, Lem. 43.11 (i)] we have

$$\lim_{|z|\to\infty} |z|^{-2}\int_{\mathbb{R}^d}\left(e^{\mathrm{i}\langle z,x\rangle} - 1 - \mathrm{i}\langle z, x\rangle \mathbb{1}_{[0,1]}(|x|)\right)\nu^{\pm}(\mathrm{d}x) = 0.$$

Since further $-\frac{1}{2}\langle z, Az\rangle \le -\frac{\lambda_0}{2}|z|^2$ for all $z \in \mathbb{R}^d$, we have $\limsup_{|z|\to\infty} |z|^{-2}\Psi_\mu(z) \le -\frac{1}{2}\lambda_0$, so there exists $K > 0$ such that $|\widehat{\mu}(z)| =$

$e^{\Re\Psi_\mu(z)} \le e^{-\lambda_0|z|^2/4}$ for all $z \in \mathbb{R}^d$ with $|z| \ge K$. Hence, we have $\int_{\mathbb{R}^d} |\widehat{\mu}(z)||z|^k \mathrm{d}z < \infty$ for all $k \in \mathbb{N}$ and the claim follows by [32, Prop. 28.1].

Now, suppose that (4.1) is satisfied. Using the fact that $\frac{1}{3}y^2 \le 1 - \cos(y) \le \frac{2}{3}y^2$ for all $y \in [-1, 1]$, we estimate

$$\begin{aligned}\int_{\mathbb{R}^d} (1 - \cos\langle z, x\rangle)\, \nu^+(\mathrm{d}x) &\ge \frac{1}{3}\int_{|x|\le 1/|z|} \langle z, x\rangle^2\, \nu^+(\mathrm{d}x) = \frac{1}{3}\int_{|x|\le 1/|z|} z^T x x^T z\, \nu^+(\mathrm{d}x)\\ &\ge \frac{1}{3}|z|^2 G^+(1/|z|)\end{aligned}$$

and similarly

$$\begin{aligned}\int_{\mathbb{R}^d} (1 - \cos\langle z, x\rangle)\, \nu^-(\mathrm{d}x) &\le \frac{2}{3}\int_{|x|\le 1/|z|} \langle z, x\rangle^2\, \nu^-(\mathrm{d}x) + 2\nu^-(\{x \in \mathbb{R}^d : |x| > 1/|z|\})\\ &\le \frac{2}{3}|z|^2 G^-(1/|z|) + 2\nu^-(\{x \in \mathbb{R}^d : |x| > 1/|z|\})\end{aligned}$$

for all $z \in \mathbb{R}^d$. Hence, for $|z| \ge 1$,

$$\begin{aligned}-(\log|z|)^{-1}\Re\Psi_\mu(z) &= (\log|z|)^{-1}\left(\frac{1}{2}\langle z, Az\rangle + \int_{\mathbb{R}^d}(1 - \cos\langle z, x\rangle)\nu(\mathrm{d}x)\right)\\ &\ge (\log|z|)^{-1}\left(\int_{\mathbb{R}^d}(1 - \cos\langle z, x\rangle)\nu(\mathrm{d}x)\right) \to \infty \text{ as } |z| \to \infty\end{aligned}$$

by assumption. As a consequence, for every $k \in \mathbb{N}$ there exists $K > 0$ such that $\Re\Psi_\mu(z) \le -(k+2)\log|z|$ for all $z \in \mathbb{R}^d$ with $|z| > K$, and therefore $|\widehat{\mu}(z)| = e^{\Re\Psi_\mu(z)} \le e^{-(k+2)\log|z|} = |z|^{-(k+2)}$ when $|z| > K$. This implies that $\int_{\mathbb{R}^d} |\widehat{\mu}(z)||z|^k \mathrm{d}z < \infty$ for every $k \in \mathbb{N}$ and the claim follows. □

Remark 4.2

(a) Observe that for fixed $r \in (0, 1)$, the matrices $\int_{|x|\le r} x x^T \nu^\pm(\mathrm{d}x) \in \mathbb{R}^{d\times d}$ are symmetric and non-negative definite and that $G^+(r)$ is the smallest eigenvalue of the matrix $\int_{|x|\le r} x x^T\, \nu^+(\mathrm{d}x)$ and $G^-(r)$ is the largest eigenvalue of $\int_{|x|\le r} x x^T\, \nu^-(\mathrm{d}x)$.
(b) If μ is infinitely divisible, then $\nu^- = 0$ and hence $G^-(r) = 0$ and (4.1) reduces to

$$\lim_{r\to 0} r^{-2}|\log r|^{-1} G^+(r) = +\infty. \tag{4.2}$$

Hence, if an infinitely divisible distribution μ on $\mathbb{R}^d$ with characteristic triplet (A, ν^+, γ) is such that A is strictly positive definitive or such that (4.2) is satisfied, then μ has an infinitely often differentiable Lebesgue density whose derivatives tend to 0 as $|x| \to \infty$. In dimension $d = 1$, this reduces to

$A > 0$ or $\lim_{r\to 0} r^{-2}|\log r|^{-1} \int_{|x|\le r} |x|^2\, \nu^+(\mathrm{d}x) = +\infty$, which is Kallenberg's classical condition [17, p. 794 f.]. The described multivariate generalisation of Kallenberg's condition seems to be new even in the case of infinitely divisible distributions.

(c) Let $\mu \sim$ q.i.d.(A, ν, γ) in $\mathbb{R}^d$. A sufficient condition for (4.1) to hold is that (4.2) is satisfied along with $G^-(r) = o(G^+(r))$ and $r^2\nu^-(\{x \in \mathbb{R}^d : |x| > r\}) = o(G^+(r))$ as $r \to 0$, where we used the "little o" Landau symbol notation.

(d) Let $\mu \sim$ q.i.d.(A, ν, γ) in $\mathbb{R}^d$. Define

$$g^-(r) := \int_{|x|\le r} |x|^2\, \nu^-(\mathrm{d}x) \quad \text{and} \quad g^+(r) := \int_{|x|\le r} |x|^2\, \nu^+(\mathrm{d}x)$$

for $r > 0$. Then

$$g^{\pm}(r) = \int_{|x|\le r} \operatorname{trace}(xx^T)\, \nu^{\pm}(\mathrm{d}x) = \operatorname{trace}\left(\int_{|x|\le r} xx^T\, \nu^{\pm}(\mathrm{d}x)\right).$$

Since the trace of a symmetric $d \times d$-matrix is the sum of its eigenvalues, we observe from (a) that

$$G^-(r) \le g^-(r) \le d\, G^-(r) \quad \text{and} \quad G^+(r) \le g^+(r).$$

So we can conveniently bound $G^-(r)$ from below and above in terms of $g^-(r)$, in particular, if we replace $G^-(r)$ in (4.1) by $g^-(r)$ (and leave the rest unchanged, in particular we do not replace $G^+(r)$ by $g^+(r)$) then we also obtain a sufficient condition for μ to have an infinitely often differentiable Lebesgue density with derivatives tending to 0 as $|x| \to \infty$. A similar remark applies to (c) above, where we can replace $G^-(r)$ by $g^-(r)$ (but not $G^+(r)$ by $g^+(r)$).

In [21, Thm. 7.1], an Orey-type condition (cf. [32, Prop. 28.3]) was given for absolute continuity of one-dimensional quasi-infinitely divisible distributions. We can now generalise this to the multivariate setting:

Corollary 4.3 *Let μ be a quasi-infinitely divisible distribution on $\mathbb{R}^d$ with characteristic triplet (A, ν, γ). With the notations of Theorem 4.1, suppose that A is strictly positive definite or that there exists some $\beta \in (0, 2)$ such that*

$$\liminf_{r\to 0} r^{-\beta} G^+(r) > \limsup_{r\to 0} r^{-\beta} G^-(r) = 0. \tag{4.3}$$

Then μ is absolutely continuous and its Lebesgue density f is infinitely often differentiable with all its derivatives tending to 0 as $|x| \to \infty$.

Proof The case when A is strictly positive definite is clear, so suppose that (4.3) is satisfied with some $\beta \in (0, 2)$. Then clearly $\lim_{r\to 0} r^{-2}|\log r|^{-1} G^+(r) = \infty$ since $\beta < 2$, and $G^-(r) = o(G^+(r))$ as $r \to 0$. By Remark 4.2 (c) it is hence sufficient to show that $r^2\nu^-(\{x \in \mathbb{R}^d : |x| > r\}) = o(G^+(r))$ as $r \to 0$. To see this, define $g^-(r) := \int_{|x|\le r} |x|^2 \, \nu^-(\mathrm{d}x)$. Using integration by parts and $g^-(r) \le d\,G^-(r)$ by Remark 4.2 (d), we then obtain

$$\begin{aligned}\nu^-(\{x \in \mathbb{R}^d : r < |x| \le 1\}) &= \int_r^1 s^{-2}\,\mathrm{d}g^-(s)\\ &= g^-(1) - r^{-2}g^-(r) - \int_r^1 g^-(s)\,\mathrm{d}s^{-2}\\ &\le d\left(G^-(1) + 2\int_r^1 G^-(s)s^{-3}\,\mathrm{d}s\right)\end{aligned}$$

for $r \in (0,1]$. By (4.3) for every $\varepsilon > 0$ we can find an $r_\varepsilon \in (0,1)$ such that $G^-(s) \le \varepsilon s^\beta$ for all $s \in (0, r_\varepsilon]$, so that we continue to estimate for $r < r_\varepsilon$

$$\nu^-(\{x \in \mathbb{R}^d : r < |x| \le 1\}) \le dG^-(1) + 2dG^-(1)r_\varepsilon^{-3}(1-r_\varepsilon) + 2\varepsilon d\int_r^{r_\varepsilon} s^{\beta-3}\,\mathrm{d}s.$$

This implies

$$r^2\frac{\nu^-(\{x \in \mathbb{R}^d : r < |x| \le 1\})}{G^+(r)} \le r^2 d\frac{G^-(1) + 2G^-(1)r_\varepsilon^{-3}}{G^+(r)} + 2\varepsilon d\frac{r^\beta}{(2-\beta)G^+(r)}.$$

Denoting the limit inferior on the left hand side of (4.3) by L, and observing that $\lim_{r\to 0} r^{-2}G^+(r) = \infty$ by (4.3), we obtain

$$\limsup_{r\to 0}\left(r^2\nu^-(\{x \in \mathbb{R}^d : r < |x| \le 1\})/G^+(r)\right) \le \frac{2\varepsilon d}{(2-\beta)L},$$

and since $\varepsilon > 0$ was arbitrary we see $\lim_{r\to 0} r^2\nu^-(\{x \in \mathbb{R}^d : r < |x| \le 1\})/G^+(r) = 0$. That $\lim_{r\to 0} r^2\nu^-(\{x \in \mathbb{R}^d : |x| > 1\})/G^+(r) = 0$ is clear so that we obtain $r^2\nu^-(\{x \in \mathbb{R}^d : |x| > r\}) = o(G^+(r))$ as $r \to 0$, finishing the proof. □

Example 4.4

(a) Let μ be a non-trivial strictly α-stable rotation invariant distribution on $\mathbb{R}^d$, where $\alpha \in (0,2)$. It is well-known that μ has a C^∞-density with all derivatives vanishing at infinity. Let us check that this can also be derived from Theorem 4.1. The Lévy measure ν of μ is given by $\nu(\mathrm{d}x) = C|x|^{-(d+\alpha)}\mathrm{d}x$ for some constant $C > 0$, see [32, Ex. 62.1]. For $r > 0$ let $G^+(r) := \inf_{\xi\in S^{d-1}} \xi^T \int_{|x|\le r} xx^T \,\nu(\mathrm{d}x)\xi$. Since μ is infinitely divisible, condition (4.1) of

Theorem 4.1 reduces to (4.2). In order to show that (4.2) is satisfied, let $r > 0$. For $k, j \in \{1, \dots, d\}$, $k \neq j$ we obtain

$$\int_{|x|\le r} x_k x_j \nu(\mathrm{d}x) = C \int_{|x|\le r} \frac{x_k x_j}{|x|^{d+\alpha}} \mathrm{d}x = 0.$$

To see that, by symmetry it suffices to consider the case $k = d$. Then we have

$$\int_{|x|\le r} \frac{x_d x_j}{|x|^{d+\alpha}} \mathrm{d}x = \int_{-r}^{r} x_d \int_{\substack{x'\in\mathbb{R}^{d-1}:\\ |x'|\le\sqrt{r^2-x_d^2}}} \frac{x_j}{\left(x_d^2 + (x')^2\right)^{(d+\alpha)/2}} \mathrm{d}x' \mathrm{d}x_d,$$

and the integrand of the outer integral is an odd function. Hence, the matrix $A_r := \int_{|x|\le r} xx^T \nu(\mathrm{d}x)$ is a diagonal matrix. We compute the trace of A_r as

$$\begin{aligned}\sum_{k=1}^{d} \int_{|x|\le r} x_k^2 \,\nu(\mathrm{d}x) &= C \int_{|x|\le r} |x|^{2-\alpha-d} \mathrm{d}x = C \int_0^r \int_{sS^{d-1}} s^{2-\alpha-d} \mathrm{d}\theta \mathrm{d}s \\ &= C \int_0^r s^{2-\alpha-d} s^{d-1} \omega_d \mathrm{d}s = C\omega_d r^{2-\alpha}/(2-\alpha),\end{aligned}$$

where ω_d denotes the $(d-1)$-dimensional volume of the surface S^{d-1}. Again by symmetry, it follows that $A_r = \frac{C\omega_d}{d(2-\alpha)} r^{2-\alpha} I_d$ with the identity matrix $I_d \in \mathbb{R}^{d\times d}$. Therefore, $G^+(r) = \frac{C\omega_d}{d(2-\alpha)} r^{2-\alpha}$ which implies that $\lim_{r\to 0} r^{-2} |\log r|^{-1} G^+(r) = \infty$, showing that μ satisfies (4.2).

(b) Now let μ be a non-trivial rotation invariant strictly α-stable distribution on $\mathbb{R}^d$ as in (a), and let σ be a probability distribution on $\mathbb{R}^d$. Since μ has a C^∞ density with all derivatives tending to 0 as $|x| \to \infty$, the same is true for the convolution $\mu' := \mu * \sigma$. When σ is additionally quasi-infinitely divisible and concentrated in $\mathbb{Z}^d$, then this can be also seen from Theorem 4.1. To see this, observe that σ has finite quasi-Lévy measure ν_σ concentrated in $\mathbb{Z}^d$ by Theorem 3.3. It follows that μ' has quasi-Lévy measure $\nu_\sigma(\mathrm{d}x) + C|x|^{-(d+\alpha)} \mathrm{d}x$. Hence the quantities $G^\pm(r)$ for μ and μ' coincide when $r < 1$ (with $G^-(r)$ being zero when $r < 1$). It follows that also μ' satisfies the assumptions of Theorem 4.1, so that μ' has a C^∞-density with derivatives tending to 0 as $|x| \to \infty$. This is of course a constructed example, but it shows that there are cases of quasi-infinitely divisible distributions that are not infinitely divisible for which the assumptions of Theorem 4.1 are applicable.

Remark 4.5 While the problem of a complete description of absolute continuity in terms of the Lévy measure remains challenging for infinitely divisible distributions, the corresponding question for continuity is completely solved: It is well-known that an infinitely divisible distribution μ on $\mathbb{R}^d$ with characteristic triplet (A, ν, γ)

is continuous if and only if $A \neq 0$ or $\nu(\mathbb{R}) = \infty$, see [32, Theorem 27.4]. The same characterisation fails however when considering quasi-infinitely divisible distributions. Berger [4, Ex. 4.7] showed that the distribution $\mu = 0.001\delta_0 + 0.999N(1, 1)$, where $N(1, 1)$ is the one-dimensional normal distribution with mean and variance 1, is quasi-infinitely divisible with infinite quasi-Lévy measure. Observe that μ is not continuous. Using Proposition 3.10 it is then also easy to construct non-continuous multivariate quasi-infinitely divisible distributions with infinite quasi-Lévy measure.

5 Topological Properties of the Class of Infinitely Divisible Distributions

Let $\mathrm{QID}(\mathbb{R}^d)$ denote the set of all quasi-infinitely divisible distributions on $\mathbb{R}^d$ and $\mathcal{P}(\mathbb{R}^d)$ the set of all distributions on $\mathbb{R}^d$. Equipped with the Prokhorov-metric π, $\mathcal{P}(\mathbb{R}^d)$ gets a metric space and the convergence in this space corresponds to the weak convergence of distributions. In this section we will always identify $\mathcal{P}(\mathbb{R}^d)$ with this metric space. The aim of this section is to state some topological properties of $\mathrm{QID}(\mathbb{R}^d)$ that were already given by [4] and [21] in one dimension. We start with some results that can be shown similarly in any dimension $d \in \mathbb{N}$. The following was shown in [4, Prop. 5.1] and [21, Sect. 4] in dimension 1.

Theorem 5.1 *Let $d \in \mathbb{N}$. The set $\mathrm{QID}(\mathbb{R}^d)$ is neither open nor closed in $\mathcal{P}(\mathbb{R}^d)$. Moreover, the set $\mathcal{P}(\mathbb{R}^d) \setminus \mathrm{QID}(\mathbb{R}^d)$ is dense in $\mathcal{P}(\mathbb{R}^d)$.*

Proof To see that $\mathcal{P}(\mathbb{R}^d) \setminus \mathrm{QID}(\mathbb{R}^d)$ is dense in $\mathcal{P}(\mathbb{R}^d)$, let μ be an arbitrary distribution on $\mathbb{R}^d$ and let σ be a distribution on $\mathbb{R}^d$ such that its characteristic function $\widehat{\sigma}$ has zeros. For $n \in \mathbb{N}$ define the distribution μ_n by $\mu_n(\mathrm{d}x) = \mu(\mathrm{d}x) * \sigma(n\mathrm{d}x)$. The characteristic function of μ_n is given by $\widehat{\mu_n}(z) = \widehat{\mu}(z)\widehat{\sigma}(z/n)$ for $z \in \mathbb{R}^d$ and has zeros, hence μ_n cannot be quasi-infinitely divisible. Furthermore, $\widehat{\mu_n}(z) \to \widehat{\mu}(z)$ as $n \to \infty$ for every $z \in \mathbb{R}^d$, implying that $\mu_n \overset{w}{\to} \mu$ as $n \to \infty$. Hence, $\mathcal{P}(\mathbb{R}^d) \setminus \mathrm{QID}(\mathbb{R}^d)$ is dense in $\mathcal{P}(\mathbb{R}^d)$, so $\mathcal{P}(\mathbb{R}^d) \setminus \mathrm{QID}(\mathbb{R}^d)$ cannot be closed, therefore $\mathrm{QID}(\mathbb{R}^d)$ cannot be an open set. In order to show that $\mathrm{QID}(\mathbb{R}^d)$ is not closed, first observe that for $n \in \mathbb{N}$ the distribution $\frac{n+1}{2n}\delta_0 + \frac{n-1}{2n}\delta_{e_1}$ is quasi-infinitely divisible due to Theorem 3.2, where $e_1 = (1, 0, \ldots, 0)^T \in \mathbb{R}^d$ is the first unit vector. We have

$$\frac{n+1}{2n}\delta_0 + \frac{n-1}{2n}\delta_{e_1} \overset{w}{\to} \frac{1}{2}\delta_0 + \frac{1}{2}\delta_{e_1} \quad \text{as } n \to \infty$$

and $\frac{1}{2}\delta_0 + \frac{1}{2}\delta_{e_1}$ is not quasi-infinitely divisible since its characteristic function has zeros. Hence, the set $\mathrm{QID}(\mathbb{R}^d)$ cannot be closed. □

In any topological space and hence in any metric space one can define the notions of connected and path-connected subsets. Observe that path-connectedness implies connectedness (see [3, Thm. 3.29]). The following result shows that $\mathrm{QID}(\mathbb{R}^d)$

is path-connected (with respect to the Prokhorov metric) and hence connected, generalising [4, Prop. 5.2] for $\mathrm{QID}(\mathbb{R})$ to arbitrary dimensions.

Theorem 5.2 *Let $d \in \mathbb{N}$. The set $\mathrm{QID}(\mathbb{R}^d)$ is path-connected and hence connected in $\mathcal{P}(\mathbb{R}^d)$.*

Proof Suppose that μ_0 and μ_1 are quasi-infinitely divisible distributions on $\mathbb{R}^d$. For $t \in (0,1)$ the distributions σ_t^0 and σ_t^1 defined by $\sigma_t^0(\mathrm{d}x) := \mu_0(1/(1-t)\mathrm{d}x)$ and $\sigma_t^1(\mathrm{d}x) := \mu_1(1/t\mathrm{d}x)$ are also quasi-infinitely divisible by Lemma 2.8. Therefore, also the distribution μ_t defined by

$$\mu_t(\mathrm{d}x) := \mu_0\left(\frac{1}{1-t}\mathrm{d}x\right) * \mu_1\left(\frac{1}{t}\mathrm{d}x\right)$$

is quasi-infinitely divisible for every $t \in (0,1)$. Note that $\widehat{\mu_t}(z) = \widehat{\mu_0}((1-t)z)\widehat{\mu_1}(tz)$ for $z \in \mathbb{R}^d$. The mapping

$$p : [0,1] \to \mathcal{P}(\mathbb{R}^d), \quad t \mapsto \mu_t$$

is continuous, because $\widehat{\mu_s}(z) \to \widehat{\mu_t}(z)$ as $s \to t$ for all $z \in \mathbb{R}^d$ and hence $\mu_s \xrightarrow{w} \mu_t$ as $s \to t$. Since $p(0) = \mu_0$ and $p(1) = \mu_1$, it follows that $\mathrm{QID}(\mathbb{R}^d)$ is path-connected and hence connected. □

It is not surprising that $\mathcal{P}(\mathbb{R}^d) \setminus \mathrm{QID}(\mathbb{R}^d)$ is dense in $\mathcal{P}(\mathbb{R}^d)$. Much more surprising is the fact that also $\mathrm{QID}(\mathbb{R})$ is dense in dimension 1. This was proved in [21, Thm. 4.1]. We state the precise result and also give a sketch of the proof along the lines of [21], in order to discuss afterwards where the obstacles arise when trying to generalise the result to higher dimensions.

Theorem 5.3 *The set of quasi-infinitely divisible distributions on $\mathbb{R}$ with Gaussian variance zero and finite quasi-Lévy measure is dense in $\mathcal{P}(\mathbb{R})$.*

Sketch of proof Denote by $\mathcal{Q}(\mathbb{R})$ the set of all probability measures on $\mathbb{R}$ whose support is a finite set contained in a lattice of the form $n^{-1}\mathbb{Z}$ for some $n \in \mathbb{N}$, and by $\mathrm{QID}^0(\mathbb{R})$ the set of all quasi-infinitely divisible distributions on $\mathbb{R}$ with Gaussian variance 0 and finite quasi-Lévy measure. It is easily seen that $\mathcal{Q}(\mathbb{R})$ is dense in $\mathcal{P}(\mathbb{R})$. Hence, it suffices to show that $\mathcal{Q}(\mathbb{R}) \cap \mathrm{QID}^0(\mathbb{R})$ is dense in $\mathcal{Q}(\mathbb{R})$. To show this, let μ be a distribution in $\mathcal{Q}(\mathbb{R})$, say $\mu = \sum_{k=-m}^m p_k \delta_{k/n}$ for some $m, n \in \mathbb{N}$ with $0 \le p_k \le 1$ for all $k \in \{-m, \dots, m\}$ and $\sum_{k=-m}^m p_k = 1$, and let X be a random variable with distribution μ. From Lemma 2.8 it is clear that $\mu = \mathcal{L}(X)$ is in the closure of $\mathcal{Q}(\mathbb{R}) \cap \mathrm{QID}^0(\mathbb{R})$ if and only if $\mathcal{L}(nX + m)$ is in the closure of $\mathcal{Q}(\mathbb{R}) \cap \mathrm{QID}^0(\mathbb{R})$. Hence it is sufficient to consider distributions μ whose support is a finite set contained in $\mathbb{N}_0$. If the support of μ is contained in $\{0, \dots, m\}$ for some $m \in \mathbb{N}_0$, then it is easily seen that μ can be approximated arbitrarily well with distributions having support exactly $\{0, \dots, m\}$, i.e. distributions σ of the form $\sigma = \sum_{k=0}^m p_k \delta_k$ with $p_0, \dots, p_m > 0$ (strictly positive). So it is sufficient to show that any such distribution σ can be approximated arbitrarily well by distributions in

$\mathcal{Q}(\mathbb{R}) \cap \text{QID}^0(\mathbb{R})$. If the characteristic function $\widehat{\sigma}$ of σ has no zeroes, then we are done by Theorem 3.3. So suppose that $\mathbb{R} \ni z \mapsto \widehat{\sigma}(z) = \sum_{k=0}^m p_k e^{izk}$ has zeroes, which corresponds to zeroes of the polynomial f given by $f(\omega) = \sum_{k=0}^m p_k \omega^k$ for $\omega \in \mathbb{C}$ on the unit circle. Factorising f, we can write $f(\omega) = p_m \prod_{k=1}^m (\omega - \xi_k)$ for $\omega \in \mathbb{C}$ with $\xi_k \in \mathbb{C}$ for $k \in \{1, \ldots, m\}$. For $h > 0$ let

$$f_h(\omega) := p_m \prod_{k=1}^m (\omega - \xi_k - h) \quad \text{for all } \omega \in \mathbb{C}. \tag{5.1}$$

If h is chosen small enough, then f_h has no zeros on the unit circle. The polynomial f has real coefficients, so the non-real roots of f appear in pairs of complex conjugates. By construction, the same is true for f_h, so there exist $\alpha_{h,k} \in \mathbb{R}$ for $k \in \{0, \ldots, m\}$ such that $f_h(\omega) = \sum_{k=0}^m \alpha_{h,k} \omega^k$ for all $\omega \in \mathbb{C}$. Moreover, $\alpha_{h,k} \to p_k > 0$ as $h \to 0$, so we can assume that h is small enough such that $\alpha_{h,k} > 0$ for all $k \in \{0, \ldots, m\}$. Let $\sigma_h := \lambda^{-1} \sum_{k=1}^m \alpha_{h,k} \delta_k$ with $\lambda := \sum_{k=0}^m \alpha_{h,k}$. Then for small enough $h > 0$, σ_h is a probability distribution having support $\{0, \ldots, m\}$ and the characteristic function of σ_h has no zeroes. By Theorem 3.3, σ_h is quasi-infinitely divisible with finite quasi-Lévy measure and Gaussian variance 0, and $\sigma_h \xrightarrow{w} \sigma$ as $h \downarrow 0$. □

It is very tempting now to assume that Theorem 5.3 also holds for $\mathbb{R}^d$-valued distributions with general $d \in \mathbb{N}$. Again it is easily seen that it would suffice to show that any distribution $\sigma = \sum_{k_1,\ldots,k_d=0}^m p_{(k_1,\ldots,k_d)} \delta_{(k_1,\ldots,k_d)}$ with $p_{k_1,\ldots,k_d} > 0$ can be approximated arbitrarily well by quasi-infinitely distributions in $\mathbb{Z}^d$. Since a distribution in $\mathbb{Z}^d$ is quasi-infinitely divisible if and only if its characteristic function has no zeroes, this might appear to be an easy task at first glance. However, it is not clear how to do a modification as in Equation (5.1), the problem being that polynomials in more than one variable do not factorise and also that there may be infinitely many zeroes of such polynomials. We have tried some time to pursue such a path, but have not succeeded. Having failed in proving Theorem 5.3 for dimensions $d \geq 2$, it is natural to wonder whether such a generalisation may be true at all. Let us pose this as an open question:

Open Question 5.4 [1] Is $\text{QID}(\mathbb{R}^d)$ dense in $\mathcal{P}(\mathbb{R}^d)$ for any dimension $d \in \mathbb{N}$, or is it dense only for $d = 1$, or is it dense for certain dimensions and not for others?

Passeggeri [27, Conjecture 4.1] conjectures that $\text{QID}(\mathbb{R}^d)$ is dense in $\mathcal{P}(\mathbb{R}^d)$ for all dimensions $d \in \mathbb{N}$. It is possible that this is true, but we are more inclined to believe that $\text{QID}(\mathbb{R}^d)$ will be dense in $\mathcal{P}(\mathbb{R}^d)$ if and only if $d = 1$. An indication

[1] An answer to this question has been obtained recently by Kutlu [19]. Kutlu shows that $\text{QID}(\mathbb{R}^d)$ is not dense in $\mathcal{P}(\mathbb{R}^d)$ if $d \geq 2$, by giving an explicit example of a distribution on $\mathbb{R}^d$ which can not be approximated by quasi-infinitely divisible distributions. In particular, it is shown that its characteristic function can not be approximated arbitrarily well by zero-free continuous functions with respect to uniform convergence on every compact set.

that this might be the case is the fact that the set of all zero-free complex valued continuous functions on $[0, 1]^d$ is dense in the set of all continuous complex valued functions on $[0, 1]^d$ with respect to uniform convergence if and only if $d = 1$, see Pears [29, Prop. 3.3.2]. Knowing this, we wonder if even the set of all probability distributions on $\mathbb{R}^d$ having zero-free characteristic function is dense in $\mathcal{P}(\mathbb{R}^d)$ if and only if $d = 1$, but again we do not know the answer. We intend to invest this question more deeply in the future.

6 Conditions for Weak Convergence

Weak convergence of infinitely divisible distributions can be characterised in terms of the characteristic triplet, see Sato [32, Thms. 8.7, 56.1]. A full characterization of the weak convergence of quasi-infinitely divisible distributions seems difficult, since the class of quasi-infinitely divisible distributions is not closed with respect to weak convergence. In [21, Thm. 4.3], some sufficient conditions for weak convergence of one-dimensional quasi-infinitely divisible distributions in terms of the characteristic pair where given; the *characteristic pair* of a one-dimensional quasi-infinitely divisible distribution with characteristic triplet $(A, \nu, \gamma)_c$ with respect to a representation function $c : \mathbb{R} \to \mathbb{R}$ is given by $(\zeta, \gamma)_c$, where ζ is a finite signed measure on $\mathbb{R}$ given by $\zeta(\mathrm{d}x) = A\,\delta_0(\mathrm{d}x) + (1 \wedge x^2)\ \nu(\mathrm{d}x)$. It is not so easy to generalise the characteristic pair to the multivariate setting and hence we will rather work with another characterisation of weak convergence, in line with the conditions given in [32, Thm. 56.1] for weak convergence of infinitely divisible distributions.

Denote by $C_\#$ the set of all bounded, continuous functions $f : \mathbb{R}^d \to \mathbb{R}$ vanishing on a neighborhood of 0. Then the following provides a sufficient condition for weak convergence of quasi-infinitely divisible distributions in terms of the characteristic triplets.

Theorem 6.1 *Let $c : \mathbb{R}^d \to \mathbb{R}^d$ be a continuous representation function. Let $(\mu_n)_{n\in\mathbb{N}}$ be a sequence of quasi-infinitely divisible distributions on $\mathbb{R}^d$ such that μ_n has characteristic triplet $(A_n, \nu_n, \gamma_n)_c$ for every $n \in \mathbb{N}$ and let μ be a quasi-infinitely divisible distribution on $\mathbb{R}^d$ with characteristic triplet $(A, \nu, \gamma)_c$. Suppose that the following conditions are satisfied.*

(i) *For all $f \in C_\#$ it holds*

$$\lim_{n\to\infty} \int_{\mathbb{R}^d} f(x)\nu_n^{\pm}(dx) = \int_{\mathbb{R}^d} f(x)\nu^{\pm}(dx).$$

(ii) *If $A_{n,\varepsilon}$ is defined by*

$$\langle z, A_{n,\varepsilon} z\rangle = \langle z, Az\rangle + \int_{|x|\le\varepsilon} \langle z, x\rangle^2 \nu_n^{+}(dx)$$

for all $n \in \mathbb{N}$ and $\varepsilon > 0$, then

$$\lim_{\varepsilon \downarrow 0} \limsup_{n\to\infty} |\langle z, A_{n,\varepsilon} z\rangle - \langle z, Az\rangle| = 0 \quad \textit{for all } z \in \mathbb{R}^d.$$

(iii) *It holds*

$$\lim_{\varepsilon \downarrow 0} \limsup_{n\to\infty} \int_{|x|\le\varepsilon} |x|^2 \nu_n^-(dx) = 0.$$

(iv) $\gamma_n \to \gamma$ *as* $n \to \infty$.

Then $\mu_n \xrightarrow{w} \mu$ *as* $n \to \infty$.

Proof Let μ^1 and μ^2 be infinitely divisible distributions with characteristic triplets $(A, \nu^+, \gamma)_c$ and $(0, \nu^-, 0)_c$ and for $n \in \mathbb{N}$ let μ_n^1 and μ_n^2 be infinitely divisible distributions with characteristic triplets $(A_n, \nu_n^+, \gamma_n)_c$ and $(0, \nu_n^-, 0)_c$, respectively. Then $\mu^2 * \mu = \mu^1$ and $\mu_n^2 * \mu_n = \mu_n^1$ for every $n \in \mathbb{N}$. Further, $\mu_n^1 \xrightarrow{w} \mu^1$ and $\mu_n^2 \xrightarrow{w} \mu^2$ by [32, Thm. 56.1] and hence $\mu_n \xrightarrow{w} \mu$ as $n \to \infty$ which follows by considering characteristic functions. □

The sufficient conditions of Theorem 6.1 are not necessary. An explicit example in dimension 1 for this fact was given in [21, Example 4.4]. There, a sequence $(\mu_n)_{n\in\mathbb{N}}$ of quasi-infinitely divisible distributions on $\mathbb{R}$ with quasi-Lévy measures ν_n was constructed with $\lim_{n\to\infty} \nu_n^-(\mathbb{R} \setminus [-1, 1]) = \infty$ but such that the limit μ was even infinitely divisible, so that its quasi-Lévy measure ν satisfies $\nu^- = 0$. In particular, condition (i) of Theorem 6.1 is violated in this case. Under the extra condition that the sequence $(\zeta_n^-)_{n\in\mathbb{N}}$ of finite (positive) measures defined by $\zeta_n^-(\mathrm{d}x) = (1 \wedge |x|^2)\, \nu_n^-(\mathrm{d}x)$ is tight and uniformly bounded, Theorem 4.3 (a,b) in [21] provides necessary and sufficient conditions for weak convergence of quasi-infinitely divisible distributions in the univariate case. In view of the above example, the sequence (ζ_n^-) will not always be tight and uniformly bounded even if the limit is infinitely divisible, and a complete characterisation without any extra conditions seems difficult. We hence refrained from any further analysis of sufficient or necessary conditions in the multivariate case, but confine ourselves to giving a sufficient condition for a weak limit of quasi-infinitely divisible distributions to be again quasi-infinitely divisible:

Theorem 6.2 *Let $c : \mathbb{R}^d \to \mathbb{R}^d$ be a continuous representation function and for $n \in \mathbb{N}$ let μ_n be a quasi-infinitely divisible distribution on $\mathbb{R}^d$ with characteristic triplet $(A_n, \nu_n, \gamma_n)_c$. Suppose that the sequence $(\mu_n)_{n\in\mathbb{N}}$ converges weakly to some distribution μ and that there exists a Lévy measure σ on $\mathbb{R}^d$ such that*

$$\lim_{n\to\infty} \int_{\mathbb{R}^d} f(x)\nu_n^-(dx) = \int_{\mathbb{R}^d} f(x)\sigma(dx)$$

for all $f \in C_\#$ *and*

$$\lim_{\varepsilon \downarrow 0} \limsup_{n \to \infty} \int_{|x| \le \varepsilon} |x|^2 \nu_n^-(dx) = 0.$$

Then μ *is quasi-infinitely divisible. If we denote its characteristic triplet by* $(A, \nu, \gamma)_c$*, then* $\nu + \sigma$ *is a Lévy measure,*

$$\lim_{n \to \infty} \int_{\mathbb{R}^d} f(x) \nu_n^+(dx) = \int_{\mathbb{R}^d} f(x)(\nu + \sigma)(dx),$$

$\gamma_n \to \gamma$ *as* $n \to \infty$ *and for* $A_{n,\varepsilon} \in \mathbb{R}^{d \times d}$ *defined by*

$$\langle z, A_{n,\varepsilon} z \rangle = \langle z, Az \rangle + \int_{|x| \le \varepsilon} \langle z, x \rangle^2 \nu_n^+(dx)$$

for all $n \in \mathbb{N}$ *and* $\varepsilon > 0$ *we have*

$$\lim_{\varepsilon \downarrow 0} \limsup_{n \to \infty} |\langle z, A_{n,\varepsilon} z \rangle - \langle z, Az \rangle| = 0 \quad \textit{for all } z \in \mathbb{R}^d.$$

Proof Let μ_n^1 and μ_n^2 be infinitely divisible distributions with characteristic triplets $(A_n, \nu_n^+, \gamma_n)_c$ and $(0, \nu_n^-, 0)_c$, respectively, and μ^2 be infinitely divisible with characteristic triplet $(0, \sigma, 0)_c$. Then [32, Thm. 56.1] implies $\mu_n^2 \xrightarrow{w} \mu^2$, and hence also $\mu_n^1 = \mu_n^2 * \mu_n \xrightarrow{w} \mu^2 * \mu$ as $n \to \infty$. Since μ_n^1 is infinitely divisible for each $n \in \mathbb{N}$, so is $\mu^2 * \mu$. If we denote its characteristic triplet by (A, η, γ), then μ is quasi infinitely divisible with characteristic triplet $(A, \nu, \gamma)_c$, where $\nu = \eta_{|\mathcal{B}_0} - \sigma_{|\mathcal{B}_0}$. The other implications now follow from [32, Thm. 56.1]. □

7 Support Properties

For infinitely divisible distributions on $\mathbb{R}$, the boundedness of the support from below can be characterised in terms of the characteristic triplet. More precisely, an infinitely divisible distribution μ with characteristic triplet (a, ν, γ) has support bounded from below if and only if $a = 0$ and ν is supported in $[0, \infty)$, c.f. Sato [32, Thm. 24.7]. If μ is only quasi-infinitely divisible (and not necessarily infinitely divisible), then it was shown in [21, Prop. 5.1] that the following two statements (i) and (ii) are equivalent:

(i) μ is bounded from below, $\operatorname{supp}(\nu^-) \subset [0, \infty)$ and $\int_{|x| \le 1} |x| \, \nu^-(\mathrm{d}x) < \infty$.
(ii) $a = 0$, $\operatorname{supp}(\nu^+) \subset [0, \infty)$ and $\int_{|x| \le 1} |x| \nu^+(\mathrm{d}x) < \infty$.

Observe that for infinitely divisible distributions we have $\nu^- = 0$ and $\nu^+ = \nu$, so that the above result reduces to the known characterisation for infinitely divisible distributions. Also observe that the condition "μ is bounded from below" can be rewritten as "there is $b \in \mathbb{R}$ such that $b + \text{supp}(\mu) \subset [0, \infty)$". Our goal now is to extend this result to higher dimensions, and we will be working immediately with cones rather than only $[0, \infty)^d$. Following [30, Def. 4.8], by a *cone* in $\mathbb{R}^d$ we mean a non-empty closed convex subset K of $\mathbb{R}^d$ which is not $\{0\}$, does not contain a straight line through the origin and is such that with $x \in K$ and $\lambda \geq 0$ also $\lambda x \in K$. Observe that this definition is more restrictive than the usual notion of cones in linear algebra, in the sense that we require additionally a cone to be closed, convex, non-trivial and one-sided (the latter being sometimes also called "proper"), but in probability this seems to be more standard and they are the only cones of interest to us. Obviously, $[0, \infty)^d$ is a cone, but there are many other examples.

Coming back to the question when a quasi-infinitely divisible distribution has support contained in a translate of a cone, let us first recall the corresponding results for infinitely divisible distributions; the equivalence of (i) and (iii) below can be found in Skorohod [33, Thm. 21 in §3.3] or Rocha-Arteaga and Sato [30, Thm. 4.11], while the equivalence of (ii) and (iii) is stated in Equation (4.35) of [30]; since the latter equivalence is only given in a remark in [30] we provide a short sketch of the proof for this fact.

Theorem 7.1 *Let $L = (L_t)_{t \geq 0}$ be a Lévy process in $\mathbb{R}^d$ with characteristic triplet (A, ν, γ) (i.e. $\mathcal{L}(L_1)$ has this characteristic triplet) and let $K \subset \mathbb{R}^d$ be a cone. Then the following are equivalent:*

(i) $\text{supp}(\mathcal{L}(L_t)) \subset K$ *for every* $t \geq 0$.
(ii) $\text{supp}(\mathcal{L}(L_t)) \subset K$ *for some* $t > 0$.
(iii) $A = 0$, $\text{supp}(\nu) \subset K$, $\int_{|x| \leq 1} |x| \, \nu(\mathrm{d}x) < \infty$ *and* $\gamma^0 \in K$, *where* γ^0 *is the drift of* $\mathcal{L}(L_1)$.

Sketch of proof of the equivalence of (i) and (ii). That (i) implies (ii) is clear. For the converse, assume that $\text{supp}(\mathcal{L}(L_t)) \subset K$ for some $t > 0$. Since $\mathcal{L}(L_t) = (\mathcal{L}(L_{t/2}))^{*2}$ we also have $\text{supp}(\mathcal{L}(L_{t/2})) \subset K$; to see that, suppose there were $y \in \text{supp}(\mathcal{L}(L_{t/2})) \setminus K$. Then also $2y \notin K$ and by the closedness of K there is an open ball U containing y with $U \cap K = \emptyset$, $(U + U) \cap K = \emptyset$ and $P(L_{t/2} \in U) > 0$. Then also $P(L_t \in U + U) > 0$, a contradiction. Hence we have $\text{supp}(L_{t/2}) \subset K$ and iterating this argument we obtain $\text{supp}(\mathcal{L}(L_{2^{-n}t})) \subset K$ for all $n \in \mathbb{N}_0$. Since $K + K \subset K$ and K is closed we conclude $\text{supp}(\mathcal{L}(L_{q2^{-n}t})) \subset K$ for all $q, n \in \mathbb{N}_0$. Since $\{q2^{-n}t : q, n \in \mathbb{N}_0\}$ is dense in $[0, \infty)$ we get (i) since L has right-continuous sample paths. □

We also need the following easy lemma for infinitely divisible distributions with existing drift. It is well known, but since we were unable to find a ready reference, we include a short proof.

Lemma 7.2 *Let μ be an infinitely divisible distribution on $\mathbb{R}^d$ with characteristic triplet (A, ν, γ) such that $\int_{|x| \leq 1} |x| \, \nu(\mathrm{d}x) < \infty$. Then the drift of μ is an element of* $\text{supp}(\mu)$.

Proof Let $L = (L_t)_{t\geq 0}$ be a Lévy process such that $\mu = \mathcal{L}(L_1)$. By the Lévy–Itô decomposition we can write $L_t = B_t + \gamma^0 t + \sum_{0<s\leq t} \Delta L_s$ for each $t \geq 0$, where $B = (B_t)_{t\geq 0}$ is a Brownian motion with drift 0 and covariance matrix A, γ^0 denotes the drift of L (i.e. of $\mathcal{L}(L_1)$) and ΔL_s denotes the jump size of L at s. Since the three components are independent and 0 is obviously in the support of both $\mathcal{L}(B_t)$ and of $\mathcal{L}(\sum_{0<s\leq t} \Delta L_s)$ we conclude that $\gamma^0 t \in \operatorname{supp}(\mathcal{L}(L_t))$ for each $t \geq 0$. □

With the aid of Theorem 7.1 and Lemma 7.2 we can now obtain the desired generalisation for multivariate quasi-infinitely divisible distributions that are supported in cones.

Theorem 7.3 *Let μ be a quasi-infinitely divisible distribution on $\mathbb{R}^d$ with characteristic triplet (A, ν, γ) and let K be a cone. Then the following statements are equivalent:*

(i) $\operatorname{supp}(\mu) \subset b+K$ *for some* $b \in \mathbb{R}^d$, $\operatorname{supp}(\nu^-) \subset K$ *and* $\int_{|x|\leq 1} |x|\, \nu^-(dx) < \infty$.
(ii) $A = 0$, $\operatorname{supp}(\nu^+) \subset K$ *and* $\int_{|x|\leq 1} |x|\nu^+(dx) < \infty$.

If the equivalent conditions (i) and (ii) are satisfied, then $\operatorname{supp}(\mu) \subset K$ *if and only if the drift of μ lies in K.*

Proof Let μ_1 and μ_2 be infinitely divisible distributions with characteristic triplets and (A, ν^+, γ) and $(0, \nu^-, 0)$, respectively, so that $\mu_1 = \mu * \mu_2$. To show that (i) implies (ii), suppose that $\operatorname{supp}(\mu) \subset b + K$ for some $b \in \mathbb{R}^d$, $\operatorname{supp}(\nu^-) \subset K$ and $\int_{|x|\leq 1} |x|\nu^-(\mathrm{d}x) < \infty$. Then $\mu * \delta_{-b}$ is supported in K and denoting the drift of μ_2 by $b_2 := -\int_{|x|\leq 1} x\nu^-(\mathrm{d}x)$, the distribution $\mu_2 * \delta_{-b_2}$ is infinitely divisible with characteristic triplet $(0, \nu^-, -b_2)$ and has drift 0. By Theorem 7.1, $\mu_2 * \delta_{-b_2}$ is supported in K as well, hence also $\mu_2 * \mu * \delta_{-b-b_2} = \mu_1 * \delta_{-b-b_2}$ is supported in K. Using Theorem 7.1 again, since $\mu_1 * \delta_{-b-b_2}$ is infinitely divisible with characteristic triplet $(A, \nu^+, \gamma - b - b_2)$, it follows that $A = 0$, $\operatorname{supp}(\nu^+) \subset K$ and $\int_{|x|\leq 1} |x|\, \nu^+(\mathrm{d}x) < \infty$. Moreover, it follows that the drift of $\mu_1 * \delta_{-b-b_2}$ is an element of K, that is,

$$\gamma - b - b_2 - \int_{|x|\leq 1} |x|\nu^+(\mathrm{d}x) \in K. \tag{7.1}$$

For the other direction, suppose that (ii) is satisfied and let $b_1 := \gamma - \int_{|x|\leq 1} x\, \nu^+(\mathrm{d}x)$ denote the drift of μ_1. By Theorem 7.1, the infinitely divisible distribution $\mu_1 * \delta_{-b_1}$ is supported in K. Using [32, Lem. 24.1] we obtain $\operatorname{supp}(\mu_2) + \operatorname{supp}(\mu) \subset b_1 + \operatorname{supp}(\mu_1 * \delta_{-b_1}) \subset b_1 + K$. Choosing arbitrary elements $u \in \operatorname{supp}(\mu_2)$ and $v \in \operatorname{supp}(\mu)$, we conclude

$$-b_1 + u + \operatorname{supp}(\mu) \subset -b_1 + \operatorname{supp}(\mu_2) + \operatorname{supp}(\mu_1) \subset K, \tag{7.2}$$

and similarly $\operatorname{supp}(\mu_2 * \delta_{-b_1+v}) = -b_1 + v + \operatorname{supp}(\mu_2) \subset K$. It follows that $\operatorname{supp}(\mu) \subset b_1 - u + K$ and by Theorem 7.1, since $\mu_2 * \delta_{-b_1+v}$ is infinitely divisible with characteristic triplet $(0, \nu^-, -b_1 + v)$, we have that $\mathrm{supp}(\nu^-) \subset K$ and $\int_{|x|\le 1} |x|\nu^-(\mathrm{d}x) < \infty$.

Now suppose that both (i) and (ii) are satisfied. With the notations above, note that the drift of μ is $b_1 - b_2$. If μ is supported in K, then in (i) we can choose $b = 0$ and (7.1) implies that $b_1 - b_2 \in K$. Conversely, if the drift $b_1 - b_2$ of μ is in K, then by Lemma 7.2 we can choose $u = b_2$ and observe from (7.2) that $\mathrm{supp}(\mu) \subset K + (b_1 - b_2) \subset K$, finishing the proof. □

8 Moments

In this section we study the finiteness of the h-moment of quasi-infinitely divisible distributions in the case of a submultiplicative function h. Recall that a function $h : \mathbb{R}^d \to \mathbb{R}$ is *submultiplicative* if it is non-negative and there exists a constant $C > 0$ such that

$$h(x + y) \le Ch(x)h(y) \quad \text{for all } x, y \in \mathbb{R}^d.$$

Given an infinitely divisible distribution μ and a locally bounded submultiplicative and measurable function h on $\mathbb{R}^d$, μ has finite h-moment if and only if the Lévy measure ν of μ restricted to $\{x \in \mathbb{R}^d : |x| > 1\}$ has finite h-moment, i.e. $\int_{\mathbb{R}^d} h(x)\,\mu(\mathrm{d}x) < \infty$ if and only if $\int_{|x|>1} h(x)\,\nu(\mathrm{d}x) < \infty$, see [32, Thm. 25.3]. This does not generalise to quasi-infinitely divisible distributions in the sense that $\int_{\mathbb{R}^d} h(x)\,\mu(\mathrm{d}x) < \infty$ is equivalent to $\int_{|x|>1} h(x)\,\nu^{\pm}(\mathrm{d}x) < \infty$ (see Example 8.2 below), but it generalises in the sense that $\int_{|x|>1} h(x)\,\nu^+(\mathrm{d}x) < \infty$ is finite if and only if both $\int_{\mathbb{R}^d} h(x)\,\mu(\mathrm{d}x)$ and $\int_{|x|>1} h(x)\,\nu^-(\mathrm{d}x)$ are finite. For univariate distributions this was shown in [21, Thm. 6.2]. The proof given there easily generalises to multivariate distributions. We have:

Theorem 8.1 *Let μ be a quasi-infinitely divisible distribution on $\mathbb{R}^d$ with standard characteristic triplet (A, ν, γ) and let $h : \mathbb{R}^d \to [0, \infty)$ be a submultiplicative, locally bounded and measurable function.*

(a) *Then $(\nu^+)_{|\{x\in\mathbb{R}^d:|x|>1\}}$ has finite h-moment if and only if both μ and $(\nu^-)_{|\{x\in\mathbb{R}^d:|x|>1\}}$ have finite h-moment, i.e. $\int_{|x|>1} h(x)\,\nu^+(\mathrm{d}x) < \infty$ if and only if $\int_{\mathbb{R}^d} h(x)\,\mu(\mathrm{d}x) + \int_{|x|>1} h(x)\,\nu^-(\mathrm{d}x) < \infty$.*

(b) *Let X be a random vector in $\mathbb{R}^d$ with distribution μ. Then the following are true:*

  (i) *If $\int_{|x|>1} |x|\nu^+(dx) < \infty$, then the expectation $\mathbb{E}(X)$ of X exists and is given by*

$$\mathbb{E}(X) = \gamma + \int_{|x|>1} x\nu(dx) = \gamma_m,$$

  which is the center of μ as defined in Remark 2.9.

(ii) *If* $\int_{|x|>1} |x|^2 \nu^+(dx) < \infty$, *then X has finite second moment and the covariance matrix* $\mathrm{Cov}(X) \in \mathbb{R}^{d\times d}$ *of X is given by*

$$\mathrm{Cov}(X) = A + \int_{\mathbb{R}^d} xx^T \nu(dx).$$

(iii) *If* $\int_{|x|>1} e^{\langle \alpha, x\rangle} \nu^+(dx) < \infty$ *for some* $\alpha \in \mathbb{R}^d$, *then* $\mathbb{E}(\mathrm{e}^{\langle \alpha, X\rangle}) < \infty$ *and*

$$\mathbb{E}(\mathrm{e}^{\langle \alpha, X\rangle}) = \exp\left(\langle \alpha, \gamma\rangle + \frac{1}{2}\langle \alpha, A\alpha\rangle + \int_{\mathbb{R}^d} \left(\mathrm{e}^{\langle \alpha, x\rangle} - 1 - \langle \alpha, x\rangle \mathbb{1}_{[0,1]}(|x|)\right) \nu(dx)\right).$$

Proof The proof given in [21, Thm. 6.2] carries over word by word to the multivariate setting. □

As shown in [21, Ex. 6.3], for univariate quasi-infinitely divisible distributions it is not true that finiteness of $\int_{\mathbb{R}} h(x)\,\mu(\mathrm{d}x)$ implies finiteness of $\int_{|x|>1} h(x)\,\nu^+(\mathrm{d}x)$. Let us give another but simpler example of this phenomenon and also remark on the multivariate setting:

Example 8.2

(a) Let $p \in (0, 1/2)$. By Theorem 3.2, the Bernoulli distribution $b(1, p)$ on $\mathbb{R}$ is quasi-infinitely divisible with quasi-Lévy measure $\nu = \sum_{k=1}^{\infty} \frac{(-1)^{k+1}}{k} \left(\frac{p}{1-p}\right)^k \delta_k$. Especially, we obtain $\nu^+ = \sum_{k=0}^{\infty} \frac{1}{2k+1} \left(\frac{p}{1-p}\right)^{2k+1} \delta_{2k+1}$. Let $c > -\log \frac{p}{1-p} > 0$. The function $g : \mathbb{R} \to \mathbb{R}$, $x \mapsto e^{cx}$ is submultiplicative and by the monotone convergence theorem it holds

$$\int_{|x|>1} \mathrm{e}^{cx}\,\nu^+(\mathrm{d}x) = \sum_{k=1}^{\infty} \frac{1}{2k+1} \left(\frac{p}{1-p}\right)^{2k+1} \mathrm{e}^{(2k+1)c} = \infty,$$

since $\frac{p}{1-p}\mathrm{e}^c > 1$. Similarly, $\int_{|x|>1} \mathrm{e}^{cx}\,\nu^-(\mathrm{d}x) = \infty$, although the Bernoulli-distribution $b(1, p)$ has finite g-moment.

(b) Let $p \in (0, 1/2)$ and $X_1, \dots, X_d$ be independent $b(1, p)$-distributed and define $X = (X_1, \dots, X_d)^T$. By Proposition 3.10, $\mathcal{L}(X)$ is a quasi-infinitely divisible distribution on $\mathbb{R}^d$ whose quasi-Lévy measure $\widetilde{\nu}$ is concentrated on the axes. Let $c > -\log \frac{p}{1-p}$ and consider the function $h : \mathbb{R}^d \to [0, \infty)$ given by $h(x) = \mathrm{e}^{\langle ce_1, x\rangle}$, where e_1 is the first unit vector in $\mathbb{R}^d$. Then h is submultiplicative and $\int_{|x|>1} h(x)\,\widetilde{\nu}^{\pm}(\mathrm{d}x) = +\infty$ by (a), although $\mathcal{L}(X)$ has bounded support and hence finite h-moment.

Acknowledgments We would like to thank the anonymous referee for careful reading and valuable suggestions which helped to improve the exposition of the manuscript.

References

1. Aoyama, T., Nakamura, T.: Behaviors of multivariable finite Euler products in probabilistic view. Math. Nachr. **286**, 1691–1700 (2013)
2. Applebaum, D.: Lévy Processes and Stochastic Calculus, 2nd edn. Cambridge University Press, Cambridge (2009)
3. Armstrong, M.A.: Basic Topology. Springer, New York (1983)
4. Berger, D.: On quasi-infinitely divisible distributions with a point mass. Math. Nachr. **292**, 1674–1684 (2018)
5. Berger, D. Lindner, A.: A Cramer–Wold device for infinite divisibility of $\mathbb{Z}^d$-valued distributions, Bernoulli, accepted for publication (2021). Preprint. Available at https://arxiv.org/abs/2011.08530v1
6. Bertoin, J.: Lévy Processes. Cambridge University Press, Cambridge (1996)
7. Bertoin, J., Doney, R.A., Maller, R.A.: Passage of Lévy processes across power law boundaries at small times. Ann. Probab. **36**, 160–197 (2008)
8. Chhaiba, H., Demni, N., Mouayn, Z.: Analysis of generalized negative binomial distributions attached to hyperbolic Landau levels. J. Math. Phys. **57**(7), 072103, 14pp (2016)
9. Cuppens, R.: Quelques nouveaux résultats en arithmétique des lois de probabilité. In: C.R. Colloqu. C.N.R.S., Les probabilités sur les structures algébraiques, pp. 97–112. C.N.R.S., Paris (1969)
10. Cuppens, R.: Decomposition of Multivariate Probabilites. Academic, New York (1975)
11. Demni, N., Mouayn, Z.: Analysis of generalized Poisson distributions associated with higher Landau leves. Infin. Dimens. Anal. Quantum Probab. Relat. Top. **18**(4), 1550028, 13pp (2015)
12. Doney, R.: Small-time behaviour of Lévy processes. Electron. J. Probab. **9**(8), 209–229 (2004)
13. Doney, R.A.: Fluctuation Theory for Lévy Processes. Lecture Notes in Mathematics, vol. 1897. Springer, Berlin (2007)
14. Doney, R.A., Maller, R.A.: Stability and attraction to normality for Lévy processes at zero and at infinitey. J. Theoret. Probab. **15**(3), 751–792 (2002)
15. Gnedenko, B.V., Kolmogorov, A.N.: Limit Distributions for Sums of Independent Random Variables, Rev. edition. Addison Wesley, Reading (1968). Translation of the Russian original of 1949
16. Kadankova, T., Simon, T., Wang, M.: On some new moments of gamma type. Statist. Probab. Lett. **165**, 108854, 7pp (2020)
17. Kallenberg, O.: Splitting at backward times in regenerative sets. Ann. Probab. **9**(5), 781–799 (1981)
18. Khartov, A.: Compactness criteria for quasi–infinitely divisible distributions on the integers. Stat. Probab. Lett. **153**, 1–6 (2019)
19. Kutlu, M.: On a denseness result for quasi-infinitely divisible distributions. Statist. Probab. Lett. **176**, 109139, (2021)
20. Kyprianou, A.E.: Fluctuations of Lévy Processes with Applications. Introductory Lectures, 2nd edn. Springer, Heidelberg (2014)
21. Lindner, A., Pan, L., Sato, K.: On quasi-infinitely divisible distributions. Trans. Am. Math. Soc. **370**, 8483–8520 (2018)
22. Lindner, A., Sato, K.: Properties of stationary distributions of a sequence of generalised Ornstein-Uhlenbeck processes. Math. Nachr. **284**, 2225–2248 (2011)
23. Linnik, Y.V.: Decomposition of Probability Distributions. Oliver and Boyd Ltd, Edingburgh (1964)
24. Linnik, Y.V., Ostrovskiǐ, I.V.: Decomposition of Random Variables and Vectors. American Mathematical Society, Providence (1977)
25. Nakamura, T.: A quasi-infinitely divisible characteristic function and its exponentation. Stat. Probab. Lett. **83**, 2256–2259 (2013)
26. Nakamura, T.: A complete Riemann zeta distribution and the Riemann hypothesis. Bernoulli **21**, 604–617 (2015)

27. Passeggeri, R.: A density property for stochastic processes (2020). Preprint. Available at arXiv:2010.07752
28. Passeggeri, R.: Spectral representations of quasi-infinitely divisible processes. Stoch. Process. Appl. **130**, 1735–1791 (2020)
29. Pears, A.R.: Dimension Theory of General Spaces. Cambridge University Press, Cambridge (1975)
30. Rocha-Arteaga, A., Sato, K.: Topics in Infinitely Divisible Distributions and Lévy Processes. Springer Briefs in Probability and Mathematical Statistics, Revised Edition. Springer, Cham (2019)
31. Samorodnitsky, G., Taqqu, M.S.: Stable Non-Gaussian Random Processes. Chapman & Hall, Boca Raton/New York (1994)
32. Sato, K.: Lévy Processes and Infinitely Divisible Distributions. Corrected Printing with Supplement. Cambridge University Press, Cambridge (2013)
33. Skorohod, A.V.: Random Processes with Independent Increments. Translated from the second Russian edition. Kluwer Academic Publishers, Dordrecht (1991)
34. Zhang, H., Liu, Y., Li, B.: Notes on discrete compound Poisson model with applications to risk theory. Insurance Math. Econ. **59**, 325–336 (2014)

Extremes and Regular Variation

Nick H. Bingham and Adam J. Ostaszewski

Abstract We survey the connections between extreme-value theory and regular variation, in one and higher dimensions, from the point of view of our recent work on general regular variation.

Keywords Extreme value · General regular variation · Generalised Pareto distribution · Peaks over thresholds · Copula · Spectral measure · D-norm · Dependence structure · Max-stable process · Spatio-temporal process

1 One Dimension

The simplest case is that of an independent and identically distributed (iid) sequence (X_n) with law F; write

$$M_n := \max\{X_1, \cdots, X_n\} \text{ or } \vee_1^n X_i.$$

If there are centering constants b_n and norming constants a_n such that

$$a_n M_n + b_n \to G \quad \text{in law} \quad (n \to \infty)$$

for some non-degenerate probability distribution G, then G is called an *extreme-value distribution (EVD)* (or *extremal law*), and F belongs to the *domain of attraction* of G, $F \in D(G)$. The EVD are also the *max-infinitely divisible (max-id)* laws [6].

N. H. Bingham
Mathematics Department, Imperial College, London, UK
e-mail: n.bingham@ic.ac.uk

A. J. Ostaszewski (✉)
Mathematics Department, London School of Economics, London, UK
e-mail: A.J.Ostaszewski@lse.ac.uk

L. Chaumont, A. E. Kyprianou (eds.), *A Lifetime of Excursions Through Random Walks and Lévy Processes*, Progress in Probability 78,
https://doi.org/10.1007/978-3-030-83309-1_7

We are working here to within an affine transformation (this would change the centering and scaling but preserve the limit), that is, to within *type* (location and scale) [71, I, IV.14]. Modulo type, the limits G (in *one* dimension) have a simple parametric description (see e.g. [13, Th. 8.13.1]):

Theorem 1.1 (Fisher-Tippett theorem, [41], 1928) *To within type, the extremal laws are exactly the following:*

$$\Phi_\xi, \quad (\xi > 0); \qquad \Psi_\xi, \quad (\xi > 0); \qquad \Lambda,$$

where the Fréchet (Φ_ξ), Weibull (Ψ_ξ) and Gumbel (Λ) laws are given by

$$\Phi_\xi := 0 \quad (x \le 0), \quad \exp\{-x^{-\xi}\} \quad (x \ge 0);$$

$$\Psi_\xi := \exp\{-(-x)^\xi\} \quad (x \le 0), \qquad 1 \quad (x \ge 0);$$

$$\Lambda(x) := \exp\{-e^{-x}\} \quad (x \in \mathbb{R}).$$

Particularly for statistical purposes, it is often better to combine these three into one parametric family, the *generalized extreme value (GEV)* laws (see e.g. [20, 3.1.3]). These have one extremal parameter $\alpha \in \mathbb{R}$ and two type parameters $\mu \in \mathbb{R}$ (location) and $\sigma > 0$ (scale):

$$G(x) := \exp\Big(-\Big[1 + \alpha\Big(\frac{x-\mu}{\sigma}\Big)\Big]^{-1/\alpha}\Big) \qquad \text{where } \Big[\cdots\Big] > 0. \tag{GEV}$$

Here $\alpha > 0$ corresponds to the Fréchet, $\alpha = 0$ to the Gumbel (using $(1 + x/n)^n \to e^x$ as $n \to \infty$; we interpret e^x as the '$n = \infty$', or '$\alpha = 0$', case, by the 'L'Hospital convention') and $\alpha < 0$ to the Weibull cases. Taking $\mu = 0$ and $\sigma = 1$ for simplicity (as we may), this gives the *extreme-value distributions*

$$G_\alpha(x) := \exp(-g_\alpha(x)), \qquad g_\alpha(x) := [1 + \alpha x]_+^{-1/\alpha}. \tag{EVD}$$

Here the parameter $\alpha \in \mathbb{R}$ is called the *extreme-value index (EVI)* or *extremal index.* The upper end-point x_+ of F is ∞ for $\alpha \ge 0$ (with a power tail for $\alpha > 0$ and an exponential tail for $\alpha = 0$); for $\alpha < 0$ $x_+ = -1/\alpha$, with a power tail to the left of x_+.

The domains of attraction in the Fréchet and Weibull cases, due to Gnedenko [46] in 1943 (see e.g. [13, Th. 8.13.2,3]) are simple: writing $\overline{F} := 1 - F$ for the tail of F and R_ρ for the class of (positive measurable) functions varying regularly at infinity with index ρ,

(i) $F \in D(\Phi_\alpha)$ iff $\overline{F} \in R_{-\alpha}$;

(ii) $F \in D(\Psi_\alpha)$ iff F has finite upper end-point x_+ and $\overline{F}(x_+ - 1/.) \in R_{-\alpha}$.
The Gumbel case is more complicated (de Haan [51, 52] in 1970–1971, [13, Th. 8.13.4]; cf. [13, Ch. 3, De Haan theory]):

(iii) $F \in D(\Lambda)$ iff

$$\overline{F}(t + xa(t))/\overline{F}(t) \to g_0(x) := e^{-x} \qquad (t \to \infty), \qquad (*)$$

for some *auxiliary function* $a > 0$, which may be taken [32, (3.34)] as

$$a(t) := \int_t^{x_+} \overline{F}(u)du/\overline{F}(t) \quad (t < x_+), \qquad (aux)$$

and satisfies (in the usual case, $x_+ = \infty$)

$$a(t + xa(t))/a(t) \to 1 \quad (t \to \infty). \qquad (Beu)$$

Such functions are called *Beurling slowly varying* (see e.g. [13, §2.11], [14] and the references cited there). If also (Beu) holds locally uniformly (i.e. uniformly on compact x-sets), a is called *self-neglecting*, $a \in SN$ (cf. [10, §2.5.2]):

$$a(t + xa(t))/a(t) \to 1 \quad (t \to \infty) \quad \text{(uniformly on compact } x-\text{sets)}. \quad (SN)$$

An alternative criterion for $D(\Lambda)$ had been given in 1968 by Marcus and Pinsky [73].

The three domain-of-attraction conditions may be unified (using the L'Hospital convention as above) as follows: $F \in D(G_\alpha)$ iff

$$\overline{F}(t + xa(t))/\overline{F}(t) \to g_\alpha(x) := (1 + \alpha x)_+^{-1/\alpha} \quad (t \to \infty) \qquad (**)$$

for some auxiliary function a, and then

$$a(t + xa(t))/a(t) \to 1 + \alpha x \quad (t \to \infty), \qquad (\alpha \text{ Beu})$$

extending the $\alpha = 0$ case (Beu) above (see e.g. [10, §2.6]).

For a *continuous* Beurling slowly varying a, $a \in SN$ and $a(x) = o(x)$ (Bloom's theorem: [13, §2.11], [14]). The relation (αBeu) with local uniformity defines the *self-equivarying* functions $a \in SE$ [77]; here $a(x)/(1 + \alpha x) \in SN$ and so allows for $a(x) = O(x)$ (cf. the case $a(x) := 1 + \alpha x$ with $\alpha > 0$).

Von Mises conditions In 1936, von Mises [76] gave *sufficient* conditions for membership of these domains of attraction, assuming that F has a density f (there is no essential loss of generality here; see below). We formulate these in terms of the *hazard rate* h of survival analysis (see e.g. [24, §2.2]):

$$h(x) := f(x)/\overline{F}(x) = f(x)/\int_x^\infty f(u)du.$$

Below, we shall also need the *inverse hazard function*

$$i(x) := 1/h(x) = \int_x^\infty f(u)du/f(x).$$

Observe that (when the density f exists, as here) the numerator and denominator in (aux) are the integrals of those here. As one may integrate (though not necessarily differentiate) asymptotic relations, we infer that when i exists it may be used as an auxiliary function a as in (aux).

Recall the Smooth Variation Theorem [13, §1.8]: in any situation in regular variation, (one is working to within asymptotic equivalence $\sim$, and so) there is no essential loss in assuming that F has a density (even a C^∞ density) f. Indeed, Balkema and de Haan [5] show that in all three cases, if $F \in D(G)$ for G an extremal law, then $\overline{F} \sim \overline{F_*}$, where F_* satisfies a von Mises condition.

The von Mises conditions in the three cases are (a)–(c) below.

(a) For Φ_α: if $x_+ = \infty$ and

$$xh(x) = x/i(x) \to \alpha > 0 \quad (x \to \infty), \qquad (vM\Phi)$$

then $F \in D(\Phi_\alpha)$.
That (a) is equivalent to (i) in the density case follows by Karamata's Theorem [13, §1.6]; [13, Th. 8.13.5].

(b) For Ψ_α: if $x_+ < \infty$ and

$$(x_+ - x)h(x) = (x_+ - x)/i(x) \to \alpha > 0 \quad (x \to \infty), \qquad (vM\Psi)$$

then $F \in \Psi(\alpha)$.
The proof uses (ii) as above [13, Th. 8.13.6].

(c) Taking $x_+ = \infty$ for simplicity: if

$$i'(x) \to 0 \quad (x \to \infty), \qquad (vM\Lambda)$$

then $F \in D(\Lambda)$ [4]. The proof [13, Th. 8.13.7] hinges on [13, Lemma 8.13.8]: if $a(.) > 0$ and $a'(t) \to 0$ as $t \to \infty$, then $a \in SN$. This actually characterises SN: the representation theorem for $a \in SN$ is [14, Th. 9]

$$a(x) = c(1 + o(1)) \int_0^x e(u)dy, \quad e \in C^1,\ e(x) \to 0\ (x \to \infty).$$

Rates of convergence in the above were studied by Falk and Marohn [38].

Von Mises functions Call a distribution function F a *von Mises function* with *auxiliary function* a if [32, §3.3.3] for some $c, d \in (0, \infty)$,

$$\overline{F}(x) = c \exp\{-\int_d^x dt/a(t)\}, \quad a'(t) \to 0 \ (t \to \infty).$$

Then (as above) one can take the auxiliary function a as the inverse hazard function i, or (see below) the mean excess function e (when it exists). One can pass to full generality by replacing the constant c above by a function $c(x) \to c$: $D(\Lambda)$ consists of von Mises functions and their tail-equivalent distributions [32, p.144]. And (from $a' \to 0$): when $x_+ = \infty$, tails in $D(\Lambda)$ decrease faster than any power [32, p.139]. Example: the standard normal law (take $a = i$ and use the Mills ratio).

Peaks over thresholds (POT) As always in extreme-value theory, one has two conflicting dangers. The maxima – the very high values – are rare, and focussing on them discards information and may leave too little data. But if one over-compensates for this by including too much data, one risks distorting things as the extra data is also informative about the distribution away from the tails. One approach is to choose a large threshold (which the statistician may choose), $u > 0$ say, and look only at the data exceeding u. These are the *peaks over thresholds (POT)*. Here one focusses on the *exceedances* $Y = X - u$ when positive, and their conditional law F_u given $X > u$. This leads to

$$\begin{aligned}\overline{F_u}(x) = P(Y > xa(u)|Y > 0) &= P\Big(\frac{X-u}{a(u)} > x|X > u\Big)\\ &= \overline{F}(u + xa(u))/\overline{F}(u)\\ &\to g_\alpha(x) := (1 + \alpha x)_+^{-1/\alpha} \quad (u \to \infty),\end{aligned}$$

as in (∗∗) above. Thus the conditional distribution of $(X - u)/a(u)|X > u$ has limit

$$H_\alpha(x) := 1 + \log G_\alpha(x) = 1 - g_\alpha(x) = 1 - (1 + \alpha x)_+^{-1/\alpha}, \qquad (GPD)$$

the *generalised Pareto distribution (GPD)* ('EVD for max, GPD for POT').

There are several ways of motivating the use of GPD:

(i) Pickands [79] showed in 1975 that F_u has GPD H_α as limit law iff it has the corresponding EVD as limit of its maxima, i.e. $F \in D(G_\alpha)$. This (in view of [5]) is the *Pickands-Balkema-de Haan theorem* [74, Th. 7.20].

(ii) There is *threshold stability*: if Y is GP and $u > 0$, then the conditional law of $Y - u|Y > u$ is also GP, and this characterises the GPD. This property is useful in applications; see e.g. [74, §7.2.2].

(iii) If N is Poisson and $(Y_1, \cdots, Y_N)|N$ are iid GP, then $\max(Y_1, \cdots, Y_N)$ has the corresponding EVD; again, this characterises GPD.

For details, see e.g. Davison and Smith [28, §2] and the references there.

Statistical work in the one-dimensional setting here centres on the estimation of the extreme-value index α. One of the commonest estimators here is *Hill's estimator* [59]; see e.g. [74, §7.2.4], [10, §9.5.2].

Mean excess function When the mean of X exists, the *mean excess* (or *mean exceedance) function* of X over the threshold u exists and is

$$e(u) := E[X - u|X > u].$$

Integrating by parts,

$$e(u) = \int_u^\infty (x-u)dF(x)/\overline{F}(u) = -\int_u^\infty (x-u)d\overline{F}(x)/\overline{F}(u) = \int_u^\infty \overline{F}(x)dx/\overline{F}(u),$$

which by (aux) is the general form of the auxiliary function a. Thus, *when e exists, one may take it as the auxiliary function a* (in preference to the inverse hazard function i, if preferred).

Self-exciting processes One way to relax the independence assumption is to allow *self-exciting processes*, where an occurrence makes other occurrences more likely. This is motivated by aftershocks of earthquakes, but also relevant to financial crises. This uses *Hawkes processes* [58]; see [74, §7.4.3]. For point processes in extremes and regular variation, see [82].

General regular variation Referring to $(*)$ and (αBeu), these can now be recognised as the relevant instances of *general regular variation*, for which see [15]. The signature is the argument $t + xa(t)$, where the auxiliary function a is self-neglecting. See the 3×3 table in [15, Th. 3] (relevant here is the top right-hand corner with $\kappa = -1$). Likewise, the limit in $(**)$ gives the (2,3) (or middle right) entry in the table, with $\kappa = -1$, and after taking logs, the (2,1) (or middle left) entry:

$$\log \overline{F}(t + xa(t)) - \log \overline{F}(t) \to -1/\alpha \log(1 + \alpha x)_+ \quad (t \to \infty), \qquad (***)$$

exactly of the form studied in [15] (there the RHS is called the *kernel*, $K(x)$).

The authors in [32] remark (e.g. their p.140) that regular variation 'does not seem to be the right tool' for describing von Mises functions. The general regular variation of [14, 15] *does* seem to be the right tool here, including as it does the Karamata, Bojanic-Karamata/de Haan and Beurling theories of regular variation.

Note. The unification that general regular variation brings to this classically important area justifies brief mention here of what it rests on: the theory of *Popa groups* and *Goldie equations*; see [15] for details.

Dependence, time series; extremal index The assumption above most often unjustified in practice is that of independence. We turn briefly now to the simplest dependent case, a stationary time series (or temporal process; cf. the spatio-temporal processes in §2 below; we do not pursue the interesting and important question of *trends*). For background here see e.g. the books [68, II], [32, §4.4, Ch. 7], Aldous [2, C], Berman [8], and the survey [69]. Comparison with the independent case is useful: let M_n, $\tilde{M}_n$ be the maxima in the time-series and independent cases (with the same F). Under suitable conditions [68, §3.7], for $u_n \to \infty$ and $\tau > 0$, the following are equivalent:

$$n\overline{F}(u_n) \to \tau, \quad P(\tilde{M}_n \leq u_n) \to e^{-\tau}, \quad P(M_n \leq u_n) \to e^{-\theta\tau},$$

for some $\theta \geq 0$, called the *extremal index*. The extremes tend to cluster [2]; $1/\theta$ is the *mean cluster size* (so $\theta \in [0, 1]$); for estimating θ, see [95].

Heat waves Often it is not the maxima as such that matter most, but extended periods of very high levels. Heat waves are notorious for causing large numbers of excess fatalities; similarly for crop loss in agriculture, disruption with flooding, etc. For studies here, see Reich et al. [81], Winter and Tawn [97].

2 Higher Dimensions

For general references for multidimensional EVT, see e.g. [32, Ch. 3,5,6], [10, Ch. 8], [55, Part II], [56], [37], [35], [36]. For multidimensional regular variation, see e.g. Basrak et al. [7].

The situation in dimensions $d > 1$ is different from and more complicated than that for $d = 1$ above, regarding both EVT and Popa theory. We hope to return to such matters elsewhere.

Copulas The theory above extends directly from dimensions 1 to d, if each X or x above is now interpreted as a d-vector, $x = (x_1, \cdots, x_d)$ etc. Then, as usual in dimension $d > 1$, we split the d-dimensional joint distribution function F into the marginals $F_1, \cdots, F_d$, and the *copula* C (a distribution function on the d-cube $[0, 1]^d$ with uniform marginals on $[0, 1]$), which encodes the dependence structure via *Sklar's theorem* ([92]; see e.g. [74, Ch. 5]):

$$F(x) = F(x_1, \cdots, x_d) = C(F_1(x_1), \cdots, F_d(x_d)). \qquad (Skla)$$

In particular, this shows that one may standardise the marginals F_i in any convenient way, changing only the joint law F but not the copula (dependence structure). One choice often made in extreme-value theory is to transform to *standard Fréchet* marginals,

$$F_i(x) = \exp\{-1/x\} \quad (x > 0).$$

When this is done, the EV law is called *simple* [10, §8.2.2].

The limit distributions that can arise are now the *multivariate extreme-value (MEV)* laws. Their copulas link MEV laws with their GEV margins. With $u^t := (u_1^t, \cdots, u_d^t)$ for $t > 0$, these, the *EV copulas*, denoted by C_0, are characterised by their *scaling relation* [74, Th. 7.44]

$$C_0(u^t) = C_0^t(u) \quad (t > 0). \qquad (Sca)$$

By analogy with stable laws for sums, a law G is *max-stable* if

$$G^n(a_n x + b_n) \equiv G(x) \quad (n \in \mathbb{N})$$

for suitable centering and scaling sequences (b_n), (a_n); these are the GEV laws.

Survival copulas In extreme-value theory, it is the upper tails that count. Taking operations on vectors componentwise and writing

$$F(x) := P(X \le x), \qquad \overline{F}(x) := P(X > x),$$

one can rewrite Sklar's theorem in terms of *survival functions* $\overline{F}$, $\overline{F_i}$: a d-dimensional survival function $\overline{F}$ has a *survival copula* C with [75, Th. 2.1]

$$\overline{F}(x) = C(\overline{F_1}(x_1), \cdots, \overline{F_d}(x_d)), \qquad C(u) = \overline{F}(\overline{F_1}^{-1}(u_1), \cdots \overline{F_d}^{-1}(u_d))$$

(F, F_i are 1 at $+\infty$; $\overline{F}, \overline{F_i}$ are 0 at $+\infty$; they are accordingly often studied for $x \le 0$, $x \to -\infty$ rather than $x \ge 0$, $x \to \infty$).

Copula convergence The question of multivariate domains of attraction (MD, or MDA) decomposes into those for the marginals and for the copula by the *Deheuvels-Galambos theorem* ([29, 43]; [74, Th. 7.48]): with F as above, $F \in MD(H)$ with

$$H(x_1, \cdots, x_d) := C_0(H_1(x_1), \cdots, H_d(x_d)),$$

an MEV law with GEV marginals H_i and EV copula C_0, iff

(i) $F_i \in D(H_i), i = 1, \cdots, d$;
(ii) $C \in CD(C_0)$ ('CD for copula domain of attraction'), i.e.

$$C^t(u_1^{1/t}, \cdots, u_d^{1/t}) \to C_0(u_1, \cdots, u_d) = C_0(u) \quad (t \to \infty) \quad (u \in [0, 1]^d).$$

Peaks over thresholds (POT) The first (and most important) two of the three properties above of POT in one dimension extend to d dimensions. The first is the d-dimensional version of the Pickands-Balkema-de Haan theorem, linking EVD and GPD; the second is threshold stability. For details, see Rootzén and Tajvidi [86], Rootzén, Segers and Wadsworth [84, 85], Kiriliouk et al. [63].

Spectral representation The scaling property (Sca) suggests using spherical polar coordinates, $x = (r, \theta)$ say ($x \in \mathbb{R}^d_+$, $r > 0$, $\theta \in \mathbb{S}^{d-1}_+$). Then the MEV law G has (with $\wedge$ for min) a *spectral representation*

$$\log G(x) = \int_{\mathbb{S}_+} \wedge_{i=1}^d \Big(\frac{\theta_i}{\|\theta\|} \log G_i(x_i) \Big)\, dS(\theta) \quad (x = (r, \theta) \in \mathbb{R}^d),$$

where the *spectral measure* S satisfies

$$\int_{\mathbb{S}_+} \frac{\theta_i}{\|\theta\|} dS(\theta) = 1 \quad (i = 1, \cdots, d)$$

(see e.g. [53], [10, §8.2.3], [72]). The regular-variation (or other limiting) properties are handled by the radial component, the dependence structure by the spectral measure.

D-norms The standard (i.e. with unit Fréchet marginals) max-stable (SMS) laws are those with survival functions of the form

$$\exp\{-\|x\|\} \quad (x \le 0 \in \mathbb{R}^d),$$

for some norm, called a *D-norm* ('D for dependence', as this norm encodes the dependence structure). For a textbook treatment, see Falk [36].

Pickands dependence function An EV copula may be specified by using the *Pickands dependence function*, B ([80]; [74, Th. 7.45]): C is a d-dimensional EV copula iff it has the representation

$$C(\mathbf{u}) = \exp\Big(B\Big(\frac{\log u_1}{\sum_1^d u_i}, \cdots, \frac{\log u_d}{\sum_1^d u_i} \Big) \sum_1^d \log u_i \Big),$$

where with S_d the d-simplex $\{x : x_i \ge 0, \sum_1^d x_i = 1\}$,

$$B(w) = \int_{S_d} \max(x_1 w_1, \cdots, x_d w_d) dH(x)$$

with H a finite measure on S_d. Of course, the d-simplex needs only $d-1$ coordinates to specify it; such a simplification is most worthwhile when $d = 2$ (below).

Two dimensions Things can be made more explicit in two dimensions. For the theory above, one obtains the representation

$$C(u_1, u_2) = \exp\{(\log u_1 + \log u_2) A(\frac{\log u_1}{\log u_1 + \log u_2})\},$$

where

$$A(w) = \int_0^1 \max((1-x)w, x(1-w))dH(x),$$

for H a measure on $[0, 1]$. The Pickands dependence function A here is characterised by the bounds

$$\max(w, 1-w) \le A(w) \le 1 \quad (0 \le w \le 1)$$

and being (differentiable and) convex.

Archimedean copulas In d dimensions, the *Archimedean copula* C with *generator* ψ is given by

$$C(u) = \psi(\psi^{-1}(u_1) + \cdots + \psi^{-1}(u_d)).$$

Here [75] $\psi(0) = 1$, $\psi(x) \to 0$ as $x \to \infty$ and ψ is d-monotone (has $d-2$ derivatives alternating in sign with $(-)^{d-2}\psi^{(d-2)}$ nonincreasing and convex); this characterises Archimedean copulas. In particular, for $d = 2$,

$$C(u, v) = \psi(\psi^{-1}(u) + \psi^{-1}(v))$$

is a copula iff ψ is convex.

An alternative to spherical polars uses the d-simplex S_d in place of $\mathbb{S}^d_+$. This leads to *ℓ_1-norm symmetric distributions*, or *simplex distributions*, and the *Williamson transform*; see [75, §3]. Here the Archimedean generator ψ is the Williamson transform of the law of the radial part R, and one can read off the domain-of-attraction behaviour of R from regular-variation conditions on ψ [66, Th. 1]. The dependence structure is now handled by the *simplex measure* [44].

This feature that one 'radial' variable handles the tail behaviour and regular-variation aspects, while the others handle the dependence structure, has led to 'one-component regular variation' in this context; see Hitz and Evans [60].

The marginals may require different normalisations; for background here, see e.g. [83, §6.5.6] ('standard v. non-standard regular variation').

Gumbel copulas For $\theta \in [1, \infty)$, the *Gumbel copula* with parameter θ is the Archimedean copula with generator $\psi(x) = \exp\{-x^{1/\theta}\}$. This is the only copula which is both Archimedean and extreme-value ([45]; [75, Cor. 1]).

Archimax copulas Call ℓ a (d-variate) *stable tail-dependence function* if for $x_i \ge 0$,

$$\ell(x_1, \cdots, x_d) = -\log C_0(e^{-x_1}, \cdots, e^{-x_d})$$

for some extreme-value copula C_0. The Archimax copulas [19] are those of the form

$$C_{\psi,\ell}(u_1, \cdots, u_d) := (\psi \circ \ell)(\psi^{-1}(u_1), \cdots, \psi^{-1}(u_d)).$$

This construction does indeed yield a copula [19], and in the case $d = 2$ gives the Archimedean copulas ($A(.) \equiv 1$ above) and the extreme-value copulas ($\psi(t) = e^{-t}$), whence the name.

Dependence structure Particularly when d is large, the spectral measure above may be too general to be useful in practice, and so special types of model are often used, the commonest being those of Archimedean type. While convenient, Archimedean copulas are *exchangeable*, which of course is often not the case in practice ('sea and wind'). The arguments of the copula typically represent covariates, and these are often related by conditional independence relationships; these may be represented graphically (see e.g. the monograph by Lauritzen [67], and for applications to extremes, Engelke and Hitz [34]; see also [60]). Hierarchical relationships between the covariates (e.g. 'phylogenetic trees') may be represented by hierarchical Archimedean copulas; see e.g. Cossette et al. [23]. Special types of graphs (vines) occur in such contexts; see e.g. Chang and Joe [18], Joe et al. [62], Lee and Joe [70].

Max-stable processes The case of infinitely many dimensions – stochastic processes – is just as important as the case $d < \infty$ above (the classic setting here is the whole of the Dutch coastline, rather than just coastal monitoring stations). For theory here, see e.g. [55, Part III].

A process Y is *max-stable* if when Y_i are independent copies of Y, $\max\{Y_1, \cdots, Y_r\}$ has the same distribution as rY for each $r \in \mathbb{N}_+$. These have a spectral representation, for which see de Haan [53]. For estimation of max-stable processes, see e.g. Chan and So [17].

Spatio-temporal processes Spectral representations have useful interpretations for modelling spatio-temporal processes, e.g. for the *storm-profile process* or *Smith process* [17, 94]:

$$Z(x) = \max_i \phi(x - X_i)\Gamma_i,$$

where $Z(x)$ represents the maximum effect at location x over an infinite number of storms centred at random points X_i (forming a homogeneous Poisson process on $\mathbb{R}^d$) of strengths Γ_i (a Poisson point process of rate 1), the effect of each being $\phi(t - X_i)\Gamma_i$ (here ϕ is a Gaussian density function with mean 0 and covariance matrix Σ, whose contours represent the decreasing effect of a storm away from its centre). Thus the process measures 'the all-time worst (storm effect), here'. Perhaps such models could also be used to describe e.g. the bush fires currently threatening Australia.

Spatio-temporal max-stable processes in which the space-time spectral function decouples into ones for time and for space given time are given by Embrechts, Koch and Robert [33] (see also [65]). This allows for the different roles of time and space given time to be modelled separately. They also allow space to be a sphere, necessary for realistic modelling on a global scale.

Tail dependence Studying asymptotic dependence in multivariate tails is important in, e.g., risk management, where one may look to diversify by introducing negative correlation. For very thin tails (e.g., Gaussian) this is not possible in view of *asymptotic independence* (Sibuya [91]). But with heavier tails, tail dependence coefficients are useful here; see e.g. [88], [66, §5].

Tail-dependence coefficients deal with asymptotic dependence between pairs of coordinates, and so can be assembled into a matrix, the *tail-dependence matrix*. This bears some analogy with the correlation matrix in multivariate analysis, which (together with the mean) is fully informative in the Gaussian case, and partially informative in general. For various developments here, see e.g. Embrechts et al. [31, 40, 96].

Applications For more on spatial processes (random fields), spatio-temporal processes and applications to such things as weather, see e.g. Smith [94], Schlather [87], Cooley et al. [21], Davison et al. [27], Davis et al. [25, 26], Huser and Davison [61], Sharkey and Winter [90], Abu-Awwad et al. [1], Cooley and Thibaud [22].

An extended study of sea and wind, applied to the North Sea flood defences of the Netherlands, is in de Haan and de Ronde [57].

Particularly with river networks, the spatial relationships between the points at which the data is sampled is crucial. For a detailed study here, see Asadi et al. [3].

For financial applications (comparison of two exchange rates), see [74, Ex. 7.53].

Statistics The great difference between one and higher dimensions in the statistics of extreme-value theory is that in the former, *parametric* methods suffice (whether one works with EVD or with GPD). In the latter, one has d such one-dimensional parametric problems (or one d-dimensional one) for the marginals, and a non-parametric one for the copula. The problem is thus *semi-parametric*, and may be treated as such (cf. [9, 64]). But our focus here is on the copula, which needs to be estimated nonparametrically; see e.g. [28, 42, 48, 49] (cf. [30, 78]), and in two dimensions, [50, 89].

For peaks over thresholds in higher dimensions, see e.g. [63]; for graphical methods, see [70].

3 Historical Comments

1. The extremal laws are known as the *Fréchet* (heavy-tailed, Φ_α), *Gumbel* (light-tailed, Λ) and *Weibull* (bounded tail, Ψ_α) distributions, after Maurice Fréchet (1878–1973), French mathematician, in 1937, Emil Julius Gumbel

(1891–1966), German statistician, in 1935 and 1958, and Waloddi Weibull (1887–1979), Swedish engineer, in 1939 and 1951.
2. The Pareto distributions are named after Vilfredo Pareto (1848–1923), Italian economist, in 1896.
3. The remarkable pioneering work of Fisher and Tippett [41] in 1928 of course pre-dated regular variation, which stems from Karamata in 1930.
4. The remarkable pioneering work of von Mises [76] in 1936 did not use regular variation, perhaps because he was not familiar with the journal Karamata published in, *Mathematica (Cluj)*, perhaps because what Karamata was then famous for was his other 1930 paper, on the (Hardy-Littlewood-)Karamata Tauberian theorem for Laplace transforms – analysis, while von Mises was an applied mathematician.
5. The pioneering work of Gnedenko [46] in 1943 on limits of maxima also did not use regular variation; nor did the classic monograph of Gnedenko and Kolmogorov [47] of 1949. As a result, the analytic aspects of both were excessively lengthy, tending to mask the essential probabilistic content.
6. The subject of extreme-value theory was made much more important by the tragic events of the night of 31 January–1 February 1953. There was great loss of life in the UK, and much greater loss in the low-lying Netherlands (see e.g. [54]).
7. The realisation that regular variation was the natural language for limit theorems in probability is due to Sudakov in 1955 (in Volume 1 of *Theory of Probability and its Applications*). But this was not picked up at the time, and was rediscovered by Feller in Volume II of his book (1966 and 1971). See e.g. [12] for details.
8. Beurling slow variation appeared in Beurling's unpublished work of 1957 on his Tauberian theorem. See e.g. [11], [14, §10.1] for details and references.
9. The first systematic application of regular variation to limit theorems in probability was de Haan's 1970 thesis [51]. This has been the thread running through his extensive and influential work for the last half-century.
10. The Balkema-de Haan paper [5] of 1974 was explicitly a study of applications of regular variation, and in 'great age' set the stage for 'high thresholds'.
11. Threshold methods were developed by hydrologists in the 1970s. Their theoretical justification stems from Pickands's result ([79], Pickands-Balkema-de Haan theorem), giving a sense in which F_u is well-approximated by some GPD iff F lies in the domain of attraction of some EVD (cf. [93, §3]). Full references up to 1990 are in [28].

Postscript

It is a pleasure to contribute to this volume, celebrating Ron Doney's 80th birthday. His long and productive career in probability theory has mainly focused on random walks and (later) Lévy processes, essentially the limit theory of sums. Extreme-value theory is essentially the limit theory of maxima. Sums and maxima have many points of contact (see e.g. [13, §8.15]), recently augmented by the fine paper by Caravenna and Doney [16] related to the Garsia-Lamperti problem (see e.g. [13,

§§8.6.3, 8.7.1]). A personal point of contact with extremes came for Ron with flooding and the partial collapse on 6 August 2019 of the dam at Whaley Bridge near his home. Many residents had to be evacuated, fortunately not including Ron and Margaret. Any such incident stands as a riposte to climate-change deniers everywhere. Of course, Australia is much in our minds at the time of writing.

We thank the editors for their kind invitation to contribute to this Festschrift. The first author thanks the organisers of Extreme Value Analysis 11 for their invitation to speak at EVA11 in Zagreb in July 2019.

Both authors send their very best wishes to Ron and Margaret.

References

1. Abu-Awwad, A.-F., Maume-Deschampts, V., Ribereau, P.: Semi-parametric estimation for space-time max-stable processes: F-madogram-based estimation approaches. Stat. Inference Stoch. Processes **24**, 241–276 (2021)
2. Aldous, D.: Probability Approximations Via the Poisson Clumping Heuristic. Springer, New York (1989)
3. Asadi, P., Davison, A.C., Engelke, S.: Extremes on river networks. Ann. Appl. Stat. **9**, 2023–2050 (2015)
4. Balkema, A.A., de Haan, L.: On R. von Mises' condition for the domain of attraction of $\exp(-e^{-x})$. Ann. Math. Statist. **43**, 1352–1354 (1972)
5. Balkema, A.A., de Haan, L.: Residual Lifetime at Great Age. Ann. Prob. **2**, 792–804 (1974)
6. Balkema, A.A., Resnick, S.I.: Max-infinite divisibility. J. Appl. Probab. **14**, 309–319 (1977)
7. Basrak, B., Davis, R.A., Mikosch, T.: A characterization of multidimensional regular variation. Ann. Appl. Prob. **12**, 908–920 (2012)
8. Berman, S.M.: Sojourns and Extremes of Stochastic Processes. Wadsworth, Pacific Grove, CA (1992)
9. Bickel, P.J., Klaassen, C.A.J., Ritov, Y., Wellner, J.A.: Efficient and Adaptive Estimation for Semiparametric Models. The Johns Hopkins University Press, Baltimore (1993) (2nd edn. Springer, New York, 1998)
10. Bierlant, J., Goegebeur, Y., Segers, J., Teugels, J.: Statistics of Extremes: Theory and Applications. Wiley, Chichester (2004)
11. Bingham, N.H.: Tauberian theorems and the central limit theorem. Ann. Probab. **9**, 221–231 (1981)
12. Bingham, N.H.: Regular variation and probability: the early years. J. Comput. Appl. Math. **200**, 357–363 (2007) (J. L. Teugels Festschrift)
13. Bingham, N.H., Goldie, C.M., Teugels, J.L.: Regular Variation. Cambridge University Press, Cambridge (1987/1989)
14. Bingham, N.H., Ostaszewski, A.J.: Beurling slow and regular variation. Trans. Lond. Math. Soc. **1**, 29–56 (2014)
15. Bingham, N.H., Ostaszewski, A.J.: General regular variation, Popa groups and quantifier weakening. J. Math. Anal. Appl. **483**, 123610 (2020). arXiv:1901.05996
16. Caravenna, F., Doney, R.A.: Local large deviations and the strong renewal theorem. Electron. J. Prob. **24**, 1–48 (2019)
17. Chan, R.K.S., So, M.K.P.: On the performance of the Bayes composite likelihood estimation of max-stable processes. J. Stat. Comput. Simul. **87**, 2869–2881 (2017)
18. Chang, B., Joe, H.: Predictions based on conditional distributions of vine copulas. Comput. Stat. Data Anal. **139**, 45–63 (2019)

19. Charpentier, A., Fougères, A.-L., Genest, C., Nešlehová, J.G.: Multivariate Archimax copulas. J. Multivariate Anal. **126**, 118–136 (2014)
20. Coles, S.: An Introduction to Statistical Modeling of Extreme Values. Springer, London (2001)
21. Cooley, D., Nychka, D., Naveau, P.: Bayesian spatial modelling of extreme precipitation return levels. J. Am. Stat. Assoc. **102**, 824–860 (2007)
22. Cooley, D., Thibaud, E.: Decompositions of dependence for high-dimensional extremes. Biometrika **106**, 587–604 (2019)
23. Cossette, H., Gadoury, S.-P., Marceau, E., Robert, C.Y.: Composite likelihood estimation methods for hierarchical Archimedean copulas defined with multivariate compound distributions. J. Multivariate Anal. **172**, 59–83 (2019)
24. Cox, D.R., Oakes, D.: Analysis of survival data. Chapman & Hall/CRC, Boca Raton (1984)
25. Davis, R.A., Klüppelberg, C., Steinkohl, C.: Max-stable processes for modelling extremes observed in space and time. J. Korean Stat. Soc. **42**, 399–413 (2013)
26. Davis, R.A., Klüppelberg, C., Steinkohl, C.: Statistial inference for max-stable processes in space and time. J. R. Stat. Soc. B **75**, 791–819 (2013)
27. Davison, A.C., Padoan, S.A., Ribatet, M.: Statistical modelloing of spatial extremes (with discussion). Stat. Sci. **27**, 161–201 (2012)
28. Davison, A.C., Smith, R.L.: Models for exceedances over high thresholds (with discussion). J. R. Stat. Soc. B **52**, 393–442 (1990)
29. Deheuvels, P.: Characterisation complète des lois extrêmes multivariées et la convergence aux types extrêmes. Publ. Inst. Sci. U. Paris **23**, 1–36 (1978)
30. Easthoe, E.F., Heffernan, J.E., Tawn, J.A.: Nonparametric estimation of the spectral measure, and associated dependence measures, for multivariate extreme values using a limiting conditional representation. Extremes **17**, 25–43 (2014)
31. Embrechts, P., Hofert, M., Wang, R.: Bernoulli and tail-dependence compatibility. Ann. Appl. Prob. **26**, 1636–1658 (2016)
32. Embrechts, P., Klüppelberg, C., Mikosch, T.: Modelling extremal events for insurance and finance. Springer, Berlin (1997)
33. Embrechts, P., Koch, E., Robert, C.: Space-time max-stable models with spectral separability. Adv. Appl. Prob. **48A**, 77–97 (2016) (N. H. Bingham Festschrift)
34. Engelke, S., Hitz, A.S.: Graphical models for extremes. With discussion. J. Roy. Stat. Soc. **B** **82**, 871–932 (2020)
35. Falk, M.: It was 30 years ago today when Laurens de Haan went the multivariate way. Extremes **11**, 55–80 (2008)
36. Falk, M.: Multivariate Extreme-Value Theory and D-Norms. Springer, Cham (2019)
37. Falk, M., Hüsler, J., Reiss, R.-D.: Laws of Small Numbers: Extremes and Rare Events, 3rd edn. Birkhäuser, Basel (2010) (2nd edn., 2004, 1st edn., 1994)
38. Falk, M., Marohn, F.: Von Mises conditions revisited. Ann. Probab. **21**, 1310–1328 (1993)
39. Falk, M., Padoan, S.A., Wisheckel, F.: Generalized Pareto copulas: a key to multivariate extremes. J. Multivariate Anal. **174**, 104538 (2019)
40. Fiebig, U.-R., Strokorb, K., Schlather, M.: The realization problem for tail correlation functions. Extremes **20**, 121–168 (2017)
41. Fisher, R.A., Tippett, L.H.C.: Limiting forms of the frequency of the largest or smallest member of a sample. Proc. Camb. Phil. Soc. **24**, 180–190 (1928) (reprinted in Collected papers of R. A. Fisher **2**, 208–219 (1972), U. Adelaide Press, Adelaide 1972)
42. de Fondeville, R., Davison, A.C.: High-dimensional peaks-over-threshold inference. Biometrika **105**, 575–592 (2018)
43. Galambos, J.: The Asymptotic Theory of Extreme Order Statistics. Wiley (1978), New York, Melbourne, FL (2nd edn., Krieger, 1987)
44. Genest, C., Nešlehová, J.G., Rivest, L.-P.: The class of multivariate max-id copulas with ℓ_1-norm symmetric exponent measure. Bernoulli **24**(4B), 3751–3790 (2018)
45. Genest, C., Rivest, L.: On the multivariate probability integral transform. Stat. Prob. Lett. **53**, 391–399 (2001)

46. Gnedenko, B.V.: Sur la distribution limite du terme maximum d'une série aléatoire. Ann. Math. **44**, 423–453 (1943)
47. Gnedenko, B.V., Kolmogorov, A.N.: Limit Distributions for Sums of Independent Random Variables. Addison-Wesley, Cambridge MA (1954) (Russian edition, Gostelkhizdat, Moscow, 1949)
48. Gudendorf, G., Segers, J.: Extreme-value Copulas. Copula Theory and Its Applications. Lecture Notes in Statistics (Proceedings), vol. 198, pp. 127–145. Springer, Heidelberg (2010)
49. Gudendorf, G., Segers, J.: Nonparametric estimation of an extreme-value copula in arbitrary dimensions. J. Multivariate Anal. **102**, 37–47 (2011)
50. Guillotte, S., Perron, F., Segers, J.: Non-parametric Bayesian inference on bivariate extremes. J. R. Stat. Soc. B **73**, 377–406 (2011)
51. de Haan, L.: On Regular Variation and Its Application to the Weak Cnvergence of Sample Extremes. Mathematical Centre Tract, vol. 32. Amsterdam (1970)
52. de Haan, L.: A form of regular variation and its application to the domain of attraction of the double exponential distribution. Z. Wahrschein. **17**, 241–258 (1971)
53. de Haan, L.: A spectral representation of max-stable processes. Ann. Prob. **12**, 1194–1204 (1984)
54. de Haan, L.: Fighting the arch-enemy with mathematics. Statistica Neerlandica **44**, 45–68 (1990)
55. de Haan, L., Ferreira, A.: Extreme Value Theory: An Introduction. Springer, New York (2006)
56. de Haan, L., Resnick, S.I.: Limit theory for multivariate sample extremes. Z. Wahrschein. **40**, 317–337 (1977)
57. de Haan, L., de Ronde, J.: Sea and wind: multivariate extremes at work. Extremes **1**, 7–45 (1998)
58. Hawkes, A.G.: Point spectra of some self-exciting point processes. J. R. Stat. Soc. B **33**, 438–443 (1971)
59. Hill, B.M.: A simple general approach to inference about the tail of a distribution. Ann. Stat. **3**, 1163–1174 (1975)
60. Hitz, A., Evans, R.: One-component regular variation and graphical modelling of extremes. J. Appl. Prob. **53**, 733–746 (2016)
61. Huser, R., Davison, A.C.: Spce-time modelling of extremal events. J. R. Stat. Soc. **76**, 439–461 (2014)
62. Joe, H., Li, H., Nikoloulopoulos, A.K.: Tail dependence functions and vine copulas. J. Multivariate Anal. **101**, 252–270 (2010)
63. Kiriliouk, A., Rootzén, H., Wadsworth, J.L., Segers, J.: Peaks-over-thresholds modeling with multivariate generalized Pareto distributions. Technometrics **61**, 123–135 (2019)
64. Klüppelberg, C., Kuhn, G., Peng, L.: Semi-parametric models for the multivariate tail-dependence function – the asymptotically dependent case. Scand. J. Stat. **35**, 701–718 (2008)
65. Koch, E., Robert, C.Y.: Geometric ergodicity for some space-time max-stable Markov chains. Stat. Probab. Letters **145**, 43–49 (2019)
66. Larsson, M., Něslehová, J.: Extremal behaviour of Archimedean copulas. Adv. Appl. Probab. **43**, 185–216 (2011)
67. Lauritzen, S.: Graphical Models. Oxford University Press, Oxford (1996)
68. Leadbetter, M.R., Lindgren, G., Rootzén, H.: Extremes and Related Properties of Random Sequences and Processes. Springer, New York (1983)
69. Leadbetter, M.R., Rootzén, H.: Extremal theory for stochastic processes. Ann. Probab. **16**, 431–478 (1988)
70. Lee, D., Joe, H.: Multivariate extreme-value copulas with factor and tree dependence structures. Extremes **21**, 147–176 (2018)
71. Loève, M.: Probability Theory, vol. I, II, 4th edn. Springer, New York (1977)
72. Mao, T., Hu, T.: Relations between the spectral measures and dependence of multivariate extreme value distributions. Extremes **18**, 65–84 (2015)
73. Marcus, M.B., Pinsky, M.: On the domain of attrraction of $\exp(-e^{-x})$. J. Math. Anal. Appl. **28**, 440–449 (1968)

74. McNeil, A.J., Frey, R., Embrechts, P.: Quantitative Risk Management: Concepts, Techniques, Tools. Princeton University Press, Princeton (2005) (2nd edn., 2015)
75. McNeil, A.J., Něslehová, J.: Multivariate Archimedean coulas, d-monotone functions and ℓ_1-norm symmetric distributions. Ann. Stat. **37**, 3059–3097 (2009)
76. von Mises, R.: La distribution de la plus grande de n valeurs. Revue Mathématique de l'Union Interbalkanique **1**, 141–160 (1936). (reprinted in Selected Papers II, Coll. Works, pp. 271–294, AMS, Providence, RI
77. Ostaszewski, A.J.: Beurling regular variation, Bloom dichotomy, and the Gołąb-Schinzel functional equation. Aequat. Math. **89**, 725–744 (2015)
78. Papastathopoulos, I., Tawn, J.A.: Dependence properties of multivariate max-stable distributions. J. Multivariate Anal. **130**, 134–140 (2014)
79. Pickands, J. III: Statistical inference using extreme order statistics. Ann. Stat. **3**, 119–131 (1975)
80. Pickands, J. III: Multivariate extreme value distributions. Bull. Inst. Int. Stat. **49**, 859–878 (1981)
81. Reich, B.J., Shaby, B.A., Cooley, D.: A hierarchical model for serially dependent extremes: a study of heat waves in the western US. J. Agric. Biol. Environ. Stat. **19**, 119–135 (2014)
82. Resnick, S.I.: Extreme Values, Regular Variation and Point Processes. Springer, New York (1987)
83. Resnick, S.I.: Heavy-Tail Phenomena: Probabilistic and Statistical Modelling. Springer, New York (2007)
84. Rootzén, H., Segers, J., Wadsworth, J.L.: Multivariate peaks over thresholds models. Extremes **21**, 115–145 (2018)
85. Rootzén, H., Segers, J., Wadsworth, J.L.: Multivariate generalized Pareto distributions: parametrizations, representations, and properties. J. Multivariate Anal. **165**, 117–131 (2018)
86. Rootzén, H., Tajvidi, N.: Multivariate generalized Pareto distributions. Bernoulli **12**, 917–930 (2006)
87. Schlather, M.: Models for stationary max-stable random fields. Extremes **5**, 33–44 (2002)
88. Schmidt, R., Stadtmüller, U.: Non-parametric estimation of tail dependence. Scand. J. Stat. **33**, 307–335 (2006)
89. Segers, J.: Non-parametric inference for bivariate extreme-value copulas. In: Ahsanullah, M., Kirmani, S.N.U.A. (eds.) Topics in Extreme Values, pp. 181–203. Nova Science, New York (2007)
90. Sharkey, P., Winter, H.C.: A Bayesian spatial hierarchical model for extreme precipitation in Great Britain. Environmetrics **30**, e2529 (2019)
91. Sibuya, M.: Bivariate extreme statistics. Ann. Inst. Stat. Math. **11**, 195–210 (1960)
92. Sklar, A.: Fonctions de répartition à n dimensions et leurs marges. Publ. Inst. Stat. U. Paris **8**, 229–231 (1959)
93. Smith, R.L.: Threshold methods for sample extremes. In: J. Tiago de Oliveira (ed.) Statistical Extremes and Applications, pp. 621–638. Reidel (1984)
94. Smith, R.L.: Max-stable processes and spatial extremes. Unpublished manuscript, U. Surrey (1990). www.stat.unc.edu/postscript/rs/spatex.pdf
95. Smith, R.L., Weissman, I.: Estimating the extremal index. J. R. Stat. Soc. **56**, 515–528 (1994)
96. Strokorb, K., Ballani, F., Schlather, M.: Tail-correlation functions of max-stable processes: construction principles, recovery and diversity of some mixing max-stable processes with identical TCF. Extremes **18**, 241–271 (2015)
97. Winter, H.C., Tawn, J.A.: Modelling heatwaves in central France: a case study in extremal dependence. J. R. Stat. Soc. C **65**, 345–365 (2016)

Some New Classes and Techniques in the Theory of Bernstein Functions

Safa Bridaa, Sonia Fourati, and Wissem Jedidi

Abstract In this paper we provide some new properties that are complementary to the book of Schilling-Song-Vondraček (Bernstein functions, 2nd edn. De Gruyter, Berlin, 2012).

Keywords Bernstein functions · Complete Bernstein functions · Thorin Bernstein functions · Subordinators · Infinite divisible distributions · Generalized Gamma convolutions · Positive stable density

1 A Unified View on Subclasses of Bernstein Functions

In the sequel all measures will be understood on the space $(0, \infty)$ and their densities, if they have one, are with respect to Lebesgue measure on $(0, \infty)$ which will be denoted by dx. We recall that the *Mellin convolution* (or multiplicative convolution)

S. Bridaa
Faculté des Sciences de Tunis, Département de Mathématiques, Laboratoire d'Analyse Mathématiques et Applications, Université de Tunis El Manar, Tunis, Tunisia

S. Fourati
LMI (Laboratoire de Mathématiques de l'INSA Rouen Normandie EA 3226 – FR CNRS 3335) and Laboratoire de Probabilités, Statistique et Modélisation (LPSM UMR 8001), Sorbonne Université, Paris, France
e-mail: sonia.fourati@upmc.fr

W. Jedidi (✉)
Department of Statistic and Operation Research, College of Sciences, King Saud University, Riyadh, Saudi Arabia

Faculté des Sciences de Tunis, Département de Mathématiques, Laboratoire d'Analyse Mathématiques et Applications, Université de Tunis El Manar, Tunis, Tunisia
e-mail: wjedidi@ksu.edu.sa; wissem.jedidi@fst.utm.tn

L. Chaumont, A. E. Kyprianou (eds.), *A Lifetime of Excursions Through Random Walks and Lévy Processes*, Progress in Probability 78,
https://doi.org/10.1007/978-3-030-83309-1_8

of two measures ν and τ on $(0, \infty)$ is defined by:

$$\nu \circledast \tau(A) = \int_{(0,\infty)^2} 1_A(xy)\nu(dx)\tau(dy), \qquad \text{if } A \text{ is a Borel set of } (0, \infty)\,.$$

If ν is absolutely continuous with density function h, then $\nu \circledast \tau$ is the function given by

$$\nu \circledast \tau(x) = h \circledast \tau(x) = \int_{(0,\infty)} h\left(\frac{x}{y}\right) \frac{\tau(dy)}{y}, \quad x > 0.$$

Another nice property of the Mellin convolution is that if a is a real number, then

$$x^a\,(\nu \circledast \tau) = (x^a \nu) \circledast (x^a \tau) \tag{1.1}$$

Notice that all the integrals above may be infinite if ν and/or τ are not finite measures. A function f defined on $(0, \infty)$ is called *completely monotone*, and we denote $f \in \mathcal{CM}$, if it is infinitely differentiable there satisfies

$$(-1)^n f^{(n)}(x) \geq 0, \quad \text{for all } n = 0, 1, 2, \cdots, \; x > 0. \tag{1.2}$$

Bernstein's theorem says that $f \in \mathcal{CM}$ if, and only if, it is the Laplace transform of some measure τ on $[0, \infty)$:

$$f(\lambda) = \int_{[0,\infty)} e^{-\lambda x}\, \tau(dx), \quad \lambda > 0.$$

Denote $\check{\tau}$ the image of $\tau_{|(0,\infty)}$ by the function $x \mapsto 1/x$ and notice that f has the representation

$$f(\lambda) = \tau(\{0\}) + \int_{(0,\infty)} e^{-\frac{\lambda}{x}}\, \check{\tau}(dx) = e^{-x} \circledast (x\check{\tau})(\lambda). \tag{1.3}$$

A function ϕ is called a **Bernstein function**, and we denote $\phi \in \mathcal{BF}$, if it has the representation

$$\phi(\lambda) = \mathrm{q} + \mathrm{d}\lambda + \int_{(0,\infty)} (1 - e^{-\lambda x})\pi(dx), \quad \lambda \geq 0\,, \tag{1.4}$$

where $\mathrm{q}, \ \mathrm{d} \geq 0$, the measure π, supported by $(0, \infty)$, satisfies

$$\int_{(0,\infty)} (x \wedge 1)\, \pi(dx) < \infty.$$

The usage is to call q the *killing* term and d the *drift* term. Any measure on $(0, \infty)$ that satisfies the preceding integrability condition is called a *Lévy measure*. As for completely monotone functions, notice that ϕ is represented by

$$\phi(\lambda) = \mathrm{q} + \mathrm{d}\lambda + (1 - e^{-x}) \circledast (x\check{\pi})(\lambda).$$

Bernstein functions are more likely called by probabilists *Laplace exponents of infinite divisible sub-probability distributions* or *Laplace exponents of* (possibly killed) *subordinators*, and the previous representation is their Lévy-Khintchine representation. See [6] for more account on subordinators and Lévy processes.

Next theorem illustrates to what extent the Mellin convolution is involved into the most popular subclasses of infinitely divisible distributions. Roughly speaking, we will see that each of these subclass **C** is associated to a Lévy measures π of the form $\pi = \mathbf{c} \circledast \nu$, where **c** is a specified function and ν is some Lévy measure.

Theorem 1.1 *Let π be a Lévy measure.*

(1) *The measure π has a non increasing density if, and only if, π is of the form*

$$\pi = 1_{(0,1]}(x)dx \circledast \nu,$$

where ν is some a Lévy measure;

(2) *The measure $x\pi(dx)$ has a non increasing density if, and only if, π is of the form*

$$\pi = 1_{(0,1]}(x)\frac{dx}{x} \circledast \nu,$$

where ν is a measure which integrates the function $g_0(x) = x\, 1_{(0,1]}(x) + \log x\, 1_{[1,\infty)}(x)$ (in particular ν is a Lévy measure);

(3) *The measure π has a density of the form $x^{a-1}\, k(x)$ with $a \in (-1, \infty)$ and k a completely monotonic function such that $\lim_{x\to+\infty} k(x) = 0$ if, and only if, π has the expression*

$$\pi = x^{a-1}\, e^{-x} dx \circledast \nu,$$

where ν is a measure which integrates the function g_a given by

$$g_a(x) := \begin{cases} x\, 1_{(0,1]}(x) + x^{-a}\, 1_{[1,\infty)}(x) & \text{if } a \in (-1, 0), \\ x\, 1_{(0,1]}(x) + \log x\, 1_{[1,\infty)}(x) & \text{if } a = 0, \\ x\, 1_{(0,1]}(x) + 1_{[1,\infty)}(x) & \text{if } a \in (0, \infty)\,. \end{cases} \tag{1.5}$$

Consequently, ν is a Lévy measure in all cases. Moreover ν may be an arbitrary Lévy measure in case (1) and in case (3) with $a > 0$.

Proof Notice that if μ has a density h, then $\mu \circledast \nu$ has a density, denoted by $h \circledast \nu$, and taking values in $[0, \infty]$:

$$h \circledast \nu(x) = \int_{(0,\infty)} \frac{1}{y} h\left(\frac{x}{y}\right) \nu(dy), \quad x > 0.$$

(1) Using the last expression for $h(x) = u_0(x) = 1_{(0,1]}(x)$, we have

$$u_0 \circledast \nu(x) = \int_x^\infty \frac{\nu(dy)}{y}.$$

Notice that any non-increasing function (taking values in $[0, \infty]$) is of the form $u_0 \circledast \nu$ and conversely. Since

$$\int_0^\infty (x \wedge 1)\,(u_0 \circledast \nu)(x)\,dx = \int_0^1 \frac{x^2}{2}\frac{\nu(dx)}{x} + \frac{1}{2}\int_1^\infty \frac{\nu(dy)}{y} + \int_1^\infty \frac{z-1}{z}\nu(dz)\,,$$

we deduce that the measure with density $\int_x^\infty \nu(dy)/y$ is a Lévy measure if, and only if, ν integrates the function $x \wedge 1$ or, in other words, ν is a Lévy measure.

(2) Using the expression of $h \circledast \nu$ with $h(x) = u_1(x) = 1_{(0,1]}(x)/x$, we have:

$$u_1 \circledast \nu(x) = \frac{\nu(x, \infty)}{x}, \quad x > 0.$$

Notice that any function π, valued in $[0, \infty]$, such that $x\pi(x)$ is non increasing is of the form $u_1 \circledast \nu$ and conversely. After that, note that

$$\int_0^\infty (x \wedge 1) u_1 \circledast \nu(x) dx = \int_0^1 x\,\nu(dx) + \nu(1, \infty) \int_1^\infty \log x\ \nu(dx)\,.$$

Thus, the measure with density $u_1 \circledast \nu(x)$ is a Lévy measure if, and only if, $\nu(dx)$ integrates $g_0(x) = x 1_{(0,1]}(x) + \log x\, 1_{(1,\infty)}(x)$.

(3) Without surprise, one is tempted to use the fact (1.1) together with representation (1.3) and write that for some measure τ

$$x^{a-1}k(x) = x^{a-1}(e^{-x} \circledast (x\check{\tau})) = (x^{a-1}e^{-x}) \circledast (x^a\check{\tau}),$$

where the transform $\check{\tau}$ of τ is given right before (1.3). We will do this in detail and provide the integrability conditions for the involved measures: let $a \in (-1, \infty)$, k be a completely monotone function such that $k(\infty) = 0$ and

$\pi(dx) = x^{a-1}\,k(x)\,1_{(0,\infty)}dx$. By Bernstein theorem, k is the Laplace transform of a measure on $(0,\infty)$, and may be written in the form

$$k(x) := \int_{(0,\infty)} e^{-xu}\; u^a \sigma(du), \quad x > 0\,. \tag{1.6}$$

Defining

$$h_a(u) := \int_0^\infty (x\wedge u)\,x^{a-1}\,e^{-x}dx, \quad u > 0,$$

and using Fubini's theorem, write

$$\int_{(0,\infty)} (x\wedge 1)\pi(dx) = \int_{(0,\infty)} (x\wedge 1)x^{a-1}\,k(x)dx = \int_{(0,\infty)} h_a(u)\frac{\sigma(du)}{u}\,.$$

We will now find the necessary and sufficient conditions on σ insuring that the last integral is finite. First, notice that $h_a(u) \nearrow \Gamma(a+1)$ when $u\to\infty$ and then h_a is bounded for any $a > -1$. Then, elementary computations give the following behavior of h_a in a neighborhood of 0,

$$\begin{aligned} \lim_{0+}\frac{h_a(u)}{u} &= \Gamma(a), \ \text{if } a > 0\,;\\ 0 < \liminf_{0+}\frac{h_a(u)-u}{u|\log u|} \le \limsup_{0+}\frac{h_a(u)-u}{u|\log u|} &< \infty, \ \text{if } a = 0\,;\\ \lim_{0+}\frac{h_a(u)}{u^{1-|a|}} = \frac{1}{|a|}+\frac{1}{1-|a|}, &\ \ \text{if } -1 < a < 0. \end{aligned}$$

and then π is a Lévy measure iff $\int_1^\infty \frac{\sigma(du)}{u}du < \infty$ and

$$\begin{aligned} \sigma([0,1]) &< \infty, \ \text{if } a > 0;\\ \int_{(0,1]} |\log u|\,\sigma(du) &< \infty, \ \text{if } a = 0;\\ \int_{(0,1]}\frac{\sigma(du)}{u^{|a|}} &< \infty, \ \text{if } -1 < a < 0. \end{aligned}$$

Notice that in each case $\sigma([0,1]) < \infty$ and then $\sigma(du)/u$ is a Lévy measure. Also notice that the measure ν, defined as the image of $\sigma(du)$ induced by the function $u\mapsto 1/u$, is also a Lévy measure, so that the integrability properties of the measure σ are equivalent to ν integrates the function g_a in (1.5). In order to conclude, write

$$x^{a-1}\,k(x) = x^{a-1}\int_{(0,\infty)} e^{-xu}\,u^a\,\sigma(du) = \int_{(0,\infty)}\left(\frac{x}{y}\right)^{a-1} e^{-\frac{x}{y}}\,\frac{\nu(dy)}{y} = (y^{a-1}e^{-y}\circledast\nu)\,(x)\,.$$

(4) After the above developments, the last assertion becomes obvious.

□

Remark 1.2 Below are some classes of Bernstein functions which can be defined via the correspondence between π and μ obtained in Theorem 1.1:

(i) The class $\mathcal{JB}$ of Bernstein function whose Lévy measure is of type (1) is often called the class *the Jurek class of Bernstein functions*. It is also characterized by those function

$$\phi \geq 0 \quad s.t. \quad \lambda \mapsto (x \mapsto x\phi(x))'(\lambda) \in \mathcal{BF}.$$

(ii) Bernstein functions whose Lévy measure is of type (2) is called *self-decomposable Bernstein functions*, and we denote $\mathcal{SDBF}$ their set. It is easy to check (see [9, Theorem 2.6 ch. VI], for instance), that

$$\phi \in \mathcal{SDBF} \quad \Longleftrightarrow \phi(0) \geq 0 \quad \text{and} \quad \lambda \mapsto \lambda\phi'(\lambda) \in \mathcal{BF}.$$

The class $\mathcal{SDBF}$ functions corresponds to self-decomposable distributions: namely, a r.v. X has a self-decomposable distribution if there exists a family of positive r.v. $(Y_c)_{0<c<1}$, each Y_c is independent from X such that the identity in distribution holds: $X \stackrel{d}{=} cX + Y_c$.

(iii) In [7, pp. 49], the class $\mathcal{CBF}$ of *complete Bernstein functions* corresponds to the Bernstein functions appearing in point (3) of Theorem 1.1 when the parameter a equals 1. In matrix analysis and operator theory, the name "operator monotone function" is more common for $\mathcal{CBF}$-functions. Another feature is that $\mathcal{CBF}$ is included into the class of $\mathcal{SBF}$ of special Bernstein functions, i.e. the class of Bernstein functions ϕ such that $\lambda \mapsto \lambda/\phi(\lambda) \in \mathcal{BF}$. The class $\mathcal{CBF}$ will be deeply investigated in next section.

(iv) The class $\mathcal{TBF}$ [7, pp. 73] of *Thorin Bernstein functions* corresponds to $a = 0$. The class $\mathcal{TBF}$ corresponds to the Laplace exponents of the generalized Gamma distributions, shortly GGC, introduced by Bondesson [1, 2] and the GGC subordinators studied by James, Roynette and Yor [5]. For more developments on $\mathcal{TBF}$, see [7].

2 Investigating the Class $\mathcal{CBF}_a$

We have seen that the well known Thorin class $\mathcal{TBF}$ corresponds to $\mathcal{CBF}_0$, and we will not go into further investigations in it. The simplest $\mathcal{CBF}$-function is given by $\lambda \mapsto \lambda/(\lambda + 1)$.

Point (3) of Theorem 1.1 suggests a generalization of the notion of $\mathcal{CBF}$ and $\mathcal{TBF}$ for any parameter $a > -1$ by introducing the set class $\mathcal{CBF}_a$ of Bernstein functions such that the corresponding Lévy measure π has a density of the form

$x^{a-1}k(x)$ such that k is a completely monotonic function and $k(\infty) = 0$. It is clear that $\mathcal{CBF}_a \subset \mathcal{CBF}_b$ for every $a \le b$, and that $\mathcal{TBF} \subset \mathcal{CBF}_a \cap \mathcal{SDBF} \subset \mathcal{CBF}$ for every $0 \le a \le 1$. The simplest functions in $\mathcal{CBF}_a$ are given when taking the complete monotonic functions k of the form $k(x) = e^{-cx}$, which is the Laplace transform of the Dirac measure at point $c > 0$. Then the associated Bernstein function is

$$\varphi_{a,b}(\lambda) = \int_0^\infty (1 - e^{-\lambda x})\, x^{a-1}\, e^{-bx}\, dx = \begin{cases} \Gamma(a)\left(\dfrac{1}{b^a} - \dfrac{1}{(b+\lambda)^a}\right) & \text{if } a \neq 0 \\ \log(1 + \frac{\lambda}{b}) & \text{if } a = 0\,. \end{cases} \tag{2.1}$$

Notice that for $a \in (-1, 0)$, these Bernstein functions are those associated to the so-called *tempered stable processes* of index $\alpha = -a$ and, for $a = 0$, it is associated to the normalized *Gamma process*. As stated in the next theorem, any $\mathcal{CBF}_a$ function is a conic combination of these simple ones. Next theorem is a straightforward consequence of Theorem 1.1:

Theorem 2.1 (Representation of $\mathcal{CBF}_a$-functions) *Let $a > -1$, $\phi : [0, \infty) \to [0, \infty)$, $\mathrm{q} = \phi(0)$ and $\mathrm{d} = \lim_{+\infty} \phi(x)/x < \infty$. Then ϕ belongs to $\mathcal{CBF}_a$ if, and only if, it $\lambda \mapsto \phi(\lambda) - \mathrm{q} - \mathrm{d}\lambda$ is the Mellin convolution of $\varphi_{a,1}$ defined in* (2.1) *with some measure. Namely,*

$$\phi(\lambda) = \begin{cases} \mathrm{q} + \mathrm{d}\lambda + \Gamma(a) \displaystyle\int_{(0,\infty)} \left(1 - \frac{u^a}{(u+\lambda)^a}\right) \sigma(du), & \text{if } a \neq 0 \\ \mathrm{q} + \mathrm{d}\lambda + \displaystyle\int_{(0,\infty)} \log\left(1 + \frac{\lambda}{u}\right) \sigma(du) & \text{if } a = 0\,, \end{cases} \tag{2.2}$$

where σ is a measure that integrates the function $g_a(1/t)$ given by (1.5). *In this case, the Lévy measure associated to ϕ has the density function*

$$x^{a-1} \int_{(0,\infty)} e^{-xt}\, t^a\, \sigma(dt), \qquad x > 0.$$

Example 2.2 The stable Bernstein function given by the power function $\lambda \mapsto \lambda^\alpha$, $\alpha \in (0, 1)$, is a trivial example of a function in $\mathcal{CBF}_\alpha$, because

$$\lambda^\alpha = \int_0^\infty (1 - e^{-\lambda x}) \frac{c_\alpha}{x^{\alpha+1}}\, dx, \quad \text{where } c_\alpha = \frac{\alpha}{\Gamma(1-\alpha)} \quad \text{and} \quad x \mapsto k(x) = x^{-2\alpha} \in \mathcal{CM},$$

In order to prove Proposition 2.4 below, we need some formalism and a Lemma. Let S_α, $\alpha \in (0,1)$, denotes a normalized positive stable random variable with density function f_α, i.e.

$$\mathbb{E}[e^{-\lambda S_\alpha}] = \int_0^\infty e^{-\lambda x} f_\alpha(x)dx = e^{-\lambda^\alpha}, \quad \lambda \geq 0, \tag{2.3}$$

and observe that for any $t > 0$,

$$e^{-t\lambda^\alpha} = t^{-1/\alpha} \int_0^\infty e^{-\lambda x} f_\alpha(xt^{-1/\alpha})dx \,. \tag{2.4}$$

Also, let γ_t denotes a normalized gamma distributed random variable with parameter $t > 0$, i.e.

$$\mathbb{E}[e^{-\lambda \gamma_t}] = \frac{1}{(1+\lambda)^t}, \quad \lambda \geq 0.$$

For any positive r.v. S satisfying $\mathbb{E}[S^s] < \infty$, $s \in \mathbb{R}$, we adopt the notation $S^{(s)}$ for a version of the size biased distribution of order s:

$$\mathbf{P}(S^{[s]} \in dx) = \frac{x^s}{\mathbb{E}[S^s]} \mathbf{P}(S \in dx). \tag{2.5}$$

Shanbhag and Sreehari [8] showed the remarkable identity in law

$$\gamma_t^{1/\alpha} \stackrel{d}{=} \frac{\gamma_{\alpha t}}{S_\alpha^{[-\alpha t]}}.$$

from which we can extract from, when taking two independent and identically distributed random variables S_α and S'_α, that

$$\gamma_1^{1/\alpha} S_\alpha \stackrel{d}{=} \gamma_1 X_\alpha, \quad \text{where} \quad X_\alpha = \frac{S_\alpha}{S'_\alpha} \stackrel{d}{=} \frac{1}{X_\alpha} \tag{2.6}$$

$$\stackrel{d}{=} \gamma_\alpha Y_\alpha, \quad \text{where} \quad Y_\alpha = \frac{S_\alpha}{(S'_\alpha)^{[-\alpha]}} \stackrel{d}{=} \frac{1}{Y_\alpha^{[-\alpha]}} \tag{2.7}$$

$$\gamma_{1/\alpha}^{1/\alpha} S_\alpha \stackrel{d}{=} \gamma_1 Z_\alpha, \quad \text{where} \quad Z_\alpha = \frac{S_\alpha}{(S'_\alpha)^{[-1]}} \stackrel{d}{=} \frac{1}{Z_\alpha^{[-1]}}, \tag{2.8}$$

where, in each product, we have used the notation of (2.5) and the r.v.'s involved in the identities in law are assumed to be independent. Last identities are used in the following lemma:

Lemma 2.3 *Let $\alpha \in (0,1)$. With the above notations, we have*

(1) *The function* $\phi_\alpha(\lambda) := \dfrac{\lambda^\alpha}{\lambda^\alpha + 1}$ *belongs to* $\mathcal{CBF}_\alpha$ *and is represented by*

$$\phi_\alpha(\lambda) = \mathbb{E}\left[\frac{\lambda X_\alpha}{\lambda + X_\alpha}\right] = \mathbb{E}\left[\left(\frac{\lambda}{1+\lambda\, Y_\alpha}\right)^\alpha\right] = 1 - \mathbb{E}\left[\frac{1}{(1+\lambda Y_\alpha)^\alpha}\right], \quad \lambda \geq 0; \tag{2.9}$$

(2) *The function* $\varphi_\alpha(\lambda) := 1 - \dfrac{1}{(\lambda^\alpha + 1)^{1/\alpha}}$ *belongs to* $\mathcal{CBF}$ *and is represented by*

$$\varphi_\alpha(\lambda) = \mathbb{E}[\frac{\lambda Z_\alpha}{1+\lambda Z_\alpha}], \quad \lambda \geq 0. \tag{2.10}$$

Proof

(1) Since

$$\frac{1}{1+\lambda^\alpha} = \mathbb{E}\left[e^{-\lambda^\alpha \gamma_1}\right] = \mathbb{E}\left[e^{-\lambda \gamma_1^{1/\alpha} S_\alpha}\right] = \mathbb{E}\left[e^{-\lambda \gamma_1 X_\alpha}\right] = \mathbb{E}\left[\frac{1}{1+\lambda X_\alpha}\right]. \tag{2.11}$$

The first equality in (2.9) comes from

$$\phi_\alpha(\lambda) = 1 - \frac{1}{1+\lambda^\alpha} = 1 - \mathbb{E}\left[\frac{1}{1+\lambda X_\alpha}\right] = \mathbb{E}\left[\frac{\lambda X_\alpha}{1+\lambda X_\alpha}\right].$$

Going back to (2.11) and using again (2.6), we obtain the second and third representations in (2.9) by writing

$$\phi_\alpha(\lambda) = \lambda^\alpha\, \mathbb{E}[e^{-\lambda\, \gamma_1^{1/\alpha}\, S_\alpha}] = \lambda^\alpha\, \mathbb{E}[e^{-\lambda\, \gamma_\alpha Y_\alpha}] = \mathbb{E}\left[\left(\frac{\lambda}{1+\lambda\, Y_\alpha}\right)^\alpha\right],$$

and also

$$\phi_\alpha(\lambda) = 1 - \frac{\phi_\alpha(\lambda)}{\lambda^\alpha} = 1 - \mathbb{E}\left[\frac{1}{(1+\lambda\, Y_\alpha)^\alpha}\right].$$

Since the third representation of ϕ_α meets the one of Theorem 2.1, we deduce that $\phi_\alpha \in \mathcal{CBF}_\alpha$.

(2) Similarly, write

$$\varphi_\alpha(\lambda) = 1 - \mathbb{E}\left[e^{-\lambda\, \gamma_{1/\alpha}^{1/\alpha}\, S_\alpha}\right] = 1 - \mathbb{E}\left[e^{-\lambda\, \gamma_1\, Z_\alpha}\right] = \mathbb{E}\left[\frac{\lambda\, Z_\alpha}{1+\lambda\, Z_\alpha}\right],$$

and deduce that $\varphi_\alpha \in \mathcal{CBF}$.

□

We are now able to exhibit additional links between $\mathcal{CBF}_a$, $a > 0$, and $\mathcal{CBF}$:

Proposition 2.4 *The following implications are true:*

(1) *If* $0 < a \le 1$ *and* $\phi \in \mathcal{CBF}$, *then* $\lambda \mapsto \phi(\lambda^a) \in \mathcal{CBF}_a$ *and* $\phi(\lambda^a)^{1/a} \in \mathcal{CBF}$.
(2) *If* $a \ge 1$ *and* $\varphi \in \mathcal{CBF}_a$, *then* $\lambda \mapsto \varphi(\lambda^{1/a}) \in \mathcal{CBF}$.

Remark 2.5 The first assertion of Proposition 2.4 is a refinement of [7, Corollary 7.15]:

$$a \ge 1 \quad \text{and} \quad \phi(\lambda^a)^{1/a} \in \mathcal{CBF} \Longrightarrow \phi \in \mathcal{CBF}.$$

The latter could be also obtained by a Pick-Nevanlinna argument as in Remark 3.6 below.

Proof of Proposition 2.4 The second assertion in (1) can be found in [7, Corollary 7.15]. In Example 2.2, we have seen that $\lambda \mapsto \lambda^\alpha \in \mathcal{CBF}_\alpha$ for every $0 < \alpha \le 1$, so, we may suppose that ϕ has no killing nor drift term. The assertions are a conic combination argument together with the result of Lemma 2.3. For the first assertion of (1), use the function $\phi_a \in \mathcal{CBF}_a$ given by (2.9), for the assertion (2), use the function $\varphi_{1/a} \in \mathcal{CBF}$ given by (2.10), and get the representations

$$\phi(\lambda^a) = \int_0^\infty \phi_a\left(\frac{\lambda}{u}\right) \nu(du) \quad \text{and} \quad \varphi(\lambda^a) = \int_0^\infty \varphi_{1/a}\left(\frac{\lambda}{u}\right) \mu(du), \quad \lambda \ge 0,$$

where ν and μ are some measure. □

3 A New Injective Mapping from $\mathcal{BF}$ onto $\mathcal{CBF}$

We recall that a $\mathcal{CBF}$ function is a Bernstein function whose Lévy measure has a density which is a completely monotonic function. We recall the connection between $\mathcal{CBF}$-functions; ϕ is a $\mathcal{CBF}$-function if, and only if, it admits the representation:

$$\phi(\lambda) = \mathrm{q} + \mathrm{d} + \int_{(0,\infty)} \frac{\lambda}{\lambda + x}\, \nu(dx), \quad \lambda \ge 0, \tag{3.1}$$

where $\mathrm{q}, \mathrm{d} \ge 0$ and ν is a measure which integrates $1 \wedge 1/x$.

Another characterization of $\mathcal{CBF}$ functions is given by the *Pick-Nevanlinna characterization of* $\mathcal{CBF}$*-functions*:

Theorem 3.1 (Theorem 6.2 [7]) *Let ϕ a non-negative continuous function on $[0, \infty)$ is a $\mathcal{CBF}$ function if, and only if, it has an analytic continuation on $\mathbb{C}(-\infty, 0]$ such that*

$$\Im\big(\phi(z)\big) \geq 0, \quad \textit{for all } z \textit{ s.t. } \Im(z) > 0.$$

Notice that any $\phi \in \mathcal{BF}$ has an analytic continuation on the half plane $\{z,\ \Re(z) > 0\}$ which can be extended by continuity to the closed half plane $\{z,\ \Re(z) \geq 0\}$ and we still denote by ϕ this continuous extension. In next theorem we state a representation similar to (3.1) and valid for any Bernstein function ϕ. Part (1) of this theorem is also quoted as [7, Proposition 3.6].[1]

Theorem 3.2

(1) *Let $\phi \in \mathcal{BF}$ represented by (1.4), then, for all $\lambda \geq 0$,*

$$\phi(\lambda) = \mathrm{d}\lambda + \int_0^\infty \frac{\lambda}{\lambda^2 + u^2}\, v(u)\, du\,, \tag{3.2}$$

where v, given by $v(u) := 2\Re\big(\phi(iu)\big)/\pi$, is a negative definite function (in the sense of [7, Definition 4.3]) satisfying the integrability condition

$$\int_1^\infty \frac{v(u)}{u^2} < \infty. \tag{3.3}$$

(2) *Conversely, let $\mathrm{d} \geq 0$ and $v : \mathbb{R} \to \mathbb{R}_+$ be a negative definite function satisfying (3.3), then*

$$\lambda \mapsto \mathrm{d}\lambda + \int_0^\infty \frac{\lambda}{\lambda^2 + u^2}\, v(u)\, du \in \mathcal{BF}\,.$$

Proof

(1) We suppose without loss of generality that $\mathrm{q} = \mathrm{d} = 0$. In this proof, we denote by $(C_t)_{t\geq 0}$ a standard Cauchy process, i.e. a Lévy process such that $\mathbb{E}[e^{iuC_t}] = e^{-t|u|}$, $u \in \mathbb{R}$. Since

$$\phi(ix) = \int_{(0,\infty)} (1 - e^{-ixs})\pi(ds), \quad x \in \mathbb{R},$$

[1]The results in Theorem 3.2 (i), Corollaries 3.4, 3.5 and Proposition 3.8 below can be found, with different proofs, in [7, Proposition 3.6, Proposition 7.22]. As is stated in [7, pp. 34 & 108 and reference entries 119, 120], *the statements of these results are due to S. Fourati and W. Jedidi and were, with a different proof, communicated by S. Fourati and W. Jedidi* in 2010.

then, for all $\lambda > 0$, we can write

$$\begin{aligned}
\phi(\lambda) &= \int_{(0,\infty)} (1 - e^{-\lambda s})\pi(ds) = \int_{(0,\infty)} \mathbb{E}[1 - e^{-isC_\lambda}]\pi(ds) \\
&= \int_{(0,\infty)} \left(\int_{\mathbb{R}} (1 - e^{-ius}) \frac{\lambda}{\pi(u^2 + \lambda^2)} du \right) \pi(ds) = \int_{\mathbb{R}} \frac{\lambda}{\pi(u^2 + \lambda^2)} \phi(iu)\, du \\
&= \int_0^\infty \frac{\lambda}{\pi(u^2 + \lambda^2)} (\phi(iu) + \phi(-iu))du = \frac{2}{\pi} \int_0^\infty \frac{\lambda}{(u^2 + \lambda^2)} \Re\big(\phi(iu)\big)\, du.
\end{aligned}$$

Notice that $v(u) := 2\Re\big(\phi(iu)\big)/\pi$ is an even function on $\mathbb{R}$, is a $[0, \infty)$-valued negative definite function in the sense of [7, Definition 4.3], and representation (3.2) proves that it necessarily satisfies (3.3).

(2) By [4, Corollary 1.1.6], every $[0, \infty)$-valued, negative definite function v, has necessarily the form

$$v(u) = \mathrm{q} + \mathrm{c}u^2 + \int_{\mathbb{R}\{0\}} (1 - \cos ux)\, \mu(dx),$$

where q, c ≥ 0 and the Lévy measure μ is symmetric and integrates $x^2 \wedge 1$. We deduce that v is an even function and necessarily c $= 0$ because of the integrability condition (3.3). So, v is actually represented by

$$v(u) = \mathrm{q} + 2 \int_{(0,\infty)} (1 - \cos ux)\, \mu(dx).$$

Then, observe that

$$\int_1^\infty \frac{v(u)}{u^2} du = \mathrm{q} + 2 \int_{(0,\infty)} \theta(x)\, \mu(dx) < \infty$$

where

$$\theta(x) = \int_1^\infty \frac{1 - \cos(xt)}{t^2} dt = x \int_x^\infty \frac{1 - \cos t}{t^2} dt \leq 2.$$

Since $\lim_{x \to 0} \theta(x)/x = \pi/2$, deduce that μ necessarily integrates $x \wedge 1$. Finally, v is the real part of some Bernstein function ϕ and conclude with part (1) of this theorem.

□

Remark 3.3

(i) Note that condition (3.3) on the negative definite function v was obtained as an immediate consequence of representation (3.2) and is equivalent, in our context, to the usual integrability condition (on the Lévy measure at 0) for a Lévy process to have finite variation paths, see the book of Breiman [3, Exercise 13 p. 316]. In Vigon's thesis, [10, Proposition 1.5.3] one can also find a nice proof of condition (3.3) based on a Fourier single-integral formula.

(ii) In (3.2), it is not clear that the constant functions belong to $\mathcal{CBF}^-$. They actually do, since for all $\mathrm{q} \ge 0$ and $\lambda > 0$,

$$\mathrm{q} = \frac{2}{\pi} \int_0^\infty \frac{\lambda\, \mathrm{q}}{\lambda^2 + u^2}\, du,$$

then $\lambda \mapsto \phi(\lambda) = \mathrm{q} \in \mathcal{CBF}^-$.

Now, it appears natural to introduce the class of functions $\mathcal{CBF}^-$ associated to negative definite functions :

$$\mathcal{CBF}^- := \left\{ \lambda \mapsto \varphi(\lambda) = \mathrm{q} + \mathrm{d}\lambda + \int_0^\infty \frac{\lambda}{\lambda + u^2}\, v(u)\, du \right\},$$

where $\mathrm{q},\ \mathrm{d} \ge 0$ and $v : [0,\infty) \to [0,\infty)$ is a negative definite function, necessarily satisfying the integrability condition (3.3). It is obvious that $\mathcal{CBF}^-$ is a (strict) subclass of $\mathcal{CBF}$.

A reformulation of last theorem gives the following two corollaries quoted as [7, Proposition 7.22 and Propositon 3.6] respectively. The reader is also addressed to the footnote before Theorem 3.2.

Corollary 3.4 (Classes $\mathcal{BF}$ and $\mathcal{CBF}^-$ are one-to-one)

(1) *If ϕ is in $\mathcal{BF}$, then $\lambda \mapsto \sqrt{\lambda}\phi(\sqrt{\lambda})$ is in $\mathcal{CBF}^-$;*
(2) *Conversely, any function in $\mathcal{CBF}^-$ is of the form $\lambda \mapsto \sqrt{\lambda}\phi(\sqrt{\lambda})$, where ϕ is in $\mathcal{BF}$.*

Corollary 3.5 *Any Bernstein function leaves globally invariant the cônes*

$$\left\{\rho e^{i\pi\theta};\quad \rho \ge 0,\, \alpha \in [-\sigma,\sigma]\right\}, \qquad \textit{for any} \quad \sigma \in [0,\frac{1}{2}].$$

Proof The function $\psi_u(\lambda) = \lambda/(\lambda^2 + u^2),\ u > 0$, maps the half-line $\left\{\rho e^{i\pi\sigma};\ \rho \ge 0\right\}$ onto in the cône $\left\{\rho e^{i\pi\theta};\ \rho \ge 0,\ \theta \in [-\sigma,\sigma]\right\}$. Since this cone is convex and closed by any conic combination of ψ_u, deduce that the integral

$$\frac{2}{\pi} \int_0^\infty \frac{\lambda}{\lambda^2 + u^2}\, \Re\big(\phi(iu)\big)\, du$$

is in the same cône. □

Remark 3.6 The property in the last corollary has to be compared with the much stronger property fulfilled by a $\mathcal{CBF}$-function: any $\mathcal{CBF}$-function has an analytic continuation on $\mathbf{C} \setminus (-\infty, 0]$ and this continuation leaves globally invariant the cônes

$$\left\{\rho e^{i\pi\theta};\ \ \rho \geq 0,\ \theta \in [0, \sigma]\right\}, \tag{3.4}$$

for any $\sigma \in [0, 1)$. Moreover this property fully characterizes the class of $\mathcal{CBF}$-functions: the Pick-Nevanlinna characterization given in Theorem 3.2 is equivalent to invariance of the cone (3.4) for $\sigma = 1$. This property is not satisfied by all Bernstein functions. For instance, $\phi(\lambda) = 1 - e^{-\lambda} \in \mathcal{BF} \setminus \mathcal{CBF}$, because $\Im\left(5e^{i\pi/4}\right) > 0$ but $\Im\left(\phi(5e^{i\pi/4})\right) < 0$.

In the following results, we give some extension of Theorem 3.4 by replacing the function $\lambda \mapsto \sqrt{\lambda}$ by other functions:

Corollary 3.7 *Let $\phi \in \mathcal{BF}$. Then,*

(1) *For any function ψ such that $\lambda \mapsto \psi(\lambda^2)$ is in $\mathcal{BF}$, we have $\psi^2 \in \mathcal{BF}$ and $\phi(\psi) \in \mathcal{CBF}^-$.*
(2) *For any function ψ such that ψ^2 is in $\mathcal{CBF}$, the following functions are in $\mathcal{CBF}$:*

$$\phi(\psi) \cdot \psi, \quad \frac{\psi}{\phi(\psi)}, \quad \phi(1/\psi) \cdot \psi, \quad \frac{\psi}{\phi(1/\psi)}$$

and also

$$\lambda \cdot \frac{\phi(\psi)}{\psi}, \quad \frac{\lambda}{\phi(\psi) \cdot \psi}, \quad \lambda \cdot \frac{\phi(1/\psi)}{\psi}, \quad \frac{\lambda}{\phi(1/\psi) \cdot \psi}.$$

Proof

(1) Since $\psi_1(\lambda) := \psi(\lambda^2) \in \mathcal{BF}$, then $\psi_2(\lambda) := \psi^2(\lambda) = \psi_1(\sqrt{\lambda})^2 \in \mathcal{BF}$. To get the last claim, just check the complete monotonicity of the derivative of ψ_2. The second assertion is seen by stability by composition of the class $\mathcal{BF}$: since $\lambda \mapsto \phi(\psi_1(\lambda)) = \phi(\psi(\lambda^2)) \in \mathcal{BF}$, then Corollary 3.4 applies on the last function.
(2) Recall $\mathcal{S}$ is the class of Stieltjes functions, i.e. the class of functions obtained by a double Laplace transform (see [7]) and observe that

$$\varphi \in \mathcal{CBF} \Longleftrightarrow \lambda \mapsto \frac{\varphi(\lambda)}{\lambda} \in \mathcal{S}.$$

As consequence of [7, (7.1), (7.2), (7.3) pp. 96], obtain that

$$\sqrt{\lambda}\phi(\sqrt{\lambda}) \in \mathcal{CBF} \Longleftrightarrow \frac{\sqrt{\lambda}}{\phi(\sqrt{\lambda})} \in \mathcal{CBF} \Longleftrightarrow \sqrt{\lambda}\phi(1/\sqrt{\lambda}) \in \mathcal{CBF} \Longleftrightarrow \frac{\sqrt{\lambda}}{\phi(1/\sqrt{\lambda})} \in \mathcal{CBF} \tag{3.5}$$

$$\Updownarrow$$

$$\frac{\phi(\sqrt{\lambda})}{\sqrt{\lambda}} \in \mathcal{S} \Longleftrightarrow \frac{1}{\sqrt{\lambda}\phi(\sqrt{\lambda})} \in \mathcal{S} \Longleftrightarrow \frac{\phi(1/\sqrt{\lambda})}{\sqrt{\lambda}} \in \mathcal{S} \Longleftrightarrow \frac{1}{\sqrt{\lambda}\phi(1/\sqrt{\lambda})} \in \mathcal{S}. \tag{3.6}$$

To get the first claim, compose the four $\mathcal{CBF}$-functions in (3.5) with $\psi^2 \in \mathcal{CBF}$, and use the stability by composition of the class $\mathcal{CBF}$ [7, Corollary 7.9.]. To obtain the last claim, also compose the four $\mathcal{S}$-functions in (3.6) with $\psi^2 \in \mathcal{CBF}$, use [7, Corollary 7.9], to get that the compositions stays in $\mathcal{S}$, and finally multiply by λ to get the announced $\mathcal{CBF}$-function in the Corollary.

□

Notice that if ψ belongs to $\mathcal{BF}$, then $\psi(\sqrt{\lambda})$ satisfies property (1). If further ψ belongs to $\mathcal{CBF}$ then $\psi(\sqrt{\lambda})$ and $\sqrt{\psi(\lambda)}$ both satisfy property (2).

Now, we summarize the properties that can be stated when composing a Bernstein function ϕ with the stable Bernstein function of Example 2.2.

Proposition 3.8 *Let $\alpha \in (0, 1]$, $\phi \in \mathcal{BF}$, π be the Lévy measure of ϕ and $\overline{\pi}$ the right tail of π: $\overline{\pi}(x) := \pi(x, \infty)$, $x > 0$. Then,*

(1) $\lambda \mapsto \phi_\alpha(\lambda) := \lambda^{1-\alpha}\phi(\lambda^\alpha) \in \mathcal{BF}$. *Further, $\phi_\alpha \in \mathcal{CBF}$ whenever*

$$x \mapsto \overline{\pi}_\alpha(x) := \alpha\, x^{\alpha-1}\,\overline{\pi}(x^\alpha) \in \mathcal{CM} \quad \textit{(which is true if } \phi \in \mathcal{CBF});$$

(2) $\lambda \mapsto \lambda^\gamma\phi(\lambda^\alpha) \in \mathcal{CBF}$ *(resp. $\mathcal{CBF}^-$) if $\alpha \le \frac{1}{2}$ and $\gamma \in (\alpha, 1-\alpha]$ (resp. $\gamma = \frac{1}{2}$).*

Proof Recall that f_α, the density function of normalized positive stable r.v., is given by (2.3).

(1) Since $\lambda \mapsto \lambda^\alpha$, $\lambda^{1-\alpha}$ are both in $\mathcal{CBF}$, there is no loss of generality to take q = d = 0 in the Lévy-Khintchine representation (1.4) of ϕ, and then to write

$$\lambda^{1-\alpha}\phi(\lambda^\alpha) = \lambda \int_0^\infty e^{-\lambda^\alpha t}\overline{\pi}(t)dt.$$

It is sufficient to prove that $\lambda \mapsto \int_0^\infty e^{-\lambda^\alpha t}\overline{\pi}(t)dt$ is the Laplace transform of a non increasing function. For that, use (2.4) and Fubini's theorem and get

$$\int_0^\infty e^{-\lambda^\alpha t}\overline{\pi}(t)dt = \int_0^\infty e^{-\lambda x}\overline{\Pi}_\alpha(x)dx,$$

where

$$\overline{\Pi}_\alpha(x) := \int_0^\infty f_\alpha\left(\frac{x}{t^{1/\alpha}}\right)\overline{\pi}(t)\,\frac{dt}{t^{1/\alpha}} = \int_0^\infty \overline{\pi}_\alpha\left(\frac{x}{z}\right) f_\alpha(z)\,\frac{dz}{z},$$

and the second representation by the change of variables $z = xt^{-\frac{1}{\alpha}}$. Since for each $z > 0$, the functions $z \mapsto \overline{\pi}_\alpha(x/z)$ are non-increasing (respectively completely monotone), deduce the same for $\overline{\Pi}_\alpha$.

(2) Observe that for $\alpha \le 1/2$, the function $\lambda \mapsto \lambda^\alpha$ satisfies the properties of Corollary 3.7: point (2) yields that the function $\lambda \mapsto \lambda^{1-\alpha}\phi(\lambda^\alpha)$ is in $\mathcal{CBF}$, and point (2) yields that $\sqrt{\lambda}\phi(\lambda^\alpha)$ is in $\mathcal{CBF}^-$. Taking representation of ϕ in Theorem 3.2, we obtain

$$\lambda^\gamma \phi(\lambda^\alpha) = \int_0^\infty \frac{\lambda^{\gamma+\alpha}}{\lambda^{2\alpha}+u^2} v(u)du, \quad \text{where } v(u) = \frac{2}{\pi}\Re\big(\phi(iu)\big).$$

Since $0 < 2\alpha \le \gamma + \alpha \le 1$, the function $\lambda \mapsto \lambda^{\gamma+\alpha}/(\lambda^{2\alpha}+u^2)$ leaves the half plane $\{\Im(\lambda) > 0\}$ globally invariant, and then, is a $\mathcal{CBF}$-function for every $u > 0$. By the argument of conic combination, this property remains true for the function $\lambda \mapsto \lambda^\gamma\phi(\lambda^\alpha)$ is $\mathcal{CBF}$. Now, the function $\psi(\lambda) = \lambda^\alpha$ satisfies property (2) of Corollary 3.7, and then, $\sqrt{\lambda}\phi(\lambda^\alpha) \in \mathcal{CBF}^-$.

□

Acknowledgments The authors would like to express their sincere gratitude to the referees for their valuable comments and acute clarifications, that helped to improve the quality of this manuscript.

References

1. Bondesson, L.: Generalized gamma convolutions and complete monotonicity. Probab. Theory Relat. Fields **85**, 181–194 (1990)
2. Bondesson, L.: Generalized Gamma Convolutions and Related Classes of Distributions and Densities. Lecture Notes in Statistics, vol. 76. Springer, New York (1992)
3. Breiman., L. Pobability, Classics in Applied Mathematics, vol. 7. Society for Industrial and Applied Mathematics (SIAM), Philadelphia (1992)
4. Farkas, W., Jacob, N., Schilling, R.L.: Function spaces related to continuous negative definite functions: Ψ-Bessel potential spaces. Dissertationes Mathematicae **393**, 1–62 (2001)
5. James, L.F., Roynette, B., Yor, M.: Generalized gamma convolutions, Dirichlet means, Thorin measures, with explicit examples. Probab. Surv **5**, 346–415 (2008)
6. Kyprianou, A.E.: Fluctuations of Lévy Processes with Applications. Introductory Lectures, 2nd edn. Springer, Berlin/Heidelberg (2006)
7. Schilling, R.L., Song, R., Vondraček, Z.: Bernstein Functions, 2nd edn. De Gruyter, Berlin (2012)
8. Shanbhag, D.N., Sreehari, M.: An extension of Goldie's result and further results in infinite divisibility. Z. Wahrscheinlichkeitstheorie verw. Gebiete **47**, 19–25 (1979)

9. Steutel, F.W., van Harn, K.: Infinite Divisibility of Probability Distributions on the Real Line. Marcel Dekker, New York (2004)
10. Vigon, V.: Lévy processes and Wiener-hopf factorization. PhD thesis, INSA Rouen (2002). https://tel.archives-ouvertes.fr/tel-00567466/document

A Transformation for Spectrally Negative Lévy Processes and Applications

Marie Chazal, Andreas E. Kyprianou, and Pierre Patie

Abstract The aim of this work is to extend and study a family of transformations between Laplace exponents of Lévy processes which have been introduced recently in a variety of different contexts, Patie (Bull Sci Math 133(4):355–382, 2009; Bernoulli 17(2):814–826, 2011), Kyprianou and Patie (Ann Inst H Poincar' Probab Statist 47(3):917–928, 2011), Gnedin (Regeneration in Random Combinatorial Structures. arXiv:0901.4444v1 [math.PR]), Patie and Savov (Electron J Probab 17(38):1–22, 2012), as well as in older work of Urbanik (Probab Math Statist 15:493–513, 1995). We show how some specific instances of this mapping prove to be useful for a variety of applications.

Keywords Spectrally negative Lévy process · Fluctuation theory · Exponential functional · Positive self-similar Markov process · Intertwining · Hypergeometric function

1 Introduction

In this paper we are interested in a Lévy process with no positive jumps, possibly independently killed at a constant rate, henceforth denoted by $\xi = (\xi_t, t \geq 0)$ with law $\mathbb{P}$. That is to say, under $\mathbb{P}$, ξ is a stochastic process which has almost

The authors would like to thank an anonymous referee for his constructive comments.

M. Chazal · P. Patie
School of Operations Research and Information Engineering, Cornell University, Ithaca, NY, USA
e-mail: mc69@cornell.edu; pp396@cornell.edu

A. Kyprianou (✉)
Department of Mathematical Sciences, University of Bath, Bath, UK
e-mail: a.kyprianou@bath.ac.uk

L. Chaumont, A. E. Kyprianou (eds.), *A Lifetime of Excursions Through Random Walks and Lévy Processes*, Progress in Probability 78,
https://doi.org/10.1007/978-3-030-83309-1_9

surely càdlàg paths, increments that are stationary and independent and killed at an independent rate $\kappa \geq 0$. The case that $\kappa = 0$ corresponds to no killing. Whilst it is normal to characterise Lévy processes by their characteristic exponent, in the case that the jumps are non-positive one may also comfortably work with the Laplace exponent which satisfies,

$$\mathbb{E}(e^{u\xi_t}) = e^{\psi(u)t}, \ t \geq 0,$$

where $u \geq 0$. It is a well established fact that the latter Laplace exponent is strictly convex on $[0, \infty)$ and admits the following Lévy-Khintchine representation

$$\psi(u) = -\kappa + au + \frac{1}{2}\sigma^2 u^2 + \int_{(-\infty,0)} (\mathrm{e}^{ux} - 1 - ux\mathbf{1}_{(|x|<1)})\Pi(\mathrm{d}x), \tag{1.1}$$

for $u \geq 0$ where $\kappa \geq 0$, $a \in \mathbb{R}$, $\sigma^2 \geq 0$ and Π is a measure concentrated on $(-\infty, 0)$ satisfying $\int_{(-\infty,0)}(1 \wedge x^2)\Pi(\mathrm{d}x) < \infty$, see for example Bertoin [1]. Note in particular that our definition includes the case that $-\xi$ is a (possibly killed) subordinator. Indeed, when Π satisfies $\int_{(-\infty,0)}(1 \wedge |x|)\Pi(\mathrm{d}x) < \infty$ and we choose $\sigma = 0$ and $a = -\mathrm{d} + \int_{(-1,0)} x\Pi(\mathrm{d}x)$ we may write for $u \geq 0$,

$$\psi(u) = -\kappa - \mathrm{d}u - \int_{(0,\infty)} (1 - \mathrm{e}^{-ux})\nu(\mathrm{d}x),$$

where $\nu(x, \infty) = \Pi(-\infty, -x)$. When $\mathrm{d} \geq 0$, writing $S_t = -\xi_t$ for $t \geq 0$, d and ν should be thought of as the drift and Lévy measure of the subordinator $S = (S_t, t \geq 0)$ respectively. Moreover, writing $\phi(u) = -\psi(u)$ for $u \geq 0$, we may think of ϕ as the Laplace exponent of S in the classical sense, namely

$$\mathbb{E}(e^{-uS_t}) = e^{-\phi(u)t}, \ t \geq 0.$$

In general we shall refer to Ψ_ℓ as the family of Laplace exponents of (possibly killed) Lévy processes with no positive jumps which are killed at rate $\kappa \geq 0$ and are well defined on (ℓ, ∞) for $\ell \leq 0$. Note that excluding the cases when $-\xi$ is not a subordinator then Ψ_ℓ boils down to the class of Laplace exponent of (possibly killed) spectrally negative Lévy processes.

Our main objective is to introduce a parametric family of linear transformations which serves as a mapping from the space of Laplace exponents of Lévy processes with no positive jumps into itself and therewith explore how a family of existing results for Lévy processes may be extrapolated further. The paper is structured as follows. In the next section we introduce our three-parameters transformation and derive some basic properties. We also describe its connection with some transformations which have already appeared in the literature. The remaining part of the paper deals with the applications of our transformation to different important issues arising in the framework of Lévy processes and related processes. More specifically, in the third section we provide some ways for getting new expressions for the so-called scale function of Lévy processes. Section 4 is devoted to the exponential functional of Lévy processes and finally in the last section we develop some applications to the study of positive self-similar Markov processes.

2 The Transformation $\mathcal{T}_{\delta,\beta}$

We begin with the definition of our new transformation and consider its properties as a mapping on Ψ_ℓ.

Definition 2.1 Suppose that $\psi \in \Psi_\ell$ where $\ell \leq 0$. Then for $\delta, \beta \geq 0$, with the additional constraint that $\psi'(0+) = \psi(0) = 0$ if $\beta = 0$, let

$$\mathcal{T}_{\delta,\beta}\psi(u) = \frac{u+\beta-\delta}{u+\beta}\psi(u+\beta) - \frac{\beta-\delta}{\beta}\psi(\beta), \quad u \geq -\beta.$$

Let us make some immediate observations on the above definition. Firstly note that $\mathcal{T}_{\delta,\beta}$ is a linear transform. In the special case that $\delta = \beta$ we shall write $\mathcal{T}_\beta$ in place of $\mathcal{T}_{\beta,\beta}$. The transform $\mathcal{T}_\beta$ was considered recently for general spectrally negative Lévy processes in Kyprianou and Patie [23] and for subordinators (as a result of a path transformation known as sliced splitting) in Gnedin [17]. Next note that, for β, γ such that $\beta + \gamma \geq 0$,

$$\mathcal{T}_\gamma \circ \mathcal{T}_\beta = \mathcal{T}_{\gamma+\beta}.$$

In the special case that $\delta = 0$ and $\beta \geq 0$ we have $\mathcal{E}_\beta := \mathcal{T}_{0,\beta}$ satisfies

$$\mathcal{E}_\beta\psi(u) = \psi(u+\beta) - \psi(\beta),\ u \geq -\beta,$$

where, as usual, $\psi \in \Psi_\ell$. This is the classical Esscher transform for Lévy processes with no positive jumps expressed in terms of Laplace exponents. It will be convenient to note for later that if $\Phi(u) := \psi(u)/u$ then we may write

$$\mathcal{T}_{\delta,\beta}\psi(u) = \mathcal{E}_\beta\psi(u) - \delta\mathcal{E}_\beta\Phi(u).$$

In particular we see that when $\beta = 0$, the assumption that $\psi'(0+) = 0$ allows us to talk safely about $\Phi(0+)$.

One may think of $\mathcal{T}_{\delta,\beta}$ as one of the many possible generalisations of the Esscher transform. For $\beta \geq 0$, the latter is a well-known linear transformation which maps Ψ_ℓ into itself and has proved to be a very effective tool in analysing many different fluctuation identities for Lévy processes with no positive jumps. It is natural to ask if $\mathcal{T}_{\delta,\beta}$ is equally useful in this respect. A first step in answering this question is to first prove that $\mathcal{T}_{\delta,\beta}$ also maps Ψ_ℓ into itself. This has already been done for the specific family of transformations $\mathcal{T}_\beta$ in Lemma 2.1 of [23] and also, for Lévy processes with two sided jumps, in Proposition 2.1 in [33] .

Proposition 2.2 *Suppose that $\psi \in \Psi_\ell$ where $\ell \leq 0$. Fix $\delta, \beta \geq 0$ with the additional constraint that $\psi'(0+) = \psi(0) = 0$ if $\beta = 0$. Then $\mathcal{T}_{\delta,\beta}\psi \in \Psi_{\ell-\beta} \subseteq \Psi_\ell$ and has no killing component. Moreover, if ψ has Gaussian coefficient σ and jump measure Π then $\mathcal{T}_{\delta,\beta}\psi$ also has Gaussian coefficient σ and its Lévy measure is given by*

$$e^{\beta x}\left(\Pi(\mathrm{d}x) + \delta\overline{\Pi}(x)\mathrm{d}x\right) + \delta\frac{\kappa}{\beta}e^{\beta x}\mathrm{d}x \quad on\ (-\infty, 0),$$

where $\overline{\Pi}(x) = \Pi(-\infty, x)$ and we understand the final term to be zero whenever $\kappa = 0$. Finally, one has $\mathcal{T}_\beta\psi \in \Psi_{\ell-\beta}$ with $l < \beta < 0$ under the additional requirements that $\kappa = 0$ and $(e^{\beta x}\overline{\Pi}(x))' = e^{\beta x}\Pi(\mathrm{d}x) + \beta e^{\beta x}\overline{\Pi}(x)\mathrm{d}x$ is a positive measure on $(-\infty, 0)$.

Proof Recall from earlier that $\mathcal{T}_{\delta,\beta}\psi(u) = \mathcal{E}_\beta\psi(u) - \delta\mathcal{E}_\beta\Phi(u)$. Moreover, from its definition, it is clear that argument of $\mathcal{T}_{\delta,\beta}$ may be taken for all u such that $u + \beta \geq \ell$. It is well understood that $\mathcal{E}_\beta\psi$ is the Laplace exponent of a spectrally negative Lévy process without killing whose Gaussian coefficient remains unchanged but whose Lévy measure is transformed from $\Pi(\mathrm{d}x)$ to $e^{\beta x}\Pi(\mathrm{d}x)$. See for example Chapter 8 of [21]. The proof thus boils down to understanding the contribution from $-\delta\mathcal{E}_\beta\phi(u)$. A straightforward computation based on integration by parts shows that

$$\Phi(u) = -\frac{\kappa}{u} + (a - \overline{\Pi}(-1)) + \frac{1}{2}\sigma^2 u + \int_{-\infty}^{0} (\mathbf{1}_{(|x|<1)} - e^{ux})\overline{\Pi}(x)\mathrm{d}x.$$

From this it follows that

$$\begin{aligned} -\delta\mathcal{E}_\beta\Phi(u) &= -\delta\frac{\kappa}{\beta}\frac{u}{u+\beta} - \frac{\delta}{2}\sigma^2 u - \delta\int_{-\infty}^{0} (1 - e^{ux})e^{\beta x}\overline{\Pi}(x)\mathrm{d}x \\ &= -\frac{\delta}{2}\sigma^2 u - \delta\int_{-\infty}^{0} (1 - e^{ux})e^{\beta x}\overline{\Pi}(x)\mathrm{d}x - \delta\frac{\kappa}{\beta}\int_0^{\infty} (1 - e^{-ux})e^{-\beta x}\mathrm{d}x. \end{aligned}$$

Here we understand the final integral above to be zero if $\kappa = 0$. In that case we see that $-\delta\mathcal{E}_\beta\psi(u)$ is the Laplace exponent of a spectrally negative Lévy process which has no Gaussian component and a jump component which is that of a negative subordinator with jump measure given by $\delta e^{\beta x}\overline{\Pi}(x)\mathrm{d}x + \beta^{-1}\kappa\delta e^{\beta x}\mathrm{d}x$ on $(-\infty, 0)$. The last claim follows readily from the previous one by choosing $\delta = \beta$ and $\kappa = 0$. □

Whilst it is now clear that the mappings $\mathcal{T}^\gamma_{\delta,\beta}$ may serve as a way of generating new examples of Lévy processes with no positive jumps from existing ones, our interest is largely motivated by how the aforesaid transformation interacts with certain path transformations and fluctuation identities associated to Lévy processes.

Indeed, as alluded to above, starting with Urbanik [38], the formalisation of these transformations is motivated by the appearance of particular examples in a number of such contexts. On account of the diversity of these examples, it is worth recalling them here briefly for interest. Kyprianou and Patie [23] use the transformation $\mathcal{T}_\beta$ to give a natural encoding for the Ciesielski-Taylor identity for a class of self similar Markov processes (pssMp). In the setting of self-similar continuous-state branching processes, [28, Proposition 4.11] uses the $\mathcal{T}_\beta$ transformation to describe a family of such processes with immigration. Finally, Gnedin [17] introduces a method of sliced splitting the path of subordinators to generate new examples of subordinators. The sliced splitting operation he used corresponds to the application of a special case of the transformation introduced here.

3 Scale Functions for Spectrally Negative Lévy Processes

Scale functions have occupied a central role in the theory of spectrally negative Lévy processes over the last ten years. They appear naturally in virtually all fluctuation identities of the latter class and consequently have also been instrumental in solving a number of problems from within classical applied probability. See Kyprianou [21] for an account of some of these applications. Despite the fundamental nature of scale functions in these settings, until recently very few explicit examples of scale functions have been found. However in the recent work of Hubalek and Kyprianou [18], Chaumont et al. [14], Patie [28], Kyprianou and Rivero [24] and the survey paper [20], many new examples as well as general methods for constructing explicit examples have been uncovered. We add to this list of contemporary literature by showing that the some of transformations introduced in this paper can be used to construct new families of scale functions from existing examples.

Henceforth we shall assume that the underlying Lévy process, ξ, is spectrally negative, but does not have monotone paths. Moreover, we allow, as above, the case of independent killing at rate $\kappa \geq 0$. For a given spectrally negative Lévy process with Laplace exponent ψ, its scale function $W_\psi : [0,\infty) \mapsto [0,\infty)$ is the unique continuous positive increasing function characterized by its Laplace transform as follows. For any $\kappa \geq 0$ and $u > \theta := \sup\{\lambda \geq 0;\ \psi(\lambda) = 0\}$,

$$\int_0^\infty e^{-ux} W_\psi(x)dx = \frac{1}{\psi(u)}.$$

In the case that $\kappa > 0$, W_ψ is also known as the κ-scale function.

Below we show how our new transformation generates new examples of scale functions from old ones; first in the form of a theorem and then with some examples.

Theorem 3.1 *Let $x, \beta \geq 0$. Then,*

$$W_{\mathcal{T}_\beta \psi}(x) = e^{-\beta x} W_\psi(x) + \beta \int_0^x e^{-\beta y} W_\psi(y) dy. \tag{3.1}$$

Moreover, if $\psi'(0+) \leq 0$, then for any $x, \beta, \delta \geq 0$, we have

$$W_{\mathcal{T}_{\delta,\theta}\psi}(x) = e^{-\theta x} \left(W_\psi(x) + \delta e^{\delta x} \int_0^x e^{-\delta y} W_\psi(y) dy \right)$$

Proof The first assertion is proved by observing that

$$\begin{aligned} \int_0^\infty e^{-ux} W_{\mathcal{T}_\beta \psi}(x) dx &= \frac{u+\beta}{u\psi(u+\beta)} \\ &= \frac{1}{\psi(u+\beta)} + \frac{\beta}{u\psi(u+\beta)}, \end{aligned} \tag{3.2}$$

which agrees with the Laplace transform of the right hand side of (3.1) for which an integration by parts is necessary. As scale functions are right continuous, the result follows by the uniqueness of Laplace transforms.

For the second claim, first note that $\mathcal{T}_{\delta,\theta}\psi = \frac{u+\theta-\delta}{u+\theta}\psi(u+\theta)$. A straightforward calculation shows that for all $u + \delta > \theta$, we have

$$\int_0^\infty e^{-ux} e^{(\theta-\delta)x} W_{\mathcal{T}_{\delta,\theta}\psi}(x) dx = \frac{u+\delta}{u\psi(u+\delta)}. \tag{3.3}$$

The result now follows from the first part of the theorem. □

When $\psi(0+) > 0$ and $\psi(0) = 0$, the first identity in the above theorem contains part of the conclusion in Lemma 2 of Kyprianou and Rivero [24]. However, unlike the aforementioned result, there are no further restrictions on the underlying Lévy processes and the expression on the right hand side is written directly in terms of the scale function W_ψ as opposed to elements related to the Lévy triple of the underlying descending ladder height process of ξ.

Note also that in the case that ψ is the Laplace exponent of an unbounded variation spectrally negative Lévy process, it is known that scale functions are almost everywhere differentiable and moreover that they are equal to zero at zero; cf. Chapter 8 of [21]. One may thus integrate by parts the expressions in the theorem above and obtain the following slightly more compact forms,

$$W_{\mathcal{T}_\beta \psi}(x) = \int_0^x e^{-\beta y} W'_\psi(y) dy \text{ and } W_{\mathcal{T}_{\delta,\theta}\psi}(x) = e^{-(\theta-\delta)x} \int_0^x e^{-\delta y} W'_\psi(y) dy.$$

We conclude this section by giving some examples.

Example 3.2 ((Tempered) Stable Processes) Let $\psi_{\kappa,c}(u) = (u + c)^\alpha - c^\alpha - \kappa$ where $1 < \alpha < 2$ and $\kappa, c \geq 0$. This is the Laplace exponent of an unbounded variation tempered stable spectrally negative Lévy process ξ killed at an independent and exponentially distributed time with rate κ. In the case that $c = 0$, the underlying Lévy process is just a regular spectrally negative α-stable Lévy process. In that case it is known that

$$\int_0^\infty e^{-ux} x^{\alpha-1} E_{\alpha,\alpha}(\kappa x^\alpha) dx = \frac{1}{u^\alpha - \kappa}$$

and hence the scale function is given by

$$W_{\psi_{\kappa,0}}(x) = x^{\alpha-1} E_{\alpha,\alpha}(\kappa x^\alpha), \quad x \geq 0,$$

where $E_{\alpha,\beta}(x) = \sum_{n=0}^\infty \frac{x^n}{\Gamma(\alpha n+\beta)}$ stands for the generalized Mittag-Leffler function. (Note in particular that when $\kappa = 0$ the expression for the scale function simplifies to $\Gamma(\alpha)^{-1} x^{\alpha-1}$). Since

$$\int_0^\infty e^{-ux} e^{-cx} W_{\psi_{\kappa+c^\alpha,0}}(x) dx = \frac{1}{(u + c)^\alpha - c^\alpha - \kappa}$$

it follows that

$$W_{\psi_{\kappa,c}}(x) = e^{-cx} W_{\psi_{\kappa+c^\alpha,0}}(x) = e^{-cx} x^{\alpha-1} E_{\alpha,\alpha}((\kappa + c^\alpha) x^\alpha).$$

Appealing to the first part of Theorem 3.1 we now know that for $\beta \geq 0$,

$$W_{\mathcal{T}_\beta \psi_{\kappa,c}}(x) = e^{-(\beta+c)x} x^{\alpha-1} E_{\alpha,\alpha}((\kappa+c^\alpha)x^\alpha) + \beta \int_0^x e^{-(\beta+c)y} y^{\alpha-1} E_{\alpha,\alpha}((\kappa+c^\alpha)y^\alpha) dy.$$

Note that $\psi'_{\kappa,c}(0+) = \alpha c^{\alpha-1}$ which is zero if and only if $c = 0$. We may use the second and third part of Theorem 3.1 in this case. Hence, for any $\delta > 0$, the scale function of the spectrally negative Lévy process with Laplace exponent $\mathcal{T}_{\delta,0}\psi_{0,0}$ is

$$\begin{aligned} W_{\mathcal{T}_{\delta,0}\psi_{0,0}}(x) &= \frac{1}{\Gamma(\alpha-1)} e^{\delta x} \int_0^x e^{-\delta y} y^{\alpha-2} dy \\ &= \frac{\delta^{\alpha-1}}{\Gamma(\alpha-1)} e^{\delta x} \Gamma(\alpha - 1, \delta x) \end{aligned}$$

where we have used the recurrence relation for the Gamma function and $\Gamma(a, b)$ stands for the incomplete Gamma function of parameters $a, b > 0$. Moreover, we have, for any $\beta > 0$,

$$W_{\mathcal{T}_{\delta,0}^{\beta}\psi_{0,0}}(x) = \frac{1}{\Gamma(\alpha-1)}\left(\frac{\beta^{\alpha}}{\beta-\delta}\Gamma(\alpha-1,\beta x) - e^{(\beta-\delta)x}\frac{\delta^{\alpha}}{\beta-\delta}\Gamma(\alpha-1,\delta x)\right).$$

Finally, the scale function of the spectrally negative Lévy process with Laplace exponent $\mathcal{T}_{\delta,0}^{\beta}\psi_{\kappa,0}$ is given by

$$W_{\mathcal{T}_{\delta,0}\psi_{\kappa,0}}(x) = (x/\delta)^{\alpha-1}E_{\alpha,\alpha-1}\left(x;\frac{\kappa}{\delta}\right)$$

$$W_{\mathcal{T}_{\delta,0}^{\beta}\psi_{\kappa,0}}(x) = \frac{\beta}{\beta-\delta}(x/\beta)^{\alpha-1}E_{\alpha,\alpha-1}\left(x;\frac{\kappa}{\beta}\right) - \frac{\delta}{\beta-\delta}e^{-(\beta-\delta)x}(x/\delta)^{\alpha-1}E_{\alpha,\alpha-1}\left(x;\frac{\kappa}{\delta}\right)$$

where we have used the notation

$$E_{\alpha,\beta}(x;\kappa) = \sum_{n=0}^{\infty}\frac{\Gamma(x;\alpha n+\beta)\kappa^{n}}{\Gamma(\alpha n+\beta)}.$$

4 Exponential Functional and Length-Biased Distribution

In this part, we aim to study the effect of the transformation to the law of the exponential functional of some Lévy processes, namely for subordinators and spectrally negative Lévy processes. We recall that this random variable is defined by

$$I_{\psi} = \int_{0}^{\infty} e^{-\xi_s}ds.$$

Note that $\lim_{t\to\infty}\xi_t = +\infty$ *a.s.* $\Leftrightarrow$ $I_{\psi} < \infty$ a.s. which is equivalent, in the spectrally negative case, to $\mathbb{E}[\xi_1] = \psi'(0^+) > 0$. We refer to the survey paper of Bertoin and Yor [7] for further discussion on this random variable. We also mention that Patie in [29, 30] and [32], provides some explicit characterizations of its law in the case ξ is a spectrally positive Lévy process. We recall that in [4], it has been proved that the law of I_{ψ} is absolutely continuous with a density denoted by f_{ψ}.

4.1 The Case of Subordinators

Let us first assume that $\tilde{\xi} = (\tilde{\xi}_t, t \geq 0)$ is a proper subordinator, that is a non-negative valued Lévy process which is conservative. Let ξ be the subordinator $\tilde{\xi}$

killed at rate $\kappa \geq 0$ and we recall that the Laplace exponent ϕ of ξ is given by $\phi(u) = -\psi(u)$, $u \geq 0$. In that case we write I_ϕ in place of I_ψ and if $\phi(0) = \kappa$, we have that

$$I_\phi = \int_0^\infty e^{-\xi_s} ds = \int_0^{\mathbf{e}_\kappa} e^{-\tilde{\xi}_s} ds$$

where $\mathbf{e}_\kappa$ stands, throughout, for an exponential random variable of mean $\kappa^{-1} > 0$, independent of ξ (we have $\mathbf{e}_0 = \infty$). Before stating our result, we recall that Carmona et al. [13] determine the law of I_ϕ through its positive entire moments as follows

$$\mathbb{E}[I_\phi^n] = \frac{n!}{\prod_{k=1}^n \phi(k)}, \; n = 0, 1 \ldots. \tag{4.1}$$

Theorem 4.1 *For any $\kappa, \beta \geq 0$, the following identity*

$$f_{\mathcal{T}_\beta \phi}(x) = \frac{x^\beta f_\phi(x)}{\mathbb{E}[I_\phi^\beta]}, \quad x > 0, \tag{4.2}$$

holds.

Proof Carmona et al. [13], see also Maulik and Zwart [27, Lemma 2.1], determine the law of I_ϕ by computing its positive entire moments which they derive from the following recursive equation, for any $s, \beta > 0$ and $\kappa \geq 0$,

$$\begin{aligned} \mathbb{E}[I_{\mathcal{T}_\beta \phi}^s] &= \frac{\mathcal{T}_\beta \phi(s)}{s} \mathbb{E}[I_{\mathcal{T}_\beta \phi}^{s-1}] \\ &= \frac{\phi(s+\beta)}{s+\beta} \mathbb{E}[I_{\mathcal{T}_\beta \phi}^{s-1}]. \end{aligned} \tag{4.3}$$

On the other hand, we also have, for any $s, \beta > 0$,

$$\mathbb{E}[I_\phi^{s+\beta}] = \frac{\phi(s+\beta)}{s+\beta} \mathbb{E}[I_\phi^{s-1+\beta}].$$

We get the first assertion by invoking uniqueness, in the space of Mellin transform of positive random variable, of the solution of such a recurrence equation, given for instance in the proof of [34, Theorem 2.4, p. 64]. □

Before providing some new examples, we note from Theorem 4.1 that if $I_\phi \overset{(d)}{=} AB$ for some independent random variables A, B then the positive entire moments of $I_{\mathcal{T}_\beta \phi}$, $\beta > 0$, admit the following expression

$$\mathbb{E}[I_{\mathcal{T}_\beta \phi}^n] = \frac{\mathbb{E}[A^{n+\beta}]}{\mathbb{E}[A^\beta]} \frac{\mathbb{E}[B^{n+\beta}]}{\mathbb{E}[B^\beta]}, \; n = 0, 1 \ldots. \tag{4.4}$$

Example 4.2 (Poisson process) Let ξ be a Poisson process with mean $c = -\log(q) > 0$ with $0 < q < 1$, i.e. $\phi(u) = -\log(q)(1 - e^{-u}), u \geq 0$. Biane et al. [3] computed the law of I_ϕ by means of $q-$calculus. More precisely, they show that its law is self-decomposable and is given by

$$f_\phi(dx) = \sum_{n=0}^{\infty}(-1)^n e^{-y/q^n} \frac{q^{\frac{n(n-1)}{2}}}{(q;q)_\infty (q;q)_n} dx, \; x > 0,$$

where

$$(a;q)_n = \prod_{k=1}^{n-1}(1 - aq^j), \; (q;q)_\infty = \prod_{k=1}^{\infty}(1 - aq^j)$$

and its Mellin transform is given, for any $s > 0$, by

$$\mathbb{E}[I_\phi^s] = \frac{\Gamma(1+s)(q^{1+s};q)_\infty}{(q;q)_\infty}.$$

The image of ξ by the mapping $\mathcal{T}_\beta$ is simply a compound Poisson process with parameter c and jumps which are exponentially distributed on $(0, 1)$ with parameter β, i.e. $\mathcal{T}_\beta\phi(u) = \frac{u}{u+\beta}(-\log(q)(1 - e^{-(u+\beta)}))$. Thus, we obtain that the law of $I_{\mathcal{T}_\beta\phi}$ has an absolute continuous density given by

$$f_{\mathcal{T}_\beta\phi}(dx) = \frac{x^\beta}{\mathbb{E}[I_\phi^\beta]} \sum_{n=0}^{\infty} e^{-x/q^n}(-1)^n \frac{q^{\frac{n(n-1)}{2}}}{(q;q)_\infty (q;q)_n} dx, \; x > 0.$$

To conclude this example, we mention that Bertoin et al. [3] show that the distribution of the random variable L_ϕ defined, for any bounded Borel function f, by

$$\mathbb{E}[f(L_\phi)] = \frac{1}{\mathbb{E}[I_\phi^{-1}]}\mathbb{E}[I_\phi^{-1} f(I'_\phi I_\phi^{-1})]$$

with I'_ϕ an independent copy of I_ϕ, shares the same moments than the log normal distribution. It is not difficult to check that such transformation applied to $I_{\mathcal{T}_\beta\phi}$ does not yield the same properties.

Example 4.3 (Killed compound Poisson process with exponential jumps) Let ξ be a compound Poisson process of parameter $c > 0$ with exponential jumps of mean $b^{-1} > 0$ and killed at a rate $\kappa \geq 0$. Its Laplace exponent has the form $\phi(u) = c\frac{u}{u+b} + \kappa$ and its Lévy measure is given by $\nu(dr) = cbe^{-br}dr, \; r > 0$. We obtain from (4.1) that

$$\mathbb{E}[I_\phi^n] = \frac{n!\Gamma(n+b+1)\Gamma(\kappa_b+1)}{((\kappa+c))^n\Gamma(b+1)\Gamma(n+\kappa_b+1)}$$

where we have set $\kappa_b = \frac{\kappa}{\kappa+c}b$. Then, noting that $b-\kappa_b>0$, we get the identity in distribution

$$I_\phi \overset{(d)}{=} ((\kappa+c))^{-1}\mathbf{e}_1 B(\kappa_b+1, b-\kappa_b)$$

where $B(a,b)$ stands for a Beta random variable of parameters $a,b>0$ and the random variables on the right-hand side are considered to be independent. The case $\kappa=0$ was considered by Carmona et al. [12]. Finally, observing that, for any $\beta\geq 0$, $\mathcal{T}_\beta\phi(u) = c\frac{u}{u+b+\beta}+\kappa\frac{u}{u+\beta}$ and its Lévy measure is $\nu_\beta(dr) = e^{-\beta r}(c(b+\beta)e^{-br}+\kappa\beta)dr,\ r>0$, we deduce from Theorem 4.1, the identity

$$I_{\mathcal{T}_\beta\phi} \overset{(d)}{=} (a(\kappa+1))^{-1}G(\beta+1)B(\kappa_b+\beta+1, b-\kappa_b)$$

where $G(a)$ is an independent Gamma random variable with parameter $a>0$.

Example 4.4 (The α-stable subordinator) Let us consider, for $0<\alpha<1$, $\phi(u)=u^\alpha,\ u\geq 0$ and in this case $\nu(dr)=\frac{\alpha r^{-(\alpha+1)}}{\Gamma(1-\alpha)}dr, r>0$. The law of I_ϕ has been characterized by Carmona et al. [12, Example E and Proposition 3.4]. More precisely, they show that the random variable $Z=\log I_\phi$ is self-decomposable and admits the following Lévy-Khintchine representation

$$\begin{aligned}\log\mathbb{E}[e^{iuZ}] &= \log(\Gamma(1+iu))^{1-\alpha}\\ &= (1-\alpha)\left(-iuC_\gamma + \int_{-\infty}^{0}(e^{ius}-ius-1)\frac{e^s}{|s|(1-e^s)}ds\right)\end{aligned}$$

where C_γ denotes the Euler constant. First, note that $\mathcal{T}_\beta\phi(u) = (u+\beta)^\alpha - \beta(u+\beta)^{\alpha-1}$ and $\nu_\beta(dr) = \frac{1}{\Gamma(1-\alpha)}e^{-\beta r}r^{-(\alpha+1)}(\alpha+\beta r)dr, r>0$. Thus, writing $Z^\beta = \log(I_{\mathcal{T}_\beta\phi})$, we obtain

$$\begin{aligned}\log\mathbb{E}[e^{iuZ^\beta}] &= \log\left(\frac{\Gamma(1+\beta+iu)}{\Gamma(1+\beta)}\right)^{1-\alpha}\\ &= (1-\alpha)\left(iu\Upsilon(1+\beta) + \int_{-\infty}^{0}(e^{ius}-ius-1)\frac{e^{(1+\beta)s}}{|s|(1-e^s)}ds\right)\end{aligned}$$

where Υ stands for the digamma function, $\Upsilon(z) = \frac{\Gamma'(z)}{\Gamma(z)}$. Observing that $\lim_{\alpha\to 0}\mathcal{T}_\beta\phi(u) = \frac{u}{u+\beta}$, by passing to the limit in the previous identity we recover the previous example.

Example 4.5 (The Lamperti-stable subordinator) Now let $\phi(u)=\phi_0(u) = (\alpha u)_\alpha, u\geq 0$, with $0<\alpha<1$. This example is treated by Bertoin and Yor in [7]. They obtain

$$I_\phi \overset{(d)}{=} \mathbf{e}_1\mathbf{e}_1^{-\alpha}.$$

Hence, with $\mathcal{T}_\beta\phi(u) = \frac{u}{u+\beta}(\alpha(u+\beta))_\alpha$, we get

$$I_{\mathcal{T}_\beta\phi} \stackrel{(d)}{=} G(\beta+1)G(\beta+1)^{-\alpha}.$$

4.2 The Spectrally Negative Case

Let us assume now that ξ is a spectrally negative Lévy process. We recall that if $\mathbb{E}[\xi_1] < 0$, then there exits $\theta > 0$ such that

$$\mathbb{E}[e^{\theta\xi_1}] = 1 \tag{4.5}$$

and we write $\psi_\theta(u) = \psi(u+\theta)$. We proceed by mentioning that, in this setting, Bertoin and Yor [6] determined the law of I_ψ by computing its negative entire moments as follows. If $\psi(0) = 0$ and $\psi'(0^+) > 0$, then, for any integer $n \geq 1$,

$$\mathbb{E}[I_\psi^{-n}] = \psi'(0^+)\frac{\prod_{k=1}^{n-1}\psi(k)}{\Gamma(n)}, \tag{4.6}$$

with the convention that $\mathbb{E}[I_\psi^{-1}] = \psi'(0^+)$. Next, it is easily seen that the strong Markov property for Lévy processes yields, for any $a > 0$,

$$I_\psi \stackrel{(d)}{=} \int_0^{T_a} e^{-\xi_s}ds + e^{-a}I'_\psi$$

where $T_a = \inf\{s > 0;\ \xi_s \geq a\}$ and I'_ψ is an independent copy of I_ψ, see e.g. [35]. Consequently, I_ψ is a positive self-decomposable random variable and thus its law is absolutely continuous with an unimodal density, see Sato [36]. We still denote its density by f_ψ. Before stating our next result, we introduce the so-called Erdélyi-Kober operator of the first kind and we refer to the monograph of Kilbas et al. [19] for background on fractional operators. It is defined, for a smooth function f, by

$$\mathbb{D}^{\alpha,\delta}f(x) = \frac{x^{-\alpha-\delta}}{\Gamma(\delta)}\int_0^x r^\alpha(x-r)^{\delta-1}f(r)dr, \quad x > 0,$$

where $\mathfrak{Re}(\delta) > 0$ and $\mathfrak{Re}(\alpha) > 0$. Note that this operator can be expressed in terms of the Markov kernel associated to a Beta random variable. Indeed, after performing a change of variable, we obtain

$$\mathbb{D}^{\alpha,\delta}f(x) = \frac{\Gamma(\alpha+1)}{\Gamma(\alpha+\delta+1)}\mathbb{E}\left[f\left(B(\alpha+1,\delta)x\right)\right]$$

which motivates the following notation

$$\mathbf{D}^{\alpha,\delta} f(x) = \frac{\Gamma(\alpha+\delta+1)}{\Gamma(\alpha+1)} \mathbb{D}^{\alpha,\delta} f(x). \tag{4.7}$$

Theorem 4.6

(1) *If $\psi'(0^+) > 0$, then for any $\beta > 0$, we have*

$$f_{\mathcal{T}_\beta \psi}(x) = \frac{x^{-\beta} f_\psi(x)}{\mathbb{E}[I_\psi^{-\beta}]}, \ x > 0. \tag{4.8}$$

In particular, I_ψ is the length-biased distribution of $I_{\mathcal{T}_1 \psi}$.

(2) *Assume that $\psi'(0^+) < 0$. Then, for any $0 < \delta < \theta$, we have*

$$I_{\mathcal{T}_{\delta,\theta}\psi} \stackrel{(d)}{=} B^{-1}(\theta-\delta, \delta) I_{\psi_\theta}. \tag{4.9}$$

This identity reads in terms of the Erdélyi-Kober operator as follows

$$f_{\mathcal{T}_{\delta,\theta}\psi}(x) = \mathbf{D}^{\theta-\delta-1,\delta} f_{\psi_\theta}(x), \ x > 0.$$

In particular, we have

$$f_{\mathcal{T}_{\delta,\theta}\psi}(x) \sim \frac{\Gamma(\theta)}{\Gamma(\delta)\Gamma(\theta-\delta)} \mathbb{E}[I_{\psi_\theta}^{\theta-\delta}]\, x^{\delta-\theta-1} \quad as \quad x \to \infty, \tag{4.10}$$

($f(t) \sim g(t)$ as $t \to a$ means that $\lim_{t\to a} \frac{f(t)}{g(t)} = 1$ for any $a \in [0,\infty]$). Combining the two previous results, we obtain, for any $0 < \delta < \theta$ and $\beta \geq 0$,

$$f_{\mathcal{T}^\beta_{\delta,\theta}\psi}(x) = \frac{x^{-\beta}}{\mathbb{E}[I^{-\beta}_{\mathcal{T}_{\delta,\theta}\psi}]} \mathbf{D}^{\theta-\delta-1,\delta} f_{\psi_\theta}(x), \ x > 0$$

and

$$f_{\mathcal{T}^\beta_{\delta,\theta}\psi}(x) \sim \frac{\Gamma(\theta)}{\Gamma(\theta-\delta)\Gamma(\delta)} \frac{\mathbb{E}[I_{\psi_\theta}^{-\beta}]}{\mathbb{E}[I_{\psi_\theta}^{\theta-\delta}]} x^{\delta-\theta-\beta-1} \quad as \quad x \to \infty.$$

Proof We start by recalling the following identity due to Bertoin and Yor [2]

$$I_\psi / I_\phi \stackrel{(d)}{=} e^{-M_{\mathbf{e}_1}}$$

where $\phi(u) = \psi(u)/u$, $u \geq 0$, that is ϕ is the Laplace exponent of the ladder height process of the dual Lévy process, $M = \sup_{t\geq 0}\{-\xi_t\}$ is the overall maximum of the dual Lévy process and the random variables are considered to be independent. Thus, recalling that $\mathbb{E}[e^{-sM}] = \psi'(0^+)/\phi(s)$, we have, for any $s > 0$,

$$\begin{aligned}
\mathbb{E}[I_\psi^{-(s+1)}] &= \frac{\psi'(0^+)\Gamma(s+2)}{\phi(s+1)\mathbb{E}[I_\phi^{s+1}]} \\
&= \frac{\psi'(0^+)\Gamma(s+1)}{\mathbb{E}[I_\phi^{s}]} \\
&= \frac{\psi(s)}{s}\mathbb{E}[I_\psi^{-s}] \qquad (4.11)
\end{aligned}$$

where we have used the recurrence relationship satisfied by I_ϕ, see (4.3). Similarly to the case of subordinators, we have, for any $s > 0$,

$$\begin{aligned}
\mathbb{E}[I_{\mathcal{T}_\beta\psi}^{-(s+1)}] &= \frac{\mathcal{T}_\beta\psi(s)}{s}\mathbb{E}[I_{\mathcal{T}_\beta\psi}^{-s}] \\
&= \frac{\psi(s+\beta)}{s+\beta}\mathbb{E}[I_{\mathcal{T}_\beta\psi}^{-s}]
\end{aligned}$$

and

$$\mathbb{E}[I_\psi^{-(s+\beta+1)}] = \frac{\psi(s+\beta)}{s+\beta}\mathbb{E}[I_\psi^{-(s+\beta+1)}].$$

The first claim follows by invoking again uniqueness, in the space of Mellin transform of positive random variables, of the solution of such a recurrence equation, given for instance in the proof of [34, Theorem 2.4, p. 64]. Next, we have both $\psi_\theta'(0^+) > 0$ and $\mathcal{T}_{\delta,\theta}\psi'(0^+) = \frac{\theta-\delta}{\theta}\psi_\theta'(0^+) > 0$ as $\delta < \theta$. Thus, the random variables I_{ψ_θ} and $I_{\mathcal{T}_{\delta,\theta}\psi}$ are well defined. Moreover, from (4.6), we get, for any integer $n \geq 1$,

$$\begin{aligned}
\mathbb{E}[I_{\mathcal{T}_{\delta,\theta}\psi}^{-n}] &= \frac{\psi'(\theta)(\theta-\delta)}{\theta}\frac{\prod_{k=1}^{n-1}\mathcal{T}_{\delta,\theta}\psi(k)}{\Gamma(n)} \\
&= \frac{\psi'(\theta)(\theta-\delta)}{\theta}\frac{\prod_{k=1}^{n-1}\frac{k+\theta-\delta}{k+\theta}\psi_\theta(k)}{\Gamma(n)} \\
&= \psi'(\theta)\frac{\Gamma(n+\theta-\delta)\Gamma(\theta)}{\Gamma(\theta-\delta)\Gamma(n+\theta)}\frac{\prod_{k=1}^{n-1}\psi_\theta(k)}{\Gamma(n)}.
\end{aligned}$$

The identity (4.9) follows by moments identification. Then, we use this identity to get, for any $x > 0$,

$$\begin{aligned} f_{\mathcal{T}_{\delta,\theta}\psi}(x) &= \frac{\Gamma(\theta)}{\Gamma(\delta)\Gamma(\theta-\delta)} \int_0^1 r^{\theta-\delta-1}(1-r)^{\delta-1} f_{\psi_\theta}(xr)dr \\ &= \frac{x^{-\theta}\Gamma(\theta)}{\Gamma(\delta)\Gamma(\theta-\delta)} \int_0^x u^{\theta-\delta-1}(x-u)^{\delta-1} f_{\psi_\theta}(u)du \\ &= \mathbf{D}^{\theta-\delta-1,\delta} f(x). \end{aligned}$$

Next, we deduce readily from (4.11) that the mapping $s \mapsto \mathbb{E}[I_{\psi_\theta}^{-s}]$ is analytic in the right-half plane $\Re\mathfrak{e}(s) > -\theta$. In particular, for any $0 < \delta < \theta$, we have $\mathbb{E}[I_{\psi_\theta}^{\theta-\delta}] < \infty$. Then, the large asymptotic behavior of the density is obtained by observing that

$$\begin{aligned} f_{\mathcal{T}_{\delta,\theta}\psi}(x) &= \frac{x^{\delta-\theta-1}\Gamma(\theta)}{\Gamma(\delta)\Gamma(\theta-\delta)} \int_0^x u^{\theta-\delta}(1-u/x)^{\delta-1} f_{\psi_\theta}(u)du \\ &\sim \frac{x^{\delta-\theta-1}\Gamma(\theta)}{\Gamma(\delta)\Gamma(\theta-\delta)} \int_0^\infty u^{\theta-\delta} f_{\psi_\theta}(u)du \quad \text{as } x \to \infty, \end{aligned}$$

which completes the proof. □

Example 4.7 (the spectrally negative Lamperti-stable process) Let us consider the Lamperti-stable process, i.e., for $1 < \alpha < 2$, $\psi(u) = ((\alpha-1)(u-1))_\alpha$, $\quad u \geq 0$. Recall that $\psi(1) = 0$, $\psi_1(u) = ((\alpha-1)u)_\alpha$ and that this example is investigated by Patie [28]. We get

$$\mathbb{E}[I_{\psi_1}^{-n}] = \psi'(0^+)\frac{\Gamma((\alpha-1)n+1)}{\Gamma(\alpha+1)}.$$

Thus, $I_{\psi_1} = \mathbf{e}_1^{-(\alpha-1)}$. Then, for any $0 < \delta < 1$,

$$I_{\mathcal{T}_{\delta,\theta}\psi} \stackrel{(d)}{=} B(1-\delta,\delta)^{-1}\mathbf{e}_1^{-(\alpha-1)},$$

and for any $\beta > 0$

$$I_{\mathcal{T}_{\delta,\theta}^{\beta}\psi} \stackrel{(d)}{=} B(1+\beta-\delta,\delta)^{-1}G(\beta+1)^{-(\alpha-1)}.$$

5 Entrance Laws and Intertwining Relations of pssMp

In this part, we show that the transformations $\mathcal{T}_{\delta,\beta}$ appear in the study of the entrance law of pssMps. Moreover, also they prove to be useful for the elaboration of intertwining relations between the semigroups of spectrally negative pssMps. We recall that the Markov kernel Λ associated to the positive random variable V is the multiplicative kernel defined, for a bounded Borelian function f, by

$$\Lambda f(x) = \mathbb{E}[f(Vx)].$$

Then, we say that two Markov semigroups P_t and Q_t are intertwined by the Markov kernel Λ if

$$P_t \Lambda = \Lambda Q_t, \quad t \geq 0. \tag{5.1}$$

We refer to the two papers of Carmona et al. [11] and [13] for a very nice account of intertwining relationships. In particular, they show, by means of the Beta-Gamma algebra, that the semigroup of Bessel processes and the one of the so-called self-similar saw tooth processes are intertwined by the Gamma kernel. Below, we provide alternative examples of such relations for a large class of pssMps with stability index 1. Recall that the latter processes were defined in Sect. 2. We also mention that for any $\alpha > -1, \delta > 0$, the linear operator $\mathbf{D}^{\alpha,\delta}$ defined in (4.7) is an instance of a Markov kernel which is, in this case, associated to the Beta random variable $B(\alpha+1,\delta)$. In what follows, when X is associated through the Lamperti mapping to a spectrally negative Lévy process with Laplace exponent ψ, we denote by P_t^{ψ} its corresponding semigroup. When $\min(\psi(0), \psi'(0^+)) < 0$ and $\theta < 1$, then P_t^{ψ} stands for the semigroup of the unique recurrent extension leaving the boundary point 0 continuously. Using the self-similarity property of X, we introduce the positive random variable defined, for any bounded borelian function f, by

$$P_t^{\psi} f(0) = \mathbb{E}[f(tJ_{\psi})].$$

Recall that Bertoin and Yor [5] showed that, when $\psi'(0^+) \geq 0$, the random variable J_{ψ} is moment-determinate with

$$\mathbb{E}\left[J_{\psi}^n\right] = \frac{\prod_{k=1}^n \psi(k)}{\Gamma(n+1)}, \quad n = 1, 2 \ldots \tag{5.2}$$

Before stating the new intertwining relations, we provide some further information concerning the entrance law of pssMps. In particular, we show that the expression (5.2) of the integer moments still holds for the entrance law of the unique continuous recurrent extension, i.e. when $\min\left(\psi(0), \psi'(0^+)\right) < 0$ with $\theta < 1$. We emphasize that we consider both cases when the process X reaches 0 continuously and by a jump.

Proposition 5.1 *Let us assume that* $\min\left(\psi(0), \psi'(0^+)\right) < 0$ *with* $\theta < 1$. *Then, we have the following identity in distribution*

$$J_{\psi} \stackrel{(d)}{=} B\left(1-\theta, \theta\right) / I_{\mathcal{T}_{1-\theta}\psi_{\theta}} \tag{5.3}$$

where $B(a, b)$ is taken independent of the random variable $I_{\mathcal{T}_{1-\theta}\psi_\theta}$. Moreover, the entrance law of the unique recurrent extension which leaves 0 continuously a.s. is determined by its positive entire moments as follows

$$\mathbb{E}\left[J_\psi^n\right] = \frac{\prod_{k=1}^n \psi(k)}{\Gamma(n+1)}, \quad n = 1, 2 \ldots$$

Proof We start by recalling that Rivero [35, Proposition 3] showed that the q-potential of the entrance law of the continuous recurrent extension is given, for a continuous function f, by

$$\int_0^\infty e^{-qt}\mathbb{E}\left[f(tJ_\psi)\right]dt = \frac{q^{-\theta}}{C_\theta}\int_0^\infty f(u)\mathbb{E}\left[e^{-quI_{\psi_\theta}}\right]u^{-\theta}du$$

where we have used the self-similarity property of X and set $C_\theta = \Gamma(1-\theta)\mathbb{E}\left[I_{\psi_\theta}^{\theta-1}\right]$. Performing the change of variable $t = uI_{\psi_\theta}$ on the right hand side of the previous identity, one gets

$$\int_0^\infty e^{-qt}\mathbb{E}\left[f(tJ_\psi)\right]dt = \frac{q^{-\theta}}{C_\theta}\int_0^\infty e^{-qt}\mathbb{E}\left[f(tI_{\psi_\theta}^{-1})I_{\psi_\theta}^{\theta-1}\right]t^{-\theta}dt.$$

Choosing $f(x) = x^s$ for some $s \in i\mathbb{R}$ the imaginary line, we get

$$\int_0^\infty e^{-qt}t^s dt\,\mathbb{E}[J_\psi^s] = \Gamma(s+1)q^{-s-1}\mathbb{E}[J_\psi^s]$$

where we have used the integral representation of the Gamma function $\Gamma(z) = \int_0^\infty e^{-t}t^z dt$, $\Re(z) > -1$. Moreover, by performing a change of variable, we obtain

$$\int_0^\infty t^s\mathbb{E}[e^{-qtI_{\psi_\theta}}]t^{-1-\theta}dt = q^{-s-\theta-1}\Gamma(s-\theta+1)\,\mathbb{E}[I_{\psi_\theta}^{-s+\theta-1}]. \tag{5.4}$$

Putting the pieces together, we deduce that

$$\mathbb{E}[J_\psi^s] = \frac{\Gamma(s-\theta+1)}{\Gamma(s+1)\Gamma\left(1-\frac{\theta}{\alpha}\right)}\frac{\mathbb{E}[I_{\psi_\theta}^{-s+\theta-1}]}{\mathbb{E}[I_{\psi_\theta}^{\theta-1}]} \tag{5.5}$$

and the proof of the first claim is completed by moments identification. Next, we have

$$\mathbb{E}\left[J_\psi^n\right] = \frac{\Gamma(n+1-\theta)}{\Gamma(n+1)\Gamma(1-\theta)\mathbb{E}\left[I_{\psi_\theta}^{\theta-1}\right]}\mathbb{E}\left[I_{\psi_\theta}^{-n+\theta-1}\right]$$

$$= \frac{\Gamma(n+1-\theta)}{\Gamma(n+1)\Gamma(1-\theta)}\mathbb{E}\left[I_{\mathcal{T}_{1-\theta}\psi_\theta}^{-n}\right]$$

$$= \frac{\Gamma(n+1-\theta)}{\Gamma(n+1)\Gamma(1-\theta)}\frac{\psi(1)}{1-\theta}\frac{\prod_{k=1}^{n-1}\frac{k}{k+1-\theta}\psi(k+1)}{\Gamma(n)}$$

$$= \frac{\Gamma(n+1-\theta)\Gamma(n)\Gamma(2-\theta)}{\Gamma(n+1-\theta)\Gamma(n+1)\Gamma(1-\theta)}\frac{\psi(1)}{1-\theta}\frac{\prod_{k=1}^{n-1}\psi(k+1)}{\Gamma(n)}$$

$$= \frac{\prod_{k=1}^{n}\psi(k)}{\Gamma(n+1)}$$

where we have used, from the second identity, successively the identities (5.5), (4.8), (4.6) and the recurrence relation of the gamma function. We point out that under the condition $\theta < 1$, $\psi(k) > 0$ for any integer $k \geq 1$. The proof of the Proposition is then completed. □

Before stating our next result, we recall a criteria given by Carmona et al. [11, Proposition 3.2] for establishing intertwining relations between pssMps. If f and g are functions of $C_0(\mathbb{R}^+)$, the space of continuous functions vanishing at infinity, satisfying the condition:

$$\forall t \geq 0, \quad P_t f(0) = P_t g(0) \quad \text{then } f = g. \tag{5.6}$$

Then the identity (5.1) is equivalent to the assertion, for all $f \in C_0(\mathbb{R}^+)$,

$$P_1 \Lambda f(0) = Q_1 f(0).$$

Finally, we introduce the following notation, for any $s \in \mathbb{C}$,

$$\mathcal{M}_\psi(s) = \mathbb{E}\left[J_\psi^{is}\right].$$

Theorem 5.2

(1) *Assume that $\psi'(0^+) < 0$ and $\mathcal{M}_{\psi_\theta}(s) \neq 0$ for any $s \in \mathbb{R}$. Then, for any $\delta < \theta + 1$, we have the following intertwining relationship*

$$P_t^{\psi_\theta}\mathbf{D}^{\theta,\delta} = \mathbf{D}^{\theta,\delta}P_t^{\mathcal{T}_{\delta,\theta}\psi}, \quad t \geq 0$$

and the following factorization

$$J_{\mathcal{T}_{\delta,\theta}\psi} \stackrel{(d)}{=} B(1+\theta-\delta,\delta)J_{\psi_\theta} \tag{5.7}$$

holds.

(2) *Assume that* $\min\left(\psi(0),\psi'(0^+)\right) < 0$ *with* $\theta < 1$, $\mathcal{M}_{\mathcal{T}_{-\theta}\psi_\theta}(s) \neq 0$ *for any* $s \in \mathbb{R}$ *and that* $\left(\int_0^\infty e^{-\theta y}\Pi(dy+x)\right)'$ *is a positive measure on* $\mathbb{R}^-$. *Then, we have the following intertwining relationship*

$$P_t^{\mathcal{T}_{-\theta}\psi_\theta}\mathbf{D}^{1,\theta} = \mathbf{D}^{1,\theta}P_t^{\psi}, \quad t \geq 0$$

and

$$J_\psi \stackrel{(d)}{=} B(1-\theta,\theta)J_{\mathcal{T}_{-\theta}\psi_\theta}.$$

(3) *Finally, assume that* $\psi'(0^+) = 0$ *and* $\mathcal{M}_\psi(s) \neq 0$ *for any* $s \in \mathbb{R}$. *Then, for any* $\delta < 1$, *we have the following intertwining relationship*

$$P_t^{\psi}\mathbf{D}^{1,\delta} = \mathbf{D}^{1,\delta}P_t^{\mathcal{T}_{\delta,0}\psi}, \quad t \geq 0$$

and

$$J_{\mathcal{T}_{\delta,0}\psi} \stackrel{(d)}{=} B(1-\delta,\delta)J_\psi.$$

Proof First, from the self-similarity property, we observe easily that the condition (5.6) is equivalent to the requirement that the kernel M_{ψ_θ} associated to the positive random variable J_{ψ_θ} is injective. Since $\mathcal{M}_{\psi_\theta}(s) \neq 0$ for any $s \in \mathbb{R}$, we deduce that the multiplicative kernel M_{ψ_θ} is indeed injective, see e.g. [8, Theorem 4.8.4]. Next, note that $\psi_\theta'(0^+) > 0$ and under the condition $\delta < \theta$, $\mathcal{T}_{\delta,\theta}\psi'(0^+) > 0$. Hence, from (5.2), we deduce that, for any $n \geq 1$,

$$\mathbb{E}[J^n_{\mathcal{T}_{\delta,\theta}\psi}] = \frac{\Gamma(n+1+\theta-\delta)\Gamma(\theta+1)}{\Gamma(1+\theta-\delta)\Gamma(n+\theta+1)}\mathbb{E}[J^n_{\psi_\theta}].$$

The identity (5.7) follows. Both processes being pssMps, the first intertwining relation follows from the criteria given above. The proof of the Theorem is completed by following similar lines of reasoning for the other claims. We simply indicate that in the case (2), we note that if

$$e^{-\theta x}\left(\int_{-\infty}^{x} e^{\theta y}\Pi(dy)\right)' = \left(\int_{-\infty}^{0} e^{\theta y}\Pi(dy+x)\right)'$$

is a positive measure on $\mathbb{R}^-$ then according to Proposition 2.2, $\mathcal{T}_{-\theta}\psi_\theta$ is the Laplace exponent of a spectrally negative Lévy process. □

A nice consequence of the previous result is some interesting relationships between the eigenfunctions of the semigroups of pssMps. Indeed, it is easily seen from the intertwining relation (5.1) that if a function f is an eigenfunction with eigenvalue 1 of the semigroup P_t then Λf is an eigenfunction with eigenvalue 1 of the semigroup Q_t. We proceed by introducing some notation taken from Patie [29]. Set $a_0(\psi) = 1$ and define for non-negative integers n

$$a_n(\psi)^{-1} = \prod_{k=1}^{n} \psi(k).$$

Next, we introduce the entire function $\mathcal{I}_\psi$ which admits the series representation

$$\mathcal{I}_\psi(z) = \sum_{n=0}^{\infty} a_n(\psi) z^n, \quad z \in \mathbb{C}.$$

In [29, Theorem 1], it is shown that

$$\mathbf{L}^\psi \mathcal{I}_\psi(x) = \mathcal{I}_\psi(x), \; x > 0, \tag{5.8}$$

where, for a smooth function f, the linear operator $\mathbf{L}^\psi$ is the infinitesimal generator associated to the semigroup P_t^ψ and takes the form

$$\mathbf{L}^\psi f(x) = \frac{\sigma}{2} x f''(x) + b f'(x) + x^{-1} \int_0^\infty f(e^{-r}x) - f(x) + x f'(x) r \mathbb{I}_{\{|r|<1\}} \nu(dr) - \kappa x f(x).$$

From the Feller property of the semigroup of X, we deduce readily that the identity (5.8) is equivalent to

$$e^{-t} P_t^\psi \mathcal{I}_\psi(x) = \mathcal{I}_\psi(x), \; t, x \geq 0,$$

that is $\mathcal{I}_\psi$ is 1-eigenfunction for P_t^ψ. Hence, we deduce from Theorem 5.2 the following interesting relationship between eigenfunctions.

Corollary 5.3

(1) *Let $\psi'(0^+) < 0$. Then, for any $\delta < \theta + 1$, we have the following identity*

$$\mathbf{D}^{\theta,\delta} \mathcal{I}_{\psi_\theta}(x) = \mathcal{I}_{\mathcal{T}_{\delta,\theta}\psi}(x).$$

(2) *If $\psi'(0^+) < 0$ and $\theta < 1$, then*

$$\mathbf{D}^{1,\theta} \mathcal{I}_{\mathcal{T}_{-\theta}\psi_\theta}(x) = \mathcal{I}_\psi(x).$$

(3) *Finally, if* $\psi'(0^+) = 0$, *then, for any* $\delta < 1$, *we have*

$$\mathbf{D}^{1,\delta}\mathcal{I}_{\psi}(x) = \mathcal{I}_{\mathcal{T}_{\delta,0}\psi}(x).$$

We illustrate this last result by detailing some interesting instances of such relationships between some known special functions.

Example 5.4 (Mittag-Leffler type functions) Let us consider, for any $1 < \alpha < 2$, the Laplace exponent $\psi(u) = (\alpha(u - 1/\alpha))_{\alpha}$. We easily check that $\theta = 1/\alpha$ and we have $\psi_{1/\alpha}(u) = (\alpha u)_{\alpha}$. Observing that

$$\prod_{k=1}^{n} \psi_{1/\alpha}(k) = \prod_{k=1}^{n} (\alpha k)_{\alpha} = \frac{\Gamma(\alpha(n+1))}{\Gamma(\alpha)}, \quad n \geq 1,$$

and using the fact that the random variable J_{ψ_θ} is moment-determinate, we readily check, from the expression (5.2), that

$$\mathcal{M}_{\psi_\theta}(s) = \frac{\Gamma(\alpha(is+1))}{\Gamma(\alpha)\Gamma(is+1)}$$

The pole of the gamma function being the negative integers, the condition $\mathcal{M}_{\psi_\theta}(s) \neq 0$ is satisfied for any $s \in \mathbb{R}$. Moreover, we obtain

$$\mathcal{I}_{\psi_{1/\alpha}}(x) = \Gamma(\alpha)\mathcal{E}_{\alpha,\alpha}(x)$$

where we recall that the Mittag-Leffler function $\mathcal{E}_{\alpha,\alpha}$ is defined in the Example 3.2. Next, for any $\delta < 1 + 1/\alpha$, we have

$$\begin{aligned}\mathcal{I}_{\mathcal{T}_{\delta,1/\alpha}\psi}(x) &= \frac{\Gamma(\alpha)\Gamma(1/\alpha+1-\delta)}{\Gamma(1/\alpha+1)} \sum_{n=0}^{\infty} \frac{\Gamma(n+1/\alpha+1)}{\Gamma(n+1/\alpha+1-\delta)\Gamma(\alpha n+\alpha)} x^n \\ &= {}_2F_2\left(\left. \begin{matrix} (1, 1/\alpha+1), (1,1) \\ (1, 1/\alpha+1-\delta), (\alpha,\alpha) \end{matrix} \right| x \right),\end{aligned}$$

where ${}_2F_2$ is the Wright hypergeometric function, see e.g. Braaksma [9, Chap. 12]. Hence, we have

$$\Gamma(\alpha)\mathbf{D}^{\theta,\delta}\mathcal{E}_{\alpha,\alpha}(x) = {}_2F_2\left(\left. \begin{matrix} (1, 1/\alpha+1), (1,1) \\ (1, 1/\alpha+1-\delta), (\alpha,\alpha) \end{matrix} \right| x \right).$$

Example 5.5 Now, for any $1 < \alpha < 2$, we set $\psi(u) = u^{\alpha}$ and we note that $\psi'(0^+) = 0$. Proceeding as in the previous example, we get

$$\mathcal{M}_{\psi}(s) = \Gamma^{\alpha-1}(is+1)$$

and hence the condition $\mathcal{M}_{\psi}(s) \neq 0$ is satisfied for any $s \in \mathbb{R}$. We have

$$\mathcal{I}_{\psi}(x) = \sum_{n=0}^{\infty} \frac{1}{\Gamma^{\alpha}(n+1)} x^{n}$$

and, for any $\delta < 1$, we write

$$\mathcal{I}_{\mathcal{T}_{\delta,0}\psi}(x) = \Gamma(1-\delta) \sum_{n=0}^{\infty} \frac{\Gamma(n+1)}{\Gamma(n+1-\delta)\Gamma^{\alpha}(n+1)} x^{n}.$$

Consequently,

$$\mathbf{D}^{1,\delta}\mathcal{I}_{\psi}(x) = \mathcal{I}_{\mathcal{T}_{\delta,0}\psi}(x).$$

References

1. Bertoin, J.: Lévy Processes. Cambridge University Press, Cambridge (1996)
2. Bertoin, J.: On the first exit time of a completely asymmetric stable process from a finite interval. Bull. Lond. Math. Soc. **28**, 514–520 (1996)
3. Bertoin, J., Biane, P., Yor, M.: Poissonian exponential functionals, q-series, q-integrals, and the moment problem for log-normal distributions. In: Seminar on Stochastic Analysis, Random Fields and Applications IV. Progress in Probability, vol. 58, pp. 45–56. Birkhäuser, Basel (2004)
4. Bertoin, J., Lindner, A., Maller, R.: On continuity properties of the law of integrals of Lévy processes. Séminaire de probabilités XLI. Lecture Notes in Mathematics, vol. 1934, pp. 137–159. Springer, Berlin (2008)
5. Bertoin, J., Yor, M.: The entrance laws of self-similar Markov processes and exponential functionals of Lévy processes. Potential Anal. **17**(4), 389–400 (2002)
6. Bertoin, J., Yor, M.: On the entire moments of self-similar Markov processes and exponential functionals of Lévy processes. Ann. Fac. Sci. Toulouse Math. **11**(1), 19–32 (2002)
7. Bertoin, J., Yor, M.: Exponential functionals of Lévy processes. Probab. Surv. **2**, 191–212 (2005)
8. Bingham, N.H., Goldie, C.M., Teugels, J.L.: Regular Variation. Encyclopedia of Mathematics and Its Applications, vol. 27. Cambridge University Press, Cambridge (1989)
9. Braaksma, B.L.J.: Asymptotic expansions and analytic continuations for a class of Barnes-integrals. Compositio Math. **15**, 239–341 (1964)
10. Caballero, M.E., Chaumont, L.: Conditioned stable Lévy processes and the Lamperti representation. J. Appl. Probab. **43**, 967–983 (2006)
11. Carmona, P., Petit, F., Yor, M.: Sur les fonctionnelles exponentielles de certains processus de Lévy. Stoch. Stoch. Rep. **47**(1–2), 71–101 (1994) (English version in Yor (2001), pp. 139–171)
12. Carmona, P., Petit, F., Yor, M.: On the distribution and asymptotic results for exponential functionals of Lévy processes. In: Yor, M. (ed.) Exponential Functionals and Principal Values Related to Brownian Motion. Biblioteca de la Rev. Mat. Iberoamericana, pp. 73–121. Madrid (1997)
13. Carmona, P., Petit, F., Yor, M.: Beta-gamma random variables and intertwining relations between certain Markov processes. Rev. Mat. Iberoamericana **14**(2), 311–368 (1998)

14. Chaumont, L., Kyprianou, A.E., Pardo, J.C.: Some explicit identities associated with positive self-similar Markov processes. Stoch. Proc. Appl. **119**, 980–1000 (2009)
15. Fitzsimmons, P.J.: On the existence of recurrent extension of positive self-similar Markov processes. Elec. Commun. Probab. **11**, 230–241 (2006)
16. Getoor, R.K., Sharpe, M.J.: Excursions of Brownian motion and Bessel processes. Z. Wahr. **47**, 83–106 (1979)
17. Gnedin, A.V.: Regeneration in random combinatorial structures. (English summary) Probab. Surv. **7**, 105–156 (2010)
18. Hubalek, F., Kyprianou, A.E.: Old and new examples of scale functions for spectrally negative Lévy processes. In: Dalang, R., Dozzi, M., Russo, F. (eds.) Stochastic Analysis, Random Fields and Applications. Progress in Probability. Birkhäuser (2010). To appear
19. Kilbas, A.A., Srivastava, H.M., Trujillo, J.J.: Theory and Applications of Fractional Differential Equations. North-Holland Mathematics Studies, vol. 204. Elsevier Science B.V., Amsterdam (2006)
20. Kuznetsov, A., Kyprianou, A.E., Rivero, V.: The Theory of Scale Functions for Spectrally Negative Levy Processes. Levy Matters II. Lecture Notes in Mathematics, vol. 2061, pp. 97–186. Springer, Heidelberg
21. Kyprianou, A.E.: Introductory Lectures on Fluctuations of Lévy Processes with Applications. Universitext. Springer, Berlin (2006)
22. Kyprianou, A.E., Palmowski, Z.: A martingale review of some fluctuation theory for spectrally negative Lévy processes. In: Séminaire de Probabilités XXXVIII. Lecture Notes in Mathematics, vol. 1857, pp. 16–29. Springer, Berlin (2005)
23. Kyprianou, A.E., Patie, P.: A Ciesielski-Taylor type identity for positive self-similar Markov processes. Ann. Inst. H. Poincar' Probab. Statist. **47**(3), 917–928 (2011)
24. Kyprianou, A.E., Rivero, V.: Special, conjugate and complete scale functions for spectrally negative Lévy processes. Electron. J. Probab. **13**, 1672–1701 (2008)
25. Lamperti, J.: Semi-stable Markov processes. I. Z. Wahrsch. Verw. Geb. **22**, 205–225 (1972)
26. Lebedev, N.N.: Special Functions and Their Applications. Dover Publications, New York (1972)
27. Maulik, K., Zwart, B.: Tail asymptotics for exponential functionals of Lévy processes. Stoch. Process. Appl. **116**, 156–177 (2006)
28. Patie, P.: Exponential functional of one-sided Lévy processes and self-similar continuous state branching processes with immigration. Bull. Sci. Math. **133**(4), 355–382 (2009)
29. Patie, P.: Infinite divisibility of solutions to some self-similar integro-differential equations and exponential functionals of Lévy processes. Ann. Inst. H. Poincaré Probab. Statist. **45**(3), 667–684 (2009)
30. Patie, P.: Law of the exponential functional of one-sided Lévy processes and Asian options. C. R. Acad. Sci. Paris, Ser. I **347**, 407–411 (2009)
31. Patie, P.: On refined factorizations of the exponential law. Bernoulli **17**(2), 814–826 (2011)
32. Patie, P.: Law of the absorption time of positive self-similar markov processes. Ann. Probab. **40**(2), 765–787 (2012)
33. Patie, P., Savov, M.: Extended factorizations of exponential functionals of Lévy processes. Electron. J. Probab. **17**(38), 1–22 (2012)
34. Patie, P., Savov, M.: Bernstein-Gamma functions and exponential functionals of Lévy processes. Electron. J. Probab. **23**(75), 1–101 (2018)
35. Rivero, V.: Recurrent extensions of self-similar Markov processes and Cramér's condition. Bernoulli **11**(3), 471–509 (2005)
36. Sato, K.: Lévy Processes and Infinitely Divisible Distributions. Cambridge University Press, Cambridge (1999)
37. Sato, K.: Density transformation in Lévy processes. Technical Report Lecture Notes 7, MaPhySto, Aarhus (2000)

38. Urbanik, K.: Infinite divisibility of some functionals on stochastic processes. Probab. Math. Statist. **15**, 493–513 (1995)
39. Yor, M.: Exponential Functionals of Brownian Motion and Related Processes. Springer Finance, Berlin (2001)
40. Zolotarev, V.M.: Mellin-Stieltjes transform in probability theory. Theory Probab. Appl. **2**, 433–460 (1957)

First-Passage Times for Random Walks in the Triangular Array Setting

Denis Denisov, Alexander Sakhanenko, and Vitali Wachtel

Abstract In this paper we continue our study of exit times for random walks with independent but not necessarily identically distributed increments. Our paper "First-passage times for random walks with non-identically distributed increments" (2018) was devoted to the case when the random walk is constructed by a fixed sequence of independent random variables which satisfies the classical Lindeberg condition. Now we consider a more general situation when we have a triangular array of independent random variables. Our main assumption is that the entries of every row are uniformly bounded by a deterministic sequence, which tends to zero as the number of the row increases.

Keywords Random walk · Triangular array · First-passage time · Central limit theorem · Moving boundary · Transition phenomena

---

A.S. and V.W. were supported by RFBR and DFG according to the research project 20-51-12007

D. Denisov (✉)
Department of Mathematics, University of Manchester, Manchester, UK
e-mail: denis.denisov@manchester.ac.uk

A. Sakhanenko
Sobolev Institute of Mathematics, Novosibirsk, Russia
e-mail: aisakh@mail.ru

V. Wachtel
Institut für Mathematik, Universität Augsburg, Augsburg, Germany
e-mail: vitali.wachtel@math.uni-augsburg.de

L. Chaumont, A. E. Kyprianou (eds.), *A Lifetime of Excursions Through Random Walks and Lévy Processes*, Progress in Probability 78,
https://doi.org/10.1007/978-3-030-83309-1_10

1 Introduction and the Main Result

1.1 Introduction

Suppose that for each $n = 1, 2, \dots$ we are given independent random variables $X_{1,n}, \dots, X_{n,n}$ such that

$$\mathbf{E}X_{i,n} = 0 \quad \text{for all } i \le n \qquad \text{and} \qquad \sum_{i=1}^{n} \mathbf{E}X_{i,n}^2 = 1. \tag{1}$$

For each n we consider a random walk

$$S_{k,n} := X_{1,n} + \cdots + X_{k,n}, \quad k = 1, 2, \dots, n. \tag{2}$$

Let $\{g_{k,n}\}_{k=1}^{n}$ be deterministic real numbers, and let

$$T_n := \inf\{k \ge 1 : S_{k,n} \le g_{k,n}\} \tag{3}$$

be the first crossing over the moving boundary $\{g_{k,n}\}$ by the random walk $\{S_{k,n}\}$. The main purpose of the present paper is to study the asymptotic behaviour, as $n \to \infty$, of the probability

$$\mathbf{P}(T_n > n) = \mathbf{P}\left(\min_{1\le k\le n} (S_{k,n} - g_{k,n}) > 0\right). \tag{4}$$

We shall always assume that the boundary $\{g_{k,n}\}$ is of a small magnitude, that is,

$$g_n^* := \max_{1\le k\le n} |g_{k,n}| \to 0. \tag{5}$$

Here and in what follows, all unspecified limits are taken with respect to $n \to \infty$.

Furthermore, to avoid trivialities, we shall assume that

$$\mathbf{P}(T_n > n) > 0 \qquad \text{for all} \quad n \ge 1. \tag{6}$$

An important particular case of the triangular array scheme is given by the following construction. Let $X_1, X_2, \dots$ be independent random variables with finite variances such that

$$\mathbf{E}X_i = 0 \quad \text{for all } i \ge 1 \qquad \text{and} \qquad B_n^2 := \sum_{i=1}^{n} \mathbf{E}X_i^2 \to \infty. \tag{7}$$

For a real deterministic sequence $\{g_1, g_2, \dots\}$ set

$$T := \inf\{k \ge 1 : S_k \le g_k\}, \quad \text{where} \quad S_k := X_1 + \cdots + X_k. \tag{8}$$

Stopping time T is the first crossing over the moving boundary $\{g_k\}$ by the random walk $\{S_k\}$. Clearly, (7)–(8) is a particular case of (1), (2), and (3). Indeed to obtain (1), (2), and (3) it is sufficient to set

$$X_{k,n} = \frac{X_k}{B_n}, \quad S_{k,n} = \frac{S_k}{B_n}, \quad g_{k,n} = \frac{g_k}{B_n}. \tag{9}$$

However, the triangular array scheme is much more general than (7), (8), and (9).

If the classical Lindeberg condition holds for the sequence $\{X_k\}$ and $g_n = o(B_n)$ then, according to Theorem 1 in [2],

$$\mathbf{P}(T > n) \sim \sqrt{\frac{2}{\pi}} \frac{U(B_n^2)}{B_n}, \tag{10}$$

where U is a positive slowly varying function with the values

$$U(B_n^2) = \mathbf{E}[S_n - g_n; T > n], \quad n \ge 1.$$

The constant $\sqrt{\frac{2}{\pi}}$ in front of the asymptotics has been inherited from the tail asymptotics of exit time of standard Brownian motion. Indeed, let $W(t)$ be the standard Brownian motion and set

$$\tau_x^{bm} := \inf\{t > 0 : x + W(t) \le 0\}, \quad x > 0.$$

Then,

$$\mathbf{P}(\tau_x^{bm} > t) = \mathbf{P}(|W(t)| \le x) = \mathbf{P}\left(|W(1)| \le \frac{x}{\sqrt{t}}\right) \sim \sqrt{\frac{2}{\pi}} \frac{x}{\sqrt{t}}, \quad \text{as} \quad \frac{x}{\sqrt{t}} \to 0.$$

The continuity of paths of $W(t)$ implies that $x + W(\tau_x^{bm}) = 0$. Combining this with the optional stopping theorem, we obtain

$$\begin{aligned} x = \mathbf{E}[x + W(\tau_x^{bm} \wedge t)] &= \mathbf{E}[x + W(t); \tau_x^{bm} > t)] + \mathbf{E}[x + W(\tau_x^{bm}); \tau_x^{bm} \le t)] \\ &= \mathbf{E}[x + W(t); \tau_x^{bm} > t)]. \end{aligned}$$

Therefore, for any fixed $x > 0$,

$$\mathbf{P}(\tau_x^{bm} > t) \sim \sqrt{\frac{2}{\pi}} \frac{x}{\sqrt{t}} = \sqrt{\frac{2}{\pi}} \frac{\mathbf{E}[x + W(t); \tau_x^{bm} > t)]}{\sqrt{t}}, \quad \text{as} \quad t \to \infty.$$

Thus, the right hand sides here and in (10) are of the same type.

1.2 Main Result

The purpose of the present note is to generalise the asymptotic relation (10) to the triangular array setting. More precisely, we are going to show that the following relation holds

$$\mathbf{P}(T_n > n) \sim \sqrt{\frac{2}{\pi}} E_n, \tag{11}$$

where

$$E_n := \mathbf{E}[S_{n,n} - g_{n,n}; T_n > n] = \mathbf{E}[-S_{T_n,n}; T_n \le n] - g_{n,n}\mathbf{P}(T_n > n). \tag{12}$$

Unexpectedly for the authors, in contrast to the described above case of a single sequence, the Lindeberg condition is not sufficient for the validity of (11), see Example 6. Thus, one has to find a more restrictive condition for (11) to hold. In this paper we show that (11) holds under the following assumption: there exists a sequence r_n such that

$$\max_{1\le i\le n} |X_{i,n}| \le r_n \to 0. \tag{13}$$

It is clear that under this assumption the triangular array satisfies the Lindeberg condition and, hence, the Central Limit Theorem holds.

At first glance, (13) might look too restrictive. However we shall construct a triangular array, see Example 7, in which the assumption (13) becomes necessary for (11) to hold. Now we state our main result.

Theorem 1 *Assume that* (5) *and* (13) *are valid. Then there exists an absolute constant* C_1 *such that*

$$\mathbf{P}(T_n > n) \ge \sqrt{\frac{2}{\pi}} E_n\big(1 - C_1(r_n + g_n^*)^{2/3}\big). \tag{14}$$

On the other hand, there exists an absolute constant C_2 *such that*

$$\mathbf{P}(T_n > n) \le \sqrt{\frac{2}{\pi}} E_n\big(1 + C_2(r_n + g_n^*)^{2/3}\big), \qquad \text{if} \quad r_n + g_n^* \le 1/24. \tag{15}$$

In addition, for $m \le n$,

$$\mathbf{P}(T_n > m) \le \frac{4E_n}{B_m^{(n)}} \tag{16}$$

provided that

$$B_m^{(n)} := \left(\sum_{k=1}^{m} \mathbf{E}X_{k,n}^2\right)^{1/2} \geq 24(r_n + g_n^*).$$

Corollary 2 *Under conditions* (5), (6) *and* (13) *relation* (11) *takes place.*

Estimates (14) and (15) can be seen as an improved version of (11), with a rate of convergence. Moreover, the fact, that the dependence on r_n and g_n is expressed in a quite explicit way, is very important for our work [3] in progress, where we analyse unbounded random variables. In this paper we consider first-passage times of walks $S_n = X_1 + X_2 + \ldots + X_n$ for which the central limit theorem is valid but the Lindeberg condition may fail. We use Theorem 1 to analyse the behaviour of triangular arrays obtained from $\{X_n\}$ by truncation.

1.3 Triangular Arrays of Weighted Random Variables

Theorem 1 and Corollary 2 can be used in studying first-passage times of weighted sums of independent random variables.

Suppose that we are given independent random variables $X_1, X_2, \ldots$ such that

$$\mathbf{E}X_i = 0 \quad \text{and} \quad \mathbf{P}(|X_i| \leq M_i) = 1 \quad \text{for all } i \geq 1, \tag{17}$$

where $M_1, M_2, \ldots$ are deterministic. For each n we consider a random walk

$$U_{k,n} := u_{1,n}X_1 + \cdots + u_{k,n}X_k, \quad k = 1, 2, \ldots, n, \tag{18}$$

and let

$$\tau_n := \inf\{k \geq 1 : U_{k,n} \leq G_{k,n}\} \tag{19}$$

be the first crossing over the moving boundary $\{G_{k,n}\}$ by the random walk $\{U_{k,n}\}$. The main purpose of the present subsection is to study the asymptotic behaviour, as $n \to \infty$, of the probability

$$\mathbf{P}(\tau_n > n) = \mathbf{P}\left(\min_{1\leq k\leq n} (U_{k,n} - G_{k,n}) > 0\right). \tag{20}$$

We suppose that $\{u_{k,n}, G_{k,n}\}_{k=1}^n$ are deterministic real numbers such that

$$M := \sup_{k,n\geq 1} \left(|u_{k,n}|M_k + |G_{k,n}|\right) < \infty \tag{21}$$

and

$$\sigma_n^2 := \sum_{k=1}^{n} u_{k,n}^2 \mathbf{E} X_k^2 \to \infty. \tag{22}$$

Moreover, we assume that

$$u_{k,n} \to u_k \quad \text{and} \quad G_{n,k} \to g_k \quad \text{for every } k \geq 1. \tag{23}$$

Corollary 3 *Assume that the distribution functions of all X_k are continuous. Then, under assumptions* (17), (21), (22) *and* (23),

$$\sigma_n \mathbf{P}(\tau_n > n) \to \sqrt{\frac{2}{\pi}} \mathbf{E}[-U_\tau] \in [0, \infty), \tag{24}$$

where

$$U_k := u_1 X_1 + \cdots + u_k X_k \quad \textit{and} \quad \tau := \inf\{k \geq 1 : U_k \leq g_k\}. \tag{25}$$

It follows from condition (23) that random walks $\{U_{k,n}\}$ introduced in (18) may be considered as perturbations of the walk $\{U_k\}$ defined in (25). Thus, we see from (24) that the influence of perturbations on the behavior of the probability $\mathbf{P}(\tau_n > n)$ is concentrated in the σ_n.

Example 4 As an example we consider the following method of summation, which has been suggested by Gaposhkin [4]. Let $f : [0, 1] \mapsto \mathbb{R}^+$ be a non-degenerate continuous function. For random variables $\{X_k\}$ define

$$U_k(n, f) := \sum_{j=1}^{k} f\left(\frac{j}{n}\right) X_j, \quad j = 1, 2, \ldots, n.$$

This sequence can be seen as a stochastic integral of f with respect to the random walk $S_k = X_1 + X_2 + \ldots X_k$ normalized by n.

We assume that the random variables $\{X_k\}$ are independent and identically distributed. Furthermore, we assume that X_1 satisfies (17) and that its distribution function is continuous. In this case

$$\sigma_n^2(f) := \frac{1}{n} \mathbf{E} X_1^2 \sum_{j=1}^{n} f^2\left(\frac{j}{n}\right) \to \sigma^2(f) := \mathbf{E} X_1^2 \int_0^1 f^2(t) dt > 0.$$

From Corollary 3 with $u_{k,n} := f\left(\frac{k}{n}\right) \to f(0) =: u_k$, $G_{k,n} \equiv 0$ and $\sigma_n := \sqrt{n}\sigma_n(f)$ we immediately obtain

$$\sqrt{n}\mathbf{P}\left(\min_{k\le n} U_k(n,f) > 0\right) \to \sqrt{\frac{2}{\pi}}\frac{f(0)}{\sigma(f)}\mathbf{E}[-S_\tau] \in [0,\infty), \tag{26}$$

where

$$S_k := X_1 + \cdots + X_k \quad \text{and} \quad \tau := \inf\{k \ge 1 : S_k \le 0\}. \tag{27}$$

◇

Clearly, (26) gives one exact asymptotics only when $f(0) > 0$. The case $f(0) = 0$ seems to be much more delicate. If $f(0) = 0$ then one needs an information on the behaviour of f near zero. If, for example, $f(t) = t^\alpha$ with some $\alpha > 0$ then, according to Example 12 in [2],

$$\mathbf{P}\left(\min_{k\le n} U_k(n,f) > 0\right) = \mathbf{P}\left(\min_{k\le n}\sum_{j=1}^{k} j^\alpha X_j > 0\right) \sim \frac{Const}{n^{\alpha+1/2}}.$$

Now we give an example of application of our results to study of transition phenomena.

Example 5 Consider an autoregressive sequence

$$U_n(\gamma) = \gamma U_{n-1}(\gamma) + X_n,\ n \ge 0, \quad n = 1, 2, \ldots, \quad \text{where} \quad U_0(\gamma) = 0, \tag{28}$$

with a non-random $\gamma = \gamma_n \in (0,1)$ and with independent, identically distributed innovations $X_1, X_2, \ldots$. As in the previous example, we assume that X_1 satisfies (17) and that its distribution function is continuous. Consider the exit time

$$T(\gamma) := \inf\{n \ge 1 : U_n(\gamma) \le 0\}.$$

We want to understand the behavior of the probability $\mathbf{P}(T(\gamma) > n)$ in the case when $\gamma = \gamma_n$ depends on n and

$$\gamma_n \in (0,1) \quad \text{and} \quad \sup_n n(1-\gamma_n) < \infty. \tag{29}$$

We now show that the autoregressive sequence $U_n(\gamma)$ can be transformed to a random walk, which satisfies the conditions of Corollary 3. First, multiplying (28) by γ^{-n}, we get

$$U_n(\gamma)\gamma^{-n} = U_n(\gamma)\gamma^{-(n-1)} + X_n\gamma^{-n} = \sum_{k=1}^{n}\gamma^{-k}X_k, \quad n \ge 1.$$

Thus, for each $n \geq 1$,

$$\{T(\gamma_n) > n\} = \left\{ \sum_{j=1}^{k} \gamma_n^{-j} X_j > 0 \quad \text{for all } k \leq n \right\}. \tag{30}$$

Comparing (30) with (18) and (20), we see that we have a particular case of the model in Corollary 3 with $u_{k,n} = \gamma_n^{-k}$ and $G_{k,n} = 0$. Clearly, $u_{k,n} \to 1$ for every fixed k. Furthermore, we infer from (29) that

$$\gamma_n^{-n} = e^{-n \log \gamma_n} = e^{O(n|\gamma_n - 1|)} = e^{O(1)}$$

and

$$\sigma_n^2(\gamma_n) := \frac{\gamma_n^{-2n} - 1}{1 - \gamma_n^2} = \gamma_n^{-2} + \gamma_n^{-4} + \cdots + \gamma_n^{-2n} = ne^{O(1)}.$$

These relations imply that (23) and (21) are fulfilled. Applying Corollary 3, we arrive at

$$\sigma_n(\gamma_n)\mathbf{P}(T(\gamma_n) > n) \to \sqrt{\frac{2}{\pi \mathbf{E}X_1^2}}\mathbf{E}[-S_\tau] \in (0, \infty), \tag{31}$$

where τ is defined in (27). ◇

1.4 Discussion of the Assumption (13)

Based on the validity of CLT and considerations in [2] one can expect that the Lindeberg condition will again be sufficient. However the following example shows that this is not the case and the situation is more complicated.

Example 6 Let $X_2, X_3, \ldots$ and $Y_2, Y_3, \ldots$ be mutually independent random variables such that

$$\mathbf{E}X_k = \mathbf{E}Y_k = 0,\ \mathbf{E}X_k^2 = \mathbf{E}Y_k^2 = 1 \quad \text{and} \quad \mathbf{P}(|X_k| \leq M) = 1 \text{ for all } k \geq 2 \tag{32}$$

for some finite constant M. It is easy to see that the triangular array

$$X_{1,n} := \frac{Y_n}{\sqrt{n}},\ X_{k,n} := \frac{X_k}{\sqrt{n}},\ k = 2, 3, \ldots, n;\ n > 1 \tag{33}$$

satisfies the Lindeberg condition. Indeed, $\sum_{i=1}^{n} \mathbf{E}X_{i,n}^2 = 1$ and for every $\varepsilon > \frac{M}{\sqrt{n}}$ one has

$$\sum_{i=1}^{n} \mathbf{E}[X_{i,n}^2; |X_{i,n}| > \varepsilon] = \mathbf{E}[X_{1,n}^2; |X_{1,n}| > \varepsilon] \le \mathbf{E}X_{1,n}^2 = \frac{\mathbf{E}Y_n^2}{n} = \frac{1}{n} \to 0 \tag{34}$$

due to the fact that $|X_{k,n}| \le \frac{M}{\sqrt{n}}$ for all $k \ge 2$.

We shall also assume that $g_{k,n} \equiv 0$. For each $n > 1$ let random variable Y_n be defined as follows

$$Y_n := \begin{cases} N_n, & \text{with probability } p_n := \frac{1}{2N_n^2}, \\ 0, & \text{with probability } 1 - 2p_n, \\ -N_n, & \text{with probability } p_n, \end{cases} \tag{35}$$

where $N_n \ge 1$. Note that $\mathbf{E}Y_n = 0$ and $\mathbf{E}Y_n^2 = 1$.

For every $n > 1$ we set

$$U_n := X_2 + X_3 + \ldots + X_n \quad \text{and} \quad \underline{U}_n := \min_{2 \le i \le n} U_i. \tag{36}$$

It is easy to see that

$$\{T_n > n\} = \{Y_n = N_n\} \cap \left\{\underline{U}_n > -N_n\right\}.$$

Noting now that $\underline{U}_n \ge -(n-1)M$, we infer that

$$\{T_n > n\} = \{Y_n = N_n\}, \quad \text{for any } N_n > (n-1)M. \tag{37}$$

In this case we have

$$\begin{aligned} E_n = \mathbf{E}[S_{n,n}; T_n > n] &= \mathbf{E}\left[\frac{Y_n + U_n}{\sqrt{n}}; Y_n = N_n\right] \\ &= \mathbf{P}(Y_n = N_n)\mathbf{E}\left[\frac{N_n + U_n}{\sqrt{n}}\right] = \mathbf{P}(Y_n = n)\frac{N_n + \mathbf{E}U_n}{\sqrt{n}} \\ &= \mathbf{P}(Y_n = n)\frac{N_n}{\sqrt{n}}. \end{aligned} \tag{38}$$

In particular, from (37) and (38) we conclude that

$$\mathbf{P}(T_n > n) = \mathbf{P}(Y_n = n) = \frac{E_n\sqrt{n}}{N_n} < \frac{E_n\sqrt{n}}{M(n-1)} = o(E_n)$$

provided that $N_n > (n-1)M$.

This example shows that (11) can not hold for all triangular arrays satisfying the Lindeberg condition. ◇

We now construct an array, for which the assumption (13) becomes necessary for the validity of (11).

Example 7 We consider again the model from the previous example and assume additionally that the variables $X_2, X_3, \ldots$ have the Rademacher distribution, that is,

$$\mathbf{P}(X_k = \pm 1) = \frac{1}{2}.$$

Finally, in order to have random walks on lattices, we shall assume that N_n is a natural number.

It is then clear that $r_n := \frac{N_n}{\sqrt{n}}$ is the minimal deterministic number such that

$$\max_{k \le n} |X_{k,n}| \le r_n.$$

As in Example 6, we shall assume that $g_{k,n} \equiv 0$.

In order to calculate E_n we note that

$$\begin{aligned} E_n = \mathbf{E}[S_{n,n}; T_n > n] &= \mathbf{P}\left(X_{1,n} = r_n\right) \mathbf{E}\left[r_n + \frac{U_n}{\sqrt{n}}; r_n + \frac{\underline{U}_n}{\sqrt{n}} > 0\right] \\ &= \mathbf{P}\left(X_{1,n} = r_n\right) \frac{1}{\sqrt{n}} \mathbf{E}\left[N_n + U_n; N_n + \underline{U}_n > 0\right]. \end{aligned}$$

It is well known that for $m \ge 1$ the sequence $(N + U_m)1_{\{N + \underline{U}_m > 0\}}$ is a martingale with $U_1 = \underline{U}_1 = 0$. This implies that

$$\mathbf{E}[N + U_m; N + \underline{U}_m > 0] = N \quad \text{for all} \quad m, N \ge 1.$$

Consequently,

$$E_n = p_n \frac{N_n}{\sqrt{n}} = p_n r_n. \tag{39}$$

Furthermore,

$$\mathbf{P}(T_n > n) = \mathbf{P}\left(X_{1,n} = r_n\right) \mathbf{P}\left(\frac{N_n}{\sqrt{n}} + \frac{\underline{U}_n}{\sqrt{n}} > 0\right) = p_n \mathbf{P}(N_n + \underline{U}_n > 0).$$

Using the reflection principle for the symmetric simple random walk, one can show that

$$\mathbf{P}\left(N + \underline{U}_m > 0\right) = \mathbf{P}(-N < U_m \le N) \quad \text{for all} \quad m, N \ge 1. \tag{40}$$

Consequently, $\mathbf{P}(T_n > n) = p_n \mathbf{P}(-N_n < U_n \le N_n)$. Combining this equality with (39), we obtain

$$\frac{\mathbf{P}(T_n > n)}{E_n} = \frac{1}{r_n}\mathbf{P}\left(-r_n < \frac{U_n}{\sqrt{n}} \le r_n\right). \tag{41}$$

Using the central limit theorem, one obtains

$$\mathbf{P}\left(-r_n < \frac{U_n}{\sqrt{n}} \le r_n\right) \sim \Psi(r_n), \tag{42}$$

where

$$\varphi(u) := \frac{1}{\sqrt{2\pi}} e^{-u^2/2} \quad \text{and} \quad \Psi(x) := 2\int_0^{x^+} \varphi(u)du. \tag{43}$$

We will postpone the proof of (40) and (42) till the end of the paper. Assuming that (40) and (42) are true, as a result we have

$$\frac{\mathbf{P}(T_n > n)}{E_n} \sim \frac{\Psi(r_n)}{r_n}.$$

Noting now that $\frac{\Psi(a)}{a} < 2\varphi(0) = \sqrt{\frac{2}{\pi}}$ for every $a > 0$, we conclude that the assumption $r_n \to 0$ is necessary and sufficient for the validity of (11). More precisely,

- $\mathbf{P}(T_n > n) \sim \sqrt{\frac{2}{\pi}} E_n$ iff $r_n \to 0$;
- $\mathbf{P}(T_n > n) \sim \frac{\Psi(a)}{a} E_n$ iff $r_n \to a > 0$;
- $\mathbf{P}(T_n > n) = o(E_n)$ iff $r_n \to \infty$.

◇

These examples show that the standard asymptotic negligibility condition

$$\max_{k \le n} \mathbf{P}(|X_{k,n}| > r_n) \to 0 \quad \text{for some} \quad r_n \downarrow 0$$

is not sufficient for the validity of (11). (It is well known that this asymptotic negligibility follows from the Lindeberg condition.) Our assumption (13) is a much stronger version of the asymptotic negligibility. This fact leads to the question, whether (11) holds under a weaker assumption. For example, one can consider arrays satisfying

$$\max_{k \le n} \mathbf{P}(|X_{k,n}| \ge r_n x) \le \mathbf{P}(\zeta > x), \quad \mathbf{E}\zeta^2 < \infty$$

or

$$\max_{k\le n}\frac{\mathbf{E}|X_{k,n}|^3}{\mathbf{E}X_{k,n}^2}\le r_n.$$

(We are grateful to the referee, who has suggested to consider the first of these assumptions.) We shall try to answer this question in our future works. In the current paper we are sticking to (13) for the following reasons. First, this assumption allows us to obtain lower and upper bounds for $\mathbf{P}(T_n > n)$, see (14) and (15). These bounds provide also a rate of convergence in (11). Another, even more important, feature of these bounds is a very clear form: they contain only g_n^* and r_n. These estimates play a crucial role in our ongoing project, where we consider exit times for random walks, which belong to the domain of attraction of the Brownian motion but the Lindeberg condition may fail. Second, the use of (13) allows us to give much simpler and shorter proofs.

2 Proofs

In this section we are going to obtain estimates, which are valid for each fixed n. For that reason we will sometimes omit the subscript n and introduce the following simplified notation:

$$T := T_n,\quad X_k := X_{k,n},\quad S_k := S_{k,n},\quad g_k := g_{k,n},\quad 1\le k<n \tag{44}$$

and

$$\rho := r_n + g_n^*,\quad B_k^2 := \sum_{i=1}^k \mathbf{E}X_i^2,\quad B_{k,n}^2 := B_n^2 - B_k^2 = 1 - B_k^2,\quad 1\le k<n. \tag{45}$$

2.1 Some Estimates in the Central Limit Theorem

For every integer $1\le k\le n$ and every real y define

$$Z_k := S_k - g_k,\ \widehat{Z}_k := Z_k\mathbf{1}\{T>k\} \text{ and } Q_{k,n}(y) := \mathbf{P}\Big(y + \min_{k\le j\le n}(Z_j - Z_k) > 0\Big). \tag{46}$$

Lemma 1 *For all $y \in R$ and for all $0 \le k < n$ with $B_{k,n} > 0$*

$$\left|Q_{k,n}(y) - \Psi\Big(\frac{y}{B_{k,n}}\Big)\right| \le \frac{C_0 \rho}{B_{k,n}} 1\{y > 0\}, \tag{47}$$

where C_0 is an absolute constant.

Proof For non-random real y define

$$q_{k,n}(y) := \mathbf{P}\Big(y + \min_{k \le j \le n}(S_j - S_k) > 0\Big), \quad n > k \ge 1. \tag{48}$$

It follows from Corollary 1 in Arak [1] that there exists an absolute constant C_A such that

$$\left|q_{k,n}(y) - \Psi\Big(\frac{y}{B_{k,n}}\Big)\right| \le \frac{C_A}{B_{k,n}} \max_{k<j\le n} \frac{\mathbf{E}|X_j|^3}{\mathbf{E}X_j^2} \le \frac{C_A r_n}{B_{k,n}}, \tag{49}$$

where maximum is taken over all j satisfying $\mathbf{E}X_j^2 > 0$. In the second step we have used the inequality $\mathbf{E}|X_j|^3 \le r_n \mathbf{E}X_j^2$ which follows from (13).

We have from (46) that $|Z_k - S_k| = |g_k| \le g_n^*$. Hence, for $Q_{k,n}$ and $q_{k,n}$ defined in (46) and (48), we have

$$q_{k,n}(y_-) \le Q_{k,n}(y) \le q_{k,n}(y_+), \quad \text{where} \quad y_\pm := y \pm 2g_n^*. \tag{50}$$

Then we obtain from (49) that

$$\left|q_{k,n}(y_\pm) - \Psi\Big(\frac{y_\pm}{B_{k,n}}\Big)\right| \le \frac{C_A r_n}{B_{k,n}}. \tag{51}$$

On the other hand, it is easy to see from (43) that

$$\left|\Psi\Big(\frac{y_\pm}{B_{k,n}}\Big) - \Psi\Big(\frac{y}{B_{k,n}}\Big)\right| \le \frac{2\varphi(0)|y_\pm - y|}{B_{k,n}} = \frac{4\varphi(0) g_n^*}{B_{k,n}}.$$

Applying this inequality together with (50) and (51) we immediately obtain (47) for $y > 0$ with $C_0 := C_A + 4\varphi(0)$. For $y \le 0$ inequality (47) immediately follows since $Q_{k,n}(y) = 0 = \Psi(y)$. □

Lemma 2 *If $1 \le m \le n$, then*

$$\mathbf{E}S_m^+ \ge \frac{3}{8} B_m - r_n. \tag{52}$$

Moreover, for all m satisfying $B_m \ge 24(r_n + g_n^*)$ *we have*

$$\mathbf{P}(T > m) \le 3\frac{\mathbf{E}\widehat{Z}_m}{B_m}. \tag{53}$$

Proof We will use the following extension of the Berry-Esseen inequality due to Tyurin [5]:

$$\sup_{x\in\mathbb{R}} |\mathbf{P}(S_m > x) - \mathbf{P}(B_m\eta > x)| \le 0.5606\frac{\sum_{j=1}^m \mathbf{E}|X_j|^3}{B_m^3} \le 0.5606\frac{r_n}{B_m},$$

when $B_m > 0$. Here η is a random variable that follows the standard normal distribution. This inequality implies that, for every $C > 0$,

$$\begin{aligned}\mathbf{E}S_m^+ &= \int_0^\infty \mathbf{P}(S_m > x)dx \ge \int_0^{CB_m} \mathbf{P}(S_m > x)dx \\ &\ge \int_0^{CB_m}\left(\mathbf{P}(B_m\eta > x) - 0.5606\frac{r_n}{B_m}\right)dx = B_m\mathbf{E}(\eta^+ \wedge C) - 0.5606Cr_n.\end{aligned}$$

Further,

$$\begin{aligned}\mathbf{E}(\eta^+ \wedge C) = \int_0^\infty (x \wedge C)\varphi(x)dx &= \int_0^C x\frac{1}{2\pi}e^{-x^2/2}dx + C\int_C^\infty \varphi(x)dx \\ &= \varphi(0) - \varphi(C) + C\int_C^\infty \varphi(x)dx.\end{aligned}$$

Taking here $C = 1/0.5606$ and using tables of the standard normal distribution we conclude that $\mathbf{E}(\eta^+ \wedge C) > 0.375 > \frac{3}{8}$ and (52) holds.

Next, according to Lemma 25 in [2],

$$\mathbf{E}Z_m^+\mathbf{P}(T > m) \le \mathbf{E}\widehat{Z}_m, \qquad 1 \le m \le n. \tag{54}$$

Therefore, it remains to derive a lower bound for $\mathbf{E}Z_m^+$. We first note that

$$S_m = Z_m + g_m \le Z_m^+ + g_m^+ \le Z_m^+ + g_n^*.$$

Hence, $S_m^+ \le Z_m^+ + g_n^*$ and, taking into account (52), we get

$$\mathbf{E}Z_m^+ \ge \mathbf{E}S_m^+ - g_n^* \ge \frac{3}{8}B_m - (r_n + g_n^*). \tag{55}$$

If m is such that $\frac{B_m}{24} \geq r_n + g_n^*$, then we infer from (54) and (55) that

$$\mathbf{E}\widehat{Z}_m \geq \mathbf{E}Z_m^+\mathbf{P}(T > m) \geq \left(\frac{3}{8}B_m - (r_n + g_n^*)\right)\mathbf{P}(T > n)$$
$$\geq \left(\frac{3}{8} - \frac{1}{24}\right)B_m\mathbf{P}(T > m) = \frac{1}{3}B_m\mathbf{P}(T > m).$$

Thus, (53) is proven. □

2.2 Estimates for Expectations of $\widehat{Z}_k$

Lemma 3 *Let α be a stopping time such that $1 \leq \alpha \leq l \leq n$ with probability one. Then*

$$\mathbf{E}\widehat{Z}_\alpha - \mathbf{E}\widehat{Z}_l \leq 2g_n^* p(\alpha, l) \quad \textit{with} \quad p(\alpha, l) := \mathbf{P}(\alpha < T, \alpha < l). \tag{56}$$

Moreover,

$$\mathbf{E}\widehat{Z}_\alpha - \mathbf{E}\widehat{Z}_l \geq \mathbf{E}[X_T; \alpha < T \leq l] - 2g_n^* p(\alpha, l) \geq -(2g_n^* + r_n)p(\alpha, l). \tag{57}$$

In addition, the equality in (12) *takes place.*

Proof Define events

$$A_1 := \{\alpha < T \leq l\} \quad \text{and} \quad A_2 := \{\alpha < l < T\}.$$

Then, clearly, $\{\alpha < T, \alpha < l\} = A_1 \cup A_2$. Using Lemma 20 from [2], we obtain

$$\begin{aligned}
\mathbf{E}\widehat{Z}_\alpha + \mathbf{E}[S_T; T \leq \alpha] &= -\mathbf{E}[g_\alpha; \alpha < T] \\
&= -\mathbf{E}[g_\alpha; A_2] - \mathbf{E}[g_l; \alpha = l < T] - \mathbf{E}[g_\alpha; A_1], \\
\mathbf{E}\widehat{Z}_l + \mathbf{E}[S_T; T \leq l] &= -\mathbf{E}[g_l; T > l] = -\mathbf{E}[g_l; A_2] - \mathbf{E}[g_l; \alpha = l < T].
\end{aligned} \tag{58}$$

Thus,

$$\mathbf{E}\widehat{Z}_\alpha - \mathbf{E}\widehat{Z}_l = \mathbf{E}[S_T - g_\alpha; A_1] + \mathbf{E}[g_l - g_\alpha; A_2]. \tag{59}$$

Next, by the definition of T,

$$g_T \geq S_T = S_{T-1} + X_T > g_{T-1} + X_T.$$

Hence,

$$\mathbf{E}[S_T - g_\alpha; A_1] \le \mathbf{E}[g_T - g_\alpha; A_1] \le 2g_n^*\mathbf{P}(A_1)$$

and

$$\begin{aligned}\mathbf{E}[S_T - g_\alpha; A_1] &\ge \mathbf{E}[g_{T-1} - g_\alpha + X_T; A_1] \\ &\ge \mathbf{E}[X_T; A_1] - 2g_n^*\mathbf{P}(A_1) \ge -(2g_n^* + r_n)\mathbf{P}(A_1).\end{aligned}$$

Furthermore,

$$|\mathbf{E}[g_n - g_\alpha; A_2]| \le 2g_n^*\mathbf{P}(A_2).$$

Plugging these estimates into (59), we arrive at desired bounds.

The equality in (12) follows from (58) with $l = n$. □

For every $h > 0$ define

$$\nu(h) := \inf\{k \ge 1 : S_k \ge g_k + h\} = \inf\{k \ge 1 : Z_k \ge h\}. \tag{60}$$

Lemma 4 *Suppose that $m \le n$ is such that the inequality* (53) *takes place,*

$$B_m \ge 24g_n^* \quad \textit{and} \quad h \ge 6g_n^*. \tag{61}$$

Then

$$2\mathbf{E}\widehat{Z}_{\nu(h)\wedge m} \le 3\mathbf{E}\widehat{Z}_m \le 4\mathbf{E}\widehat{Z}_n = 4E_n, \quad \mathbf{P}(\widehat{Z}_{\nu(h)\wedge m} > 0) \le \varkappa E_n, \tag{62}$$

$$2\varkappa g_n^* E_n \ge \mathbf{E}\widehat{Z}_{\nu(h)\wedge m} - E_n \ge \delta(h) - 2\varkappa g_n^* E_n, \tag{63}$$

where

$$0 \ge \delta(h) := \mathbf{E}[X_T; n \ge T > \nu(h) \wedge m] \ge -\varkappa r_n E_n \quad \textit{and} \quad \varkappa := \frac{2}{h} + \frac{4}{B_m}. \tag{64}$$

In particular, (16) *takes place.*

Proof First, we apply Lemma 3 with $l = m$ and $\alpha = \nu(h) \wedge m$. For this choice of the stopping time one has

$$\begin{aligned}p(\nu(h) \wedge m, m) &= \mathbf{P}\left(\nu(h) \wedge m < T, \nu(h) \wedge m < m\right) \\ &\le \mathbf{P}(\widehat{Z}_{\nu(h)\wedge m} \ge h) \le \frac{\mathbf{E}\widehat{Z}_{\nu(h)\wedge m}}{h}.\end{aligned}$$

Plugging this bound into (56) and using the inequality $h \geq 6g_n^*$, we get

$$\mathbf{E}\widehat{Z}_{\nu(h)\wedge m} - \mathbf{E}\widehat{Z}_m \leq \frac{2g_n^*}{h}\mathbf{E}\widehat{Z}_{\nu(h)\wedge m} \leq \frac{\mathbf{E}\widehat{Z}_{\nu(h)\wedge m}}{3}$$

and hence

$$\frac{2}{3}\mathbf{E}\widehat{Z}_{\nu(h)\wedge m} \leq \mathbf{E}\widehat{Z}_m. \tag{65}$$

Next, we apply Lemma 3 with $l = n$ and $\alpha = m$. In this case $p(m,n) = \mathbf{P}(T > m)$ and we may use (53). Substituting these estimates into (56) and using (61), we obtain

$$\mathbf{E}\widehat{Z}_m - \mathbf{E}\widehat{Z}_n \leq 2g_n^*\mathbf{P}(T > m) \leq \frac{6g_n^*}{B_m}\mathbf{E}\widehat{Z}_m \leq \frac{1}{4}\mathbf{E}\widehat{Z}_m.$$

Therefore,

$$\frac{3}{4}\mathbf{E}\widehat{Z}_m \leq \mathbf{E}\widehat{Z}_n. \tag{66}$$

We conclude from (65) and (66) that the first relation in (62) takes place. In particular, from (53) and (66) we get that (16) holds under assumptions of Lemma 4.

At last, we are going to apply Lemma 3 with $l = n > m$ and $\alpha = \nu(h) \wedge m$. For this choice of the stopping time one has

$$\begin{aligned} p(\nu(h) \wedge m, n) &= \mathbf{P}\left(T > \nu(h) \wedge m\right) = \mathbf{P}(\widehat{Z}_{\nu(h)\wedge m} > 0) \\ &\leq \mathbf{P}(\widehat{Z}_{\nu(h)\wedge m} \geq h) + \mathbf{P}\left(T > m\right) \\ &\leq \frac{\mathbf{E}\widehat{Z}_{\nu(h)\wedge m}}{h} + \frac{3\mathbf{E}\widehat{Z}_m}{B_m} \leq \frac{2E_n}{h} + \frac{4E_n}{B_m} = \varkappa E_n. \end{aligned} \tag{67}$$

Plugging this bound into (56) and (57), we immediately obtain (63). The second inequality in (62) also follows from (67); and using (13) together with (67) we find (64).

Thus, all assertions of Lemma 4 are proved.

2.3 *Proof of Theorem* 1

According to the representation (36) in [2],

$$\begin{aligned} \mathbf{P}(T > n) &= \mathbf{E}\left[Q_{\nu(h)\wedge m,n}(Z_{\nu(h)\wedge m}); T > \nu(h) \wedge m\right] \\ &= \mathbf{E}Q_{\nu(h)\wedge m,n}(\widehat{Z}_{\nu(h)\wedge m}). \end{aligned} \tag{68}$$

Lemma 5 *Suppose that all assumptions of Lemma* 4 *are fulfilled and that* $B_{m,n} > 0$. *Then one has*

$$\left|\mathbf{P}(T > n) - \mathbf{E}\Psi\Big(\frac{\widehat{Z}_{\nu(h)\wedge m}}{B_{\nu(h)\wedge m,n}}\Big)\right| \le \frac{C_0\rho}{B_{m,n}}\mathbf{P}(\widehat{Z}_{\nu(h)\wedge m} > 0)$$
$$\le 2\varphi(0)\frac{1.3C_0\varkappa\rho E_n}{B_{m,n}}. \tag{69}$$

In addition,

$$\mathbf{E}\Psi\Big(\frac{\widehat{Z}_{\nu(h)\wedge m}}{B_{\nu(h)\wedge m,n}}\Big) \le \frac{2\varphi(0)E_n(1+2\varkappa g_n^*)}{B_{m,n}}, \tag{70}$$

$$\mathbf{E}\Psi\Big(\frac{\widehat{Z}_{\nu(h)\wedge m}}{B_{\nu(h)\wedge m,n}}\Big) \ge 2\varphi(0)E_n\Big(1 - \frac{(r_n+h)^2}{6} - 2\varkappa g_n^* - \varkappa r_n\Big). \tag{71}$$

Proof Using (47) with $y = \widehat{Z}_{\nu(h)\wedge m}$, we obtain the first inequality in (69) as a consequence of (68). The second inequality in (69) follows from (62).

Next, it has been shown in [2, p. 3328] that

$$2\varphi(0)a \ge \Psi(a) \ge 2\varphi(0)a(1-a^2/6) \quad \text{for all } a \ge 0. \tag{72}$$

Recall that $0 \le z := \widehat{Z}_{\nu(h)\wedge m} \le r_n + h$ and $B_n = 1$. Hence, by (72),

$$\Psi\Big(\frac{z}{B_{\nu(h)\wedge m,n}}\Big) \le \Psi\Big(\frac{z}{B_{m,n}}\Big) \le \frac{2\varphi(0)z}{B_{m,n}}, \tag{73}$$

$$\Psi\Big(\frac{z}{B_{\nu(h)\wedge m,n}}\Big) \ge \Psi\Big(\frac{z}{B_n}\Big) \ge \frac{2\varphi(0)z}{B_n}\Big(1 - \frac{z^2}{6B_n^2}\Big) \ge 2\varphi(0)z\Big(1 - \frac{(r_n+h)^2}{6}\Big). \tag{74}$$

Taking mathematical expectations in (73) and (74) with $z = \widehat{Z}_{\nu(h)\wedge m}$, we obtain:

$$\frac{2\varphi(0)\mathbf{E}\widehat{Z}_{\nu(h)\wedge m}}{B_{m,n}} \ge \mathbf{E}\Psi\Big(\frac{\widehat{Z}_{\nu(h)\wedge m}}{B_{\nu(h)\wedge m,n}}\Big) \ge 2\varphi(0)\mathbf{E}\widehat{Z}_{\nu(h)\wedge m}\Big(1 - \frac{(r_n+h)^2}{6}\Big). \tag{75}$$

Now (70) and (71) follow from (75) together with (62) and (63).

Lemma 6 *Assume that* $\rho \le 1/64$. *Then inequalities* (14) *and* (15) *take place with some absolute constants* C_1 *and* C_2.

Proof Set

$$m := \min\{j \le n : B_j \ge \frac{3}{2}\rho^{1/3}\} \quad \text{and} \quad h := \rho^{1/3}. \tag{76}$$

Noting that $r_n \le \rho \le \rho^{1/3}/4^2$ we obtain

$$B_m^2 = B_{m-1}^2 + \mathbf{E}X_m^2 < \left(\frac{3}{2}\rho^{1/3}\right)^2 + r_n^2 \le \frac{9}{4}\rho^{2/3} + \frac{1}{4^6} < \frac{1}{7}. \tag{77}$$

Consequently, $B_{m,n}^2 = 1 - B_m^2$ and we have from (76) that

$$B_{m,n}^2 > \frac{6}{7}, \quad 24\rho \le \frac{24}{4^2}\rho^{1/3} = \frac{3}{2}\rho^{1/3} \le B_m, \quad 6g_n < \frac{6}{4^2}\rho^{1/3} < \rho^{1/3} = h. \tag{78}$$

Thus, all assumptions of Lemmas 4 and 5 are satisfied. Hence, Lemma 5 implies that

$$2\varphi(0)E_n(1 - \rho_1 - \rho_2 - 2\varkappa\rho) \le \mathbf{P}(T > n), \tag{79}$$

$$\mathbf{P}(T > n) \le 2\varphi(0)E_n(1 + \rho_1)(1 + 2\varkappa\rho)(1 + \rho_3), \tag{80}$$

where we used that $2g_n^* + r_n \le 2\rho$ and

$$\rho_1 := 1.3C_0\varkappa\rho, \quad \rho_2 := \frac{(r_n + h)^2}{6}, \quad \rho_3 := \frac{1}{B_{m,n}} - 1. \tag{81}$$

Now from (64) and (76) with $\rho^{1/3} \le 1/4$ we have

$$\rho\varkappa = \frac{2\rho}{h} + \frac{4\rho}{B_m} \le 2\rho^{2/3} + \frac{4\rho^{2/3}}{3/2} < 4.7\rho^{2/3}, \quad r_n + h \le \frac{1}{4^2}\rho^{1/3} + \rho^{1/3}.$$

Then, by (77),

$$\frac{1}{B_{m,n}} = \frac{B_{m,n}}{B_{m,n}^2} = \frac{\sqrt{1 - B_m^2}}{1 - B_m^2} \le \frac{1 - B_m^2/2}{1 - B_m^2} = 1 + \frac{B_m^2}{2B_{m,n}^2} < 1 + 1.4\rho^{2/3}.$$

So, these calculations and (81) yield

$$\rho_1 < 5C_0\rho^{2/3}, \quad \rho_2 < 0.2\rho^{2/3}, \quad \rho_3 < 1.4\rho^{2/3}, \quad 2\varkappa\rho < 9.4\rho^{2/3}. \tag{82}$$

Substituting (82) into (79) we obtain (14) with any $C_1 \ge 5C_0 + 9.6$. On the other hand from (82) and (80) we may obtain (15) with a constant C_2 which may be calculated in the following way:

$$C_2 = \sup_{\rho^{1/3} \le 1/4} [5C_0(1 + 2\varkappa\rho)(1 + \rho_3) + 9.4(1 + \rho_3) + 1.4] < \infty.$$

□

Thus, when $\rho \le 1/4^3$, the both assertions of Theorem 1 immediately follow from Lemma 6. But if $\rho > 1/4^3$ then (15) is valid with any $C_1 \ge 4^2 = 16$ because in this case right-hand side in (15) is negative.

Let us turn to the upper bound (15). If $\rho \le \frac{1}{24}$ but $\rho > \frac{1}{64}$ then (16) holds for $m = n$; and as a result we have from (16) with any $C_2 \ge 32/\varphi(0)$ that

$$\mathbf{P}(T_n > n) \le 4E_n \le 4^3 E_n \rho^{2/3} \le 2\varphi(0)E_n(1 + C_2\rho^{2/3}) \quad \text{for} \quad \rho^{1/3} > 1/4.$$

So, we have proved all assertions of Theorem 1 in all cases.

2.4 *Proof of Corollary* 3

In order to apply Corollary 2 we introduce the following triangular array:

$$X_{j,n} := \frac{u_{j,n}X_j}{\sigma_n}, \quad g_{j,n} := \frac{G_{j,n}}{\sigma_n}, \quad 1 \le j \le n,\ n \ge 1. \tag{83}$$

The assumptions in (21) and (22) imply that the array introduced in (83) satisfies (13) and (5). Thus,

$$\begin{aligned}\mathbf{P}(\tau_n > n) = \mathbf{P}(T_n > n) &\sim \sqrt{\frac{2}{\pi}}\mathbf{E}[S_{n,n} - g_{n,n}; T_n > n]\\ &= \sqrt{\frac{2}{\pi}}\Big(\mathbf{E}[S_{n,n}; T_n > n] - g_{n,n}\mathbf{P}(T_n > n)\Big).\end{aligned}$$

Here we also used (12). Since $g_{n,n} \to 0$, we conclude that

$$\mathbf{P}(\tau_n > n) \sim \sqrt{\frac{2}{\pi}}\mathbf{E}[S_{n,n}; T_n > n].$$

Noting that $S_{n,n} = U_{n,n}/\sigma_n$, we get

$$\mathbf{P}(\tau_n > n) \sim \sqrt{\frac{2}{\pi}}\frac{1}{\sigma_n}\mathbf{E}[U_{n,n}; \tau_n > n]. \tag{84}$$

By the optional stopping theorem,

$$\mathbf{E}[U_{n,n}; \tau_n > n] = -\mathbf{E}[U_{\tau_n,n}; \tau_n \le n].$$

It follows from (23) that, for every fixed $k \ge 1$,

$$U_{k,n} \to U_k \text{ a.s.} \tag{85}$$

and, taking into account the continuity of distribution functions,

$$\begin{aligned}\mathbf{P}(\tau_n > k) &= \mathbf{P}(U_{1,n} > G_{1,n}, U_{2,n} > G_{2,n}, \ldots, U_{k,n} > G_{k,n}) \\ &\to \mathbf{P}(U_1 > g_1, U_2 > g_2, \ldots, U_k > g_k) = \mathbf{P}(\tau > k). \qquad (86)\end{aligned}$$

Obviously, (86) implies that

$$\mathbf{P}(\tau_n = k) \to \mathbf{P}(\tau = k) \quad \text{for every } k \ge 1. \qquad (87)$$

Furthermore, it follows from the assumptions (17) and (21) that

$$|U_{\tau_n,n}| \le M \quad \text{on the event } \{\tau_n \le n\}. \qquad (88)$$

Then, combining (85), (87) and (88), we conclude that

$$\mathbf{E}[U_{\tau_n,n}; \tau_n \le k] = \sum_{j=1}^{k} \mathbf{E}[U_{j,n}; \tau_n = j] \to \sum_{j=1}^{k} \mathbf{E}[U_j; \tau = j] = \mathbf{E}[U_\tau; \tau \le k]. \qquad (89)$$

Note also that, by (88) and (86),

$$\limsup_{n\to\infty} |\mathbf{E}[U_{\tau_n,n}; k < \tau_n \le n]| \le M \limsup_{n\to\infty} \mathbf{P}(\tau_n > k).$$

Therefore,

$$\begin{aligned}\limsup_{n\to\infty} \mathbf{E}[U_{\tau_n,n}; \tau_n \le n] &\le \limsup_{n\to\infty} \mathbf{E}[U_{\tau_n,n}; \tau_n \le k] + \limsup_{n\to\infty} |\mathbf{E}[U_{\tau_n,n}; k < \tau_n \le n]| \\ &= \mathbf{E}[U_\tau; \tau \le k] + M\mathbf{P}(\tau > k) \qquad (90)\end{aligned}$$

and

$$\begin{aligned}\liminf_{n\to\infty} \mathbf{E}[U_{\tau_n,n}; \tau_n \le n] &\ge \liminf_{n\to\infty} \mathbf{E}[U_{\tau_n,n}; \tau_n \le k] - \limsup_{n\to\infty} |\mathbf{E}[U_{\tau_n,n}; k < \tau_n \le n]| \\ &= \mathbf{E}[U_\tau; \tau \le k] - M\mathbf{P}(\tau > k). \qquad (91)\end{aligned}$$

Letting $k \to \infty$ in (90) and (91), and noting that τ is almost surely finite, we infer that

$$\mathbf{E}[U_{\tau_n,n}; \tau_n \le n] \to \mathbf{E}[U_\tau].$$

Consequently, by the optional stopping theorem,

$$\mathbf{E}[U_{\tau_n,n}; \tau_n > n] = -\mathbf{E}[U_{\tau_n,n}; \tau_n \le n] \to \mathbf{E}[-U_\tau].$$

Plugging this into (84), we obtain the desired result.

2.5 Calculations Related to Example 7

Lemma 7 *For the simple symmetric random walk $\{U_m\}$ one has*

$$\mathbf{P}\left(N+\underline{U}_m>0\right)=\mathbf{P}(-N<U_m\le N)\quad \textit{for all}\quad m,N\ge 1$$

and

$$\sup_{N\ge 1}\left|\frac{\mathbf{P}(-N<U_n\le N)}{\Psi(N/\sqrt{n})}-1\right|\to 0.$$

Proof By the reflection principle for symmetric simple random walks,

$$\mathbf{P}\left(N+U_m=k, N+\underline{U}_m\le 0\right)=\mathbf{P}(U_m=N+k)\quad \text{for every } k\ge 1.$$

Thus, by the symmetry of the random walk U_m,

$$\mathbf{P}\left(N+U_m>0, N+\underline{U}_m\le 0\right)=\mathbf{P}(U_m<-N)=\mathbf{P}(U_m>N).$$

Therefore,

$$\begin{aligned}\mathbf{P}\left(N+\underline{U}_m>0\right)&=\mathbf{P}\left(N+U_m>0\right)-\mathbf{P}\left(N+U_m>0, N+\underline{U}_m\le 0\right)\\&=\mathbf{P}(U_m>-N)-\mathbf{P}(U_m>N)=\mathbf{P}(-N<U_m\le N).\end{aligned}$$

We now prove the second statement. Recall that U_n is the sum of $n-1$ independent, Rademacher distributed random variables. By the central limit theorem, $U_n/\sqrt{n-1}$ converges to the standard normal distribution. Therefore, $U_n/\sqrt{n}$ has the same limit. This means that

$$\varepsilon_n^2:=\sup_{x>0}|\mathbf{P}(-x\sqrt{n}<U_n\le x\sqrt{n})-\Psi(x)|\to 0.$$

Taking into account that $\Psi(x)$ increases, we conclude that, for every $\delta>0$,

$$\sup_{x\ge\delta}\left|\frac{\mathbf{P}(-x\sqrt{n}<U_n\le x\sqrt{n})}{\Psi(x)}-1\right|\le\frac{\varepsilon_n^2}{\Psi(\delta)}.$$

Choose here $\delta=\varepsilon_n$. Noting that $\Psi(\varepsilon_n)\sim 2\varphi(0)\varepsilon_n$, we obtain

$$\sup_{N\ge\varepsilon_n\sqrt{n}}\left|\frac{\mathbf{P}(-N<U_n\le N)}{\Psi(N/\sqrt{n})}-1\right|\le\frac{\varepsilon_n^2}{\Psi(\varepsilon_n)}\sim\frac{\varepsilon_n}{2\varphi(0)}\to 0.$$

It remains to consider the case $N \le \varepsilon_n \sqrt{n}$. Here we shall use the local central limit theorem. Since U_n is 2-periodic,

$$\sup_{k:\, k \equiv n-1(\mathrm{mod}2)} |\sqrt{n-1}\mathbf{P}(U_n = k) - 2\varphi(k/\sqrt{n-1})| \to 0.$$

Noting that

$$\sup_{k \le \varepsilon_n \sqrt{n}} |\varphi(k/\sqrt{n-1}) - \varphi(0)| \to 0,$$

we obtain

$$\sup_{N \le \varepsilon_n \sqrt{n}} \left| \frac{\sqrt{n-1}\mathbf{P}(-N < U_n \le N)}{2\varphi(0) m(n,N)} - 1 \right| \to 0,$$

where

$$m(n,N) = \#\{k \in (-N, N] : \ k \equiv n-1(\mathrm{mod}2)\}.$$

Since the interval $(-N, N]$ contains N even and N odd lattice points, $m(n,N) = N$ for all $n, N \ge 1$. Consequently,

$$\sup_{N \le \varepsilon_n \sqrt{n}} \left| \frac{\sqrt{n-1}\mathbf{P}(-N < U_n \le N)}{2\varphi(0) N} - 1 \right| \to 0,$$

It remains now to notice that

$$\Psi(N/\sqrt{n}) \sim \frac{2\varphi(0)N}{\sqrt{n}}$$

uniformly in $N \le \varepsilon_n \sqrt{n}$. □

References

1. Arak, T.V.: On the distribution of the maximum of the succesive partial sums of independent random variables. Theory Probab Appl. **19**, 245–266 (1975)
2. Denisov, D., Sakhanenko, A., Wachtel, V.: First-passage times for random walks with non-identically distributed increments. Ann. Probab. **46**(6), 3313–3350 (2018)
3. Denisov, D., Sakhanenko, A., Wachtel, V.: First-passage times for random walks without Lindeberg condition. (2022)
4. Gaposhkin, V.F.: The law of the iterated logarithm for Cesaro's and Abel's methods of summation. Theory Probab. Appl. **10**, 411–420 (1965)
5. Tyurin, I.S.: Refinement of the upper bounds of the constants in Lyapunov's theorem. Russian Math. Surv. **65**(3), 586–588 (2010)

On Local Times of Ornstein-Uhlenbeck Processes

Nathalie Eisenbaum

Abstract We establish expressions of the local time process of an Ornstein-Uhlenbeck process in terms of the local times on curves of a Brownian motion.

Keywords Local time · Ornstein-Uhlenbeck process

1 Introduction and Main Result

The purpose of this note is to provide expressions of the local time process of an Ornstein-Uhlenbeck process in terms of the local times of a Brownian motion. Indeed we have recently been confronted to the lack of such expression. More precisely, we have recently solved a stochastic differential equation modeling macrophage dynamics in atherosclerotic plaques [4]. The solution is a functional of the local time process of an Ornstein-Uhlenbeck process and surprisingly we could not find in the literature a tractable expression of this local time. Our contribution below (Theorem 1.1) is based on the notion of local time on curves for Brownian motion. Various definitions of this notion (see [1, 3, 5]) exist and all coincide. We remind in Sect. 2, the ones that we need to establish Theorem 1.1. Theorem 1.1 is established in Sect. 3. Finally in Sect. 4, we make some remarks on the asymptotic behavior of the local time of the Ornstein-Uhlenbeck process.
Fix $\lambda > 0$. Let Y be an Ornstein-Uhlenbeck process solution of

$$Y_t = y_0 + W_t - \frac{\lambda}{2} \int_0^t Y_s ds, \tag{1.1}$$

where $(W_t, t \geq 0)$ is a real Brownian motion starting from 0 and y_0 a fixed real number.

N. Eisenbaum (✉)
Laboratoire MAP5, Université de Paris, Paris, France
e-mail: nathalie.eisenbaum@parisdescartes.fr

L. Chaumont, A. E. Kyprianou (eds.), *A Lifetime of Excursions Through Random Walks and Lévy Processes*, Progress in Probability 78,
https://doi.org/10.1007/978-3-030-83309-1_11

Doob [2] has established the following expression for Y

$$(Y_t, t \geq 0) = (e^{-\lambda t/2}(Z_{u(t)} + y_0), t \geq 0), \tag{1.2}$$

where $Z = (Z_t, t \geq 0)$ is a real Brownian motion starting from 0 and $u(t) = \frac{e^{\lambda t}-1}{\lambda}$.
We will denote the local time process of Z by $(L_t^x(Z), x \in I\!R, t \geq 0)$.
As a continuous semimartingale Y admits a local time process $(L_t^y(Y), y \in I\!R, t \geq 0)$, satisfying the following occupation time formula ([8] Chap VI - Corollary 1.6)

$$\int_0^t f(Y_s)ds = \int_{I\!R} f(y)L_t^y(Y)dy. \tag{1.3}$$

Under the assumption that the Ornstein-Uhlenbeck process Y starts at 0, the expression of its local time at 0 has already been noticed in [6]

$$(L_t^0(Y), t \geq 0) = (\int_0^{\frac{e^{\lambda t}-1}{\lambda}} \frac{1}{\sqrt{1+\lambda s}} dL_s^0(Z), t \geq 0).$$

In [6], the result is given without proof. In their introduction, the authors first mention that the local time at 0 of a continuous semimartingale $M = (\varphi(t)Z_t)_{t\geq 0}$ with Z real Brownian motion starting from 0, is given by $(L_t^0(M))_{t\geq 0} = (\int_0^t \varphi(s)d_sL_s^0(Z))_{t\geq 0}$. This can be actually obtained as a consequence of the extended occupation formula

$$\int_0^t h(s, Z_s)ds = \int_{I\!R} \int_0^t h(s, x)\, d_sL_s^x(Z)\, dx, \tag{1.4}$$

where $d_sL_s^x(Z)$ denotes integration with respect to the time variable.
Using (1.2), Y can be written under the form $(M_{u(t)})_{t\geq 0}$, with $u(t) = \frac{e^{\lambda t}-1}{\lambda}$ and $\varphi(t) = \frac{1}{\sqrt{1+\lambda t}}$. One then obtains immediately the expression of its local time at 0

$$L_t^0(Y) = L_{u(t)}^0(M) = \int_0^{u(t)} \varphi(s)d_sL_s^0(Z),\ t \geq 0.$$

The problem is that this argument does not lead to a tractable expression of $L_t^a(Y)$ for a distinct from 0. The reason is that one has to deal with more general functional of Z than functionals of type $\int_0^t h(s, Z_s)ds$. More precisely, the key notion to handle $L_t^a(Y)$ is stochastic integration over the plane with respect to the local time process $(L_t^x(Z), x \in I\!R, t \geq 0)$ as a doubly-indexed process. We give a brief exposure of that notion in Sect. 2.

Theorem 1.1 *Fix a real number y_0. The Ornstein-Uhlenbeck process Y starting from y_0, admits a local time process $(L_t^y(Y), y \in I\!R, t \geq 0)$, related to the local time process of Z, as follows*

$$(L_t^a(Y), a \in I\!R, t \geq 0) = (\int_0^{\frac{e^{\lambda t}-1}{\lambda}} \frac{1}{\sqrt{1+\lambda s}} d_s L_s^{F_a(.)}(Z), a \in I\!R, t \geq 0) \qquad (1.5)$$

where $F : a \to F_a$ is a functional from $I\!R$ into the continuous path from $I\!R_+$ into $I\!R$, defined as follows

$$F_a(t) = a\sqrt{1+\lambda t} - y_0,$$

and $(L_t^{F_a(.)}(Z), t \geq 0)$ denotes the local time process of Z along the curve F_a. In particular, one has

$$(L_t^0(Y), t \geq 0) = (\int_0^{\frac{e^{\lambda t}-1}{\lambda}} \frac{1}{\sqrt{1+\lambda s}} d_s L_s^{-y_0}(Z), t \geq 0).$$

Remark 1.2 *Note that* (1.5) *is equivalent to*

$$L_t^a(Y) = a \int_0^{u(t)} \frac{1}{Z_s + y_0} d_s L_s^{F_a(.)}(Z).$$

Besides, to avoid integration with respect to $(L_t^{F_a(.)}(Z), t \geq 0)$, one can make an integration by parts, and obtain

$$L_t^a(Y) = \frac{L_{u(t)}^{F_a(.)}(Z)}{(1+\lambda u(t))^{1/2}} + \lambda/2 \int_0^{u(t)} L_s^{F_a(.)}(Z)(1+\lambda s)^{-3/2} ds, \quad a \in I\!R, t \geq 0. \qquad (1.6)$$

Remark 1.3 *As a diffusion, the Ornstein-Uhlenbeck process Y admits a bicontinuous local time $\ell = (\ell_t^x(Y), x \in I\!R, t \geq 0)$ with respect to its speed measure m $(m(dy) = 2e^{-\frac{\lambda}{2}y^2}dy)$. One obviously has*

$$(\ell_t^x(Y), x \in I\!R, t \geq 0) = (\frac{e^{\frac{\lambda}{2}x^2}}{2} L_t^x(Y), x \in I\!R, t \geq 0).$$

But ℓ can also be related to the local time process of the Brownian motion W. Indeed denote by $(L_t^x(W), x \in I\!R, t \geq 0)$ the local time process of W. For simplicity assume that Y starts from 0*. Following Orey* [7]*, one obtains*

$$(\ell_t^x(Y), x \in I\!R, t \geq 0) = (L_{\beta(t)}^{\rho(x)}(W), x \in I\!R, t \geq 0),$$

where ρ is the inverse of S the scale function of Y ($S(x) = \int_0^x e^{\frac{\lambda}{2}y^2}dy$, $x \in I\!R$) and β is the inverse of the function α defined by

$$\alpha(t) = \int_{I\!R} L_t^x(W)(m \circ \rho)(dx) = \int_{I\!R} L_t^{S(x)}(W)m(dx).$$

One finally obtains

$$(L_t^a(Y), a \in I\!R, t \geq 0) = (2e^{-\frac{\lambda}{2}a^2} L_{\beta(t)}^{\rho(a)}(W), a \in I\!R, t \geq 0). \tag{1.7}$$

Some questions (as questions of stochastic structure) on the local time of Y might be easier to handle using (1.7) *than* (1.5).

2 Local Times on Curves

Let Z be a real Brownian motion starting from 0. If one needs a local time of Z along a deterministic measurable curve $b = (b(t), t \geq 0)$, the first idea that comes to mind is to define it as follows

$$L_t^{b(.)}(Z) = \lim_{\varepsilon \to 0} \frac{1}{2\varepsilon} \int_0^t 1_{[b(s)-\varepsilon, b(s)+\varepsilon]}(Z_s)ds, \tag{2.1}$$

which is precisely the definition introduced by Bass and Burdzy in [1]. They show that the limit is uniform in t on compact sets, in L^2.
When b is such that $(Z_t - b(t))_{t \geq 0}$ is a semi-martingale, then it admits a local time process and in particular a local time process at 0 which is given by Tanaka's formula

$$L_t^0(Z_. - b(.)) = |Z_t - b(t)| - |b(0)| - \int_0^t sgn(Z_s - b(s))d(Z_s - b(s)). \tag{2.2}$$

As noticed in [1], in that case the two definitions (2.1) and (2.2) coincide a.s.

$$L_t^{b(.)}(Z) = L_t^0(Z_. - b(.)),\ t \geq 0.$$

In [3], we have defined a local time for Z along any mesurable curve by using integration with respect to local time over the plane. We first remind this notion of integration of deterministic functions with respect to local time over the plane. Consider the space $\mathcal{H}$

$$\mathcal{H} = \{F : I\!R_+ \times I\!R \to I\!R : ||F|| < \infty\},$$

where the norm $||.||$ is defined by

$$||F|| = 2(\int_0^\infty \int_{\mathbb{R}} F^2(s,x)\exp(-\frac{x^2}{2s})\frac{dsdx}{\sqrt{2\pi s}})^{1/2} + \int_0^\infty \int_{\mathbb{R}} |xF(s,x)|\exp(-\frac{x^2}{2s})\frac{dsdx}{s\sqrt{2\pi s}}.$$

For F elementary function i.e. such that there exist a finite sequence of $\mathbb{R}_+$, $(s_i)_{1\le i\le n}$, and a finite sequence of $\mathbb{R}$, $(x_j)_{1\le j\le n}$, with

$$F(s,x) = \sum_{1\le i,j\le n} F(s_i,x_j)1_{[s_i,s_{i+1})}(s)1_{[x_j,x_{j+1})}(x),$$

one sets

$$\int_0^t \int_{\mathbb{R}} F(s,x)dL_s^x = \sum_{1\le i,j\le n} F(s_i,x_j)(L_{s_{i+1}}^{x_{j+1}} - L_{s_i}^{x_{j+1}} - L_{s_{i+1}}^{x_j} + L_{s_i}^{x_j}).$$

We have shown in [3] that for every F in $\mathcal{H}$, and every sequence of elementary functions converging to F for the norm $||.||$, $(F_n)_{n\ge 0}$, the sequence $(\int_0^t \int_{\mathbb{R}} F_n(s,x)dL_s^x)_{n\ge 0}$ converges in L^1 uniformly in t on compact sets. The limit does not depend on the choice of the sequence (F_n) and represents

$$\int_0^t \int_{\mathbb{R}} F(s,x)dL_s^x.$$

Moreover

$$\mathbb{E}[|\int_0^t \int_{\mathbb{R}} F(s,x)dL_s^x|] \le ||F||. \tag{2.3}$$

We remind that for every $t > 0$, the process $(Z_{t-s}, 0 \le s \le t)$ is a semimartingale. Extending a result of [5], we have established in [3], that

$$\int_0^v \int_{\mathbb{R}} F(s,x)dL_s^x = \int_0^v F(s,Z_s)dZ_s + \int_{t-v}^t F(t-s,Z_{t-s})dZ_{t-s},\ 0 \le v \le t. \tag{2.4}$$

We have also shown (Corollary 3.2 (ii) in [3]) that if $\frac{\partial F}{\partial x}(s,x)$ exists, then

$$\int_0^t \int_{\mathbb{R}} F(s,x)dL_s^x = -\int_0^t \frac{\partial F}{\partial x}(s,Z_s)ds. \tag{2.5}$$

Finally, it has been established in [3] that for any measurable curve $(b(s), s \ge 0)$, one has a.s.

$$(L_t^{b(.)}(Z), t \ge 0) = (\int_0^t \int_{\mathbb{R}} 1_{\{x<b(s)\}}dL_s^x(Z), t \ge 0). \tag{2.6}$$

Using (2.4), one obtains

$$(L_v^{b(.)}(Z), 0 \le v \le t) = (\int_0^v 1_{\{Z_s < b(s)\}} dZ_s + \int_{t-v}^t 1_{\{Z_{t-s} < b(t-s)\}} dZ_{t-s},\ 0 \le v \le t).$$

3 Proof of Theorem 1.1

As a continuous semimartingale, the process Y admits a local time process $(L_t^y(Y), y \in I\!R, t \ge 0)$ which satisfies the occupation time formula (1.3). This local time process admits a bicontinuous modification. Indeed, according to Theorem 1.7 in [8], the local time process of Y is a.s. continuous in t and cadlag in y with the following discontinuities

$$L_t^y(Y) - L_t^{y-}(Y) = -\frac{\lambda}{2} \int_0^t 1_{(Y_s = y)} Y_s ds = -\frac{\lambda y}{2} \int_0^t 1_{(Y_s = y)} ds = 0,$$

using (1.3). We work with a bicontinuous modification of the local time process of Y.

Using (1.2), we have

$$\int_0^t f(Y_s) ds = \int_0^t f(e^{-\lambda s/2}(Z_{u(s)} + y_0)) ds = \int_0^{\frac{e^{\lambda t}-1}{\lambda}} f(\frac{1}{\sqrt{1+\lambda u}}(Z_u + y_0)) \frac{du}{1+\lambda u}. \tag{3.1}$$

We now choose the function f: $f(x) = f_{a,\varepsilon}(x) = \frac{1}{2\varepsilon} 1_{[a-\varepsilon, a+\varepsilon]}(x)$.

On one hand, using the occupation time formula and the continuity of the local time process of Y, a.s. for every $t \ge 0$, every real a

$$\lim_{\varepsilon \to 0} \int_0^t f_{a,\varepsilon}(Y_s) ds = L_t^a(Y). \tag{3.2}$$

On the other hand, defining the function $F_{a,\varepsilon}$ on $I\!R_+ \times I\!R$ by

$$F_{a,\varepsilon}(s, x) = \int_x^\infty \frac{1}{2\varepsilon} 1_{[a-\varepsilon, a+\varepsilon]}(\frac{y + y_0}{\sqrt{1+\lambda s}}) \frac{1}{1+\lambda s} dy,$$

one notes that for every $T > 0$, the function $F_{a,\varepsilon}(s, x) 1_{[0,T]}(s)$ belongs to $\mathcal{H}$. Using (2.5), one obtains

$$\int_0^{\frac{e^{\lambda t}-1}{\lambda}} f_{a,\varepsilon}(\frac{1}{\sqrt{1+\lambda s}}(Z_s + y_0)) \frac{ds}{1+\lambda s} = \int_0^{\frac{e^{\lambda t}-1}{\lambda}} \int_{I\!R} F_{a,\varepsilon}(s, x) dL_s^x(Z),$$

where $dL_s^x(Z)$ refers to integration over the plane $\mathbb{R}_+ \times \mathbb{R}$ wrt local times (see Sect. 2).
For every $T > 0$, as ε tends to 0, the function $F_{a,\varepsilon}(s,x)1_{[0,T]}(s)$ converges for the norm $||.||$ to $\frac{1}{\sqrt{1+\lambda s}} 1_{[x,+\infty)}(a\sqrt{1+\lambda s} - y_0)1_{[0,T]}(s)$. Hence, thanks to (2.3), one obtains

$$\int_0^{\frac{e^{\lambda t}-1}{\lambda}} \int_{\mathbb{R}} F_{a,\varepsilon}(s,x) dL_s^x(Z)$$

tends in L^1 as ε tends to 0, to

$$\int_0^{\frac{e^{\lambda t}-1}{\lambda}} \int_{\mathbb{R}} \frac{1}{\sqrt{1+\lambda s}} 1_{\{x<a\sqrt{1+\lambda s}-y_0\}} dL_s^x(Z).$$

One easily shows that for any measurable bounded function h from $\mathbb{R}_+$ into $\mathbb{R}$

$$\int_0^T h(s) d_s L_s^{b(.)}(Z) = \int_0^T \int_{\mathbb{R}} h(s) 1_{\{x<b(s)\}} dL_s^x(Z).$$

Consequently, $\int_0^{\frac{e^{\lambda t}-1}{\lambda}} \int_{\mathbb{R}} F_{a,\varepsilon}(s,x) dL_s^x(Z)$ tends in L^1 to

$$\int_0^{\frac{e^{\lambda t}-1}{\lambda}} \frac{1}{\sqrt{1+\lambda s}} d_s L_s^{F_a(.)}(Z).$$

Hence, thanks to (3.1), $\int_0^t f_{a,\varepsilon}(Y_s) ds$ tends to $\int_0^{\frac{e^{\lambda t}-1}{\lambda}} \frac{1}{\sqrt{1+\lambda s}} d_s L_s^{F_a(.)}(Z)$ in L^1 uniformly in t on compact sets. This is still true finite-dimensionnaly in a. Using (3.2), one obtains then (1.5) by continuity arguments. □

4 Some Asymptotics of Local Times

Proposition 4.1 *For Y Ornstein-Uhlenbeck process starting from any fixed real number, the following properties hold for every real a.*

(i) $\limsup_{t\to\infty} \frac{L_t^a(Y)}{e^{\frac{\lambda t}{2}}\sqrt{\log t}} \leq 2\sqrt{\frac{2}{\lambda}}$ *a.s.*

and in particular: $\limsup_{t\to\infty} \frac{L_t^0(Y)}{e^{\frac{\lambda t}{2}}\sqrt{\log(t)}} \leq \sqrt{\frac{2}{\lambda}}$ *a.s.*

(ii) $\limsup_{t\to\infty} \frac{L_t^a(Y)}{\sqrt{\log t}} \geq \sqrt{\frac{2}{\lambda}}$ *a.s.*

Proof Denote by y_0 the starting point of Y. Since $F_a(s)$ is an increasing continuous function of s, $M = (Z_t - F_a(t), t \geq 0)$ is a continuous semimartingale. Consequently (2.2) leads to

$$L_t^{F_a(.)}(Z) = |Z_t + y_0 - a\sqrt{1+\lambda t}| - |a - y_0| - \int_0^t sgn(Z_s + y_0 - a\sqrt{1+\lambda s})dZ_s$$
$$+\frac{a\lambda}{2}\int_0^t sgn(Z_s + y_0 - a\sqrt{1+\lambda s})\frac{ds}{\sqrt{1+\lambda s}}. \tag{4.1}$$

Besides, Theorem 1.1 gives the following bounds

$$\frac{L_{u(t)}^{F_a(.)}(Z)}{\sqrt{1+\lambda u(t)}} \leq L_t^a(Y) \leq L_{u(t)}^{F_a(.)}(Z). \tag{4.2}$$

Set: $H(t) = \sqrt{2t \log_2 t}$. One has: $H(u(t)) \sim_{t\to\infty} \sqrt{\frac{2}{\lambda}} e^{\frac{\lambda t}{2}} \sqrt{\log t}$.

(i) According to the well-known Strassen law of iterated logarithm [9], one has for B real Brownian motion

$$\limsup_{t\to\infty} \frac{B_t}{H(t)} = 1 \; a.s. \tag{4.3}$$

Applying it to the two Brownian motions Z and $(\int_0^t sgn(Z_s + y_0 - a\sqrt{1+\lambda s})dZ_s)_{t\geq 0}$, one obtains immediately from (4.1): $\limsup_{t\to\infty} \frac{L_{u(t)}^{F_a(.)}}{e^{\frac{\lambda t}{2}}\sqrt{\log t}} \leq 2\sqrt{\frac{2}{\lambda}}$, which with the right hand inequality of (4.2) leads to (i).

From (4.3) one easily deduces the following law of iterated logarithm for local time: $\limsup_{t\to\infty} \frac{L_t^{-y_0}(Z)}{H(t)} = 1 \; a.s.$ It leads to: $\limsup_{t\to\infty} \frac{L_{u(t)}^{-y_0}(Z)}{e^{\frac{\lambda t}{2}}\sqrt{\log(t)}} = \sqrt{\frac{2}{\lambda}} \; a.s.$ Using the right hand inequality of (4.2) for $a = 0$, one obtains: $\limsup_{t\to\infty} \frac{L_t^0(Y)}{e^{\frac{\lambda t}{2}}\sqrt{\log(t)}} \leq \sqrt{\frac{2}{\lambda}} \; a.s.$

(ii) Using the left hand inequality of (4.2) one has: $\frac{L_t^a(Y)}{\sqrt{\frac{2}{\lambda}}\sqrt{\log t}} \geq \frac{L_{u(t)}^{F_a(.)}(Z)}{\sqrt{\frac{2}{\lambda}}e^{\frac{\lambda t}{2}}\sqrt{\log t}}$.

Besides, denoting by B the Brownian motion $(-\int_0^t sgn(Z_s + y_0 - a\sqrt{1+\lambda s})dZ_s, t \geq 0)$, one obtains with (4.1)

$$\limsup_{t\to\infty} \frac{L_{u(t)}^{F_a(.)}(Z)}{\sqrt{\frac{2}{\lambda}}e^{\frac{\lambda t}{2}}\sqrt{\log t}} \geq \limsup_{t\to\infty} \frac{B_{u(t)}}{H(u(t))} + \frac{|Z_{u(t)}|}{H(u(t))} \geq \limsup_{t\to\infty} \frac{B_{u(t)}}{H(u(t))} = 1 \; a.s.,$$

which leads to the result. □

References

1. Bass, R., Burdzy, K.: The supremum of Brownian local times on hölder curves. Ann. I.H. Poincaré **37**(6), 627–642 (2001)
2. Doob, J.L.: The Brownian movement and stochastic equations. Ann. Math. **43**(2), 351–369 (1942)
3. Eisenbaum, N.: Integration with respect to local time. Potential Anal. **13**, 303–328 (2000)
4. Eisenbaum, N., Meunier, N.: A lipid structured model for macrophage dynamics in atherosclerotic plaques. Work in progress
5. Föllmer, H., Protter, P., Shiryayev, A.N.: Quadratic covariation and an extension of Itô's formula. Bernoulli **1**, 149–169 (1995)
6. Gradinaru, M., Roynette, B., Vallois, P., Yor, M.: The laws of Brownian local time integrals. Comput. Appl. Math. **3**, 259–331 (1999)
7. Orey, S.: Conditions for the absolute continuity of two diffusions. Trans. Am. Math. Soc. **193**, 413–426 (1974)
8. Revuz, D., Yor, M.: Continuous Martingales and Brownian Motion, 3rd edn. Springer (1999)
9. Strassen, V.: An invariance principle for the law of the iterated logarithm. Z. Wahrscheinlichkeitstheorie und Verw. Gebiete **3**, 211–226 (1964)

Two Continua of Embedded Regenerative Sets

Steven N. Evans and Mehdi Ouaki

Abstract Given a two-sided real-valued Lévy process $(X_t)_{t\in\mathbb{R}}$, define processes $(L_t)_{t\in\mathbb{R}}$ and $(M_t)_{t\in\mathbb{R}}$ by $L_t := \sup\{h \in \mathbb{R} : h - \alpha(t-s) \le X_s \text{ for all } s \le t\} = \inf\{X_s + \alpha(t-s) : s \le t\}$, $t \in \mathbb{R}$, and $M_t := \sup\{h \in \mathbb{R} : h - \alpha|t-s| \le X_s \text{ for all } s \in \mathbb{R}\} = \inf\{X_s + \alpha|t-s| : s \in \mathbb{R}\}$, $t \in \mathbb{R}$. The corresponding contact sets are the random sets $\mathcal{H}_\alpha := \{t \in \mathbb{R} : X_t \wedge X_{t-} = L_t\}$ and $\mathcal{Z}_\alpha := \{t \in \mathbb{R} : X_t \wedge X_{t-} = M_t\}$. For a fixed $\alpha > \mathbb{E}[X_1]$ (resp. $\alpha > |\mathbb{E}[X_1]|$) the set $\mathcal{H}_\alpha$ (resp. $\mathcal{Z}_\alpha$) is non-empty, closed, unbounded above and below, stationary, and regenerative. The collections $(\mathcal{H}_\alpha)_{\alpha>\mathbb{E}[X_1]}$ and $(\mathcal{Z}_\alpha)_{\alpha>|\mathbb{E}[X_1]|}$ are increasing in α and the regeneration property is compatible with these inclusions in that each family is a continuum of embedded regenerative sets in the sense of Bertoin. We show that $(\sup\{t < 0 : t \in \mathcal{H}_\alpha\})_{\alpha>\mathbb{E}[X_1]}$ is a càdlàg, nondecreasing, pure jump process with independent increments and determine the intensity measure of the associated Poisson process of jumps. We obtain a similar result for $(\sup\{t < 0 : t \in \mathcal{Z}_\alpha\})_{\alpha>|\beta|}$ when $(X_t)_{t\in\mathbb{R}}$ is a (two-sided) Brownian motion with drift β.

Keywords Lévy process · Fluctuation theory · Subordinator · Lipschitz minorant

1 Introduction

Let $X = (X_t)_{t\in\mathbb{R}}$ be a two-sided, real-valued Lévy process on a complete probability space $(\Omega, \mathcal{F}, \mathbb{P})$. That is, X has càdlàg paths and stationary, independent increments. Assume that $X_0 = 0$. Let $(\mathcal{F}_t)_{t\in\mathbb{R}}$ be the natural filtration of X

SNE supported in part by NSF grant DMS-1512933.

S. N. Evans (✉) · M. Ouaki
Department of Statistics, University of California at Berkeley, Berkeley, CA, USA
e-mail: evans@stat.berkeley.edu; mouaki@berkeley.edu

L. Chaumont, A. E. Kyprianou (eds.), *A Lifetime of Excursions Through Random Walks and Lévy Processes*, Progress in Probability 78,
https://doi.org/10.1007/978-3-030-83309-1_12

augmented by the $\mathbb{P}$-null sets. Suppose that $\mathbb{E}[X_1^+] < \infty$ so that $\mathbb{E}[X_1]$ is well-defined (but possibly $-\infty$).

For $\alpha > \mathbb{E}[X_1]$ define a process $(L_t)_{t\in\mathbb{R}}$ by

$$L_t := \sup\{h \in \mathbb{R} : h-\alpha(t-s) \le X_s \text{ for all } s \le t\} = \inf\{X_s+\alpha(t-s) : s \le t\}, \quad t \in \mathbb{R},$$

and set

$$\mathcal{H}_\alpha := \{t \in \mathbb{R} : X_t \wedge X_{t-} = L_t\}.$$

Equivalently,

$$\mathcal{H}_\alpha := \left\{t \in \mathbb{R} : X_t \wedge X_{t-} - \alpha t = \inf_{u\le t}(X_u - \alpha u)\right\}.$$

By the strong law of large numbers for Lévy processes (see, for example, [10, Example 7.2])

$$\lim_{t\to+\infty} \frac{X_t}{t} = \lim_{t\to-\infty} \frac{X_t}{t} = \mathbb{E}[X_1] \quad \text{a.s.}$$

so that

$$\lim_{t\to+\infty} X_t - \alpha t = -\infty \quad \text{a.s.}$$

and

$$\lim_{t\to-\infty} X_t - \alpha t = +\infty \quad \text{a.s.}$$

It follows from Lemma 7.1 below that $\mathcal{H}_\alpha$ is almost surely a non-empty, closed set that is unbounded above and below.

We show in Theorem 2.5 that $\mathcal{H}_\alpha$ is a *regenerative set* in the sense of [8]. Moreover, we observe in Lemma 5.1 that

$$\mathcal{H}_{\alpha_1} \subseteq \mathcal{H}_{\alpha_2} \subseteq \cdots \subseteq \mathcal{H}_{\alpha_n}$$

for

$$\mathbb{E}[X_1] < \alpha_1 < \alpha_2 < \cdots < \alpha_n.$$

In Sect. 4 we recall from [4] the notion of *regenerative embeddings* and establish in Proposition 5.2 that these embeddings are regenerative. As a consequence, we derive the following result which we prove in Sect. 5.

Theorem 1.1 *For $\alpha > \mathbb{E}[X_1]$ set*

$$G_\alpha := \sup\{t < 0 : t \in \mathcal{H}_\alpha\}.$$

Then $(G_\alpha)_{\alpha > \mathbb{E}[X_1]}$ is a nondecreasing, càdlàg, pure jump process with independent increments. The point process

$$\{(\alpha, G_\alpha - G_{\alpha-}) : G_\alpha - G_{\alpha-} > 0\}$$

is a Poisson point process on $(\mathbb{E}[X_1], \infty) \times (0, \infty)$ with intensity measure

$$\gamma(dx \times dt) = t^{-1}\mathbb{P}\left\{\frac{X_t}{t} \in dx\right\} dt \ \mathbb{1}_{\{t>0,\, x>\mathbb{E}[X_1]\}}.$$

The set $\mathcal{H}_\alpha$ is obviously closely related to the *ladder time*

$$\mathcal{R}_\alpha := \left\{t \in \mathbb{R} : X_t - \alpha t = \inf_{u \le t}(X_u - \alpha u)\right\}$$

of the Lévy process $(X_t - \alpha t)_{t\in\mathbb{R}}$. We clarify the connection with the following result which is proved in Sect. 3.

Proposition 1.2 *The following hold almost surely.*

(i) $\mathcal{R}_\alpha \subseteq \mathcal{H}_\alpha$.
(ii) $\mathcal{R}_\alpha$ *is closed from the right.*
(iii) $\mathbf{cl}(\mathcal{R}_\alpha) = \mathcal{H}_\alpha$.
(iv) $\mathcal{H}_\alpha \setminus \mathcal{R}_\alpha$ *consists of points in $\mathcal{H}_\alpha$ that are isolated on the right and so, in particular, this set is countable.*

Remark 1.3 The embedded regenerative sets structure for the sets $\mathcal{H}_\alpha = \mathcal{R}_\alpha$ when X is Brownian motion with drift has already been noted in [5] in relation to the additive coalescent of Aldous and Pitman (see also [6]). This is further related to the Burgers turbulence (see [11] and the references therein).

Given $\alpha > 0$, denote by $(M_t)_{t\in\mathbb{R}}$ be the α-Lipschitz minorant of the two-sided Lévy process $(X_t)_{t\in\mathbb{R}}$; that is, $t \mapsto M_t$ is the greatest α-Lipschitz function dominated by $t \mapsto X_t$ (our notation suppresses the dependence of M on α). We refer the reader to [1] and [7] for extensive investigations of the Lipschitz minorant of a Lévy process. The α-Lipschitz minorant exists if

$$\mathbb{E}[|X_1|] < \infty \text{ and } \alpha > |\mathbb{E}[X_1]|$$

and we suppose that these conditions hold when discussing $(M_t)_{t\in\mathbb{R}}$. Then,

$$M_t = \sup\{h \in \mathbb{R} : h-\alpha|t-s| \le X_s \text{ for all } s \in \mathbb{R}\} = \inf\{X_s+\alpha|t-s| : s \in \mathbb{R}\}, \quad t \in \mathbb{R}.$$

Set

$$\mathcal{Z}_\alpha := \{t \in \mathbb{R} : X_t \wedge X_{t-} = M_t\}.$$

It is shown in [1][Theorem 2.6] that this set is closed, unbounded above and below, stationary, and regenerative. We establish in Proposition 6.1 that

$$\mathcal{Z}_{\alpha_1} \subseteq \cdots \subseteq \mathcal{Z}_{\alpha_n}$$

for $|\mathbb{E}[X_1]| < \alpha_1 < \cdots < \alpha_n$ and that these embeddings are regenerative. As a consequence, we derive the following result which is proved in Sect. 6.

Theorem 1.4 *Suppose that $(X_t)_{t\in\mathbb{R}}$ is a two-sided standard Brownian motion with drift β For $\alpha > |\beta|$ set*

$$Y_\alpha := \sup\{t < 0 : t \in \mathcal{Z}_\alpha\}.$$

Then $(Y_\alpha)_{\alpha>|\beta|}$ is a nondecreasing, càdlàg, pure jump process with independent increments. The point process

$$\{(\alpha, Y_\alpha - Y_{\alpha-}) : Y_\alpha - Y_{\alpha-} > 0\}$$

is a Poisson point process on $(|\beta|, \infty) \times (0, \infty)$ with intensity measure

$$\gamma(ds \times dr) = \frac{\phi(\frac{\sqrt{r}}{s-\beta}) + \phi(\frac{\sqrt{r}}{s+\beta})}{\sqrt{r}}\, ds\, dr\, \mathbb{1}_{\{s>|\beta|, r>0\}},$$

where $\phi(x) := \frac{e^{-\frac{x^2}{2}}}{\sqrt{2\pi}}$, for $x > 0$.

2 Regenerative Sets

We introduce the notion of a regenerative set in the sense of [8]. For simplicity, we specialize the definition by only considering random sets defined on probability spaces (rather than general σ-finite measure spaces).

Notation 2.1 *Let $\Omega^{\leftrightarrow}$ denote the class of closed subsets of $\mathbb{R}$. For $t \in \mathbb{R}$ and $\omega^{\leftrightarrow} \in \Omega^{\leftrightarrow}$, define*

$$d_t(\omega^{\leftrightarrow}) := \inf\{s > t : s \in \omega^{\leftrightarrow}\}, \quad r_t(\omega^{\leftrightarrow}) := d_t(\omega^{\leftrightarrow}) - t,$$

and

$$\tau_t(\omega^{\leftrightarrow}) := \mathbf{cl}\{s - t : s \in \omega^{\leftrightarrow} \cap (t, \infty)\} = \mathbf{cl}\left((\omega^{\leftrightarrow} - t) \cap (0, \infty)\right).$$

Here **cl** *denotes closure and we adopt the convention* $\inf \emptyset = +\infty$. *Note that* $t \in \omega^{\leftrightarrow}$ *if and only if* $\lim_{s \uparrow t} r_s(\omega^{\leftrightarrow}) = 0$, *and so* $\omega^{\leftrightarrow} \cap (-\infty, t]$ *can be reconstructed from* $r_s(\omega^{\leftrightarrow})$, $s \le t$, *for any* $t \in \mathbb{R}$. *Set* $\mathcal{G}^{\leftrightarrow} := \sigma\{r_s : s \in \mathbb{R}\}$ *and* $\mathcal{G}_t^{\leftrightarrow} := \sigma\{r_s : s \le t\}$. *Clearly,* $(d_t)_{t \in \mathbb{R}}$ *is an increasing càdlàg process adapted to the filtration* $(\mathcal{G}_t^{\leftrightarrow})_{t \in \mathbb{R}}$, *and* $d_t \ge t$ *for all* $t \in \mathbb{R}$.

Let $\Omega^{\rightarrow}$ *denote the class of closed subsets of* $\mathbb{R}_+$. *Define a* σ*-field* $\mathcal{G}^{\rightarrow}$ *on* $\Omega^{\rightarrow}$ *in the same manner that the* σ*-field* $\mathcal{G}^{\leftrightarrow}$ *was defined on* $\Omega^{\leftrightarrow}$.

Definition 2.2 *A* random closed set *is a measurable mapping* S *from a measurable space* $(\Omega, \mathcal{F})$ *into* $(\Omega^{\leftrightarrow}, \mathcal{G}^{\leftrightarrow})$.

Definition 2.3 *A probability measure* $\mathbb{Q}^{\leftrightarrow}$ *on* $(\Omega^{\leftrightarrow}, \mathcal{G}^{\leftrightarrow})$ *is regenerative with regeneration law* $\mathbb{Q}^{\rightarrow}$ *a probability measure on* $(\Omega^{\rightarrow}, \mathcal{G}^{\rightarrow})$ *if*

(i) $\mathbb{Q}^{\leftrightarrow}\{d_t = +\infty\} = 0$, *for all* $t \in \mathbb{R}$;
(ii) *for all* $t \in \mathbb{R}$ *and for all* $\mathcal{G}^{\rightarrow}$*-measurable nonnegative functions* F,

$$\mathbb{Q}^{\leftrightarrow}\left[F(\tau_{d_t}) \mid \mathcal{G}_{t+}^{\leftrightarrow}\right] = \mathbb{Q}^{\rightarrow}[F], \tag{2.1}$$

where we write $\mathbb{Q}^{\leftrightarrow}[\cdot]$ *and* $\mathbb{Q}^{\rightarrow}[\cdot]$ *for expectations with respect to* $\mathbb{Q}^{\leftrightarrow}$ *and* $\mathbb{Q}^{\rightarrow}$.

A random set S *defined on a probability space* $(\Omega, \mathcal{F}, \mathbb{P})$ *is a regenerative set if the push-forward of* $\mathbb{P}$ *by the map* S *(that is, the distribution of* S*) is a regenerative probability measure.*

Remark 2.4 Suppose that the probability measure $\mathbb{Q}^{\leftrightarrow}$ on $(\Omega^{\leftrightarrow}, \mathcal{G}^{\leftrightarrow})$ is stationary; that is, if $S^{\leftrightarrow}$ is the identity map on $\Omega^{\leftrightarrow}$, then the random set $S^{\leftrightarrow}$ on $(\Omega^{\leftrightarrow}, \mathcal{G}^{\leftrightarrow}, \mathbb{Q}^{\leftrightarrow})$ has the same distribution as $u + S^{\leftrightarrow}$ for any $u \in \mathbb{R}$ or, equivalently, that the process $(r_t)_{t \in \mathbb{R}}$ has the same distribution as $(r_{t-u})_{t \in \mathbb{R}}$ for any $u \in \mathbb{R}$. Then, in order to check conditions (i) and (ii) of Definition 2.3, it suffices to check them for the case $t = 0$.

Theorem 2.5 *The random set* $\mathcal{H}_\alpha$ *is stationary and regenerative.*

Proof We first show that $\mathcal{H}_\alpha$ is stationary. Let $a \in \mathbb{R}$. Define the process $(X_t^{(a)})_{t \in \mathbb{R}} := (X_{t-a} - X_{-a})_{t \in \mathbb{R}}$. This process is a Lèvy process that has the same distribution as $(X_t)_{t \in \mathbb{R}}$, and we have

$$\begin{aligned}
t \in \mathcal{H}_\alpha^X + a &\Leftrightarrow t - a \in \mathcal{H}_\alpha^X \\
&\Leftrightarrow X_{t-a} \wedge X_{(t-a)-} - \alpha(t-a) = \inf_{u \le t-a} (X_u - \alpha u) \\
&\Leftrightarrow X_{t-a} \wedge X_{(t-a)-} - X_{-a} - \alpha(t-a) = \inf_{u \le t} (X_{u-a} - X_{-a} - \alpha(u-a)) \\
&\Leftrightarrow X_t^{(a)} \wedge X_{t-}^{(a)} - \alpha t = \inf_{u \le t} (X_u^{(a)} - \alpha u) \\
&\Leftrightarrow t \in \mathcal{H}_\alpha^{X^{(a)}}.
\end{aligned}$$

Hence, $\mathcal{H}_\alpha^X + a = \mathcal{H}_\alpha^{X^{(a)}} \stackrel{d}{=} \mathcal{H}_\alpha^X$ for all $a \in \mathbb{R}$, and the stationarity is proved.

Now, because of Remark 2.4, to prove the regeneration property it suffices to check that conditions (i) and (ii) of Definition 2.3 hold for $t = 0$. As pointed out in the Introduction, the random set $\mathcal{H}_\alpha$ is almost surely unbounded from above, hence condition (i) is verified.

For $t \in \mathbb{R}$ introduce the random times

$$D_t := \inf\{s > t : s \in \mathcal{H}_\alpha\} = \inf\left\{s > t : X_s \wedge X_{s-} - \alpha s = \inf_{u \le s}(X_u - \alpha u)\right\}$$

and put

$$R_t := D_t - t.$$

It is clear from the début theorem that $D := D_0$ is a stopping time with respect to the filtration $(\mathcal{F}_t)_{t\in\mathbb{R}}$. To prove condition (ii), it suffices to show that the random set

$$\tau_D(\mathcal{H}_\alpha) = \mathbf{cl}\left\{t > 0 : X_{t+D} \wedge X_{(t+D)-} - \alpha(t+D) = \inf_{u \le t+D}(X_u - \alpha u)\right\}$$

is independent of the σ-field $\bigcap_{\epsilon>0} \sigma\{R_s : s \le \epsilon\}$.

We shall prove first that

$$\bigcap_{\epsilon>0} \sigma\{R_s : s \le \epsilon\} \subseteq \mathcal{F}_D. \tag{2.2}$$

It is clear that

$$\bigcap_{\epsilon>0} \sigma\{R_s : s \le \epsilon\} \subseteq \bigcap_{n\in\mathbb{N}} \mathcal{F}_{D_{\frac{1}{n}}}. \tag{2.3}$$

Moreover, for a sequence of nonincreasing stopping times T_n converging almost surely to a stopping time T, we have

$$\bigcap_{n\in\mathbb{N}} \mathcal{F}_{T_n} = \mathcal{F}_T. \tag{2.4}$$

To see this, take $\epsilon > 0$ and consider a random variable Z that is $\bigcap_{n\in\mathbb{N}} \mathcal{F}_{T_n}$–measurable. We have almost surely the convergence $Z\mathbb{1}_{\{T_n \le T+\epsilon\}} \to Z$. Note that $Z\mathbb{1}_{\{T_n \le T+\epsilon\}}$ is $\mathcal{F}_{T+\epsilon}$–measurable. Thus Z is $\mathcal{F}_{T+\epsilon}$–measurable. It follows from the strong Markov property and the Blumenthal zero–one law that

$$\bigcap_{\epsilon>0} \mathcal{F}_{T+\epsilon} = \mathcal{F}_T$$

and so Z is $\mathcal{F}_T$–measurable.

In order to establish (2.2), it follows from (2.3) and (2.4) that it is enough to conclude that

$$D_+ := \lim_{n\to\infty} D_{\frac{1}{n}} = D, \quad \text{a.s.} \tag{2.5}$$

To see this, suppose to the contrary that $\mathbb{P}\{D < D+\} > 0$. We claim that $D > 0$ on the event $\{D < D_+\}$. This is so, because on the event $\{0 = D < D_+\}$ the point 0 is a right accumulation point of $\mathcal{H}_\alpha$ and then $D_{\frac{1}{n}}$ must converge to zero, which is not possible. On the event $\{0 < D\}$ we have that $D_+ \le D_{\frac{1}{N}} \le D$ as soon as N is large enough so that $\frac{1}{N} < D$. Thus, $\mathbb{P}\{D < D+\} = 0$ and (2.5) holds, implying that (2.2) also holds.

With (2.2) in hand, it is enough to prove that the set $\tau_D(\mathcal{H}_\alpha)$ is independent of $\mathcal{F}_D$. Observe that

$$\begin{aligned}\tau_D(\mathcal{H}_\alpha) = \mathbf{cl}\Big\{t > 0 : X_{t+D} \wedge X_{(t+D)-} - X_D - \alpha t \\ = (X_D \wedge X_{D-} - X_D) \wedge \inf_{0\le u\le t} (X_{u+D} - X_D - \alpha u)\Big\}.\end{aligned}$$

Because D is a stopping time, the process $(X_{t+D} - X_D)_{t\ge 0}$ is independent of $\mathcal{F}_D$. It therefore suffices to prove that $X_D \le X_{D-}$ a.s.

Suppose that the event $\{X_D > X_{D-}\}$ has positive probability. Because $X_0 = X_{0-}$ almost surely, $D > 0$ on this event.

Introduce the nondecreasing sequence $(D^{(n)})_{n\in\mathbb{N}}$ of stopping times

$$D^{(n)} := \inf\left\{t > 0 : X_t \wedge X_{t-} - \alpha t \le \inf_{u\le t}(X_u - \alpha u) + \frac{1}{n}\right\}$$

and put $D^{(\infty)} := \sup_{n\in\mathbb{N}} D^{(n)}$. By Lemma 7.1,

$$D = \inf\left\{t > 0 : X_t \wedge X_{t-} - \alpha t \le \inf_{u\le t}(X_u - \alpha u)\right\},$$

and so $D^{(\infty)} \le D$. Because X has càdlàg paths, for all $n \in \mathbb{N}$ we have that $X_{D^{(n)}} \wedge X_{D^{(n)}-} - \alpha D^{(n)} \le \inf_{u\le D^{(n)}}(X_u - \alpha u) + \frac{1}{n}$. Sending n to infinity and again using the fact that X has càdlàg paths, we get that $X_{D^{(\infty)}} \wedge X_{D^{(\infty)}-} - \alpha D^{(\infty)} \le \inf_{u\le D^{(\infty)}}(X_u - \alpha u)$, and so $D^{(\infty)} \in \mathcal{H}_\alpha$. By definition of D, we conclude that $D^{(\infty)} = D$.

Set $N := \inf\{n \in \mathbb{N} : D^{(n)} = D\}$ with the usual convention that $\inf\emptyset = \infty$. Suppose we are on the event $\{X_D > X_{D-}\} \cap \{N < \infty\}$. Recall that $D > 0$ on this event. For all $0 < s < D$ we have that $X_s \wedge X_{s-} - \alpha s > \inf_{u\le s}(X_u - \alpha u) + \frac{1}{N}$ so by sending $s \uparrow D$ we get that: $X_{D-} - \alpha D \ge \inf_{u\le D}(X_u - \alpha u) + \frac{1}{N}$, which contradicts $X_{D-} < X_D$. Hence $N = \infty$ almost surely on the event $\{X_D > X_{D-}\}$ and so

$D^{(n)} < D$ for all $n \in \mathbb{N}$ on the event $\{X_D > X_{D-}\}$. By the quasi-left continuity of X we thus have on the event $\{X_D > X_{D-}\}$ that

$$X_{D-} = \lim_{n\to\infty} X_{D^{(n)}} = X_D, \quad \text{a.s.}$$

Therefore $\mathbb{P}\{X_D > X_{D-}\} = 0$ as claimed. □

3 Relationship with the Set of Ladder Times

Proof of Proposition 1.2 (i) If $t \in \mathcal{R}_\alpha$, then $X_t - \alpha t = \inf_{u\le t}(X_u - \alpha u)$ and so $X_t \wedge X_{t-} - \alpha t \le \inf_{u\le t}(X_u - \alpha u)$. It follows from Lemma 7.1 that $t \in \mathcal{H}_\alpha$.

(ii) Because the process $(X_t)_{t\in\mathbb{R}}$ is right-continuous, it is clear that $\mathcal{R}_\alpha$ is closed from the right; that is, for every sequence $t_n \downarrow t$ such that $t_n \in \mathcal{R}_\alpha$ we have $t \in \mathcal{R}_\alpha$.

(iii) As the set $\mathcal{H}_\alpha$ is closed and $\mathcal{R}_\alpha \subseteq \mathcal{H}_\alpha$ we certainly have $\mathbf{cl}(\mathcal{R}_\alpha) \subseteq \mathcal{H}_\alpha$. We showed in the proof of Theorem 2.5 that $X_D \le X_{D-}$ a.s. and so $D \in \mathcal{R}_\alpha$ a.s. By stationarity, $D_t \in \mathcal{R}_\alpha$ a.s. for any $t \in \mathbb{R}$. Therefore, almost surely for all $r \in \mathbb{Q}$ we have $D_r \in \mathcal{R}_\alpha$. Suppose that $t \in \mathcal{H}_\alpha$. Take a sequence of rationals $\{r_n\}_{n\in\mathbb{N}}$ such that $r_n \uparrow t$. Then, for all $n \in \mathbb{N}$, we have $r_n \le D_{r_n} \le t$ and $D_{r_n} \in \mathcal{R}_\alpha$. It follows that $t \in \mathbf{cl}(\mathcal{R}_\alpha)$ and so $\mathbf{cl}(\mathcal{R}_\alpha) = \mathcal{H}_\alpha$.

(iv) Take $t \in \mathcal{H}_\alpha$ that is not isolated on the right so that there exists a sequence $\{t_n\}_{n\in\mathbb{N}}$ of point in $\mathcal{H}_\alpha$ such that $t_n \downarrow t$ and $t_n > t$. Consider a sequence $(r_n)_{n\in\mathbb{N}}$ of rational numbers such that for every $n \in \mathbb{N}$ we have $t \le r_n \le t_n$. We then have $t \le r_n \le D_{r_n} \le t_n$. Thus, $D_{r_n} \downarrow t$ and, as we have already observed, $D_{r_n} \in \mathcal{R}_\alpha$ for all $n \in \mathbb{N}$. Since $\mathcal{R}_\alpha$ is closed from the right, we must have $t \in \mathcal{R}_\alpha$. Finally, as the set of points isolated on the right is countable, the set $\mathcal{H}_\alpha \setminus \mathcal{R}_\alpha$ consists of at most countably many points. □

Remark 3.1 The ladder time set $\mathcal{R}_\alpha$ has been thoroughly studied in the fluctuation theory of Lévy processes. From Proposition VI.1 in [2], we know that the process $(X_t - \alpha t - \inf_{u\le t}\{X_u - \alpha u\})_{t\in\mathbb{R}}$ is a strong Markov process with cádlág paths and hence, by the strong Markov property, the closure of its zero set is a regenerative set in the sense of the Definition 2.3. This result together with Proposition 1.2 proves that $\mathcal{H}_\alpha = \mathbf{cl}(\mathcal{R}_\alpha)$ is a regenerative set.

4 Regenerative Embedding Generalities

We recall the notion of a regenerative embedding of a sequence of regenerative sets from [4]. We modify it slightly to encompass the whole real line instead of the set of nonnegative real numbers. For ease of notation we restrict our definition to the case of two sets. The generalization to a greater number of sets is straightforward.

Definition 4.1 *Recall that* $\Omega^{\leftrightarrow}$ *is the set of closed subsets of* $\mathbb{R}$ *and that* $\Omega^{\rightarrow}$ *is the set of closed subsets of* $\mathbb{R}_+$*). Set*

$$\bar{\Omega} := \{\omega = (\omega^{(1)}, \omega^{(2)}) \in \Omega^{\leftrightarrow} \times \Omega^{\leftrightarrow} : \omega^{(1)} \subseteq \omega^{(2)}\}.$$

and

$$\bar{\Omega}^{\rightarrow} := \{\omega = (\omega^{(1)}, \omega^{(2)}) \in \Omega^{\rightarrow} \times \Omega^{\rightarrow} : \omega^{(1)} \subseteq \omega^{(2)}\}.$$

Write $M^{(1)}(\omega) = \omega^{(1)}$ *and* $M^{(2)}(\omega) = \omega^{(2)}$ *for the canonical projections on* $\bar{\Omega}$, $M = (M^{(1)}, M^{(2)})$*. For* $t \in \mathbb{R}$ *put*

$$d_t^{(1)}(\omega) = d_t(\omega^{(1)})$$

and, with a slight abuse of notation,

$$\tau_t(\omega) = (\tau_t(\omega^{(1)}), \tau_t(\omega^{(2)})).$$

Denote by $\mathcal{G}_t$ *the sigma-field generated by* $d_t^{(1)}$, $M^{(1)} \cap (-\infty, d_t^{(1)}]$*, and* $M^{(2)} \cap (-\infty, d_t^{(1)}]$*. It is easy to check that* $(\mathcal{G}_t)_{t\in\mathbb{R}}$ *is a filtration. A probability measure* $\mathcal{P}$ *is called a regenerative embedding law with regeneration law* $\mathcal{P}^{\rightarrow}$ *if for each* $t \in \mathbb{R}$ *and each bounded measurable function* $f : \bar{\Omega}^{\rightarrow} \to \mathbb{R}$

$$\mathcal{P}[f(M \circ \tau_{d_t^{(1)}}) \,|\, \mathcal{G}_t] = \mathcal{P}^{\rightarrow}[f(M)] \text{ on } \{d_t^{(1)} < \infty\}. \tag{4.1}$$

We denote such an embedding by the notation $M^{(1)} \prec M^{(2)}$.

Remark 4.2

(i) If under the probability measure $\mathcal{P}$, the canonical pair (M_1, M_2) of random sets is *jointly stationary*, in the sense that for all $t \in \mathbb{R}$ the pair $(M_1 + t, M_2 + t)$ has the same distribution as (M_1, M_2), then to check that there is a regenerative embedding it suffices to verify (4.1) for $t = 0$.
(ii) A similar definition holds for subsets of $\mathbb{R}_+$ that contain zero almost surely, which is the version present in [4].

The following theorem follows straightforwardly from the results in [4].

Theorem 4.3 *Let:*

$$\mathcal{S}^{(1)} \prec \mathcal{S}^{(2)} \prec \dots \mathcal{S}^{(n)}$$

be a jointly stationary sequence of subsets of $\mathbb{R}$ *that are regeneratively embedded in the sense of the Definition* 4.1*. Let* Φ_i *be the Laplace exponent of the subordinator* *associated with each $\mathcal{S}^{(i)}$. Introduce the measures $\mu_1, \dots, \mu_n$ on $\mathbb{R}_+$, defined by their Laplace transforms*

$$\int_{\mathbb{R}_+} e^{-\lambda x}\,\mu_i(dx) := \frac{\Phi_i(\lambda)}{\Phi_{i+1}(\lambda)}, \quad \lambda > 0,\ 1 \le i \le n,$$

where we adopt the convention $\Phi_{n+1}(\lambda) := \lambda$, $\lambda > 0$. Put

$$c_i := \frac{1}{\mu_i(\mathbb{R}_+)} = \lim_{\lambda \downarrow 0} \frac{\Phi_{i+1}(\lambda)}{\Phi_i(\lambda)}, \quad 1 \le i \le n.$$

Define the age processes A_t^i for each set $\mathcal{S}^{(i)}$ by

$$A_t^i := \inf\{s \ge 0 : t - s \in \mathcal{S}^{(i)}\}.$$

Then, for any $t \in \mathbb{R}$,

$$(A_t^1 - A_t^2, \dots, A_t^{n-1} - A_t^n, A_t^n) \stackrel{d}{=} c_1\mu_1 \otimes c_2\mu_2 \otimes \cdots \otimes c_n\mu_n.$$

Remark 4.4 We elaborate here on the relationship between subordinators and regenerative sets. If $(\sigma_t)_{t \ge 0}$ is a subordinator (i.e an increasing Lévy process) then the closure of its range $\mathbf{cl}\{\sigma_t : t \ge 0\}$ has the distribution of a regeneration law on $(\Omega^{\rightarrow}, \mathcal{G}^{\rightarrow})$. Conversely, if $\mathcal{S}$ is a regenerative set and we define $\mathcal{S}^{\rightarrow} := \tau_{d_0}(\mathcal{S})$. There exists a continuous nondecreasing process $(L_s)_{s \ge 0}$ which increases exactly on $\mathcal{S}^{\rightarrow}$. We call L the local time on $\mathcal{S}$. Its right continuous inverse defined by $\sigma_t = \inf\{s \ge 0 : L_s > t\}$ is a subordinator, and $\mathcal{S}^{\rightarrow}$ coincides almost surely with the closed range of σ.

Remark 4.5 If $\mathcal{S}^{(1)}$ and $\mathcal{S}^{(2)}$ are regenerative sets in the sense of the Definition 2.3 such that almost surely $\mathcal{S}^{(1)} \subseteq \mathcal{S}^{(2)}$, and Φ_1 (resp Φ_2) is the Laplace exponent of the subordinator associated with $\mathcal{S}^{(1)}$ (resp $\mathcal{S}^{(2)}$). Then from a result of Bertoin (see Theorem 1 in [3]), we have that $\mathcal{S}^{(1)} \prec \mathcal{S}^{(2)}$ iff $\frac{\Phi_1}{\Phi_2}$ is a completely monotone function. As any completely monotone function is a Laplace transform of a nonnegative measure, that proves the existence of the measures μ_i in the statement of Theorem 4.3.

5 A Continuous Family of Embedded Regenerative Sets

For this section, we suppose that X has a Brownian component or infinite Lévy measure. That is, we suppose that X is not a compound Poisson process with drift. The latter case is trivial to study.

Lemma 5.1 *For*

$$\mathbb{E}[X_1] < \alpha_1 < \alpha_2 < \cdots < \alpha_n.$$

we have

$$\mathcal{H}_{\alpha_1} \subseteq \mathcal{H}_{\alpha_2} \subseteq \cdots \subseteq \mathcal{H}_{\alpha_n}.$$

Proof By part (i) of Lemma 7.1,

$$\mathcal{H}_\alpha := \left\{ t \in \mathbb{R} : X_t \wedge X_{t-} - \alpha t \le \inf_{u \le t}(X_u - \alpha u) \right\}.$$

Hence, if $\mathbb{E}[X_1] < \alpha' < \alpha''$, $t \in \mathcal{H}_{\alpha'}$, and $u \le t$, then

$$X_t \wedge X_{t-} - \alpha'' t \le X_u - \alpha' u - (\alpha'' - \alpha')t \le X_u - \alpha' u - (\alpha'' - \alpha')u = X_u - \alpha'' u,$$

so that $t \in \mathcal{H}_{\alpha''}$. Thus $\mathcal{H}_{\alpha'} \subseteq \mathcal{H}_{\alpha''}$ for $\mathbb{E}[X_1] < \alpha' < \alpha''$.

Proposition 5.2 *For* $\mathbb{E}[X_1] < \alpha_1 < \alpha_2 < \cdots < \alpha_n$ *we have*

$$\mathcal{H}_{\alpha_1} \prec \mathcal{H}_{\alpha_2} \prec \cdots \prec \mathcal{H}_{\alpha_n}.$$

□

Proof For ease of notation, we restrict our proof to the case $n = 2$.

By Lemma 5.1 we have $\mathcal{H}_{\alpha_1} \subseteq \mathcal{H}_{\alpha_2}$ when $\mathbb{E}[X_1] < \alpha_1 < \alpha_2$.

By stationarity, we only need to verify (4.1) for $t = 0$. It is clear that

$$D_0^{(1)} := \inf \left\{ s > 0 : X_s \wedge X_{s-} - \alpha_1 s = \inf_{u \le s}(X_u - \alpha_1 u) \right\}$$

is an $(\mathcal{F}_t)_{t \in \mathbb{R}}$-stopping time. From the proof of Theorem 2.5, we have that almost surely

$$X_{D_0^{(1)}} \le X_{D_0^{(1)}-}.$$

Now $D_0^{(1)} \in \mathcal{H}_{\alpha_2}$ and hence

$$\begin{aligned}\mathcal{H}_{\alpha_i} \circ \tau_{D_0^{(1)}} = \mathbf{cl}\Big\{ s > 0 : & X_{s+D_0^{(1)}} \wedge X_{s+D_0^{(1)}-} - X_{D_0^{(1)}} - \alpha_i s \\ & = \inf_{u \le s} \left(X_{u+D_0^{(1)}} - X_{D_0^{(1)}} - \alpha_i u \right) \Big\}\end{aligned}$$

for $i = 1, 2$. Now each of $D_0^{(1)}$, $\mathcal{H}_{\alpha_1} \cap (-\infty, D_0^{(1)}]$, and $\mathcal{H}_{\alpha_2} \cap (-\infty, D_0^{(1)}]$ is $\mathcal{F}_{D_0^{(1)}}$-measurable, so it remains to note that $(X_{s+D_0^{(1)}} - X_{D_0^{(1)}})_{s \geq 0}$ is independent of $\mathcal{F}_{D_0^{(1)}}$. □

Proof of Theorem 1.1 It is clear that G is nondecreasing.

As for the right-continuity, consider $\beta > \mathbb{E}[X_1]$ and a sequence $\{\beta_n\}_{n \in \mathbb{N}}$ with $\beta_n \downarrow \beta$ and $\beta_n > \beta$. Suppose that $G_{\beta+} := \lim_{n \to \infty} G_{\beta_n} > G_\beta$. For any $u \leq G_{\beta+} \leq G_{\beta_n}$ we have

$$X_{G_{\beta_n}} \wedge X_{G_{\beta_n}-} - \beta_n G_{\beta_n} \leq X_u - \beta_n u.$$

Taking the limit as n goes to infinity gives

$$X_{G_{\beta+}} - \beta G_{\beta+} \leq X_u - \beta u$$

and hence

$$X_{G_{\beta+}} \wedge X_{G_{\beta+}-} - \beta G_{\beta+} \leq X_u - \beta u.$$

It follows from Lemma 7.1 that $G_\beta < G_{\beta+} \in \mathcal{H}_\beta$, but this contradicts the definition of G_β.

Corollary VI.10 in [2] gives that the Laplace exponent of the subordinator associated with the ladder time set of the process $(\alpha t - X_t)_{t \geq 0}$ (the subordinator is the right-continuous inverse of the local time associated with this set) is

$$\Phi_\alpha(\lambda) = \exp\left(\int_0^\infty (e^{-t} - e^{-\lambda t}) t^{-1} \mathbb{P}\{X_t \geq \alpha t\} dt\right).$$

Fix $\mathbb{E}[X_1] < \alpha_1 < \alpha_2 < \cdots < \alpha_n$. Introduce the measures $\mu_1, \ldots, \mu_n$ on $\mathbb{R}_+$, defined by their Laplace transforms

$$\int_{\mathbb{R}_+} e^{-\lambda x} \mu_i(dx) := \frac{\Phi_{\alpha_i}(\lambda)}{\Phi_{\alpha_{i+1}}(\lambda)}, \quad \lambda > 0, \ 1 \leq i \leq n,$$

where we adopt the convention $\Phi_{\alpha_{n+1}}(\lambda) := \lambda$, $\lambda > 0$. Put

$$c_i := \frac{1}{\mu_i(\mathbb{R}_+)} = \lim_{\lambda \downarrow 0} \frac{\Phi_{\alpha_{i+1}}(\lambda)}{\Phi_{\alpha_i}(\lambda)}, \quad 1 \leq i \leq n.$$

Set $\nu_i = c_i \mu_i$, $1 \le i \le n$, so that

$$\int_{\mathbb{R}_+} e^{-\lambda x}\, \nu_i(dx) = \exp\left(- \int_0^\infty (1 - e^{-\lambda t}) t^{-1} \mathbb{P}\{\alpha_i t \le X_t \le \alpha_{i+1} t\}\, dt \right), \quad 1 \le i \le n-1, \tag{5.1}$$

and

$$\int_{\mathbb{R}_+} e^{-\lambda x}\, \nu_i(dx) = \exp\left(- \int_0^\infty (1 - e^{-\lambda t}) t^{-1} \mathbb{P}\{X_t \ge \alpha_n t\}\, dt \right). \tag{5.2}$$

Then, by Theorem 4.3,

$$(G_{\alpha_2} - G_{\alpha_1}, \ldots, G_{\alpha_n} - G_{\alpha_{n-1}}, -G_{\alpha_n}) \stackrel{d}{=} \nu_1 \otimes \nu_2 \otimes \cdots \otimes \nu_n.$$

It follows that the process G has independent increments and that $\lim_{\alpha \to \infty} G_\alpha = 0$ almost surely. That $(G_\alpha)_{\alpha > \mathbb{E}[X_1]}$ is a pure jump process (that is, the process is a sum of its jumps and there is no deterministic drift component) along with the Poisson description of $\{(\alpha, G_\alpha - G_{\alpha-}) : G_\alpha - G_{\alpha-} > 0\}$ follows from (5.1), (5.2), and standard Lévy–Khinchin–Itô theory: for example, from [9, p 146], the process $(G_\alpha)_{\alpha > \mathbb{E}[X_1]}$ can be written as:

$$G_\alpha = - \int_0^\infty l \mathfrak{p}([\alpha, \infty) \times dl)$$

where $\mathfrak{p}$ is a Poisson random measure with intensity measure γ. □

Remark 5.3 Taking the concatenation of the lines with slopes α between G_α and $G_{\alpha-}$ for every jump time α constructs the graph of the convex minorant of the Lévy process $(-X_{t-})_{t \ge 0}$. The conclusion of Theorem 1.1 thus agrees with the study of the convex minorant of a Lévy process carried out in [12].

6 Another Continuous Family of Embedded Regenerative Sets

Proposition 6.1 *For* $|\mathbb{E}[X_1]| < \alpha_1 < \cdots < \alpha_n$, *we have that*

$$\mathcal{Z}_{\alpha_1} \prec \cdots \prec \mathcal{Z}_{\alpha_n}.$$

Proof We shall just prove the result for the case $n = 2$. It is very clear that $\mathcal{Z}_{\alpha_1} \subseteq \mathcal{Z}_{\alpha_2}$, as any α_1-Lipschitz function is also an α_2-Lipschitz function. Moreover, the sets $(\mathcal{Z}_{\alpha_1}, \mathcal{Z}_{\alpha_2})$ are obviously jointly stationary, and thus it suffices to check the independence condition for $t = 0$. Note that $D_{\alpha_1} \in \mathcal{Z}_{\alpha_2}$. Using [7, Lemma 7.2] gives that

$$(\mathcal{Z}_{\alpha_1} \circ \tau_{D_{\alpha_1}}, \mathcal{Z}_{\alpha_2} \circ \tau_{D_{\alpha_1}})$$

is measurable with respect to $\sigma\{X_{t+D_{\alpha_1}} - X_{D_{\alpha_1}} : t \geq 0\}$. The same argument yields

$$\mathcal{G}_0 = \sigma\{\mathcal{Z}_{\alpha_1} \cap (-\infty, D_{\alpha_1}], \mathcal{Z}_{\alpha_2} \cap (-\infty, D_{\alpha_1}]\} \subseteq \sigma\{X_t : t \leq D_{\alpha_1}\}$$

An appeal to [7, Theorem 3.5] completes the proof. □

Proof of Theorem 1.4 As in the proof of Theorem 1.1, it is clear that the process $(Y_\alpha)_{\alpha > |\beta|}$ is nondecreasing and has independent increments. We leave to the reader the straightforward proof of that this process is càdlàg.

We compute the Laplace exponent Φ_α of the subordinator associated with the regenerative set $\mathcal{Z}_\alpha$. From [1, Proposition 8.1] we have

$$\Phi_\alpha(\lambda) = \frac{4(\alpha^2 - \beta^2)\lambda}{(\sqrt{2\lambda + (\alpha - \beta)^2} + \alpha - \beta)(\sqrt{2\lambda + (\alpha + \beta)^2} + \alpha + \beta)}.$$

Thus, for $|\beta| < \alpha_1 < \alpha_2$, we have

$$\begin{aligned}\mathbb{E}[e^{-\lambda(Y_{\alpha_2} - Y_{\alpha_1})}] &= c\frac{\Phi_{\alpha_1}(\lambda)}{\Phi_{\alpha_2}(\lambda)} \\ &= c\frac{(\sqrt{2\lambda + (\alpha_2 - \beta)^2} + \alpha_2 - \beta)(\sqrt{2\lambda + (\alpha_2 + \beta)^2} + \alpha_2 + \beta)}{(\sqrt{2\lambda + (\alpha_1 - \beta)^2} + \alpha_1 - \beta)(\sqrt{2\lambda + (\alpha_1 + \beta)^2} + \alpha_1 + \beta)},\end{aligned}$$

where

$$c = \lim_{\lambda \downarrow 0} \frac{(\sqrt{2\lambda + (\alpha_1 - \beta)^2} + \alpha_1 - \beta)(\sqrt{2\lambda + (\alpha_1 + \beta)^2} + \alpha_1 + \beta)}{(\sqrt{2\lambda + (\alpha_2 - \beta)^2} + \alpha_2 - \beta)(\sqrt{2\lambda + (\alpha_2 + \beta)^2} + \alpha_2 + \beta)};$$

that is,

$$c = \frac{\alpha_1^2 - \beta^2}{\alpha_2^2 - \beta^2}.$$

Hence,

$$\log\left(\mathbb{E}\left[e^{-\lambda(Y_{\alpha_2} - Y_{\alpha_1})}\right]\right) = f(a_3) + f(a_4) - f(a_1) - f(a_2),$$

where $a_1 = (\alpha_1 + \beta)^{-1}$, $a_2 = (\alpha_1 - \beta)^{-1}$, $a_3 = (\alpha_2 + \beta)^{-1}$ and $a_4 = (\alpha_2 - \beta)^{-1}$, and

$$f(x) = -\log(1 + \sqrt{2\lambda x^2 + 1}).$$

It remains to observe that

$$f(x) = -\int_0^\infty (1 - e^{-\lambda r}) r^{-\frac{1}{2}} \int_0^x t^{-2} \phi(t\sqrt{r})\, dt\, dr$$

and do a change of variables inside the integral to finish the proof. □

7 Some Real Analysis

Lemma 7.1 *Fix a càdlàg function $f : \mathbb{R} \mapsto \mathbb{R}$ and consider the set*

$$\mathcal{H} := \{t \in \mathbb{R} : f(t) \wedge f(t-) = \inf_{u \le t} f(u)\}.$$

(i) *The set $\mathcal{H}$ coincides with*

$$\{t \in \mathbb{R} : f(t) \wedge f(t-) \le \inf_{u \le t} f(u)\}.$$

(ii) *The set $\mathcal{H}$ is closed.*
(iii) *If $\lim_{t\to-\infty} f(t) = +\infty$ and $\lim_{t\to+\infty} f(t) = -\infty$, then the set $\mathcal{H}$ is nonempty and unbounded from above and below.*

Proof

(i) Note that $\{t \in \mathbb{R} : f(t) \wedge f(t-) \le \inf_{u \le t} f(u)\}$ is the disjoint union $\{t \in \mathbb{R} : f(t) \wedge f(t-) = \inf_{u \le t} f(u)\} \sqcup \{t \in \mathbb{R} : f(t) \wedge f(t-) < \inf_{u \le t} f(u)\}$. Clearly, $f(t) \wedge f(t-) \ge \inf_{u \le t} f(u)$ for all $t \in \mathbb{R}$ and so the second set on the right hand side is empty.
(ii) We want to show that if $\{t_n\}_{n\in\mathbb{N}}$ is a sequence of elements of $\mathcal{H}$ converging to some $t^* \in \mathbb{R}$, then $t^* \in \mathcal{H}$. The result is clear if $t_n = t^*$ infinitely often, so we may suppose that $t^* \notin \{t_n\}_{n\in\mathbb{N}}$.
Suppose to begin with that there are only finitely many $n \in \mathbb{N}$ such that $t_n < t^*$. Then, for n large enough, we have that $t_n > t^*$ and thus $f(t_n) \wedge f(t_n-) \le f(u)$ for all $u \le t^*$. Now $\lim_{n\to\infty} f(t_n) = \lim_{n\to\infty} f(t_n-) = f(t^*)$. Hence, $f(t^*) \wedge f(t^*-) \le f(t^*) \le f(u)$ for all $u \le t^*$ and so $t^* \in \mathcal{H}$ by part (i).
Suppose on the other hand, that the set $\mathcal{N}$ of $n \in \mathbb{N}$ such that $t_n < t^*$ is infinite. For $u < t^*$ we have for large $n \in \mathcal{N}$ sufficiently large that $u \le t_n$ and thus $f(t_n) \wedge f(t_n-) \le f(u)$. Now the limit as $n \to \infty$ with $n \in \mathcal{N}$ of

$f(t_n) \wedge f(t_n-)$ is $f(t^*-)$. Hence, $f(t^*) \wedge f(t^*-) \le f(t^*-) \le \inf_{u<t^*} f(u)$. This implies that $f(t^*) \wedge f(t^*-) \le \inf_{u\le t^*} f(u)$ and so $t^* \in \mathcal{H}$ by part (i).

(iii) Fix $M \in \mathbb{R}$, put $I = \inf_{t\le M} f(t)$, and let $\{t_n\}_{n\in\mathbb{N}}$ be a sequence of elements of $(-\infty, M]$ such that $\lim_{n\to\infty} f(t_n) = I$. Because $f(t)$ goes to $+\infty$ as $t \to -\infty$, the sequence $\{t_n\}_{n\in\mathbb{N}}$ is bounded and thus admits a subsequence $\{t_{n_k}\}_{k\in\mathbb{N}}$ that converges to some $t^* \in (-\infty, M]$. By the argument in part (ii), $I \in \{f(t^*), f(t^*-)\}$. Moreover, $I \le f(t^*)$ and $I \le f(t^*-)$. Thus, $f(t^*) \wedge f(t^*-) = I = \inf_{u\le M} f(u) \le \inf_{u\le t} f(u)$ and $t \in \mathcal{H}$ by part (i). Since $M \in \mathbb{R}$ is arbitrary it follows that $\mathcal{H}$ is not only nonempty but also unbounded below. □

Because $f(t)$ goes to $+\infty$ as $t \to -\infty$ and $f(t)$ goes to $-\infty$ as $t \to +\infty$, for each $n \in \mathbb{N}$ we have that the set $\{t \in \mathbb{R} : f(t) \le -n\}$ is nonempty and bounded below and so $s_n := \inf\{t \in \mathbb{R} : f(t) \le -n\} \in \mathbb{R}$. The sequence $\{s_n\}_{n\in\mathbb{N}}$ is clearly nondecreasing and unbounded above. Now $f(s_n) \wedge f(s_n-) = f(s_n) = \inf\{f(u) : u \le s_n\}$ for all $n \in \mathbb{N}$ so that $s_n \in \mathcal{H}$ for all $n \in \mathbb{N}$ and hence $\mathcal{H}$ is unbounded above.

Acknowledgments We thank the anonymous referee for several helpful suggestions.

References

1. Abramson, J., Evans, S.N.: Lipschitz minorants of Brownian motion and Lévy processes. Probab. Theory Relat. Fields **158** (3–4), 809–857 (English) (2014)
2. Bertoin, J.: Lévy processes, Cambridge Tracts in Mathematics, vol. 121. Cambridge University Press, Cambridge (1996). MR1406564 (98e:60117)
3. Bertoin, J.: Regenerative embedding of Markov sets. Probab. Theory Relat. Fields **108**(4), 559 (1997)
4. Bertoin, J.: Renewal theory for embedded regenerative sets. Ann. Probab. **27**(3), 1523–1535 (1999). MR1733158
5. Bertoin, J.: A fragmentation process connected to Brownian motion. Probab. Theory Relat. Fields **117**(2), 289–301 (2000). MR1771665
6. Chassaing, P., Louchard, G.: Phase transition for parking blocks, Brownian excursion and coalescence. Random Struct. Algoritm. **21**(1), 76–119 (2002). MR1913079
7. Evans, S.N., Ouaki, M.: Excursions away from the Lipschitz minorant of a Lévy process (2019). arXiv:1905.07038
8. Fitzsimmons, P.J., Taksar, M.: Stationary regenerative sets and subordinators. Ann. Probab. **16**(3), 1299–1305 (1988). MR942770 (89m:60176)
9. Itō, K., McKean, H.P.: Diffusion processes and their sample paths. Grundlehren Der Mathematischen Wissenschaften. Springer (1974)
10. Kyprianou, A.E.: Introductory lectures on fluctuations of Lévy processes with applications, Universitext. Springer, Berlin (2006). MR2250061
11. Menon, G., Pego, R.L.: Universality classes in Burgers turbulence. Commun. Math. Phys. **273**(1), 177–202 (2007). MR2308754
12. Pitman, J., Uribe Bravo, G.: The convex minorant of a Lévy process. Ann. Probab. **40**(4), 1636–1674 (2012). MR2978134

No-Tie Conditions for Large Values of Extremal Processes

Yuguang Ipsen and Ross Maller

Abstract We give necessary and sufficient conditions for there to be no ties, asymptotically, among large values of a space-time Poisson point process evolving homogeneously in time. The convergence is at small times, in probability or almost sure.

Keywords Extremal processes · Poisson point processes · No tie conditions · Extreme values · Lévy processes

1 Introduction

The possibility of tied (equal) large values among the increments of a random walk (in discrete time) or a Lévy process (in continuous time) arises whenever we want to place in order of magnitude the jumps of the process, occurring up to some time. We may want to study them in their own right, as in extreme value theory, or as components of the original random walk or Lévy process – perhaps with a view to trimming extreme values from the process. In some studies the issue is dealt with simply by assuming that tied values do not occur (occur with probability 0); in others the possibility of ties is included as part of the formulation, usually with some consequent complications in the analysis.

If the distribution function describing the increments in a random walk is continuous, or the Lévy measure of a Lévy process has no atoms, then ties among large increments occur with probability 0, at all times. Assumptions like this of course would rule out the use of many common distributions and processes. But it may also be that the probability of a tie becomes negligible asymptotically, for large times (or, in the case of a continuous time process, alternatively, for small times),

---

Y. Ipsen · R. Maller (✉)
College of Business and Economics, The Australian National University, Canberra, ACT, Australia
e-mail: yuguang.ipsen@anu.edu.au; ross.maller@anu.edu.au

L. Chaumont, A. E. Kyprianou (eds.), *A Lifetime of Excursions Through Random Walks and Lévy Processes*, Progress in Probability 78,
https://doi.org/10.1007/978-3-030-83309-1_13

where the asymptotic may be in probability or almost sure. In this paper we give some necessary and sufficient conditions for these possibilities, in the more general setting of an extremal process associated with a Poisson point process, drawing on the general methodology of Buchmann, Fan & Maller [4] for the relevant point process-extremal process setup.

In the following, conditions for no ties in probability, asymptotically at small times, are in Sect. 2, and almost sure results are in Sect. 3. A brief discussion of sufficient conditions for no ties and an example are in Sect. 4. For the remainder of this section we recap the necessary point process and extremal theory formulations.

1.1 Poisson Point Processes

In this subsection we establish the framework of Poisson point processes needed for the results that follow. Our results were motivated by thinking of the jump process of a subordinator, but apply more generally to a Poisson point process (PPP) evolving homogeneously in time (an "evolutionary" or "space-time" process, [5], Ch.14, 15). Let N be a point process on a probability space $(\Omega, \mathcal{F}, P)$ with intensity measure $dt \times \Pi(dx)$, $t \geq 0$, $x > 0$, where Π is a Borel measure on $(0, \infty)$, locally finite at infinity. So the numbers of points in disjoint Borel subsets of $[0, \infty) \times (0, \infty)$ are independent, and, for A and B Borel subsets of $[0, \infty)$ and $(0, \infty)$ respectively, the number of points, $N(A, B)$, in $A \times B$ has distribution

$$P\big(N(A, B) = k\big) = \frac{e^{-\lambda}\lambda^k}{k!}, \; k = 0, 1, 2\ldots, \tag{1.1}$$

where $\lambda = m(A) \times \Pi(B)$, with $m(A)$ denoting the Lebesgue measure of A. The measure Π has finite-valued tail function $\overline{\Pi} : (0, \infty) \to (0, \infty)$, defined by

$$\overline{\Pi}(x) := \Pi\{(x, \infty)\}, \; x > 0,$$

a nonnegative, right-continuous, non-increasing function with $\overline{\Pi}(+\infty) = 0$. Π need not be a Lévy measure but as a special case the points of the process evolving in time could be the jump process $(X_t - X_{t-})_{t>0}$ of a subordinator X on $(0, \infty)$ with Lévy canonical measure Π.

The process is "simple" in the time parameter in the sense that, by (1.1), $N(\{t\}, B) = 0$ a.s. for any $t > 0$ and Borel $B \subseteq (0, \infty)$; and we denote the (a.s. unique) value (magnitude) of a point of N at time t by Δ_t. We allow the possibility of atoms in Π, so there may be coincident points in the space component, with positive probability. In general it is possible to write the spatial components of the points of the process occurring in $[0, t]$ in decreasing order of magnitude, possibly with ties, as

$$\infty > \Delta_t^{(1)} \geq \cdots \geq \Delta_t^{(r)} \geq \cdots \geq 0, \; r \in \mathbb{N} := \{1, 2, \ldots\},$$

for a fixed $t > 0$. For a fixed $r \in \mathbb{N}$, the $\Delta_t^{(r)}$ are nondecreasing in t; whenever $0 \le s \le t$, we have $0 \le \Delta_s^{(r)} \le \Delta_t^{(r)}$.

We use this notation to set up the extremal processes in the next section.

1.2 Extremal Processes

The distribution of the points ordered by their magnitudes can be conveniently written in terms of Gamma random variables (rvs). Let $(\mathfrak{E}_i)$ be an i.i.d. sequence of exponentially distributed rvs with common parameter $\mathbf{E}\mathfrak{E}_i = 1$. Then $\Gamma_j := \sum_{i=1}^{j} \mathfrak{E}_i$ is a Gamma$(j, 1)$ random variable, $j \in \mathbb{N}$. Let

$$\overline{\Pi}^{\leftarrow}(x) = \inf\{y > 0 : \overline{\Pi}(y) \le x\},\ x > 0,$$

be the right-continuous inverse of $\overline{\Pi}$. A basic property of this function is

$$\overline{\Pi}(\overline{\Pi}^{\leftarrow}(x)) \le x \le \overline{\Pi}(\overline{\Pi}^{\leftarrow}(x)-),\ x > 0. \tag{1.2}$$

Buchmann, Fan, and Maller [4] use a randomisation procedure to define the ordered points $(\Delta_t^{(r)})$ at time $t > 0$, and give the representation

$$\left\{\Delta_t^{(r)}\right\}_{r\ge 1} \stackrel{\mathrm{D}}{=} \left\{\overline{\Pi}^{\leftarrow}(\Gamma_r/t)\right\}_{j\ge 1},\ t > 0, \tag{1.3}$$

for their joint distribution at a fixed time $t > 0$. Considered as a process in continuous time, the process $(\Delta_t^{(r)})_{t\ge 0}$ is the rth-*order extremal process*, $r \in \mathbb{N}$. We refer to [4] for background information on the properties of $(\Delta_t^{(r)})_{t\ge 0}$. (Some other related literature is referenced at the end of Sect. 4.)

We observe a *tied value* for the r-th largest point at time $t > 0$, $r \in \mathbb{N}$, if the event $\{\Delta_t^{(r)} = \Delta_t^{(r+1)}\}$ occurs. More generally, we could have a number of points tied for r-th largest at time t, that is, if $\Delta_t^{(r)} = \Delta_t^{(r+n)}$ for some $n \in \mathbb{N}$. We can consider the probability that ties occur or not, asymptotically as $t \downarrow 0$ (small time) or as $t \to \infty$ (large time). We will concentrate on the small time case $t \downarrow 0$ herein.

2 No Ties in Probability

We assume throughout that $\Pi\{(0,\infty)\} = \overline{\Pi}(0+) = \infty$ and $\overline{\Pi}(x) > 0$ for all $x > 0$. Then also $\overline{\Pi}^{\leftarrow}$ is nonincreasing with $\overline{\Pi}^{\leftarrow}(x) > 0$ for all $x > 0$, $\overline{\Pi}^{\leftarrow}(0+) = \infty$ and $\overline{\Pi}^{\leftarrow}(\infty) = 0$. The assumption $\overline{\Pi}(0+) = \infty$ means there are infinitely many non-zero points of the process in any right neighbourhood of 0, a.s., so $\Delta_t^{(r)} > 0$ a.s.

for each $t > 0$ and $r \in \mathbb{N}$. But we have $\lim_{t\downarrow 0} \Delta_t^{(r)} = 0$ a.s., because for any $\varepsilon > 0$, $r \in \mathbb{N}$,

$$P(\Delta_t^{(r)} > \varepsilon) \le P(\Delta_t^{(1)} > \varepsilon) = 1 - e^{-t\overline{\Pi}(\varepsilon)} \to 0, \text{ as } t \downarrow 0.$$

So $\Delta_t^{(r)} \stackrel{\mathrm{P}}{\longrightarrow} 0$ as $t \downarrow 0$ and hence by monotonicity, $\lim_{t\downarrow 0} \Delta_t^{(r)} = 0$ a.s. for each $r \in \mathbb{N}$.

Let $\Delta\Pi(x) = \Pi\{x\} = \overline{\Pi}(x-) - \overline{\Pi}(x)$, $x > 0$, be the mass attributed by Π to x, if any. Our first theorem characterises "no ties" in probability.

Theorem 2.1 *For $r, n \in \mathbb{N}$,*

$$\lim_{t\downarrow 0} P\big(\Delta_t^{(r)} = \Delta_t^{(r+n)}\big) = 0 \tag{2.1}$$

if and only if

$$\lim_{x\downarrow 0} \frac{\overline{\Pi}(x-)}{\overline{\Pi}(x)} = 1, \text{ or, equivalently, } \lim_{x\downarrow 0} \frac{\Delta\Pi(x)}{\overline{\Pi}(x)} = 0. \tag{2.2}$$

Proof of Theorem 2.1 Fix $t, x > 0$, $r, n \in \mathbb{N}$, and write $\Gamma_{r+n} = \Gamma_r + \widetilde{\Gamma}_n$, as independent components. Calculate, using (1.3), the main identity

$$\begin{aligned}
P\big(\Delta_t^{(r)} = \Delta_t^{(r+n)} < x\big) &= P\big(\overline{\Pi}^{\leftarrow}(\Gamma_r/t) = \overline{\Pi}^{\leftarrow}((\Gamma_r + \widetilde{\Gamma}_n)/t) < x\big) \\
&= \int_{\overline{\Pi}^{\leftarrow}(y/t)<x} P\left(\overline{\Pi}^{\leftarrow}(y/t) = \overline{\Pi}^{\leftarrow}((y+\Gamma_n)/t)\right) P\left(\Gamma_r \in \mathrm{d}y\right) \\
&= \int_{\overline{\Pi}^{\leftarrow}(y/t)<x} P\left(y + \Gamma_n \le t\overline{\Pi}\left(\overline{\Pi}^{\leftarrow}(y/t)-\right)\right) P\left(\Gamma_r \in \mathrm{d}y\right).
\end{aligned} \tag{2.3}$$

Here we used that $\overline{\Pi}^{\leftarrow}(a) = \overline{\Pi}^{\leftarrow}(a+b)$ iff $\overline{\Pi}(\overline{\Pi}^{\leftarrow}(a)-) \ge a + b$ for $a, b > 0$. Taking $x = \infty$ in (2.3) gives

$$P\big(\Delta_t^{(r)} = \Delta_t^{(r+n)}\big) = \int_{y>0} P\left(y + \Gamma_n \le t\overline{\Pi}\left(\overline{\Pi}^{\leftarrow}(y/t)-\right)\right) P\left(\Gamma_r \in \mathrm{d}y\right). \tag{2.4}$$

Clearly the two conditions in (2.2) are equivalent. Assume the first condition in (2.2). Then, given $\delta > 0$, we can choose $x_0 = x_0(\delta)$ small enough for $\overline{\Pi}(x-) \le (1+\delta)\overline{\Pi}(x)$ whenever $0 < x \le x_0$. Fix $\eta > 0$ and keep $y \ge \eta$. Since $\overline{\Pi}^{\leftarrow}(\infty) = 0$, we have $\overline{\Pi}^{\leftarrow}(y/t) \le \overline{\Pi}^{\leftarrow}(\eta/t) \le x_0(\delta)$ if $0 < t \le$ some $t_0(\delta, \eta)$. This implies (see

(1.2) for the left and right hand inequalities)

$$y \le t\overline{\Pi}\left(\overline{\Pi}^{\leftarrow}(y/t)-\right) \le (1+\delta)t\overline{\Pi}\left(\overline{\Pi}^{\leftarrow}(y/t)\right) \le (1+\delta)y.$$

Hence we deduce that $t\overline{\Pi}((\overline{\Pi}^{\leftarrow}(y/t)-) \to y$ as $t \downarrow 0$, for each $y \ge \eta$, and so by dominated convergence the component of the integral in (2.4) over $y \ge \eta$ tends to 0 as $t \downarrow 0$. The component of the integral in (2.4) over $0 < y < \eta$ is no larger than $P(\Gamma_r \le \eta)$ and hence can be made arbitrarily small with η. Thus (2.2) implies (2.1).

Conversely, we show that (2.1) implies (2.2). Assume (2.1). Define $g_t(y) := t\overline{\Pi}\left(\overline{\Pi}^{\leftarrow}(y/t)-\right)$, a nondecreasing function on $(0,\infty)$ for each $t>0$. Then for any sequence $t_k \downarrow 0$ as $k \to \infty$, we have by (2.4)

$$\lim_{k\to\infty} \int_{y>0} P\big(\Gamma_n \le g_{t_k}(y) - y\big) P\left(\Gamma_r \in \mathrm{d}y\right) = 0. \tag{2.5}$$

Suppose there is a $y_0 > 0$ such that $\lim_{k\to\infty} g_{t_k}(y_0) = \infty$. Then $\lim_{k\to\infty} g_{t_k}(y) = \infty$ for all $y \ge y_0$ and (2.5) with Fatou's lemma gives the contradiction $P(\Gamma_r > y_0) = 0$. So the sequence $(g_{t_k}(y))_{k\ge1}$ is bounded for each $y > 0$ and by Helly's selection theorem we can take a subsequence of t_k if necessary so that $\lim_{k\to\infty} g_{t_k}(y) = g(y)$, a finite nondecreasing function on $(0,\infty)$. Another application of Fatou's lemma to (2.5) gives

$$\int_{y>0} P\big(\Gamma_n \le g(y) - y\big) P\left(\Gamma_r \in \mathrm{d}y\right) = 0.$$

Since Γ_n and Γ_r are continuous random variables this implies that the nondecreasing function $g(y) = y$ for almost all $y > 0$. We conclude that $g(y) = y$ for all $y > 0$. Thus the limit through the subsequence t_k does not depend on the choice of t_k. Recalling the definition of $g(y)$ this means that $\lim_{t\downarrow 0} t\overline{\Pi}(\overline{\Pi}^{\leftarrow}(y/t)-) = y$ for all $y > 0$. Fix a $y_0 > 0$ so that $\lim_{t\downarrow 0} t\overline{\Pi}(\overline{\Pi}^{\leftarrow}(y_0/t)-) = y_0$ and take any sequence $(y_k)_{k\ge1}$ such that $y_k \downarrow 0$. We claim that

$$\lim_{k\to\infty} \frac{\overline{\Pi}(y_k-)}{\overline{\Pi}(y_k)} = 1.$$

To prove this we can restrict consideration to points y_k for which $\overline{\Pi}(y_k-) > \overline{\Pi}(y_k)$. Define $t_k = y_0/\overline{\Pi}(y_k)$. Then $\overline{\Pi}(y_k) = y_0/t_k$ and $y_k = \overline{\Pi}^{\leftarrow}(y_0/t_k)$; note that the latter holds because y_k is a discontinuity point of $\overline{\Pi}$. Finally

$$1 \le \frac{\overline{\Pi}(y_k-)}{\overline{\Pi}(y_k)} = \frac{t_k\overline{\Pi}(\overline{\Pi}^{\leftarrow}(y_0/t_k)-)}{y_0} \to 1,$$

as $k \to \infty$, as required. So we have proved (2.2). □

3 No Ties Almost Surely

For almost sure convergence we need to strengthen (2.2) to an integral criterion.

Theorem 3.1 *For* $n \in \mathbb{N}$,

$$P\big(\Delta_t^{(1)} = \Delta_t^{(n+1)} \text{ i.o., as } t \downarrow 0\big) = 0 \tag{3.1}$$

("i.o" means "infinitely often") if and only if

$$\int_0^1 \left(\frac{\Delta\Pi(x)}{\overline{\Pi}(x)}\right)^n \frac{\Pi(dx)}{\overline{\Pi}(x)} < \infty. \tag{3.2}$$

Either of (3.1) *or* (3.2) *implies* (2.1) *and* (2.2).

Remark 3.1 The complete a.s. analogue of Theorem 2.1 would have $\Delta_t^{(r)} = \Delta_t^{(r+n)}$, $r \in \mathbb{N}$, replacing $\Delta_t^{(1)} = \Delta_t^{(n+1)}$ in Theorem 3.1. If there are n values of the Δ_t tied for r-th largest as $t \downarrow 0$ then there will be n values tied for largest as $t \downarrow 0$, because the lower ranked points will ultimately become the largest as $t \downarrow 0$. Thus (3.1) and (3.2) are sufficient for $P\big(\Delta_t^{(r)} = \Delta_t^{(n+r)} \text{ i.o., as } t \downarrow 0\big) = 0$, for any $r \in \mathbb{N}$. But currently we do not have a necessary and sufficient condition for the general case.

Proof of Theorem 3.1: Let (3.2) hold for some fixed $n \in \mathbb{N}$. First we prove, for $t \in (0, 1]$,

$$P\big(\Delta_{s-}^{(n+1)} = \Delta_{s-}^{(1)} \text{ for some } s \le t\big) \le P\big(\Delta_{s-}^{(n+1)} = \Delta_{s-}^{(1)} < \Delta_s \text{ for some } s \le t\big). \tag{3.3}$$

To see this, let

$$E_t = E_t(n) := \left\{\Delta_{s-}^{(n+1)} = \Delta_{s-}^{(1)} < \Delta_s \text{ for some } s \le t\right\} \tag{3.4}$$

be the event on the RHS of (3.3). Suppose sample point ω is in the event on the LHS of (3.3). Then ω is in E_t or else ω is in the event $\big\{\Delta_{s-}^{(1)} = \Delta_s \text{ for all } s \le t\big\}$. But this event has probability 0 because $\lim_{t\downarrow 0} \Delta_t^{(1)} = 0$ a.s. Thus (3.3) holds, in fact with equality.

We aim to show that $\lim_{t\downarrow 0} P(E_t) = 0$ under the assumption (3.2), and this will establish (3.1) via (3.3). To this end, define

$$N_t = N_t(n) := \int_{(0,t]\times(0,\infty)} \mathbf{1}\left\{\Delta_{s-}^{(n+1)} = \Delta_{s-}^{(1)} < x\right\} N(\mathrm{d}s \times \mathrm{d}x), \tag{3.5}$$

the number of points (s, Δ_s) which satisfy $\Delta_{s-}^{(n+1)} = \Delta_{s-}^{(1)} < \Delta_s$ with $s \le t$. Then, for $0 < t < 1$,

$$\begin{aligned} P(E_t) &= P(N_t \ge 1) \le \mathbf{E}(N_t) \qquad \text{(by Markov's inequality)} \\ &= \int_0^t \mathrm{d}s \int_{x>0} E\mathbf{1}\left\{\Delta_{s-}^{(n+1)} = \Delta_{s-}^{(1)} < x\right\} \Pi(\mathrm{d}x) \quad \text{(by the compensation formula)} \\ &= \int_0^t \mathrm{d}s \int_{x>0} \int_{\overline{\Pi}^{\leftarrow}(y/s)<x} P\left(y + \Gamma_n \le s\overline{\Pi}\left(\overline{\Pi}^{\leftarrow}(y/s)-\right)\right) P\left(\Gamma_1 \in \mathrm{d}y\right) \Pi(\mathrm{d}x), \end{aligned} \tag{3.6}$$

where the last equality holds by (2.3). Let $a(y) = \overline{\Pi}\left(\overline{\Pi}^{\leftarrow}(y)-\right) - y \ge 0$ and write the RHS of (3.6) as

$$\int_0^t \mathrm{d}s \int_{x>0} \int_{\overline{\Pi}^{\leftarrow}(y/s)<x} P\left(\Gamma_n \le sa(y/s)\right) P\left(\Gamma_1 \in \mathrm{d}y\right) \Pi(\mathrm{d}x). \tag{3.7}$$

Consider first the component over $x > 1$ of the integral in (3.7). That component is bounded above by

$$\int_0^t \mathrm{d}s \int_{x>1} \Pi(\mathrm{d}x) = t\overline{\Pi}(1). \tag{3.8}$$

Next consider the component over $0 < x \le 1$ of the integral in (3.7). That component can be written as

$$\int_{0<x\le 1} \int_{y>\overline{\Pi}(x)} \int_0^t sP\left(\Gamma_n \le sa(y)\right) e^{-sy} \mathrm{d}s\, \mathrm{d}y\, \Pi(\mathrm{d}x). \tag{3.9}$$

We can estimate, for any $a > 0$,

$$\begin{aligned} P(\Gamma_n \le a) &= P\Big(\sum_{i=1}^n \mathfrak{E}_i \le a\Big) \le P\Big(\max_{1\le i\le n} \mathfrak{E}_i \le a\Big) \\ &= P^n\left(\mathfrak{E}_1 \le a\right) = \left(1 - e^{-a}\right)^n \le a^n. \end{aligned} \tag{3.10}$$

Applying this to (3.9) with $a = sa(y)$ gives an upper bound of

$$\begin{aligned} &\int_{0<x\le 1} \int_{y>\overline{\Pi}(x)} a^n(y) \int_0^t s^{n+1} e^{-sy} \mathrm{d}s\, \mathrm{d}y\, \Pi(\mathrm{d}x) \\ &= \int_{0<x\le 1} \int_{y>\overline{\Pi}(x)} \frac{a^n(y)}{y^{n+2}} \int_0^{ty} s^{n+1} e^{-s} \mathrm{d}s\, \mathrm{d}y\, \Pi(\mathrm{d}x). \end{aligned} \tag{3.11}$$

The innermost integral here is bounded above by $\Gamma(n+2)$ for all $y, t > 0$. Interchanging the order of the remaining integrals, we have the following bound for the RHS of (3.11) with the innermost integral removed:

$$\int_{0<x\le 1}\int_{y>\overline{\Pi}(x)} \frac{a^n(y)}{y^{n+2}}\mathrm{d}y\,\Pi(\mathrm{d}x) = \int_{y>\overline{\Pi}(1)}\int_{\overline{\Pi}^{\leftarrow}(y)\le x\le 1} \frac{a^n(y)}{y^{n+2}}\mathrm{d}y\,\Pi(\mathrm{d}x)$$
$$\le \int_{y>\overline{\Pi}(1)} \overline{\Pi}(\overline{\Pi}^{\leftarrow}(y)-)\frac{a^n(y)}{y^{n+2}}\mathrm{d}y. \tag{3.12}$$

Recalling the definition of $a(y)$, and that $y \ge \overline{\Pi}\big(\overline{\Pi}^{\leftarrow}(y)\big)$, we have

$$a(y) = \overline{\Pi}\big(\overline{\Pi}^{\leftarrow}(y)-\big) - y \le \overline{\Pi}\big(\overline{\Pi}^{\leftarrow}(y)-\big) - \overline{\Pi}\big(\overline{\Pi}^{\leftarrow}(y)\big) = \Delta\Pi\big(\overline{\Pi}^{\leftarrow}(y)\big),$$

so the function $a(y)$ and hence the integrands in (3.12) are positive only when $y \in \widetilde{D} := \{y > \overline{\Pi}(1) : \overline{\Pi}(\overline{\Pi}^{\leftarrow}(y)-) > \overline{\Pi}(\overline{\Pi}^{\leftarrow}(y))\}$. The points in $\widetilde{D}$ are the points of discontinuity of $\overline{\Pi}$, so $\widetilde{D} = D := \{0 < x \le 1 : \overline{\Pi}(x-) > \overline{\Pi}(x)\} = \{0 < x \le 1 : \Delta\Pi(x) > 0\}$. Thus the RHS of (3.12) can be bounded above by

$$\sum_{x\in D}\int_{\overline{\Pi}(x)\le y\le\overline{\Pi}(x-)} \overline{\Pi}(\overline{\Pi}^{\leftarrow}(y)-)\frac{a^n(y)}{y^{n+2}}\mathrm{d}y$$
$$\le \sum_{x\in D}\frac{\overline{\Pi}(x-)}{\overline{\Pi}(x)^{n+2}}\int_{\overline{\Pi}(x)\le y\le\overline{\Pi}(x-)} \big(\overline{\Pi}(x-)-y\big)^n\mathrm{d}y$$
$$= \frac{1}{n+1}\sum_{x\in D}\frac{\overline{\Pi}(x-)}{\overline{\Pi}(x)^{n+2}}\big(\overline{\Pi}(x-)-\overline{\Pi}(x)\big)^{n+1}$$
$$= \frac{1}{n+1}\sum_{x\in D}\frac{\overline{\Pi}(x-)}{\overline{\Pi}(x)^{n+2}}\big(\Delta\Pi(x)\big)^{n+1}$$
$$= \frac{1}{n+1}\int_{0<x\le 1}\frac{\overline{\Pi}(x-)}{\overline{\Pi}(x)^{n+2}}\big(\Delta\Pi(x)\big)^{n}\,\Pi(\mathrm{d}x). \tag{3.13}$$

Now note that (3.2) implies

$$\sum_{x\in D}\left(\frac{\Delta\Pi(x)}{\overline{\Pi}(x)}\right)^{n+1} < \infty,$$

hence that $\lim_{x\downarrow 0}\Delta\Pi(x)/\overline{\Pi}(x) = 0$, i.e., (2.2) holds. So $\overline{\Pi}(x-) \sim \overline{\Pi}(x)$ as $x \downarrow 0$ and (3.2) implies that the last integral in (3.13) is finite.

It follows that the component over $0 < x \le 1$ of the integral in (3.7) tends to 0 as $t \downarrow 0$, by dominated convergence applied to (3.11) (recall the bound of $\Gamma(n+2)$ for the innermost integral). We also have the bound (3.8). Hence, overall, the integral in

(3.7) tends to 0 as $t \downarrow 0$. So we have shown that (3.2) implies $\lim_{t\downarrow 0} P(E_t(n)) = 0$, $n \in \mathbb{N}$, which by (3.4) and (3.7) proves (3.1).

At this stage we note that (2.1) and (2.2) follow from (3.1) and (3.2), because

$$P\big(\Delta_t^{(1)} = \Delta_t^{(n+1)} \text{ i.o., as } t \downarrow 0\big) = \lim_{t\downarrow 0} P\big(\Delta_s^{(1)} = \Delta_s^{(n+1)} \text{ for some } s \le t\big)$$

$$\ge \lim_{t\downarrow 0} P\big(\Delta_t^{(n+1)} = \Delta_t^{(1)}\big)$$

shows that (3.1) implies (2.1). (2.2) then follows by Theorem 2.1, and we showed already that (3.2) implies (2.2).

Now we prove the converse result, that (3.1) implies (3.2). Assume (3.1) holds for some $n \in \mathbb{N}$. We want to define a sequence of random times $(\tau_k)_{k\ge 0}$ in $(0, 1]$ at which the events $F_s := \{\Delta_{s-}^{(n+1)} = \Delta_{s-}^{(1)} < \Delta_s\}$ occur. As in, e.g., Sato [17] p.131 we can enumerate the $(\Delta_t)_{0<t\le 1}$ (not in order) as $(\Delta_{t_k})_{k\ge 1}$ for some random times $t_1 > t_2 > \cdots > 0$. Then we can define: $\tau_0 = 1$, and for $k = 0, 1, 2, \ldots$

$$\begin{aligned}\tau_{k+1} &:= \sup\{0 < t < \tau_k : \Delta_t^{(1)} = \Delta_t^{(n+1)} < \Delta_{\tau_k-}^{(1)}\}\\ &= \sup\{t_\ell \in (0, \tau_k) : \Delta_{t_\ell}^{(1)} = \Delta_{t_\ell}^{(n+1)} < \Delta_{\tau_k-}^{(1)}\}.\end{aligned} \tag{3.14}$$

These are well defined as long as the event on the RHS of (3.14) is nonempty. Because we assumed (3.1), N_t as defined in (3.5) is finite a.s., and as t approaches 0 there will be a last occurrence of E_s in $[0, t]$, a.s. We let $\tau_{k+j} = 0$ for all $j \ge 1$ if $\Delta_t^{(1)} \ne \Delta_t^{(n+1)}$ for all $0 < t < \tau_k$.

With this setup let $A_k := \{\tau_k > 0\}$. Then $N_1 = \sum_{k\ge 1} \mathbf{1}_{A_k}$ and

$$\sum_{k\ge 1} P(A_k) = \mathbf{E}\sum_{k\ge 1} \mathbf{1}_{A_k} = \mathbf{E}\#\{t \in (0, 1) : \Delta_{t-}^{(1)} = \Delta_{t-}^{(n+1)} < \Delta_t\} = \mathbf{E}N_1.$$

By (3.1), $P\big(A_k \text{ i.o. as } k \to \infty\big) = 0$, and we want to deduce from this that $\sum_{k\ge 1} P(A_k) < \infty$. Take $k > j \ge 1$, and calculate

$$\begin{aligned}P(A_k \cap A_j) &= P(\tau_k > 0, \tau_j > 0) = \int_{\{t>0\}} P\big(\tau_k > 0 \big| \tau_j = t\big) P(\tau_j \in \mathrm{d}t)\\
&\le \int_{\{t>0\}} P\big(\text{there are } k-j \text{ occurrences of } F_s \text{ among } (\Delta_s)_{\tau_k < s \le t} \big|\\
&\quad \text{there are } j \text{ occurrences of } F_s \text{ among } (\Delta_s)_{t<s\le 1}, \text{ and the } j\text{th occurs at } t\big)\\
&\qquad \times P(\tau_j \in \mathrm{d}t)\\
&\le \int_{\{t>0\}} P\big(\text{there are } k-j \text{ occurrences of } F_s \text{ among } (\Delta_s)_{0<s\le t}\big) P(\tau_j \in \mathrm{d}t)\\
&\le P(A_{k-j})P(\tau_j > 0) = P(A_{k-j})P(A_j).\end{aligned} \tag{3.15}$$

With (3.15) we can apply Spitzer's converse Borel-Cantelli lemma ([18], p.317) and deduce that $\sum P(A_k)$ converges, and conclude that $EN_1 < \infty$.

Thus the expression on the RHS of (3.7) is finite. Now we want to find a lower bound for this expression. Complementary to (3.10), we can estimate

$$P(\Gamma_n \le x) = P\Big(\sum_{i=1}^{n} \mathfrak{E}_i \le x\Big) \ge P\Big(\max_{1\le i\le n} \mathfrak{E}_i \le x/n\Big)$$
$$= P^n\big(\mathfrak{E}_1 \le x/n\big) = \big(1 - e^{-x/n}\big)^n \ge x^n e^{-x}/n^n. \tag{3.16}$$

Interchanging the order of integration in (3.7) we can write

$$\mathbf{E}(N_1) = \int_0^1 \mathrm{d}s \int_{y>0} \overline{\Pi}(\overline{\Pi}^{\leftarrow}(y)-) P\left(\Gamma_n \le sa(y)\right) P\left(\Gamma_1 \in s\mathrm{d}y\right)$$
$$= \int_0^1 \mathrm{d}s \int_{y>0} \overline{\Pi}(\overline{\Pi}^{\leftarrow}(y)-) P\left(\Gamma_n \le sa(y)\right) s e^{-sy} \mathrm{d}y.$$

Now use (3.16) to get the lower bound

$$\mathbf{E}(N_1) \ge \frac{1}{n^n} \int_{y>0} \overline{\Pi}(\overline{\Pi}^{\leftarrow}(y)-) \int_0^1 s(sa(y))^n e^{-sa(y)} e^{-sy} \mathrm{d}s\, \mathrm{d}y$$
$$\ge \frac{1}{n^n} \int_{y>y_0} \overline{\Pi}(\overline{\Pi}^{\leftarrow}(y)-) \frac{(a(y))^n}{y^{n+2}} \int_0^y s^{n+1} e^{-sa(y)/y} e^{-s} \mathrm{d}s\, \mathrm{d}y. \tag{3.17}$$

Given $\varepsilon > 0$, we can by (2.2) choose $y_0 = y_0(\varepsilon) > 0$ so large that $y > y_0$ implies

$$\Delta\overline{\Pi}(\overline{\Pi}^{\leftarrow}(y))/\overline{\Pi}(\overline{\Pi}^{\leftarrow}(y)) \le \varepsilon.$$

Then we get for $y > y_0$

$$\frac{a(y)}{y} \le \frac{\Delta\Pi\big(\overline{\Pi}^{\leftarrow}(y)\big)}{\overline{\Pi}(\overline{\Pi}^{\leftarrow}(y))} \le \varepsilon,$$

and a lower bound for the RHS of (3.17) is

$$\frac{c_n e^{-\varepsilon y_0}}{n^n} \int_{y>y_0} \overline{\Pi}(\overline{\Pi}^{\leftarrow}(y)-) \frac{(a(y))^n}{y^{n+2}} \mathrm{d}y, \tag{3.18}$$

where $c_n := \int_0^{y_0} s^{n+1}e^{-s}\mathrm{d}s$. Convergence of the integral in (3.18) implies the convergence of the integral on the RHS of (3.12), and the same sorts of manipulations as in (3.13) give a lower bound for (3.18) of

$$\frac{c_n e^{-\varepsilon y_0}}{n^{n+1}} \int_{0<x\leq x_0} \frac{\left(\Delta\Pi(x)\right)^n}{\overline{\Pi}(x-)^{n+1}} \Pi(\mathrm{d}x),$$

for some $x_0 > 0$. Because of (2.2), we see from this that the integral in (3.2) is finite, completing the proof of the converse part of Theorem 3.1. □

4 Sufficient Conditions and an Example

In view of Theorem 2, we call (2.2) a "no-ties" condition (asymptotic as $x \downarrow 0$).

Proposition 4.1 *Recall we assume* $\overline{\Pi}(0+) = \infty$ *and* $\overline{\Pi}(x) > 0$ *for all* $x > 0$*. We have that* (2.2) *holds when*

$$\lim_{\lambda\uparrow 1} \limsup_{x\downarrow 0} \frac{\overline{\Pi}(x\lambda)}{\overline{\Pi}(x)} = 1. \tag{4.1}$$

Proof of Proposition 4.1: For $x > 0$ and $\varepsilon \in (0, 1)$ write

$$0 \leq \frac{\Delta\overline{\Pi}(x)}{\overline{\Pi}(x)} = \frac{\overline{\Pi}(x-) - \overline{\Pi}(x)}{\overline{\Pi}(x)} = \frac{\overline{\Pi}(x-)}{\overline{\Pi}(x)} - 1 \leq \frac{\overline{\Pi}((1-\varepsilon)x)}{\overline{\Pi}(x)} - 1. \tag{4.2}$$

Assuming (4.1), when $x \downarrow 0$ the last term in (4.2) is bounded above by a quantity which tends to 0 as $\varepsilon \to 0$. Hence (2.2). □

Equation (4.1) holds when $\overline{\Pi}(x)$ is regularly varying with index $-\alpha$, $\alpha \geq 0$ (including, when $\overline{\Pi}$ is slowly varying), at 0. More generally, we could consider subsequential convergence of a normed and centered Lévy process. Maller and Mason [13] give conditions for the *stochastic compactness* at 0 or ∞ of a normed and centered Lévy process. They take the form of a version of the dominated variation (cf. Bingham, Goldie and Teugels [1], p.54) of a relevant function (the truncated variance, or tail function) of the Lévy measure. Sufficient for stochastic compactness is

$$\limsup_{x\downarrow 0} \frac{\overline{\Pi}(x\lambda)}{\overline{\Pi}(x)} \leq c\lambda^{-\alpha} \tag{4.3}$$

for some finite $c > 0$ and α in $(0, 2)$, and the measure Λ_Y of any subsequential limit rv satisfies an analogous condition. A subclass of the corresponding limit processes (but not all)[1] will satisfy (4.1) in addition.

Here is an example where (2.2) holds but (3.2) does not. Take a sequence $p_j = j^{-1/2}e^{\sqrt{j}}/2$, $j \in \mathbb{N}$. Then $\sum_1^j p_k \sim e^{\sqrt{j}}$ as $j \to \infty$. Let $\overline{\Pi}$ have masses p_j at x_j, $j \in \mathbb{N}$, where $1 = x_1 > x_2 > \cdots$, and $x_j \downarrow 0$ as $j \to \infty$. Then for $x_{j+1} < x \le x_j$ we have

$$\frac{\Delta\Pi(x)}{\overline{\Pi}(x)} = \frac{\Delta\overline{\Pi}(x_j)}{\overline{\Pi}(x_j)} = \frac{p_j}{\sum_1^j p_k} \sim \frac{1}{2\sqrt{j}} \to 0, \text{ as } j \to \infty.$$

Thus (2.2) holds and there are no ties in probability, asymptotically, as $t \downarrow 0$.

But we can check

$$\begin{aligned}\int_{(0,1]} \left(\frac{\Delta\Pi(x)}{\overline{\Pi}(x)}\right)^n \frac{\Pi(\mathrm{d}x)}{\overline{\Pi}(x)} &= \sum_{j\ge 1} \int_{x_{j+1}<x\le x_j} \left(\frac{\Delta\Pi(x)}{\overline{\Pi}(x)}\right)^n \frac{\Pi(\mathrm{d}x)}{\overline{\Pi}(x)} \\ &= \sum_{j\ge 1} \left(\frac{p_j}{\sum_1^j p_k}\right)^n \left(\frac{p_j}{\sum_1^j p_k}\right) \\ &\asymp \sum_{j\ge 1} \frac{1}{j^{(n+1)/2}}.\end{aligned}$$

The series diverges for $n = 1$, so infinitely often there will be ties for the largest, a.s. as $t \to 0$, but the series converges for $n \ge 2$, so there will almost surely be no more than one tie for largest at time t, as $t \downarrow 0$.

Sufficient conditions for (3.2) would likely involve ideas connected with slow variation with remainder; e.g., Goldie and Smith [6]. We do not pursue this further here.

For other results on rth-order extremal processes and related results, especially at small times, see Buchmann, Fan and Maller [2], Buchmann, Maller and Resnick [3], Maller and Schmidli [12], Ipsen, Kevei and Maller [7], Ipsen, Maller and Resnick [8], Ipsen, Maller and Resnick [9], Kevei and Mason [10], Kevei and Mason [11]. For other results on extremal processes see Resnick [14], Resnick [15]), Resnick and Rubinovitch [16].

Acknowledgments We are grateful to a referee for helpful comments.

[1] The function $e^{\lceil|\log x|\rceil}$ satisfies (4.3) but not (4.1).

References

1. Bingham, N.H., Goldie, C.M., Teugels, J.L.: Regular Variation. Cambridge University Press (1987)
2. Buchmann, B., Fan, Y., Maller, R.A.: Functional laws for trimmed Lévy processes. J. Appl. Probab. **54**, 873–889 (2017)
3. Buchmann, B., Maller, R.A., Resnick, S.I.: Processes of rth largest. Extremes **21**, 485–508 (2018). http://dx.doi.org/10.1007/s10687-018-0308-x
4. Buchmann, B., Fan, Y., Maller, R.A.: Distributional representations and dominance of a Lévy process over its maximal jump processes. Bernoulli **22**, 2325–2371 (2016)
5. Daley, D.J., Vere-Jones, D.: An Introduction to the Theory of Point Processes, Vol II: General Theory and Structure, 2nd edn. Springer (2008)
6. Goldie, C.M., Smith, R.L.: Slow variation with remainder: Theory and applications. Q. J. Math. **38**, 45–71 (1987)
7. Ipsen, Y., Kevei, P., Maller, R.A.: Convergence to stable limits for ratios of trimmed Lévy processes and their jumps. Markov Proces. Relat. Fields **24**, 539–562 (2018)
8. Ipsen, Y., Maller, R.A., Resnick, S.I.: Ratios of ordered points of point processes with regularly varying intensity measures. Stochastic Proces. Appl. **129**, 205–222 (2019)
9. Ipsen, Y., Maller, R.A., Resnick, S.I.: Trimmed Lévy processes and their extremal components. Stoch. Proces. Appl. **130**, 2228–2249 (2020)
10. Kevei, P., Mason, D.M.: Randomly weighted self-normalized Lévy processes. Stoch. Proc. Appl. **123**, 490–522 (2013)
11. Kevei, P., Mason, D.M.: The limit distribution of ratios of jumps and sums of jumps of subordinators. Lat. Am. J. Probab. Math. Stat. **11**, 631–642 (2014)
12. Maller, R.A., Schmidli, P.C.: Small-time almost-sure behaviour of extremal processes. Adv. Appl. Probab. **49**, 1–19 (2017)
13. Maller, R.A., Mason, D.M.: Stochastic compactness of Lévy processes. IMS Collections, High Dimensional Probability V. The Luminy Volume **5**, 239–257 (2009)
14. Resnick, S.I.: Inverses of extremal processes. Adv. Appl. Probab. **6**, 392–406 (1974)
15. Resnick, S.I.: Extreme Values, Regular Variation, and Point Processes. Springer (1987)
16. Resnick, S.I., Rubinovitch, M.: The structure of extremal processes. Adv. Appl. Prob. **5** 287–307 (1973)
17. Sato, K.I.: Lévy Processes and Infinitely Divisible Distributions. Cambridge University Press, Cambridge (1999)
18. Spitzer, F.: Principles of Random Walk, 2nd edn. Springer (1976)

Slowly Varying Asymptotics for Signed Stochastic Difference Equations

Dmitry Korshunov

Abstract For a stochastic difference equation $D_n = A_n D_{n-1} + B_n$ which stabilises upon time we study tail distribution asymptotics for D_n under the assumption that the distribution of $\log(1 + |A_1| + |B_1|)$ is heavy-tailed, that is, all its positive exponential moments are infinite. The aim of the present paper is three-fold. Firstly, we identify the asymptotic behaviour not only of the stationary tail distribution but also of D_n. Secondly, we solve the problem in the general setting when A takes both positive and negative values. Thirdly, we get rid of auxiliary conditions like finiteness of higher moments introduced in the literature before.

Keywords Stochastic difference equations · Heavy tails · Long-tailed and subexponential distributions · Slowly varying tail asymptotics

1 Introduction

Let (A, B) be a random vector in $\mathbb{R}^2$ such that $\mathbb{E}\log|A| = -a < 0$. Let (A_k, B_k), $k \in \mathbb{Z}$, be independent copies of (A, B). Consider the following stochastic difference equation

$$
\begin{aligned}
D_n &= A_n D_{n-1} + B_n \\
&= \Pi_1^n D_0 + \sum_{k=1}^{n} \Pi_{k+1}^n B_k, \quad n \geq 1,
\end{aligned}
\tag{1}
$$

D. Korshunov (✉)
Department of Mathematics and Statistics, Lancaster University, Lancaster, UK
e-mail: d.korshunov@lancaster.ac.uk

L. Chaumont, A. E. Kyprianou (eds.), *A Lifetime of Excursions Through Random Walks and Lévy Processes*, Progress in Probability 78,
https://doi.org/10.1007/978-3-030-83309-1_14

where D_0 is independent of (A_k, B_k)'s, $\Pi_k^n := A_k \cdot \ldots \cdot A_n$ for $k \le n$ and $\Pi_{n+1}^n = 1$. The process D_n clearly constitutes a Markov chain and satisfies the following equality in distribution

$$D_n =_{st} \Pi_{-n}^{-1} D_0 + \sum_{k=-n}^{-1} \Pi_{k+1}^{-1} B_k.$$

If $a < \infty$ then, by the strong law of large numbers applied to the logarithm of $|\Pi|$, with probability 1, $e^{-2an} \le \Pi_1^n \le e^{-an/2}$ ultimately in n, hence the process D_n, $n \ge 1$, is stochastically bounded if and only if $\mathbb{E}\log(1+|B|) < \infty$. If $\mathbb{P}\{A = 0\} > 0$ which implies $a = \infty$, then the process D_n is always stochastically bounded. In both cases, the Markov chain D_n is stable, its stationary distribution is given by the following random series

$$D_\infty := \sum_{k=-\infty}^{-1} \Pi_{k+1}^{-1} B_k =_{st} \sum_{k=1}^{\infty} \Pi_1^{k-1} B_k$$

and D_n weakly converges to the stationary distribution as $n \to \infty$; in the context of financial mathematics such random variables are called stochastic perpetuities. Stability results for D_n are dealt with in [17], see also [2]; the case where $\mathbb{E}\log|A|$ is not necessarily finite is treated in [9].

Both perpetuities and stochastic difference equations have many important applications, among them life insurance and finance, nuclear technology, sociology, random walks and branching processes in random environments, extreme-value analysis, one-dimensional ARCH processes, etc. For particularities, we refer the reader to, for instance, Embrechts and Goldie [5], Rachev and Samorodnitsky [15] and Vervaat [17] for a comprehensive survey of the literature.

If $A \ge 0$ and $\mathbb{P}\{A > 1\} > 0$, then $\mathbb{E}A^\gamma \to \infty$ as $\gamma \to \infty$, so $\mathbb{E}A^\beta > 1$ for some $\beta < \infty$. If in addition $B \ge 0$, then it follows from the stationary version of the recursion (1) that $\mathbb{E}D_\infty^\beta \ge \mathbb{E}D_\infty^\beta \mathbb{E}A^\beta$ which implies that $\mathbb{E}D_\infty^\beta = \infty$, in other words, with necessity, not all moments of D_∞ are finite; see [8] for a similar conclusion for signed A and B. It was proven in the seminal paper by Kesten [11, Theorem 5], see also [7], that if $\mathbb{E}|A|^\beta = 1$ for some $\beta > 0$, then a power tail asymptotics for the stationary distribution holds, $\mathbb{P}\{|D_\infty| > x\} \sim c/x^\beta$ as $x \to \infty$, for some $c > 0$.

The problem we address in this paper is about the tail asymptotic behaviour of D_n and of its stationary version D_∞ in the case where the distribution of $\log|A|$ is heavy-tailed, that is, all positive exponential moments of $\log|A|$ are infinite, in other words, $\mathbb{E}|A|^\gamma = \infty$ for all $\gamma > 0$. It can only happen if the random variable $|A|$ has right unbounded support.

The only result in that direction we are aware of is that by Dyszewski [4] where in the context of iterated random functions it is proven that the stationary tail distribution is asymptotically equivalent to

$$\frac{1}{a}\int_x^{\infty}\mathbb{P}\{\log C > y\}dy \quad \text{as } x \to \infty,$$

where $C := \max(A, B)$, provided $A, B \geq 0$, the integrated tail distribution of $\log C$ is subexponential and under additional moment condition that $\mathbb{E}\log^{\gamma} C < \infty$ for some $\gamma > 1$. In the case of a signed B, only lower and upper asymptotic bounds are derived in [4]. An alternative approach to lower and upper bounds for the tail distribution of D_∞ is developed in [3] in the case of positive A and B.

The aim of the present paper is three-fold. Firstly, we identify the asymptotic behaviour not only of the stationary tail distribution but also for D_n in the heavy-tailed case. Secondly, we solve the problem in the general setting when A takes both positive and negative values. Thirdly, we get rid of auxiliary conditions like finiteness of higher moments.

Our approach to the problem is based on reduction of D_n – roughly speaking by taking the logarithm of it – to an asymptotically homogeneous in space Markov chain with heavy-tailed jumps and on further analysis of such chains. Namely, we define a Markov chain X_n on $\mathbb{R}$ as follows

$$X_n := \begin{cases} \log(1 + D_n) & \text{if } D_n \geq 0, \\ -\log(1 + |D_n|) & \text{if } D_n < 0, \end{cases} \tag{2}$$

hence the distribution tail of D_n may be computed as

$$\mathbb{P}\{D_n > x\} = \mathbb{P}\{X_n > \log(1 + x)\} \quad \text{for } x > 0. \tag{3}$$

At any state $x \geq 0$, the jump of the Markov chain X_n is a random variable distributed as

$$\xi(x) = \begin{cases} \log(1 + A(e^x - 1) + B) - x & \text{if } A(e^x - 1) + B \geq 0, \\ -\log(1 + |A(e^x - 1) + B|) - x & \text{if } A(e^x - 1) + B < 0, \end{cases} \tag{4}$$

and at any state $x \leq 0$,

$$\xi(x) = \begin{cases} \log(1 + A(1 - e^{-x}) + B) - x & \text{if } A(1 - e^{-x}) + B \geq 0, \\ -\log(1 + |A(1 - e^{-x}) + B|) - x & \text{if } A(1 - e^{-x}) + B < 0. \end{cases} \tag{5}$$

Also define a sequence of independent random fields $\xi_n(x)$, $x \in \mathbb{R}$, which are independent copies of $\xi(x)$. Then the recursion (1) may be rewritten as

$$X_{n+1} = X_n + \xi_n(X_n).$$

The Markov chain X_n is *asymptotically homogeneous in space*, that is, the distribution of its jump $\xi(x)$ weakly converges to that of $\xi := \log A$ as $x \to \infty$; it is particularly emphasised in [7, Section 2]. Let us underline that, in general, $\log(A + (1 - A + B)e^{-x})$ may not converge to ξ as $x \to \infty$ in total variation norm.

Asymptotically homogeneous in space Markov chains are studied in detail in [1, 13] from the point of view of their asymptotic tail behaviour in subexponential case. However, that results for general asymptotically homogeneous in space Markov chains are not directly applicable to stochastic difference equations as it is formally assumed in [1, Theorem 3] that the distribution of a Markov chain X_n converges to the invariant distribution in the total variation norm which is not always true for stochastic difference equations. Secondly, stochastic difference equations possess some specific properties that allow us to find tail asymptotics in a simpler way than it is done in [1, Theorem 3] or in [4, Theorem 3.1]; we explore that below however our approach still follows some ideas of the proof for Markov chains in [1].

Let us recall some relevant classes of distributions needed in our analysis of the heavy-tailed case.

Definition 1 A distribution H with right unbounded support is called *long-tailed*, $H \in \mathscr{L}$, if, for each fixed y, $\overline{H}(x+y) \sim \overline{H}(x)$ as $x \to \infty$; hereinafter $\overline{H}(x) = H(x, \infty)$ is the tail of H.

A random variable A has slowly varying at infinity distribution if and only if the distribution of $\xi := \log(A^+)$ is long-tailed.

Definition 2 A distribution H on $\mathbb{R}^+$ with unbounded support is called *subexponential*, $H \in \mathscr{S}$, if $\overline{H * H}(x) \sim 2\overline{H}(x)$ as $x \to \infty$. Equivalently, $\mathbb{P}\{\zeta_1 + \zeta_2 > x\} \sim 2\mathbb{P}\{\zeta_1 > x\}$, where random variables ζ_1 and ζ_2 are independent with common distribution H. A distribution H of a random variable ζ on $\mathbb{R}$ with right-unbounded support is called *subexponential* if the distribution of ζ^+ is so.

As well-known (see, e.g. [6, Lemma 3.2]) the subexponentiality of H on $\mathbb{R}^+$ implies long-tailedness of H. In particular, if the distribution of a random variable $\zeta \ge 0$ is subexponential then ζ is heavy-tailed.

For a distribution H with finite mean, we define the *integrated tail distribution* H_I generated by H as follows:

$$\overline{H}_I(x) := \min\Big(1, \int_x^\infty \overline{H}(y)dy\Big).$$

Definition 3 A distribution H on $\mathbb{R}^+$ with unbounded support and finite mean is called *strong subexponential*, $H \in \mathscr{S}^*$, if

$$\int_0^x \overline{H}(x-y)\overline{H}(y)dy \sim 2m\overline{H}(x) \quad \text{as } x \to \infty,$$

where m is the mean value of H. It is known that if $H \in \mathscr{S}^*$ then both H and H_I are subexponential distributions, see e.g. [6, Theorem 3.27].

In what follows we use the following notation for distributions: we denote

(i) the distribution of $\log(1 + |A| + |B|)$ by H;
(ii) the distribution of $\log(1 + |A|)$ by F;
(iii) the distribution of $\log(1 + |B|)$ by G;
(iv) the distribution of $\log(1 + B^+)$ by G^+;
(v) the distribution of $\log(1 + B^-)$ by G^-.

The paper is organised as follows. In Sects. 2, 4 and 5 we assume that $\log |A|$ has finite negative mean and successively investigate three different cases in the order of increasing difficulty: (i) both A and B are positive, see Theorem 1; (ii) A is positive and B is a signed random variable, see Theorem 6; (iii) both A and B are signed, see Theorem 7. In the case (i) we also explain in Theorem 4 the most probable way by which large deviations of D_n can occur – it is a version of the principle of a single big jump playing the key role in the theory of subexponential distributions. The aim of Sect. 3 is to explain what happens if the distribution of A has an atom at zero; in that case the tail asymptotics of D_n is essentially different from what we observe if A has no atom at zero.

2 Positive Stochastic Difference Equation

In this section we consider a positive D_n, so $A > 0$, $B \ge 0$ – we exclude the case where A has an atom at zero as then the tail asymptotics of D_n are essentially different, see the next section. Then the Markov chain $X_n := \log(1+D_n)$ is positive too. As above, we denote $\xi := \log A$ and the distribution of the random variable $\log(1 + A + B)$ by H.

Theorem 1 *Suppose that* $A > 0$, $B \ge 0$, $\mathbb{E}\xi = -a \in (-\infty, 0)$ *and* $\mathbb{E}\log(1+B) < \infty$, *so that* D_n *is positive recurrent.*

If the integrated tail distribution H_I *is long-tailed, then*

$$\mathbb{P}\{D_\infty > x\} \ge (a^{-1} + o(1))\overline{H_I}(\log x) \quad \text{as } x \to \infty. \tag{6}$$

If, in addition, the distribution H *is long-tailed itself, then*

$$\mathbb{P}\{D_n > x\} \ge \frac{1+o(1)}{a} \int_{\log x}^{\log x + na} \overline{H}(y)dy \text{ as } x \to \infty \text{ uniformly for all } n \ge 1. \tag{7}$$

If the integrated tail distribution H_I *is subexponential then*

$$\mathbb{P}\{D_\infty > x\} \sim a^{-1}\overline{H}_I(\log x) \quad \text{as } x \to \infty. \tag{8}$$

If moreover the distribution H is strong subexponential then

$$\mathbb{P}\{D_n > x\} \sim \frac{1}{a}\int_{\log x}^{\log x+na} \overline{H}(y)dy \quad \textit{as } x \to \infty \textit{ uniformly for all } n \geq 1. \tag{9}$$

The main contribution of Theorem 1 is (9) that states uniform asymptotic behaviour for all $n \geq 1$. It is much stronger than a rather simple conclusion that (9) holds for a fixed n demonstrated by Dyszewski in [4, Theorem 3.3] by induction argument that clearly does not work for the tail asymptotics for the entire range of $n \geq 1$.

In [4], a sufficient condition for the asymptotics (8) is formulated in terms of the distribution of $\log \max(A, B)$ instead of H. Let us show that these two approaches are equivalent. Indeed, for any two positive random variables A and B, since

$$\begin{aligned}\max(\log(1+A), \log(1+B)) &\leq \log(1+A+B)\\ &\leq \log(1+2\max(A, B))\\ &< \log 2 + \max(\log(1+A), \log(1+B)),\end{aligned}$$

it follows that

(i) the distribution H is long-tailed/subexponential/strong subexponential if and only if the distribution of $\max(\log(1+A), \log(1+B))$ is long-tailed/subexponential/strong subexponential respectively;
(ii) the distribution H_I is subexponential if and only if the integrated tail distribution of $\max(\log(1+A), \log(1+B))$ is so.

Denote the distribution of $\log(1+A)$ by F and that of $\log(1+B)$ by G. In the next result we discuss some sufficient conditions for subexponentiality and related properties of H.

Lemma 2 *Let A and B be any two positive random variables such that either of the following two conditions holds:*

(i) *the distribution H of* $\log(1+A+B)$ *is long-tailed or*
(ii) *the random variables A and B are independent.*

Then if the distribution $(F+G)/2$ *is subexponential or strong subexponential, then the distribution H is subexponential or strong subexponential respectively.*

If the integrated tail distribution $(F_I+G_I)/2$ *is subexponential, then* H_I *is subexponential too.*

Proof First assume that (i) holds. On the one hand,

$$\begin{aligned}\overline{H}(x) &= \mathbb{P}\{\log(1+A+B) > x\} \\ &\geq \frac{\mathbb{P}\{\log(1+A) > x\} + \mathbb{P}\{\log(1+B) > x\}}{2} \\ &= \big(\overline{F}(x) + \overline{G}(x)\big)/2 \end{aligned} \tag{10}$$

and thus, for all sufficiently large x,

$$\overline{H_I}(x) \geq \big(\overline{F_I}(x) + \overline{G_I}(x)\big)/3. \tag{11}$$

On the other hand,

$$\begin{aligned}\overline{H}(x) &\leq \mathbb{P}\{\log(1+2A) > x\} + \mathbb{P}\{\log(1+2B) > x\} \\ &\leq \overline{F}(x-\log 2) + \overline{G}(x-\log 2). \end{aligned} \tag{12}$$

If $(F+G)/2$ is subexponential then it is long-tailed and hence

$$\overline{H}(x) \leq (1+o(1))\big(\overline{F}(x) + \overline{G}(x)\big) \quad \text{as } x \to \infty. \tag{13}$$

If $(F_I + G_I)/2$ is subexponential then similarly

$$\overline{H_I}(x) \leq (1+o(1))\big(\overline{F_I}(x) + \overline{G_I}(x)\big) \quad \text{as } x \to \infty. \tag{14}$$

The two bounds (13) and (10) in the case of long-tailed H allow us to apply Theorem 3.11 or 3.25 from [6] and to conclude subexponentiality or strong subexponentiality of H respectively provided $(F+G)/2$ is so.

The two bounds (14) and (11) in the case of long-tailed H_I allow us to apply Theorem 3.11 from [6] and to conclude subexponentiality of H_I provided $(F_I + G_I)/2$ is so.

Now let us consider the case where A and B are independent which yields the following improvement on the lower bound (10). For all $x > 0$,

$$\begin{aligned}\overline{H}(x) &\geq \mathbb{P}\{\log(1+A) > x\} + \mathbb{P}\{\log(1+A) \leq x\}\mathbb{P}\{\log(1+B) > x\} \\ &= \overline{F}(x) + F(x)\overline{G}(x) \\ &\sim \overline{F}(x) + \overline{G}(x) \quad \text{as } x \to \infty. \end{aligned}$$

Therefore, H inherits the tail properties of the distribution $(F+G)/2$, and H_I the tail properties of $(F_I + G_I)/2$. □

Proof of Theorem 1 At any state $x \geq 0$, the Markov chain X_n has jump

$$\begin{aligned} \xi(x) &= \log(1 + A(e^x - 1) + B) - x \\ &= \log(A + e^{-x}(1 - A + B)) \\ &\geq \log(A - e^{-x} A), \end{aligned}$$

as $B \geq 0$. Fix an $\varepsilon > 0$. Choose x_0 sufficiently large such that $\log(1-e^{-x_0}) \geq -\varepsilon/2$. Then the family of jumps $\xi(x)$, $x \geq x_0$, possesses an integrable minorant

$$\begin{aligned} \xi(x) &\geq \xi + \log(1 - e^{-x_0}) \\ &\geq \xi - \varepsilon/2 =: \eta. \end{aligned} \tag{15}$$

On the other hand, since $A > 0$ and $B \geq 0$, the family of jumps $\xi(x)$, $x \geq x_0$, possesses an integrable majorant $\zeta(x_0) := \log(A + e^{-x_0}(1 + B))$. For a sufficiently large x_0,

$$\mathbb{E} \log(A + e^{-x_0}(1 + B)) \leq \mathbb{E}\xi + \varepsilon, \tag{16}$$

owing to the dominated convergence theorem which applies because firstly $\log(A + e^{-x_0}(1 + B)) \to \log A = \xi$ a.s. as $x_0 \to \infty$ and secondly, by the concavity of the function $\log(1 + z)$,

$$\begin{aligned} \log(A + e^{-x_0}(1 + B)) &< \log(1 + A + e^{-x_0}(1 + B)) \\ &\leq \log(1 + A) + \log(1 + e^{-x_0}(1 + B)), \end{aligned}$$

which is integrable by the finiteness of $\mathbb{E}\xi$ and $\mathbb{E}\log(1 + B)$.

Let us first prove the lower bound (6) following the single big jump technique known from the theory of subexponential distributions. Since D_n is assumed convergent, the associated Markov chain X_n is stable, so there exists a $c > 2$ such that

$$\mathbb{P}\{X_n \in (1/c, c]\} \geq 1 - \varepsilon \quad \text{for all } n \geq 0.$$

Let us consider an event

$$\Omega(k, n, c) := \{\eta_{k+1} + \ldots + \eta_{k+j} \geq -c - n(a + \varepsilon) \text{ for all } j \leq n\}, \tag{17}$$

where η_k are independent copies of η defined in (15). By the strong law of large numbers, there exists a sufficiently large c such that

$$\mathbb{P}\{\Omega(k, n, c)\} \geq 1 - \varepsilon \quad \text{for all } k \text{ and } n. \tag{18}$$

It follows from (15) that any of the events

$$\{X_{k-1} \le c,\ X_k > x + c + (n-k)(a+\varepsilon),\ \Omega(k, n-k, c)\} \tag{19}$$

implies $X_n > x$ and they are pairwise disjoint. Therefore, by the Markov property and (18),

$$\begin{aligned}
&\mathbb{P}\{X_n > x\} \\
&\ge \sum_{k=1}^{n} \mathbb{P}\{X_{k-1} \le c,\ X_k > x + c + (n-k)(a+\varepsilon)\}\mathbb{P}\{\Omega(k, n-k, c)\} \\
&\ge (1-\varepsilon)\sum_{k=1}^{n} \mathbb{P}\{X_{k-1} \in (1/c, c],\ X_k > x + c + (n-k)(a+\varepsilon)\}.
\end{aligned}$$

The kth probability on the right hand side equals

$$\begin{aligned}
&\int_{1/c}^{c} \mathbb{P}\{X_{k-1} \in dy\}\mathbb{P}\{y + \xi(y) > x + c + (n-k)(a+\varepsilon)\} \\
&= \int_{1/c}^{c} \mathbb{P}\{X_{k-1} \in dy\}\mathbb{P}\{\log(1 + A(e^y - 1) + B) > x + c + (n-k)(a+\varepsilon)\}.
\end{aligned}$$

For all $y > 1/c$,

$$\begin{aligned}
\log(1 + A(e^y - 1) + B) &\ge \log(1 + A(e^{1/c} - 1) + B) \\
&\ge \log(1 + A + B) + \log(e^{1/c} - 1),
\end{aligned}$$

because $e^{1/c} - 1 < \sqrt{e} - 1 < 1$. Therefore, the value of the last integral is not less than

$$\mathbb{P}\{X_{k-1} \in (1/c, c]\}\mathbb{P}\{\log(1 + A + B) > x + c_1 + (n-k)(a+\varepsilon)\},$$

where $c_1 := c - \log(e^{1/c} - 1)$. Hence, due to the choice of c,

$$\mathbb{P}\{X_n > x\} \ge (1-\varepsilon)^2 \sum_{k=1}^{n} \overline{H}(x + c_1 + (n-k)(a+\varepsilon)).$$

Since the tail is a decreasing function, the last sum is not less than

$$\frac{1}{a+\varepsilon}\int_0^{n(a+\varepsilon)} \overline{H}(x + c_1 + y)dy. \tag{20}$$

Letting $n \to \infty$ we obtain that the tail at point x of the stationary distribution of the Markov chain X is not less than

$$\begin{aligned}\frac{(1-\varepsilon)^2}{a+\varepsilon}\int_0^\infty \overline{H}(x+c_1+y)dy &= \frac{(1-\varepsilon)^2}{a+\varepsilon}\overline{H_I}(x+c_1)\\ &\sim \frac{(1-\varepsilon)^2}{a+\varepsilon}\overline{H_I}(x) \quad \text{as } x\to\infty,\end{aligned}$$

due to the long-tailedness of the integrated tail distribution H_I. Summarising altogether we deduce that, for every fixed $\varepsilon > 0$,

$$\liminf_{x\to\infty}\frac{\mathbb{P}\{D_\infty > x\}}{\overline{H_I}(\log x)} \geq \frac{(1-\varepsilon)^2}{a+\varepsilon},$$

which implies the lower bound (6) due to the arbitrary choice of $\varepsilon > 0$.

If the distribution H is long-tailed itself, then the integral in (20) is asymptotically equivalent to the integral

$$\int_x^{x+n(a+\varepsilon)} \overline{H}(y)dy \quad \text{as } x\to\infty \text{ uniformly for all } n\geq 1,$$

which implies the second lower bound (7).

Now let us turn to the asymptotic upper bound under the assumption that the integrated tail distribution H_I is subexponential. Fix an $\varepsilon \in (0, a)$. Let x_0 be defined as in (16), so $\mathbb{E}\zeta(x_0) \leq -a+\varepsilon$. Let J be the distribution of $\zeta(x_0)$. Since

$$\log(1+A+B) - x_0 \;\leq\; \zeta(x_0) \;\leq\; \log(1+A+B),$$

we have $\overline{H}(x+x_0) \leq \overline{J}(x) \leq \overline{H}(x)$. Then subexponentiality of H_I yields subexponentiality of the integrated tail distribution J_I and $\overline{J}_I(x) \sim \overline{H}_I(x)$ as $x \to \infty$.

By the construction of $\zeta(x_0)$,

$$x+\xi(x) \leq y+\zeta(x_0) \quad \text{for all } y\geq x\geq x_0. \tag{21}$$

Also, by the positivity of A,

$$\begin{aligned} x+\xi(x) &= \log(1+A(e^x-1)+B)\\ &\leq \log(1+A(e^{x_0}-1)+B)\\ &= x_0+\xi(x_0) \;\leq\; x_0+\zeta(x_0) \quad \text{for all } x\leq x_0. \end{aligned}\tag{22}$$

Consider a random walk Z_n delayed at the origin with jumps $\zeta(x_0)$:

$$Z_0 := 0, \quad Z_n := (Z_{n-1} + \zeta_n(x_0))^+,$$

where $\zeta_n(x_0)$ are independent copies of $\zeta(x_0)$. The upper bounds (21) and (22) yield that the two chains X_n and Z_n can be constructed on a common probability space in such a way that, with probability 1,

$$X_n \le x_0 + Z_n \quad \text{for all } n, \tag{23}$$

so X_n is dominated by a random walk on $[x_0, \infty)$ delayed at point x_0. Since the integrated tail distribution J_I is subexponential, the tail of the invariant measure of the chain Z_n is asymptotically equivalent to $\overline{J}_I(x)/(a-\varepsilon) \sim \overline{H}_I(x)/(a-\varepsilon)$ as $x \to \infty$, see, for example, [6, Theorem 5.2]. Thus, the tail of the invariant measure of X_n is asymptotically not greater than $\overline{H}_I(x-x_0)/(a-\varepsilon)$ which is equivalent to $\overline{H}_I(x)/(a-\varepsilon)$, since H_I is long-tailed by subexponentiality. Hence,

$$\limsup_{x\to\infty} \frac{\mathbb{P}\{D_\infty > x\}}{\overline{H}_I(\log x)} \le \frac{1}{a-\varepsilon}.$$

Due to arbitrary choice of $\varepsilon > 0$ and the lower bound proven above this completes the proof of the first asymptotics (8).

The same arguments with the same majorant (23) allow us to conclude the finite time horizon asymptotics for D_∞ if we apply Theorem 5.3 from [6] instead of Theorem 5.2. □

Theorem 1 makes it possible to identify a moment of time after which the tail distribution of D_n is equivalent to that of D_∞, in some particular strong subexponential cases.

Corollary 3 *Suppose that* $\mathbb{E}\log A = -a < 0$, $B > 0$ *and* $\mathbb{E}\log(1+B) < \infty$.

If the distribution H *of* $\log(1+A+B)$ *is regularly varying at infinity with index* $\alpha < -1$, *then* $\mathbb{P}\{D_n > x\} \sim \mathbb{P}\{D_\infty > x\}$ *as* $n, x \to \infty$ *if and only if* $n/\log x \to \infty$.

If $\overline{H}(x) \sim e^{-x^\beta}$ *for some* $\beta \in (0,1)$, *then* $\mathbb{P}\{D_n > x\} \sim \mathbb{P}\{D_\infty > x\}$ *as* n, $x \to \infty$ *if and only if* $n/\log^{1-\beta} x \to \infty$.

We conclude this section by a version of the principle of a single big jump for D_n. For any $c > 1$ and $\varepsilon > 0$ consider events

$$\Omega_k := \{1/c < X_{k-1} \le c,\ X_k > \log x + c + (n-k)(a+\varepsilon),$$
$$|X_{k+j} - X_k + aj| \le c + j\varepsilon \text{ for all } j \le n-k\Big\}$$

or, in terms of D_n,

$$\Omega_k^D := \{1/c < D_{k-1} \le c,\ A_k/c + B_k > xe^{c+(n-k)(a+\varepsilon)},$$
$$e^{-c-j(a+\varepsilon)} \le D_{k+j}/D_k \le e^{c-j(a-\varepsilon)} \text{ for all } j \le n-k\Big\}.$$

Roughly speaking, it describes a trajectory such that, for large x, the D_{k-1} is neither too far away from zero nor too close, then a single big jump occurs, both A_k and B_k may contribute to that big jump, and then the logarithm of D_{k+j}, $j \le n-k$, moves down according to the strong law of large numbers with drift $-a$. As stated in the next theorem, the union of all these events describes more precisely than the lower bound of Theorem 1 the most probable way by which large deviations of D_n do occur.

Theorem 4 *Let the distribution H of* $\log(1+A+B)$ *be strong subexponential. Then, for any fixed* $\varepsilon > 0$,

$$\lim_{c\to\infty}\lim_{x\to\infty}\inf_{n\ge 1}\mathbb{P}\{\cup_{k=0}^{n-1}\Omega_k \mid D_n > x\} = 1.$$

Proof The events $\Omega(k)$, $k \le n$, are pairwise disjoint and any of them implies $\{X_n > \log x\}$. Then similar arguments as in the proof of lower bound in Theorem 1 apply. □

3 Impact of Atom at Zero

In this section we demonstrate what happens if the distribution of A has an atom at zero. It turns out that then the tail asymptotics of D_n are essentially different – they are proportional to the tail of H which is lighter than given by the integrated tail distribution H_I in the case where $A > 0$ – because the chain satisfies Doeblin's condition, see e.g. [14, Ch. 16]. As above, we denote by H the distribution of the random variable $\log(1+A+B)$. For simplicity, we assume that $B > 0$.

Theorem 5 *Suppose that $A \ge 0$, $B > 0$ and $p_0 := \mathbb{P}\{A = 0\} \in (0, 1)$. If the distribution H is long-tailed and $D_0 > 0$, then*

$$\mathbb{P}\{D_n > x\} \ge \Big(\frac{1-(1-p_0)^n}{p_0} + o(1)\Big)\overline{H}(\log x) \tag{24}$$

as $x \to \infty$ uniformly for all $n \ge 1$. In particular,

$$\mathbb{P}\{D_\infty > x\} \ge (p_0^{-1} + o(1))\overline{H}(\log x) \quad \text{as } x \to \infty. \tag{25}$$

If the distribution H is subexponential, $D_0 > 0$ and $\{D_0 > x\} = o(\overline{H}(x))$ then

$$\mathbb{P}\{D_n > x\} \sim \frac{1-(1-p_0)^n}{p_0}\overline{H}(\log x) \tag{26}$$

as $x \to \infty$ uniformly for all $n \ge 1$. In particular,

$$\mathbb{P}\{D_\infty > x\} \sim p_0^{-1}\overline{H}(\log x) \quad \text{as } x \to \infty. \tag{27}$$

Proof Let H_0 be the distribution of $\log(1+A+B)$ conditioned on $A > 0$ and G_0 be the distribution of $\log(1+B)$ conditioned on $A = 0$, then $H = p_0 G_0 + (1-p_0) H_0$.

Let us decompose the event $X_n > x$ according to the last zero value of A_k, which gives equality

$$\begin{aligned}
\mathbb{P}\{X_n > x\} &= \mathbb{P}\{A_1, \dots, A_n > 0, X_n > x\} \\
&\qquad + \sum_{k=1}^{n} \mathbb{P}\{A_k = 0, A_{k+1} > 0, \dots, A_n > 0, X_n > x\} \\
&= (1-p_0)^n \mathbb{P}\{X_n > x \mid A_1, \dots, A_n > 0\} \\
&\quad + p_0 \sum_{k=1}^{n} (1-p_0)^{n-k} \mathbb{P}\{X_n > x \mid A_k = 0, A_{k+1}, \dots, A_n > 0\} \\
&= (1-p_0)^n \mathbb{P}\{X_n > x \mid A_1, \dots, A_n > 0\} \\
&\quad + p_0 \sum_{k=0}^{n-1} (1-p_0)^k \mathbb{P}\{X_{k+1} > x \mid A_1 = 0, A_2, \dots, A_{k+1} > 0\},
\end{aligned} \tag{28}$$

by the Markov property. In particular, the sum from 0 to $n-1$ on the right hand side is increasing as n grows as all terms are positive. For that reason, for the lower bounds for $\mathbb{P}\{D_n > x\}$ it suffices to prove by induction that, for any fixed $k \ge 0$ and $\gamma > 0$, there exists a $c < \infty$ such that

$$\begin{aligned}
\mathbb{P}\{X_{k+1} > x \mid A_1 = 0, A_2, \dots, A_{k+1} > 0\} \\
\ge (1-\gamma)\big(\overline{G_0}(x+c) + k\overline{H_0}(x+c)\big),
\end{aligned} \tag{29}$$

$$\mathbb{P}\{X_{k+1} > x \mid A_1, \dots, A_{k+1} > 0\} \ge (1-\gamma)(k+1)\overline{H_0}(x+c) \tag{30}$$

for all sufficiently large x, because then

$$\begin{aligned}
\mathbb{P}\{X_n > x\} &\ge (1-\gamma)\Bigg((1-p_0)^n n\overline{H_0}(x+c) \\
&\qquad + p_0 \sum_{k=0}^{n-1} (1-p_0)^k \big(\overline{G_0}(x+c) + k\overline{H_0}(x+c)\big)\Bigg) \\
&= (1-\gamma)((1-(1-p_0)^n)\Big(\overline{G_0}(x+c) + \frac{1-p_0}{p_0}\overline{H_0}(x+c)\Big) \\
&= (1-\gamma)\frac{1-(1-p_0)^n}{p_0}\overline{H}(x+c),
\end{aligned}$$

with further application of long-tailedness of H.

To prove (29), first let us note that the induction basis $k = 0$ is immediate, since the distribution of X_1 conditioned on $A_1 = 0$ is G_0. Now let us assume that (29) is true for some k. Denote

$$G_k(dy) := \mathbb{P}\{X_{k+1} \in dy \mid A_1 = 0, A_2, \ldots, A_{k+1} > 0\}, \quad k \geq 0,$$

which is a distribution on $(0, \infty)$. Then

$$\begin{aligned}
\overline{G_{k+1}}(x) &= \int_0^\infty \mathbb{P}\{\log(1 + A(e^y - 1) + B) > x \mid A > 0\} G_k(dy) \\
&\geq \int_\varepsilon^{1/\varepsilon} \mathbb{P}\{\log(1 + A\delta + B) > x \mid A > 0\} G_k(dy) \\
&\qquad + \int_{x+1/\varepsilon}^\infty \mathbb{P}\{\log(A(e^y - 1)) > x \mid A > 0\} G_k(dy) \\
&=: I_1 + I_2,
\end{aligned}$$

for any $\varepsilon \in (0, 1/2]$ where $\delta = e^\varepsilon - 1 < \sqrt{e} - 1 < 1$. Let us observe that then

$$\begin{aligned}
\mathbb{P}\{\log(1 + A\delta + B) > x \mid A > 0\} &= \mathbb{P}\{\log(1/\delta + A + B/\delta) > x - \log\delta \mid A > 0\} \\
&\geq \overline{H}_0(x - \log\delta).
\end{aligned}$$

Therefore,

$$I_1 \geq \overline{H}_0(x - \log\delta) G_k(\varepsilon, 1/\varepsilon].$$

The second integral may be bounded below as follows:

$$\begin{aligned}
I_2 &\geq \mathbb{P}\{\log(A(e^{x+1/\varepsilon} - 1)) > x \mid A > 0\} \overline{G}_k(x + 1/\varepsilon) \\
&\geq \mathbb{P}\{\log(Ae^{x+1/2\varepsilon}) > x \mid A > 0\} \overline{G}_k(x + 1/\varepsilon) \\
&= \mathbb{P}\{A > e^{-1/2\varepsilon} \mid A > 0\} \overline{G}_k(x + 1/\varepsilon),
\end{aligned}$$

for all sufficiently large x. Letting $\varepsilon \to 0$ we obtain that, for any fixed $\gamma > 0$, there exists a $c < \infty$ such that the following lower bound holds

$$\overline{G_{k+1}}(x) \geq (1 - \gamma)\big(\overline{H_0}(x + c) + \overline{G_k}(x + c)\big)$$

for all sufficiently large x, which implies the induction step.

The second lower bound, (30), follows by similar arguments provided $D_0 > 0$.

Let us now proceed with a matching upper bound under the assumption that H is a subexponential distribution. Since $A, B \geq 0$,

$$\xi(x) = \log(A + e^{-x}(1 - A + B)) \tag{31}$$

$$\leq \log(1 + A + B) \quad \text{for all } x > 0. \tag{32}$$

Let η and ζ be random variables with the following tail distributions

$$\mathbb{P}\{\eta > x\} = \min\left(1, \frac{\mathbb{P}\{\log(1+A+B) > x\}}{\mathbb{P}\{A=0\}}\right),$$
$$\mathbb{P}\{\zeta > x\} = \min\left(1, \frac{\mathbb{P}\{\log(1+A+B) > x\}}{\mathbb{P}\{A>0\}}\right), \quad x > 0.$$

Both are subexponential random variables provided $\log(1 + A + B)$ is so, see e.g. [6, Corollary 3.13]. It follows from (31) that, for all $x > 0$,

$$\mathbb{P}\{\xi(x) > y \mid A = 0\} \le \mathbb{P}\{\eta > y\},$$
$$\mathbb{P}\{\xi(x) > y \mid A > 0\} \le \mathbb{P}\{\zeta > y\},$$

which implies that

$$\mathbb{P}\{X_{k+1} > x \mid A_1 = 0, A_2, \ldots, A_{k+1} > 0\} \le \mathbb{P}\{\eta + \zeta_1 + \ldots + \zeta_k > x\},$$

where ζ_i's are independent copies of ζ independent of η. Then standard technique based on Kesten's bound for convolutions of subexponential distributions, see e.g. Theorem 3.39 in [6], allows us to deduce from (28) that, for any fixed $\gamma > 0$,

$$\mathbb{P}\{X_n > x\}$$
$$\le (1+\gamma)\Big((1-p_0)^n n\overline{G}_0(x) + p_0\sum_{k=0}^{n-1}(1-p_0)^k(\overline{G}_0(x) + k\overline{H}_0(x)\Big)$$

for all $n \ge 1$ and sufficiently large x. Therefore,

$$\mathbb{P}\{X_n > x\} \le (1+\gamma)\frac{1-(1-p_0)^n}{p_0}\overline{H}(x),$$

which together with the lower bound proves (26). □

4 The Case of Positive A and Signed B

In this section we consider the case where D_n takes both positive and negative values because of singed B, while A is still assumed positive in this section, $A > 0$. The Markov chain X_n is defined as in (2).

As B is no longer assumed positive, it makes the tail behaviour of D quite different if no further assumptions are made on dependency between A and B. For example, in the extreme case where $B = -cA$ for some $c > 0$, so $D_{n+1} =$

$A_{n+1}(D_n - c)$, we have that D_n is eventually negative if stable, hence $D_\infty < 0$ with probability 1.

More generally, if $B = A\eta$ where η is independent of A and takes values of both signs, then we conclude similar to (6) that, as $x \to \infty$,

$$\mathbb{P}\{D_\infty > x\} \geq \left(\frac{1}{a}\int_{\mathbb{R}} \mathbb{P}\{\eta > -c\}\mathbb{P}\{D_\infty \in dc\} + o(1)\right)\overline{F_I}(\log x),$$

provided the distribution F_I is long-tailed. However, the technique used in Sect. 2 for proving the matching upper bound does not work in such cases as the Lindley majorant returns the coefficient a^{-1} which is greater than that in the lower bound above. For that reason we restrict further considerations to the case where A and B are independent.

Theorem 6 *Suppose that $A > 0$, A and B are independent, $\mathbb{E}\xi = -a \in (-\infty, 0)$ and $\mathbb{E}\log(1 + |B|) < \infty$.*

If the integrated tail distributions F_I and G_I^+ are long-tailed, then

$$\mathbb{P}\{D_\infty > x\} \geq (a^{-1}+o(1))\left(\mathbb{P}\{D_\infty > 0\}\overline{F_I}(\log x) + \overline{G_I^+}(\log x)\right) \text{ as } x \to \infty. \tag{33}$$

If, in addition, the distributions F and G^+ are long-tailed itself, then, as $x, n \to \infty$,

$$\mathbb{P}\{D_n > x\} \geq \frac{1+o(1)}{a}\left(\mathbb{P}\{D_\infty > 0\}\int_{\log x}^{\log x+na} \overline{F}(y)dy + \int_{\log x}^{\log x+na} \overline{G^+}(y)dy\right). \tag{34}$$

If $\mathbb{P}\{D_\infty = 0\} = 0$, the integrated tail distributions F_I, G_I^+ and G_I^- are long-tailed, $\overline{G_I^-}(z) = O(\overline{F_I}(z) + \overline{G_I^+}(z))$ and H_I is subexponential then

$$\mathbb{P}\{D_\infty > x\} \sim a^{-1}\left(\mathbb{P}\{D_\infty > 0\}\overline{F_I}(\log x) + \overline{G_I^+}(\log x)\right) \text{ as } x \to \infty. \tag{35}$$

If moreover the distributions F, G^+ and G^- are long-tailed, $\overline{G^-}(z) = O(\overline{F}(z) + \overline{G^+}(z))$ and H is strong subexponential then, as $x, n \to \infty$,

$$\mathbb{P}\{D_n > x\} \sim \frac{1}{a}\left(\mathbb{P}\{D_\infty > 0\}\int_{\log x}^{\log x+na} \overline{F}(y)dy + \int_{\log x}^{\log x+na} \overline{G^+}(y)dy\right). \tag{36}$$

Proof Fix an $\varepsilon > 0$. As follows from (4), for $x \geq 0$,

$$\xi(x) \geq \begin{cases} \log(A(1-e^{-x}) - e^{-x}B^-) & \text{if } A(e^x - 1) + B \geq 0, \\ -\log(1 + A + |B|) & \text{if } A(e^x - 1) + B < 0, \end{cases}$$

where the second line follows due to $A > 0$. The minorant on the right hand side is stochastically increasing as x grows, therefore, there exists a sufficiently large x_0 and a random variable η such that

$$\xi(x) \geq \eta \quad \text{for all } x \geq x_0 \text{ and } \mathbb{E}\eta > -a - \varepsilon/2. \tag{37}$$

As in the last proof, we start with the lower bound (33) following the single big jump technique. Since D_n is assumed to be convergent, the associated Markov chain X_n is stable, so there exist n_0 and $c > 2$ such that

$$\mathbb{P}\{X_n \in (1/c, c]\} \geq (1-\varepsilon)\mathbb{P}\{D_\infty > 0\} \quad \text{for all } n \geq n_0,$$
$$\mathbb{P}\{|X_n| \leq c\} \geq 1 - \varepsilon \quad \text{for all } n,$$

and also $\mathbb{P}\{A \leq c\} \geq 1 - \varepsilon$, $\mathbb{P}\{|B| \leq c\} \geq 1 - \varepsilon$. For all k, n and c, let us consider the events $\Omega(k, n, c)$ defined in (17) and satisfying (18). It follows from (37) that any of the events (19) implies $X_n > x$ and they are pairwise disjoint. Therefore, by the Markov property and (18),

$$\begin{aligned} &\mathbb{P}\{X_n > x\} \\ &\geq \sum_{k=1}^{n} \mathbb{P}\{X_{k-1} \leq c,\ X_k > x + c + (n-k)(a+\varepsilon)\}\mathbb{P}\{\Omega(k, n-k, c)\} \\ &\geq (1-\varepsilon)\sum_{k=1}^{n} \mathbb{P}\{X_{k-1} \leq c,\ X_k > x + c + (n-k)(a+\varepsilon)\}, \end{aligned} \tag{38}$$

The kth term of the sum is not less than

$$\begin{aligned} &\left(\int_{-c}^{0} + \int_{0}^{c}\right)\mathbb{P}\{X_{k-1} \in dy\}\mathbb{P}\{y + \xi(y) > z_{n-k}\} \\ &= \int_{-c}^{0} \mathbb{P}\{X_{k-1} \in dy\}\mathbb{P}\{\log(1 + A(1-e^{-y}) + B) > z_{n-k}\} \\ &\quad + \int_{0}^{c} \mathbb{P}\{X_{k-1} \in dy\}\mathbb{P}\{\log(1 + A(e^{y}-1) + B) > z_{n-k}\} \\ &=: I_1 + I_2, \end{aligned}$$

where $z_k = x + c + k(a + \varepsilon)$. For all $y \in [-c, 0]$ and $z > 0$, owing to the condition $A > 0$ and independence of A and B

$$\begin{aligned}\mathbb{P}\{\log(1 + A(1 - e^{-y}) + B) > z\} &\ge \mathbb{P}\{\log(1 - Ae^c + B) > z\}\\ &\ge \mathbb{P}\{A \le c\}\mathbb{P}\{\log(1 - ce^c + B) > z\}\\ &\ge \mathbb{P}\{A \le c\}\overline{G^+}(z + 1)\end{aligned}$$

for all sufficiently large z which yields that

$$\begin{aligned}I_1 &\ge \mathbb{P}\{A \le c\}\mathbb{P}\{X_{k-1} \in [-c, 0]\}\overline{G^+}(z_{n-k} + 1)\\ &\ge (1 - \varepsilon)\mathbb{P}\{X_{k-1} \in [-c, 0]\}\overline{G^+}(z_{n-k} + 1),\end{aligned} \tag{39}$$

due to the choice of c. For all $y > 0$,

$$\begin{aligned}&\mathbb{P}\{\log(1 + A(e^y - 1) + B) > z\}\\ &\quad\ge \mathbb{P}\{|B| \le c\}\mathbb{P}\{\log(1 + A(e^y - 1) - c) > z\} + \mathbb{P}\{\log(1 + B) > z\},\end{aligned}$$

which yields that

$$\begin{aligned}I_2 &\ge \mathbb{P}\{|B| \le c\}\int_{1/c}^{c} \mathbb{P}\{\log(1 + A(e^y - 1) - c) > z_{n-k}\}\mathbb{P}\{X_{k-1} \in dy\}\\ &\qquad\qquad + \overline{G^+}(z_{n-k})\mathbb{P}\{X_{k-1} \in (0, c]\}\\ &\ge (1 - \varepsilon)\mathbb{P}\{\log(1 + A(e^{1/c} - 1) - c) > z_{n-k}\}\mathbb{P}\{X_{k-1} \in (1/c, c]\}\\ &\qquad\qquad + \overline{G^+}(z_{n-k})\mathbb{P}\{X_{k-1} \in (0, c]\}.\end{aligned}$$

Therefore, by the choice of c, for all sufficiently large x and $k > n_0$,

$$I_2 \ge (1 - \varepsilon)^2\mathbb{P}\{D_\infty > 0\}\overline{F}(z_{n-k} + 1) + \overline{G^+}(z_{n-k})\mathbb{P}\{X_{k-1} \in (0, c]\}. \tag{40}$$

Substituting (39) and (40) into (38) we deduce that

$$\begin{aligned}\mathbb{P}\{X_n > x\} \ge (1 - \varepsilon)^2 \sum_{k=n_0+1}^{n} \Big(&\mathbb{P}\{D_\infty > 0\}\overline{F}(x + c + 1 + (n - k)(a + \varepsilon))\\ &+ \overline{G^+}(x + c + 1 + (n - k)(a + \varepsilon))\Big)\end{aligned}$$

Since the tail is a non-increasing function, the last sum is not less than

$$\frac{1}{a+\varepsilon}\int_0^{(n-n_0-1)(a+\varepsilon)}\Big(\mathbb{P}\{D_\infty>0\}\overline{F}(x+c+1+y)+\overline{G^+}(x+c+1+y)\Big)dy. \tag{41}$$

Letting $n \to \infty$ we obtain that the tail at point x of the stationary distribution of the Markov chain X is not less than

$$\begin{aligned}\frac{(1-\varepsilon)^2}{a+\varepsilon}&\int_0^{\infty}\Big(\mathbb{P}\{D_\infty>0\}\overline{F}(x+c+1+y)+\overline{G^+}(x+c+1+y)\Big)dy\\ &=\frac{(1-\varepsilon)^2}{a+\varepsilon}\Big(\mathbb{P}\{D_\infty>0\}\overline{F_I}(x+c+1)+\overline{G_I^+}(x+c+1)\Big) \qquad (42)\\ &\sim\frac{(1-\varepsilon)^2}{a+\varepsilon}\Big(\mathbb{P}\{D_\infty>0\}\overline{F_I}(x)+\overline{G_I^+}(x)\Big)\quad \text{as } x\to\infty,\end{aligned}$$

due to the long-tailedness of the integrated tail distributions F_I and G_I^+. Summarising altogether we deduce that, for every fixed $\varepsilon > 0$,

$$\liminf_{x\to\infty}\frac{\mathbb{P}\{D_\infty>x\}}{\mathbb{P}\{D_\infty>0\}\overline{F_I}(\log x)+\overline{G_I^+}(\log x)}\ge\frac{(1-\varepsilon)^2}{a+\varepsilon},$$

which implies the lower bound (33) due to the arbitrary choice of $\varepsilon > 0$.

If the distributions F and G^+ are long-tailed itself, then the integral in (41) is asymptotically equivalent to the integral

$$\int_x^{x+n(a+\varepsilon)}\Big(\mathbb{P}\{D_\infty>0\}\overline{F}(y)+\overline{G^+}(y)\Big)dy\quad \text{as } x,\ n\to\infty,$$

and the second lower bound (34) follows too.

To prove matching upper bounds let us first observe that

$$|D_{n+1}|\le A_n|D_n|+|B_n|\quad \text{for all } n, \tag{43}$$

where the right hand side is increasing in D_n. Hence, $|D_n| \le \widetilde{D}_n$, where $\widetilde{D}_n$ is a positive stochastic difference recursion,

$$\widetilde{D}_{n+1}=A_n\widetilde{D}_n+|B_n|.$$

Since H_I is subexponential, Theorem 1 applies to $\widetilde{D}_n$, so

$$\mathbb{P}\{\widetilde{D}_\infty>x\}\sim a^{-1}\overline{H_I}(\log x)\quad \text{as } x\to\infty,$$

and hence

$$\mathbb{P}\{|D_\infty| > x\} \le (a^{-1} + o(1))\overline{H_I}(\log x) \quad \text{as } x \to \infty,$$

It follows from (12) that

$$\overline{H}(x) \le \mathbb{P}\{\log(1 + A) > x - 1\} + \mathbb{P}\{\log(1 + |B|) > x - 1\}.$$

Integrating the last inequality we get an upper bound

$$\begin{aligned} \overline{H_I}(x) &\le \overline{F_I}(x-1) + \overline{G_I^-}(x-1) + \overline{G_I^+}(x-1) \\ &\sim \overline{F_I}(x) + \overline{G_I^-}(x) + \overline{G_I^+}(x) \quad \text{as } x \to \infty, \end{aligned} \tag{44}$$

because all three distributions, F_I, G_I^- and G_I^+ are assumed long-tailed. Hence the following upper bound holds for the tail of $|D_\infty|$, as $x \to \infty$:

$$\mathbb{P}\{|D_\infty| > x\} \le (a^{-1}+o(1))\big(\overline{F_I}(\log x) + \overline{G_I^-}(\log x) + \overline{G_I^+}(\log x)\big). \tag{45}$$

The long-tailedness of F_I and G_I^- similarly to (33) implies that

$$\mathbb{P}\{D_\infty < -x\} \ge (a^{-1} + o(1))\big(\mathbb{P}\{D_\infty < 0\}\overline{F_I}(\log x) + \overline{G_I^-}(\log x)\big),$$

and the two lower bounds together imply that, as $x \to \infty$,

$$\mathbb{P}\{|D_\infty| > x\} \ge (a^{-1} + o(1))\big(\overline{F_I}(\log x) + \overline{G_I^+}(\log x) + \overline{G_I^-}(\log x)\big),$$

because $\mathbb{P}\{D_\infty = 0\} = 0$. Together with the upper bound (45) it yields that

$$\mathbb{P}\{D_\infty > x\} = a^{-1}\big(\mathbb{P}\{D_\infty > 0\}\overline{F_I}(\log x) + \overline{G_I^+}(\log x)\big) + o(\overline{H_I}(\log x)),$$

and the first asymptotics (35) follows by the condition $\overline{G_I^-}(z) = O(\overline{F_I}(z)+\overline{G_I^+}(z))$.

The second asymptotics (36) follows along similar arguments. □

5 Balance of Negative and Positive Tails in the Case of Signed A

In this section we turn to the general case where D_n takes both positive and negative values, with A taking values of both signs. Denote $\xi := \log|A|$ and the distribution of $\log(1 + |A|)$ by F. Recall that the distribution of $\log(1 + |B|)$ is denoted by G and the distribution of $\log(1 + |A| + |B|)$ by H.

The Markov chain X_n is defined as above in (2).

Theorem 7 *Suppose that* $\mathbb{P}\{D_\infty = 0\} = 0$,

$$0 \ < \ \mathbb{P}\{A > 0\} \ < \ 1, \tag{46}$$

A and B are independent, $\mathbb{E}\xi = -a \in (-\infty, 0)$ *and* $\mathbb{E}\log(1 + |B|) < \infty$.
If the integrated tail distribution H_I *is long-tailed, then*

$$\mathbb{P}\{D_\infty > x\} \geq (1/2a + o(1))\overline{H_I}(\log x) \quad \text{as } x \to \infty. \tag{47}$$

If, in addition, the distribution H is long-tailed itself, then

$$\mathbb{P}\{D_n > x\} \geq \frac{1 + o(1)}{2a} \int_{\log x}^{\log x + na} \overline{H}(y)dy \quad \text{as } n,\ x \to \infty. \tag{48}$$

If the integrated tail distribution H_I *is subexponential then*

$$\mathbb{P}\{D_\infty > x\} \sim \frac{1}{2a}\overline{H_I}(\log x) \quad \text{as } x \to \infty. \tag{49}$$

If moreover the distribution H is strong subexponential then

$$\mathbb{P}\{D_n > x\} \sim \frac{1}{2a} \int_{\log x}^{\log x + na} \overline{H}(y)dy \quad \text{as } n,\ x \to \infty. \tag{50}$$

Proof The same arguments based on the single big jump technique used in the last section for proving (42) show that, for any fixed $\varepsilon > 0$, there exists a $c < \infty$ such that

$$\mathbb{P}\{|X_\infty| > x\} \geq \frac{1 - \varepsilon}{a}\big(\mathbb{P}\{D_\infty \neq 0\}\overline{F_I}(x + c + 1) + \overline{G_I}(x + c + 1)\big)$$

for all sufficiently large x. Similar to (44),

$$\overline{H_I}(x) \leq \overline{F_I}(x - 1) + \overline{G_I}(x - 1)$$

for all sufficiently large x, which together with the condition $\mathbb{P}\{D_\infty = 0\} = 0$ implies that

$$\begin{aligned}\mathbb{P}\{|X_\infty| > x\} &\geq \frac{1 - \varepsilon}{a}\overline{H_I}(x + c + 2) \\ &\sim \frac{1 - \varepsilon}{a}\overline{H_I}(x) \quad \text{as } x \to \infty,\end{aligned}$$

due to the long-taileness of the distribution H_I. Therefore,

$$\mathbb{P}\{|X_\infty| > x\} \ge (a^{-1} + o(1))\overline{H_I}(x) \quad \text{as } x \to \infty. \tag{51}$$

At any time epoch n large absolute value of X_n changes its sign with asymptotic (as $x \to \infty$) probability $p^- = \mathbb{P}\{A < 0\}$ and keeps its sign with asymptotic probability $p^+ = \mathbb{P}\{A > 0\}$, so sign change may be asymptotically described as a Markov chain with transition probability matrix

$$\begin{pmatrix} p^+ & p^- \\ p^- & p^+ \end{pmatrix},$$

whose asymptotic distribution is $(1/2, 1/2)$, owing to the condition (46). For that reason, the probability of a large positive value of X_n is approximately at least one half of the right hand side of (51), and the proof of (47) is complete. The proof of (48) follows the same lines.

To prove the upper bound (49), similar to (43) we first note that

$$|D_{n+1}| \le |A_n||D_n| + |B_n| \quad \text{for all } n,$$

which allows to conclude the proof as it was done in the last section. □

References

1. Borovkov, A.A., Korshunov, D.: Large-deviation probabilities for one-dimensional Markov chains. Part 3: Prestationary distributions in the subexponential case. Theory Probab. Appl. **46**, 603–618 (2002)
2. Buraczewski, D., Damek, E., Mikosch, T.: Stochastic Models with Power-Law Tails. The Equation $X = AX + B$. Springer (2016)
3. Chen, B., Rhee, C.-H., Zwart, B.: Importance sampling of heavy-tailed iterated random functions. Adv. Appl. Probab. **50**, 805–832 (2018)
4. Dyszewski, P.: Iterated random functions and slowly varying tails. Stochastic Process. Appl. **126**, 392–413 (2016)
5. Embrechts, P., Goldie, C.M.: Perpetuities and random equations. In: Asymptotic Statistics (Prague, 1993). Contrib. Statist. 75–86. Heidelberg: Physica (1994)
6. Foss, S., Korshunov, D., Zachary, S.: An Introduction to Heavy-Tailed and Subexponential Distributions. Springer, New York (2011)
7. Goldie, C.M.: Implicit renewal theory and tails of solutions of random equations. Ann. Appl. Probab. **1**, 126–166 (1991)
8. Goldie, C.M., Grübel, R.: Perpetuities with thin tails. Adv. Appl. Probab. **28**, 463–480 (1996)
9. Goldie, C.M., Maller, R.A.: Stability of perpetuities. Ann. Probab. **28**, 1195–1218 (2000)
10. Grey, D.: (1994) Regular variation in the tail behaviour of solutions of random difference equations. Ann. Appl. Probab. **4**, 169–183
11. Kesten, H.: Random difference equations and renewal theory for products of random matrices. Acta Math. **131**, 207–248 (1973)

12. Konstantinides, D.G., Mikosch, T.: Large deviations and ruin probabilities for solutions to stochastic recurrence equations with heavy-tailed innovations. Ann. Probab. **33**, 1992–2035 (2005)
13. Korshunov, D.: Large-deviation probabilities for maxima of sums of independent random variables with negative mean and subexponential distribution. Theory Probab. Appl. **46**, 355–366 (2002)
14. Meyn, S., Tweedie, R.: Markov Chains and Stochastic Stability, 2nd edn. Cambridge University Press (2009)
15. Rachev, S.T., Samorodnitsky, G.: Limit laws for a stochastic process and random recursion arising in probabilistic modeling. Adv. Appl. Probab. **27**, 185–202 (1995)
16. de Saporta, B.: Tail of the stationary solution of the stochastic equation $Y_{n+1} = a_n Y_n + b_n$ with Markovian coefficients. Stochastic Process. Appl. **115**, 1954–1978 (2005)
17. Vervaat, W.: On a stochastic difference equation and a representation of nonnegative infinitely divisible random variables. Adv. Appl. Probab. **11**, 750–783 (1979)

The Doob–McKean Identity for Stable Lévy Processes

Andreas E. Kyprianou and Neil O'Connell

Abstract We re-examine the celebrated Doob–McKean identity that identifies a conditioned one-dimensional Brownian motion as the radial part of a 3-dimensional Brownian motion or, equivalently, a Bessel-3 process, albeit now in the analogous setting of isotropic α-stable processes. We find a natural analogue that matches the Brownian setting, with the role of the Brownian motion replaced by that of the isotropic α-stable process, providing one interprets the components of the original identity in the right way.

Keywords Cauchy Processes · Doob h-transform · Radial process

1 Introduction

A now-classical result in the theory of Markov processes due to Doob [8] and McKean [19] equates the law of a Brownian motion conditioned to stay positive with that of a Bessel-3 process; see also [21, 24, 25]. A precise statement of this identity can be made in a number of different ways as each of the two processes that are equal in law have several different representations. For the purpose of this exposition, it is worth reminding ourselves of them.

Research supported by the European Research Council (Grant No. 669306).

A. E. Kyprianou (✉)
Department of Mathematical Sciences, University of Bath, Bath, UK
e-mail: a.kyprianou@bath.ac.uk

N. O'Connell
School of Mathematics and Statistics, University College Dublin, Dublin, Ireland
e-mail: neil.oconnell@ucd.ie

L. Chaumont, A. E. Kyprianou (eds.), *A Lifetime of Excursions Through Random Walks and Lévy Processes*, Progress in Probability 78,
https://doi.org/10.1007/978-3-030-83309-1_15

Denote by $\mathbb{D}(\mathbb{R})$ the space of càdlàg paths $\omega : [0, \infty) \to \mathbb{R} \cup \Delta$ with lifetime $\zeta = \inf\{t > 0 : \omega_t = \Delta\}$, where Δ is a cemetery state. The space $\mathbb{D}(\mathbb{R})$ will be equipped with the Skorokhod topology and its natural Borel σ-algebra into which is embedded the natural filtration $(\mathcal{F}_s, s \geq 0)$. On this space, we will denote by $B = (B_t, t \geq 0)$ the coordinate process whose probabilities $\mathbb{P} = (\mathbb{P}_x, x \in \mathbb{R})$ are those of a standard one dimensional Brownian motion. For each $t \geq 0$, $x > 0$, the limit

$$\mathbb{P}^{\uparrow}_x(A, t < \zeta) := \lim_{\varepsilon \to 0} \mathbb{P}_x(A, t < \mathbf{e}/\varepsilon \mid \tau_0^-(B) > \mathbf{e}/\varepsilon), \tag{1}$$

where $\mathbf{e}$ is an independent exponentially distributed random variable with unit mean, $\tau_0^-(B) = \inf\{t > 0 : B_t < 0\}$, defines a new family of probabilities on $\mathbb{D}(\mathbb{R}_{\geq 0}) := \{\omega \in \mathbb{D}(\mathbb{R}) : \omega \in (0, \infty) \cup \Delta\}$. It turns out that $\mathbb{P}^{\uparrow} = (\mathbb{P}^{\uparrow}_x, x > 0)$ defines a conservative (i.e. $\zeta = \infty$) Markov process on $[0, \infty)$. As such, $(B, \mathbb{P}^{\uparrow})$ is the sense in which we can understand Brownian motion conditioned to stay positive.

Thanks to the well known fact that the probability $\mathbb{P}_x(\tau_0^-(B) > t) \sim x/\sqrt{2\pi t}$, as $t \to \infty$, it is easy to verify by taking its Laplace transform followed by an integration by parts, then an application of the classical Tauberian Theorem, that, up to an constant $c > 0$, $\mathbb{P}_x(\tau_0^-(B) > \mathbf{e}/\varepsilon) \sim cx\sqrt{\varepsilon}$. One thus easily verifies from (1), with the help of an easy dominated convergence argument, that $(B, \mathbb{P}^{\uparrow})$ satisfies

$$\left.\frac{\mathrm{d}\mathbb{P}^{\uparrow}_x}{\mathrm{d}\mathbb{P}_x}\right|_{\mathcal{F}_t} = \frac{B_t}{x}\mathbf{1}_{\{t < \tau_0^-(B)\}}, \qquad x, t > 0. \tag{2}$$

The change of measure (2) presents a second definition of the Brownian motion conditioned to stay positive via a Doob h-transform with respect to Brownian motion killed on exiting $[0, \infty)$, using the harmonic function $h(x) = x$. Suppose we write $p_t(x, y)$ and $p_t^{\dagger}(x, y)$, $t \geq 0$, $x, y > 0$, for the transition density of Brownian motion and of Brownian motion killed on exiting $[0, \infty)$, respectively. Then another way of expressing (2) is via the harmonic transformation

$$p^{\uparrow}(x, y) := \frac{y}{x} p_t^{\dagger}(x, \mathrm{d}y) = \frac{y}{x}(p_t(x, y) - p_t(x, -y)), \qquad x, y > 0. \tag{3}$$

As alluded to above, the so-called *Doob–McKean identity* states that the process $(B, \mathbb{P}^{\uparrow})$ is equal in law to a Bessel-3 process. There are also several ways that one may define the latter processes. Among the many, there are three that we mention here.

As a parametric family indexed by $\nu \geq 0$, Bessel-ν processes are defined as non-negative valued, conservative, one-dimensional diffusions which can be identified via the action of their generator L^ν, which satisfies

$$L^\nu = \frac{1}{2}\left(\frac{\mathrm{d}^2}{\mathrm{d}x^2} + \frac{\nu - 1}{x}\frac{\mathrm{d}}{\mathrm{d}x}\right), \qquad x > 0, \tag{4}$$

such that the point 0 is treated as an absorbing boundary if $\nu = 0$, as a reflecting boundary if $\nu \in (0, 2)$ and as an entrance boundary if $\nu \geq 2$. As such, the associated transition density can be identified as a non-zero solution to the backward equation given by L^ν. In general, the transition density can be identified explicitly with the help of Bessel functions (hence the name of the family of processes). In the special case that $\nu = 3$, it turns out that the transition density can be more simply identified by the right-hand side of (3).

In the setting that ν is a natural number, in particular, in the case that $\nu = 3$, the generator (4) is also the radial component of the ν-dimensional Laplacian. Noting that the latter is the generator of a ν-dimensional Brownian motion, we also see that, for positive integer values of ν, the Bessel-ν process is also the radial distance from the origin of a ν-dimensional Brownian motion; cf. [12]. This also illuminates the need for the point 0 to be either reflecting or an entrance point when $\nu > 0$, at least for $\nu \in \mathbb{N}$.

The Doob–McKean identity is present-day nested in a much bigger dialogue concerning the representation of conditioned, path-segment-sampled and time-reversed stochastic processes, including general diffusions, random walks and Lévy processes; see e.g. [1, 2, 5–8, 19, 21, 22, 24, 25] and others. In this article we add to the list of extensions to the Doob–McKean identity by looking at the setting in which the role of the Brownian motion is replaced by an isotropic α-stable process.

2 Doob-McKean for Isotropic α-Stable Processes

We recall that an isotropic α-stable process (henceforth sometimes referred to as a stable process or a symmetric stable process in one dimension) in dimension $d \in \mathbb{N}$, with coordinate process say $X = (X_t, t \geq 0)$ and probabilities $\mathrm{P}^{\alpha,d} = (\mathrm{P}^{\alpha,d}_x, x \in \mathbb{R}^d)$, is a Lévy process which is also a self-similar Markov process, which has self-similarity index α. More precisely, as a Lévy process, its transitions are uniquely described by its characteristic exponent given by the identity

$$\mathrm{E}^{\alpha,d}_0[\exp(\mathrm{i}\theta X_t)] = \exp(-|\theta|^\alpha t), \qquad t \geq 0,$$

where we interpret θX_t as an inner product in the setting that $d \geq 2$. For the pure jump case that we are interested in, it is necessary that $\alpha \in (0, 2)$. As a self-similar Markov process with index α, it satisfies the scaling property that, for all $c > 0$,

$$(cX_{c^{-\alpha}t}, t \geq 0) \text{ under } \mathrm{P}_x^{\alpha,d} \text{is equal in law to } (X, \mathrm{P}_{cx}^{\alpha,d}). \tag{5}$$

In any dimension, $(X, \mathrm{P}^{\alpha,d})$ has a transition density and, for example, in the setting $d = 1$, if we denote it by $q_t^{(\alpha)}(x, y)$, $x, y \in \mathbb{R}$, then the scaling property (5) manifests in the form

$$cq_t^{(\alpha)}(cx, cy) = q_{c^{-\alpha}t}^{(\alpha)}(x, y), \qquad x, y \geq 0, t > 0. \tag{6}$$

We note that the Cauchy process has a symmetric distribution in one dimension and is isotropic in higher dimensions. As a Lévy process, its jump measure is given by

$$\Pi(\mathrm{d}z) = 2^{\alpha}\pi^{-d/2}\frac{\Gamma((d+\alpha)/2)}{\left|\Gamma(-\alpha/2)\right|}\frac{1}{|z|^{\alpha+d}}\,\mathrm{d}z, \qquad z \in \mathbb{R}^d \tag{7}$$

where B is a Borel set in $\mathbb{R}^d$. A special case of interest will be when $\alpha = 1$ and when $d = 1$, in which case, (7) takes the form

$$\Pi(\mathrm{d}x) = \frac{1}{\pi}\frac{1}{x^2}\mathrm{d}x, \qquad x \in \mathbb{R}.$$

Moreover, the transition density, more conveniently written as $(q_t, t \geq 0)$ rather than $(q_t^{(1)}, t \geq 0)$, is given by

$$q_t(x, y) = \frac{1}{\pi}\frac{t}{(y-x)^2 + t^2}, \qquad x, y \in \mathbb{R}, t > 0, \tag{8}$$

from which we can verify the scaling property (5) directly.

Given the summary of the Doob–McKean identity for the Brownian setting above, the stable-process analogue we present as our main result below matches perfectly the Brownian setting providing one interprets the components in the identity in the right way.

Theorem 1 *The kernel*

$$q_t^{(\alpha),*}(x, y) = \frac{y}{x}\left(q_t^{(\alpha)}(x, y) - q_t^{(\alpha)}(x, -y)\right) \qquad x, y \geq 0, t > 0 \tag{9}$$

defines a conservative Feller semigroup, say $Y = (Y_t, t \geq 0)$, on $[0, \infty)$ which is self-similar with index α. Moreover, Y is equal in law to the radial part of a three-dimensional isotropic α-stable process.

An easy corollary of the above result is the following.

Corollary 1 *The transition density of the radial part of a 3-dimensional Cauchy process is given by*

$$q_t^{(1),*}(x,y) = \frac{1}{\pi}\frac{4y^2t}{(y^2-x^2)^2+2t^2(y^2+x^2)+t^4}, \qquad x,y\geq 0, t>0. \tag{10}$$

Proof of Theorem 1 The proof is a relatively elementary consequence of the classical Doob–McKean identity once one takes account of the following basic fact; cf. e.g. Chapter 3 of [17].

Lemma 1 *If* $(B_t^{(d)}, t\geq 0)$ *is a standard* d*-dimensional Brownian motion* $(d\geq 1)$ *and* $\Lambda = (\Lambda_t, t\geq 0)$ *is an independent stable subordinator with index* $\alpha/2$*, where* $\alpha\in(0,2)$*, then* $(\sqrt{2}B_{\Lambda_t}^{(d)}, t\geq 0)$ *is an isotropic* d*-dimensional stable process with index* α*.*

An immediate consequence of Lemma 1 is that, e.g. in one dimension, we can identify the semigroup of a symmetric stable process with index α via

$$q_t^{(\alpha)}(x,y) = \int_0^\infty \gamma_t^{(\alpha/2)}(s)\frac{1}{2^{d/2}}p_s^{(d)}(x,y/\sqrt{2})\mathrm{d}s$$

where $p_t^{(d)}(x,y)$, $x,y\in\mathbb{R}^d$ is the transition density of a standard Brownian motion in $\mathbb{R}^d$ (and for consistency we have $p_t^{(1)} = p_t$, $t\geq 0$.)

$$\gamma_t^{(\alpha/2)}(s) = \frac{1}{\pi}\sum_{n\geq 1}(-1)^{n-1}\frac{\Gamma(1+\frac{\alpha n}{2})}{n!}\sin\left(\frac{n\pi\alpha}{2}\right)t^n s^{-\frac{n\alpha}{2}-1}, \qquad x>0,$$

is the transition density of the stable subordinator with index $\alpha/2$.

Replacing y by $y/\sqrt{2}$ in (3) and dividing through by $\sqrt{2}$, by integrating against the kernel $\gamma^{(\alpha/2)}$ we see with the help of Lemma 1 that

$$\frac{1}{\sqrt{2}}\int_0^\infty \gamma_t^{(\alpha/2)}(s)p_s^{\uparrow}(x,y/\sqrt{2})\mathrm{d}s = \frac{y}{x}\left(q_t^{(\alpha)}(x,y) - q_t^{(\alpha)}(x,-y)\right), \qquad x,y\geq 0, t\geq 0.$$

Writing $\mathbb{P}^{(3)}$ for the law of 3-dimensional Brownian motion with coordinate process $(B_t^{(3)}, t\geq 0)$ as a coordinate process on $\mathbb{D}(\mathbb{R})$. Since $(p_t^{\uparrow}, t\geq 0)$ is the transition density of a Bessel-3 process, which is also the transition density of the radius of a 3-dimensional standard Brownian motion, we know that

$$\frac{1}{\sqrt{2}}p_s^{\uparrow}(x,y/\sqrt{2})\mathrm{d}y = \mathbb{P}^{(3)}_{(x,0,0)}(|\sqrt{2}B_t^{(3)}|\in\mathrm{d}y), \qquad y,t\geq 0.$$

As such, it follows that

$$\frac{1}{\sqrt{2}}\int_0^\infty \gamma_t^{(\alpha/2)}(s)p_s^\uparrow(x,y/\sqrt{2})\mathrm{d}s = \mathbb{P}^{(3)}_{(x,0,0)}(|\sqrt{2}B^{(3)}_{\Lambda_t}|\in \mathrm{d}y),$$

where Λ is an independent stable subordinator with index $\alpha/2$. Lemma 1 now allows us to conclude that (9) agrees with the transition semigroup of the radial component of a 3-dimensional stable process. On account of the fact that the radial component of an isotropic stable process is a conservative self-similar Markov process (and in particular a Feller process), we see that the semigroup in (9) must also offer the same properties. This also includes the existence of an entrance law at zero which is affirmed by the representation given in Lemma 1. □

3 The Special Case of Cauchy Processes

The special case of the Doob–McKean identity for $\alpha = 1$, i.e. the Cauchy process, reveals a few more details that we can explore further. In the subsections below, we look at the Doob–McKean identity in terms of the Lamperti representation of self-similar Markov processes, its relation with the Cauchy process conditioned to stay positive and in terms of a pathwise interpretation.

3.1 *Lamperti Representation of the Doob-McKean Identity*

As a self-similar Markov process with index 1, the process Y in Theorem 1 when $\alpha = 1$ enjoys a Lamperti representation. Specifically,

$$Y_t = \mathrm{e}^{\xi_{\varphi(t)}}, \qquad t \le \int_0^\infty \mathrm{e}^{\xi_u}\,du, \tag{11}$$

where $\varphi(t) = \inf\{s > 0 : \int_0^s \exp(\xi_u)\mathrm{d}u > t\}$ and $(\xi_t, t \ge 0)$ is a Lévy process, which is possibly killed at an independent and exp

Another way of understanding the statement in the second part of Theorem 1 is that the Lévy process ξ agrees with the one that underlies the Lamperti representation of the radial part of a three-dimensional Cauchy process. The reason why the latter is a positive self-similar Markov process was examined in [4]; see also Chapter 5 of [17]. Indeed, there it was shown that the radial part of a 3-dimensional Cauchy process has underlying Lévy process, say $(\eta_t, t \ge 0)$, with probabilities $(\mathbb{P}^\eta_x, x \in \mathbb{R})$, which is identified via its characteristic exponent $\Psi(z) = -\log\int_{\mathbb{R}} \mathrm{e}^{\mathrm{i}zx}\mathbb{P}^\eta_0(\eta_1 \in \mathrm{d}x)$, where

$$\Psi(z) = 2\frac{\Gamma(\frac{1}{2}(-\mathrm{i}z+1))}{\Gamma(-\frac{1}{2}\mathrm{i}z)}\frac{\Gamma(\frac{1}{2}(\mathrm{i}z+3))}{\Gamma(\frac{1}{2}(\mathrm{i}z+2))} = (z-\mathrm{i})\tanh(\pi z/2), \qquad z \in \mathbb{R}. \tag{12}$$

An equivalent way of identifying η is as a pure jump process, with no killing (note that $\Psi(0) = 0$) and with Lévy measure having density taking the form

$$\mu(x) = \frac{4}{\pi} \frac{e^{3x}}{(e^{2x} - 1)^2}, \qquad x \in \mathbb{R}. \tag{13}$$

Note that for small $|x|$ the density above behaves like $O(|x|^{-2})$, for large positive x, it behaves like $O(e^{-x})$ and for large negative x, it behaves like $O(e^{-3|x|})$. As such, the process η has paths of unbounded variation and its law enjoys exponential moments; in particular η has a finite first moment.

The long term linear growth of η (in the sense of the Strong Law of Large Numbers) is given by the mean $\mathbb{E}_0^{\eta}[\eta_1] = \pi/2$ which can also be computed from the value of $i\Psi'(0)$; see also Proposition 1 of [15]. Not surprisingly this implies that $\lim_{t\to\infty} \eta_t = \infty$ almost surely. This is consistent with the fact that a three-dimensional Cauchy process is transient and hence, its radial component drifts to $+\infty$, which implies its underlying Lévy process must too. Note, in the latter observation, we are also using the fact that positive self-similar Markov processes are either: Transient to infinity, corresponding to the underlying Lévy process drifting to $+\infty$; Interval recurrent, corresponding to the underlying Lévy process oscillating; Continuously absorbed at the origin, corresponding to the case that the underlying Lévy process drifts to $-\infty$; Absorbed at the origin by a jump; corresponding to the case that the underlying Lévy process is killed at an independent and exponentially distributed time. See [16–18] for further details.

Because η has a finite first moment, we can relate (13) to (12) via the particular arrangement of the Lévy–Khintchine formula

$$\Psi(z) = -\frac{\pi}{2} iz + \int_{\mathbb{R}} \left(1 - e^{izx} + izx\right) \mu(x) dx, \qquad z \in \mathbb{R}. \tag{14}$$

This arrangement will prove to be convenient in the following Corollary.

Corollary 2 *Suppose that $C^2(\mathbb{R}_{\geq 0})$ is the space of twice continuously integrable functions on $\mathbb{R}_{\geq 0}$. On $C^2(\mathbb{R}_{\geq 0})$, the action of the generator $\mathcal{L}$ associated to the process Y in Theorem 1 is given by*

$$\mathcal{L}f(x) = \frac{\pi}{2} f'(x) + \frac{4}{\pi x} \int_0^{\infty} \left(f(xu) - f(x) - xf'(x) \log u\right) \frac{u^2}{(u^2 - 1)^2} du, \qquad x > 0 \tag{15}$$

which agrees with the representation

$$\mathcal{L}f(x) = \frac{4}{\pi x} (PV) \int_0^{\infty} (f(xu) - f(x)) \frac{u^2}{(u^2 - 1)^2} du, \qquad x > 0, \tag{16}$$

where $(PV)\int$ is understood as a principal value integral.

Proof Because of the arrangement of the characteristic exponent in (14), from [3], we know that its generator can be accordingly arranged to have action on $f \in \mathcal{C}^2(\mathbb{R}_{\geq 0})$ given by

$$\mathcal{L}f(x) = \frac{\pi}{2}f'(x) + \frac{4}{\pi x}\int_0^\infty \left(f(xu) - f(x) - xf'(x)\log u\right)\frac{u^2}{(u^2-1)^2}\mathrm{d}u, \qquad x > 0. \tag{17}$$

For the second statement of the corollary, we need to show that

$$I := (PV)\int_0^\infty \frac{u^2 \log u}{(u^2-1)^2}\mathrm{d}u = \frac{\pi^2}{8} \tag{18}$$

and that

$$(PV)\int_0^\infty (f(xu) - f(x))\frac{u^2}{(u^2-1)^2}\mathrm{d}u$$

is well defined. The latter is easily done on account of the fact that, near the singularity $u = 1$, $f(ux) - f(x) \approx (u-1)xf'(x) + O((u-1)^2)$, $x, u > 0$, so that we can estimate the principal value of the integral there using partial fractions.

To see why the equality in (18) holds, note that after a change of variable $u = \mathrm{e}^x$ we see

$$I = (PV)\int_{-\infty}^\infty \frac{x\mathrm{e}^x}{(\mathrm{e}^x - \mathrm{e}^{-x})^2}\mathrm{d}x = -(PV)\int_{-\infty}^\infty \frac{x\mathrm{e}^{-x}}{(\mathrm{e}^x - \mathrm{e}^{-x})^2}\mathrm{d}x, \tag{19}$$

where in the second equality we have noted the simple change of variables $x \mapsto -x$. It thus follows by adding the two integrals in (19) together that

$$I = \frac{1}{2}\int_{-\infty}^\infty \frac{x}{(\mathrm{e}^x - \mathrm{e}^{-x})}\mathrm{d}x = \frac{1}{2}\int_0^\infty \frac{x}{\sinh x}\mathrm{d}x = \frac{\pi^2}{8}.$$

where the final equality follows from equation 3.521.1 of [11].

Note, another way to approach the second part of the corollary is to use the standard definition of a Feller generator on $C_c^\infty(\mathbb{R}_{\geq}0)$, the space of compactly supported smooth functions; cf [13]. We have

$$\mathcal{L}f(x) = \lim_{t\to 0}\frac{1}{t}\left(\int_0^\infty f(y)q_t^{(1),*}(x,y) - f(x)\right), \qquad x > 0.$$

Making use of (9) and monotone convergence, again taking note that the singularity in the integral can be dealt with in a similar manner, we see that

$$\begin{aligned}\mathcal{L}f(x) &= \lim_{t\to 0}\frac{4}{\pi}(PV)\int_0^\infty (f(y)-f(x))\frac{y^2}{(y^2-x^2)^2+2t^2(y^2+x^2)+t^4}\mathrm{d}y\\ &= \frac{4}{\pi}(PV)\int_0^\infty (f(y)-f(x))\frac{y^2}{(y^2-x^2)^2}\mathrm{d}y, \qquad x>0, \end{aligned} \tag{20}$$

which agrees with (16) after a simple change of variables. □

3.2 Connection to Cauchy Process Conditioned to Stay Positive

It is also worthy of note in the general case $\alpha \in (0,2)$ that the process Y does not agree with the law of a one-dimensional symmetric stable process conditioned to stay positive. The latter can be understood via the exact same limiting process in (1), again replacing the role of Brownian motion by that of the one-dimensional stable process, inducing a new family of probabilities $(\mathrm{P}_x^{1,1,\uparrow}, x>0)$ on $\mathbb{D}(\mathbb{R}_{\geq 0})$. Rather than corresponding to the change of measure (2), the law of the Cauchy process conditioned to stay positive is related to that of the Cauchy process via

$$\left.\frac{\mathrm{dP}_x^{1,1,\uparrow}}{\mathrm{dP}_x^{1,1}}\right|_{\mathcal{F}_t} = \left(\frac{X_t}{x}\right)^{1/2}\mathbf{1}_{\{t<\tau_0^-(X)\}}, \qquad x>0, t\geq 0, \tag{21}$$

where $\tau_0^-(X) = \inf\{t>0 : X_t<0\}$.

There is nonetheless a close relationship between $(\mathrm{P}_x, x>0)$ and $(\mathrm{P}_x^{1,1,\uparrow}, x>0)$, which is best seen through the Lamperti representation (11). Suppose we write $\Psi^\uparrow$ for the characteristic exponent of the Lévy process that underlies the Cauchy process conditioned to stay positive. It is known from [3] (see also Chapter 5 of [17]) that

$$\Psi^\uparrow(z) = \Psi(2z), \qquad z\in\mathbb{R}. \tag{22}$$

If we write $\mu^\uparrow$ for the Lévy measure associated to $\Psi^\uparrow$. This is equivalent to saying that $2\mu^\uparrow(x) = \mu(x/2)$, or indeed that the Lévy process underlying the Cauchy process conditioned to stay positive is equal in law to 2η. This is a curious relationship which is clearly related to the fact that the Doob h-transform in the definition (9) uses $h(x)=x$, whereas the Doob h-transform in (21) uses $h(x)=\sqrt{x}$. It is less clear if or how this relationship extends to other values of α. From Lemma 2.2 in [20] we can now identify the following simple relationship.

Corollary 3 *Denote by $Y^{\uparrow} = (Y_t^{\uparrow}, t \geq 0)$ is the co-ordinate process of a one-dimensional Cauchy process conditioned to stay positive. Then with Y denoting the process in Theorem* 1*, we have space-time path transformation relating Y to $Y^{\uparrow}$,*

$$(Y_t^{\uparrow}, t \geq 0) \overset{law}{=} \left((Y_{\chi(t)})^2, t \geq 0\right), \text{ where } \chi(t) = \inf\{s > 0 : \int_0^t Y_u^{-1} \mathrm{d}u > t\}, \qquad t \geq 0.$$

3.3 Pathwise Representation

One way to understand the Doob-McKean in the Cauchy setting is to consider it via a path transformation which mirrors the proof of Theorem 1. Think of a two-dimensional Brownian motion $\mathbb{P}^{(2)}$ on the x-y plane which is stopped when hits the line $x = t$, that is at the time $\Gamma_t = \inf\{s > 0 : \pi_x(\sqrt{2}B_s^{(2)}) = t\}$, where π_x is the projection of $\sqrt{2}B^{(2)}$ onto the x-axis. It is well known that Γ_t is a 1/2-stable subordinator and that $(\pi_y(\sqrt{2}B_{\Gamma_t}^{(2)}), t \geq 0)$ is a Cauchy process where π_y is the projection on to the y-axis.

Suppose now we replace $B^{(2)}$ by the x-y planar process (B, R), where B is a one-dimensional Brownian motion and R is an independent Bessel-3 process. Noting that R is a Doob h-transform of $\pi_y(\sqrt{2}B^{(2)})$ killed on hitting the x-axis, the independence of B and R, and hence the independence of $(\Gamma_t, t \geq 0)$ and R means that the process $(\sqrt{2}R_{\Gamma_t}, t \geq 0)$ agrees precisely with the transformation on the right-hand side of (9) with $\alpha = 1$.

3.4 Generators

We know that the generator of the process Y in Theorem 1 is given by (16). The pathwise representation in the previous section, captured e.g. in Fig. 1 also gives us some insight into the structure of the generator (16).

As alluded to above, if B is a one-dimensional Brownian motion, then $(\sqrt{2}B_{\Gamma_t}, t \geq 0)$ is a Cauchy process. Its generator $\mathcal{C}$ is written

$$\mathcal{C}f(x) = \frac{1}{\pi}(PV)\int_{-\infty}^{\infty} \frac{f(y) - f(x)}{(y - x)^2}\mathrm{d}y, \qquad f \in C_c^\infty(\mathbb{R}_{\geq}0). \tag{23}$$

We want to connect the generator $\mathcal{C}$ with the processes Y we see in the path decomposition, in particular with the process $(R_{\Gamma_t}, t \geq 0)$.

We know from (3) that a Bessel-3 process is the result of Doob h-transforming the law of a Brownian motion killed on entry to $(-\infty, 0)$. We have also seen e.g. in the proof of Theorem 1 that subordination with the 1/2-stable process $(\Gamma_t, t \geq 0)$ preserves the effect of the Doob h-transform. What we would like to understand

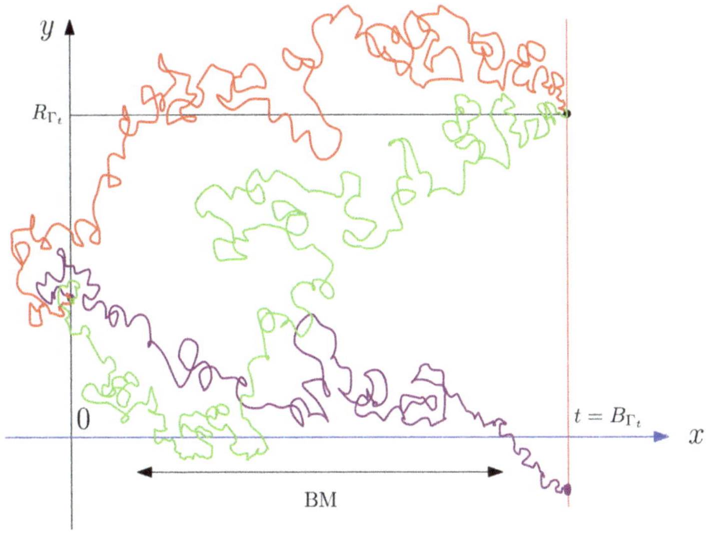

Fig. 1 A pathwise representation of the Doob-McKean transformation for Cauchy processes. The red path depicts a sample path from the process (B, R), where B is a Brownian motion in the direction of the x axis and R is a Bessel-3 process in the direction of the y axis, until it hits the vertical line $x = t$. The green and purple paths are sample paths from the two dimensional Brownian motion $B^{(2)}$ until first hitting of the vertical line $x = t$

is how the ½-stable subordination of killed Brownian motion, i.e. $(q_t^{(1/2)}(x, y) - q_t^{(1/2)}(x, -y))$, plays out in (23).

To this end, we can think of jump rate from $x \geq 0$ to $y \geq 0$ of the sub-Markov process with semigroup $(q_t^{(1/2)}(x, y) - q_t^{(1/2)}(x, -y))$, as being derived from a principal of 'path counting' using jump rates of the Cauchy process. The generator of a Cauchy process killed on exiting the upper half line is given by

$$\mathcal{C}_+ f(x) - \frac{1}{\pi x} \quad \text{where} \quad \mathcal{C}_+ f(x) := \frac{1}{\pi} \int_0^\infty \frac{f(y) - f(x)}{(y-x)^2} \mathrm{d}y, \qquad f \in C_c^\infty(\mathbb{R}_{\geq} 0).$$

Indeed, the aforesaid process jumps from $x \geq 0$ to $y \geq 0$ at rate $1/\pi(y - x)^2 \mathrm{d}y$, however, we must subtract from this rate, the rate at which killing occurs by jumping from x into the negative half line. The latter is

$$\frac{1}{\pi} \int_{-\infty}^0 \frac{1}{(y-x)^2} \mathrm{d}y = \frac{1}{\pi} \int_x^\infty \frac{1}{z^2} \mathrm{d}z = \frac{1}{\pi x}.$$

The combined effect of reflection principal and ½-stable subordination, suggests we must also subtract the rate at which jumps from $x \geq 0$ to $y \geq 0$ occur as the reflection of jumps from x to $-y$, with the additional effect of killing on the lower half line, i.e.

$$\mathcal{C}_- f(x) - \frac{1}{\pi x} \quad \text{where} \quad \mathcal{C}_- f(x) = \frac{1}{\pi}\int_0^\infty \frac{f(y) - f(x)}{(x+y)^2}\mathrm{d}y, \qquad f \in C_c^\infty(\mathbb{R}_{\geq} 0).$$

We can thus identify the generator of Y, $\mathcal{L}$, as the following Doob h-transform.

Lemma 2 *We have*

$$\mathcal{L}f(x) = \frac{1}{x}\mathcal{D}(xf(x)), \qquad f \in C_c^\infty(\mathbb{R}_{\geq} 0), x > 0,$$

where

$$\mathcal{D} = \mathcal{C}_+ - \mathcal{C}_- - \frac{2}{\pi x}.$$

Proof We compute (all integrals are Cauchy principal value integrals):

$$\begin{aligned}
\frac{1}{x}\mathcal{D}(xf(x)) &= \frac{1}{\pi x}\int_0^\infty \frac{yf(y) - xf(x)}{(y-x)^2}\mathrm{d}y - \frac{1}{\pi x}\int_0^\infty \frac{yf(y) - xf(x)}{(y+x)^2}\mathrm{d}y - \frac{2}{\pi x}f(x) \\
&= \frac{1}{\pi x}\int_0^\infty \frac{4xy(yf(y) - xf(x))}{(y^2 - x^2)^2}\mathrm{d}y - \frac{2}{\pi x}f(x) \\
&= \frac{4}{\pi}\int_0^\infty \frac{y^2(f(y) - f(x))}{(y^2 - x^2)^2}\mathrm{d}y + \frac{4}{\pi}\int_0^\infty \frac{(y^2 - xy)f(x)}{(y^2 - x^2)^2}\mathrm{d}y - \frac{2}{\pi x}f(x) \\
&= \mathcal{L}f(x),
\end{aligned}$$

where the last identity follows from the definition of $\mathcal{L}$ and the fact that

$$(PV)\int_0^\infty \frac{2xy}{(y-x)(y+x)^2}\mathrm{d}y = 1.$$

□

Note that the 'reflected' Cauchy process has generator $\mathcal{C}_R = \mathcal{C}_+ + \mathcal{C}_-$, and we earlier identified the Cauchy process killed on going negative as having generator $\mathcal{C}_A = \mathcal{C}_+ - 1/(\pi x)$. These are related to the generator $\mathcal{D}$ via $\mathcal{C}_A = (\mathcal{D} + \mathcal{C}_R)/2$. The spectral problem associated with the Cauchy process on the half-line with 'reflecting' boundary is equivalent to the so-called 'sloshing problem' in the theory of linear water waves, and this has been extensively studied [10]. The spectral problem associated with the Cauchy process on the half-line with absorbing boundary conditions has been completely solved in [14].

4 Concluding Remarks

Elliot and Feller [9] consider various examples of Cauchy processes constrained to stay in a compact interval $[0, a]$. One of the examples they consider (Example (d) in their paper), has transition density

$$p_t(x, y) = \sum_{n=-\infty}^{\infty} [q_t(x, 2an + y) - q_t(x, 2an - y)], \tag{24}$$

where $q_t(x, y)$ is the transition density of the one-dimensional Cauchy process. They remark that (24) defines *'a transition semi-group and determines a Markovian process, but it is not the absorbing barrier process.* $[\cdots]$ *It is not clear whether and how the process is related to the Cauchy process.'* In fact, the process considered in [9] is a Brownian motion in $[0, a]$ with Dirichlet boundary conditions, time-changed by an independent stable subordinator of index $1/2$. Moreover, it may be interpreted in terms of the Cauchy process via a similar pathwise interpretation to the one outlined above for the half-line.

It is also natural to consider multi-dimensional versions. For example, Dyson Brownian motion is a Brownian motion in $\mathbb{R}^n$ conditioned never to exit the Weyl chamber $C = \{x \in \mathbb{R}^n : x_1 > \cdots > x_n\}$. Its transition density is given by

$$d_t(x, y) = h(x)^{-1} h(y) \sum_{\sigma \in S_n} \operatorname{sgn}(\sigma) p_t(x, \sigma y),$$

where the sum is over permutations, σy is the vector y with components permuted by σ, $h(x) = \prod_{i<j}(x_i - x_j)$ is the Vandermonde determinant, and $p_t(x, y)$ is the standard Gaussian heat kernel in $\mathbb{R}^n$. If we time-change this process by an independent stable subordinator of index $\alpha/2$, and multiply by a factor of $\sqrt{2}$, then the resulting process in C has transition density

$$D_t(x, y) = h(x)^{-1} h(y) \sum_{\sigma \in S_n} \operatorname{sgn}(\sigma) P_t^{(\alpha)}(x, \sigma y),$$

where $P_t^{(\alpha)}(x, y)$ is the transition density of the isotropic n-dimensional stable process with index α. We note that, in the case $\alpha = 1$, this time-changed process may be interpreted as the 'radial part' of a Cauchy process in $\mathbb{R}^n$, as discussed in Section 5 of the paper [23].

Acknowledgments Both authors would like to thank an anonymous referee for their remarks which lead to an improved version of this paper.

References

1. Bertoin, J.: Lévy processes, volume 121 of Cambridge Tracts in Mathematics. Cambridge University Press, Cambridge (1996)
2. Bertoin, J., Doney, R.A.: On conditioning a random walk to stay nonnegative. Ann. Probab. **22**(4), 2152–2167 (1994)
3. Caballero, M.E., Chaumont, L.: Conditioned stable Lévy processes and the Lamperti representation. J. Appl. Probab. **43**(4), 967–983 (2006)
4. Caballero, M.E., Pardo, J.C., Pérez, J.L.: Explicit identities for Lévy processes associated to symmetric stable processes. Bernoulli **17**(1), 34–59 (2011)
5. Chaumont, L.: Conditionings and path decompositions for Lévy processes. Stochastic Process. Appl. **64**(1), 39–54 (1996)
6. Chaumont, L., Doney, R.A.: On Lévy processes conditioned to stay positive. Electron. J. Probab. **10**(28), 948–961 (2005)
7. Chaumont, L., Doney, R.A.: Corrections to: "On Lévy processes conditioned to stay positive" [Electron J. Probab. **10** (2005)(28), 948–961; mr2164035]. Electron. J. Probab. **13**(1), 1–4 (2008)
8. Doob, J.L.: Conditional Brownian motion and the boundary limits of harmonic functions. Bull. Soc. Math. France **85**, 431–458 (1957)
9. Elliott, J., Feller, W.: Stochastic processes connected with harmonic functions. Trans. Am. Math. Soc. **82**, 392–420 (1956)
10. Fox, D.W., Kuttler, J.R.: Sloshing frequencies. Z. Angew. Math. Phys. **34**(5), 668–696 (1983)
11. Gradshteyn, I.S., Ryzhik, I.M.: Table of Integrals, Series, and Products, 8th edn. Elsevier/Academic Press, Amsterdam (2015)
12. Karlin, S., McGregor, J.: Classical diffusion processes and total positivity. J. Math. Anal. Appl. **1**, 163–183 (1960)
13. Kühn, F.: Schilling, R.L.: On the domain of fractional Laplacians and related generators of Feller processes. J. Funct. Anal. **276**(8), 2397–2439 (2019)
14. Kulczycki, T., Kwaśnicki, M., Mał ecki, J., Stos, A.: Spectral properties of the Cauchy process on half-line and interval. Proc. Lond. Math. Soc. (3) **101**(2), 589–622 (2010)
15. Kuznetsov, A., Pardo, J.C.: Fluctuations of stable processes and exponential functionals of hypergeometric Lévy processes. Acta Appl. Math. **123**, 113–139 (2013)
16. Kyprianou, A.E.: Fluctuations of Lévy processes with applications, 2nd edn. Universitext. Springer, Heidelberg (2014) Introductory lectures
17. Kyprianou, A.E., Pardo, J.C.: Stable Lévy processes via Lamperti-type representations. Cambridge University Press (2020)
18. Lamperti, J.: Semi-stable Markov processes. I. Z. Wahrscheinlichkeitstheorie und Verw. Gebiete **22**, 205–225 (1972)
19. McKean, H.P. Jr.: Excursions of a non-singular diffusion. Z. Wahrscheinlichkeitstheorie und Verw. Gebiete **1**, 230–239 (1962/1963)
20. Patie, P.: Exponential functional of a new family of Lévy processes and self-similar continuous state branching processes with immigration. Bull. Sci. Math. **133**(4):355–382 (2009)
21. Pitman, J.W.: One-dimensional Brownian motion and the three-dimensional Bessel process. Adv. Appl. Prob. **7**(3):511–526 (1975)
22. Rogers, L.C.G., Williams, D.: Diffusions, Markov processes, and martingales. Vol. 2. Cambridge Mathematical Library. Cambridge University Press, Cambridge (2000) Itô calculus, Reprint of the second (1994) edition
23. Rösler, M., Voit, M.: Markov processes related with Dunkl operators. Adv. Appl. Math. **21**(4):575–643 (1998)
24. Williams, D.: Decomposing the Brownian path. Bull. Am. Math. Soc. **76**, 871–873 (1970)
25. Williams, D.: Path decomposition and continuity of local time for one-dimensional diffusions. I. Proc. Lond. Math. Soc. (3) **28**, 738–768 (1974)

Oscillatory Attraction and Repulsion from a Subset of the Unit Sphere or Hyperplane for Isotropic Stable Lévy Processes

Mateusz Kwaśnicki, Andreas E. Kyprianou, Sandra Palau, and Tsogzolmaa Saizmaa

Abstract Suppose that S is a closed set of the unit sphere $\mathbb{S}^{d-1} = \{x \in \mathbb{R}^d : |x| = 1\}$ in dimension $d \geq 2$, which has positive surface measure. We construct the law of absorption of an isotropic stable Lévy process in dimension $d \geq 2$ conditioned to approach S continuously, allowing for the interior and exterior of $\mathbb{S}^{d-1}$ to be visited infinitely often. Additionally, we show that this process is in duality with the unconditioned stable Lévy process. We can replicate the aforementioned results by similar ones in the setting that S is replaced by D, a closed bounded subset of the hyperplane $\{x \in \mathbb{R}^d : (x, v) = 0\}$ with positive surface measure, where v is the unit orthogonal vector and where $(\cdot, \cdot)$ is the usual Euclidean inner product. Our results complement similar results of the authors [17] in which the stable process was further constrained to attract to and repel from S from either the exterior or the interior of the unit sphere.

Keywords Stable process · Time reversal · Duality

M. Kwaśnicki
Department of Pure Mathematics, Wrocław University of Science and Technology, Wrocław, Poland
e-mail: Mateusz.Kwasnicki@pwr.edu.pl

A. E. Kyprianou (✉)
Department of Mathematical Sciences, University of Bath, Bath, UK
e-mail: a.kyprianou@bath.ac.uk

S. Palau
Instituto de Investigaciones en Matemáticas Aplicadas y en Sistemas, Universidad Nacional Autónoma de México, Coyoacán, Mexico
e-mail: sandra@sigma.iimas.unam.mx

T. Saizmaa
Department of Mathematical Sciences, University of Bath, Bath, UK

National University of Mongolia, Baga-toiruu, Ulaanbaatar, Mongolia
e-mail: t.saizmaa@bath.ac.uk; tsogzolmaa@num.edu.mn

L. Chaumont, A. E. Kyprianou (eds.), *A Lifetime of Excursions Through Random Walks and Lévy Processes*, Progress in Probability 78,
https://doi.org/10.1007/978-3-030-83309-1_16

1 Introduction

Let $X = (X_t, t \geq 0)$ be a d-dimensional stable Lévy process ($d \geq 2$) with probabilities $(\mathbb{P}_x, x \in \mathbb{R}^d)$. This means that X has càdlàg paths with stationary and independent increments, and there exists an $\alpha > 0$ such that, for $c > 0$, and $x \in \mathbb{R}^d$,

$$\text{under } \mathbb{P}_x \text{ the law of } (cX_{c^{-\alpha}t}, t \geq 0) \text{ is equal to } \mathbb{P}_{cx}.$$

The latter is the property of so-called self-similarity. It turns out that stable Lévy processes necessarily have $\alpha \in (0, 2]$. The case $\alpha = 2$ is that of standard d-dimensional Brownian motion, thus has a continuous path. All other $\alpha \in (0, 2)$ have no Gaussian component and are pure jump processes. In this article we are specifically interested in phenomena that can only occur when jumps are present. We thus restrict ourselves henceforth to the setting $\alpha \in (0, 2)$.

Although Brownian motion is isotropic, this need not be the case in the stable case when $\alpha \in (0, 2)$. *Nonetheless, we will restrict to the isotropic setting.* To be more precise, this means, for all orthogonal transformations $U : \mathbb{R}^d \mapsto \mathbb{R}^d$ and $x \in \mathbb{R}^d$,

$$\textit{the law of} \quad (UX_t, t \geq 0) \textit{ under} \quad \mathbb{P}_x \textit{ is equal to } (X_t, t \geq 0) \textit{ under } \mathbb{P}_{Ux}.$$

For convenience, we will henceforth refer to X as a *stable process.*

As a Lévy process, our stable process of index $(0, 2)$ has a characteristic triplet $(0, 0, \Pi)$, where the jump measure Π satisfies

$$\Pi(B) = \frac{2^\alpha \Gamma((d+\alpha)/2)}{\pi^{d/2}|\Gamma(-\alpha/2)|} \int_B \frac{1}{|y|^{\alpha+d}} \ell_d(\mathrm{d}y), \quad B \subseteq \mathcal{B}(\mathbb{R}^d), \tag{1}$$

where ℓ_d is d-dimensional Lebesgue measure.[1] This is equivalent to identifying its characteristic exponent as

$$\Psi(\theta) = -\frac{1}{t} \log \mathbb{E}(\mathrm{e}^{\mathrm{i}\theta \cdot X_t}) = |\theta|^\alpha, \quad \theta \in \mathbb{R}^d,$$

where we write $\mathbb{P}$ in preference to $\mathbb{P}_0$.

In this article, we characterise the law of a stable process conditioned to continuously approach a closed subdomain of the surface of a unit sphere, say $\mathsf{S} \subseteq \mathbb{S}^{d-1} = \{x \in \mathbb{R}^d : |x| = 1\}$, which has non-zero surface measure. Moreover,

[1] We will distinguish integrals with respect to one-dimensional Lebesgue measure as taking the form $\int \cdot \, \mathrm{d}x$, where as higher dimensional integrals will always indicate the dimension, for example $\int \cdot \, \ell_d(\mathrm{d}x)$.

our conditioning will allow the stable process to approach S by visiting the exterior and interior of $\mathbb{S}^{d-1}$ infinitely often. We note that when $\alpha \in (1, 2)$, stable processes will hit the unit sphere with positive probability and otherwise, when $\alpha \in (0, 1]$ it hits the unit sphere with probability zero; see e.g. [25] or [16]. The aforesaid conditioning is thus only of interest when $\alpha \in (0, 1]$.

In addition to constructing the conditioned process, we develop an expression for the law of the limiting point of contact on S. Moreover, we show that, when time reversed from the strike point on S, the resulting process can be described as nothing more than the stable process itself.

It turns out that the methodology we use here is robust enough to cover a similar suite of results for the case of an isotropic stable process conditioned to a closed subdomain of an arbitrary hyperplane in $\mathbb{R}^d$ that is orthogonal to an arbitrary unit-length vector $\upsilon \in \mathbb{R}^d$.

Our results naturally complement those of the recent paper [17], which considers a similar type of conditioning, albeit requiring the stable process to additionally remain either inside or outside of the unit ball. Other related works include [9] and [14], who considered a real valued stable process conditioned to hit 0 continuously and a real valued stable process conditioned to continuously approach the boundary of the interval $[-1, 1]$ from the outside, respectively. In order to make our results pertinent, we restrict ourselves to the case that $d \geq 2$.

2 Oscillatory Attraction Towards S

Let $\mathbb{D}(\mathbb{R}^d)$ denote the space of càdlàg paths $\omega : [0, \infty) \to \mathbb{R}^d \cup \partial$ with lifetime $\zeta(\omega) = \inf\{s > 0 : \omega(s) = \partial\}$, where ∂ is a cemetery point. The space $\mathbb{D}(\mathbb{R}^d)$ will be equipped with the Skorokhod topology, with its closed σ-algebra $\mathcal{F}$ and natural filtration $(\mathcal{F}_t, t \geq 0)$. The reader will note that we will also use a similar notion for $\mathbb{D}(E)$ later on in this text in the obvious way for an E-valued Markov process. We will always work with $X = (X_t, t \geq 0)$ to mean the coordinate process defined on the space $\mathbb{D}(\mathbb{R}^d)$. Hence, the notation of the introduction indicates that $\mathbb{P} = (\mathbb{P}_x, x \in \mathbb{R}^d)$ is such that $(X, \mathbb{P})$ is our stable process.

We want to construct the law of the stable process conditioned to continuously limit to $\mathsf{S} \in \mathbb{S}^{d-1}$ whilst visiting both $\mathbb{B}_d := \{x \in \mathbb{R}^d : |x| < 1\}$ and $\bar{\mathbb{B}}_d^c := \mathbb{R}^d \setminus \bar{\mathbb{B}}_d$ infinitely often at arbitrarily small times prior to striking S. We shall denote the associated probabilities by $\mathbb{P}^{\mathsf{S}} = (\mathbb{P}^{\mathsf{S}}_x, x \in \mathbb{R}^d \setminus \mathsf{S})$. For a more precise definition of what is meant by this form of conditioning, let us introduce the stopping times,

$$\tau_\beta = \inf\{t > 0 : \beta^{-1} < |X_t| < \beta\}, \qquad \text{for } \beta > 1. \tag{2}$$

Whenever it is well defined, we will write, for $t \geq 0$, $\Lambda \in \mathcal{F}_t$ and $x \notin \mathsf{S}$,

$$\mathbb{P}_x^{\mathsf{S}}(\Lambda,\ t < \zeta) = \lim_{\beta \to 1} \lim_{\varepsilon \to 0} \mathbb{P}_x \left(\Lambda, t < \tau_\beta \,\middle|\, \tau_{\mathsf{S}_\varepsilon} < \infty\right), \tag{3}$$

where

$$\tau_{\mathsf{S}_\varepsilon} = \inf\{t > 0 : X_t \in \mathsf{S}_\varepsilon\} \qquad \text{and} \qquad \mathsf{S}_\varepsilon := \{x \in \mathbb{R}^d : 1-\varepsilon \leq |x| \leq 1+\varepsilon \text{ and } \arg(x) \in \mathsf{S}\}.$$

Our first main result clarifies that the process $(X, \mathbb{P}^{\mathsf{S}})$ is well defined. In the theorem below, and thereafter, we will understand σ_1 to mean the Lebesgue surface measure on $\mathbb{S}^{d-1}$ normalised to have unit mass, i.e. $\sigma_1(\mathbb{S}^{d-1}) = 1$.

Theorem 1 *Suppose that $\alpha \in (0, 1]$ and the closed set $S \subseteq \mathbb{S}^{d-1}$ is such that $\sigma_1(S) > 0$. For $\alpha \in (0, 1]$, the limit (3) makes sense. Therefore, the process $(X, \mathbb{P}^{\mathsf{S}})$ is well defined such that*

$$\left.\frac{\mathrm{d}\mathbb{P}_x^S}{\mathrm{d}\mathbb{P}_x}\right|_{\mathcal{F}_t} = \frac{H_S(X_t)}{H_S(x)}, \qquad t \geq 0, x \notin S, \tag{4}$$

where

$$H_S(x) = \int_S |x - \theta|^{\alpha-d} \sigma_1(\mathrm{d}\theta), \qquad x \notin S.$$

Although excluded from the conclusion of Theorem 1, it is worth dwelling for a moment on the extreme case $\mathsf{S} = \{\theta\}$, for $\theta \in \mathbb{S}^{d-1}$. It has been shown in [20] that, when $\alpha \in (0, 1)$, conditioning a stable process to continuously limit to a point (which, by stationary and independent increments, can always be arranged to be $\theta \in \mathbb{S}^{d-1}$) results in a family of probability measures $(\mathbb{P}_x^{\{\theta\}}, x \neq \theta)$ which can be identified via a Doob h-transform with $h_\theta(x) = |x - \theta|^{\alpha-d}$. Although the sense in which the conditioning is performed cannot be contextualised via (3), we see that the resulting h-transformation is consistent with the use of the harmonic function H_{S}.

The way in which we will prove Theorem 1 will be to prove the following subtle result which establishes the leading order behaviour of the probability of hitting the set S_ε.

Theorem 2 *Let $S \subseteq \mathbb{S}^{d-1}$ be a closed subset such that $\sigma_1(S) > 0$.*

(i) *Suppose $\alpha \in (0, 1)$. For $x \notin S$,*

$$\lim_{\varepsilon \to 0} \varepsilon^{\alpha-1} \mathbb{P}_x(\tau_{S_\varepsilon} < \infty) = 2^{1-2\alpha} \frac{\Gamma((d+\alpha-2)/2)}{\pi^{d/2}\Gamma(1-\alpha)} \frac{\Gamma((2-\alpha)/2)}{\Gamma(2-\alpha)} H_S(x). \tag{5}$$

(ii) *When $\alpha = 1$, we have that, for $x \notin \mathsf{S}$,*

$$\lim_{\varepsilon \to 0} |\log \varepsilon|\, \mathbb{P}_x(\tau_{\mathsf{S}_\varepsilon} < \infty) = \frac{\Gamma((d-1)/2)}{\pi^{(d-1)/2}} H_{\mathsf{S}}(x). \tag{6}$$

Theorem 2 also gives us the opportunity to understand the strike position of the conditioned stable process. Indeed, let S' be a closed subset of S. Define $\mathsf{S}'_\varepsilon = \{x \in \mathbb{R}^d : 1 - \varepsilon \le |x| \le 1 + \varepsilon$ and $\arg(x) \in \mathsf{S}'\}$ and $\tau_{\mathsf{S}'_\varepsilon} := \inf\{t > 0 : X_t \in \mathsf{S}'_\varepsilon\}$. Then, $\{\tau_{\mathsf{S}'_\varepsilon} < \infty\} \subseteq \{\tau_{\mathsf{S}_\varepsilon} < \infty\}$ and thanks to Theorem 2, when $\alpha \in (0, 1)$, we have

$$\lim_{\varepsilon \to 0} \mathbb{P}_x(\tau_{\mathsf{S}'_\varepsilon} < \infty | \tau_{\mathsf{S}_\varepsilon} < \infty) = \lim_{\varepsilon \to 0} \frac{\varepsilon^{\alpha-1}\mathbb{P}_x(\tau_{\mathsf{S}'_\varepsilon} < \infty)}{\varepsilon^{\alpha-1}\mathbb{P}_x(\tau_{\mathsf{S}_\varepsilon} < \infty)} = \frac{H_{\mathsf{S}'}(x)}{H_{\mathsf{S}}(x)}, \qquad x \notin \mathsf{S}.$$

A similar statement also holds when $\alpha = 1$ by changing the scaling in ε to $|\log \varepsilon|$. This gives us the following result.

Corollary 1 *For a closed $\mathsf{S} \subseteq \mathbb{S}^{d-1}$ such that $\sigma_1(\mathsf{S}) > 0$ and $\alpha \in (0, 1]$, we have that for all closed $\mathsf{S}' \subseteq \mathsf{S}$,*

$$\mathbb{P}_x^{\mathsf{S}}(X_{\zeta-} \in \mathsf{S}') = \frac{H_{\mathsf{S}'}(x)}{H_{\mathsf{S}}(x)}, \qquad x \notin \mathsf{S}. \tag{7}$$

In light of the above Corollary, it is worth remarking that we can also see the probabilities $\mathbb{P}^{\mathsf{S}}$ as the result of first conditioning to continuously hit $\mathbb{S}^{d-1}$ and then conditioning the strike point to be in S. Indeed, we note that, for $A \in \mathcal{F}_t$ and $t \ge 0$,

$$\begin{aligned}
\mathbb{P}_x^{\mathbb{S}^{d-1}}(A | X_{\zeta-} \in \mathsf{S}) &= \mathbb{E}_x^{\mathbb{S}^{d-1}}\left[\mathbf{1}_A \frac{\mathbb{P}_{X_t}^{\mathbb{S}^{d-1}}(X_{\zeta-} \in \mathsf{S})}{\mathbb{P}_x^{\mathbb{S}^{d-1}}(X_{\zeta-} \in \mathsf{S})}\right] \\
&= \mathbb{E}_x\left[\mathbf{1}_A \frac{H_{\mathbb{S}^{d-1}}(X_t)}{H_{\mathbb{S}^{d-1}}(x)} \frac{H_{\mathsf{S}}(X_t)}{H_{\mathbb{S}^{d-1}}(X_t)} \frac{H_{\mathbb{S}^{d-1}}(x)}{H_{\mathsf{S}}(x)}\right] \\
&= \mathbb{E}_x\left[\mathbf{1}_A \frac{H_{\mathsf{S}}(X_t)}{H_{\mathsf{S}}(x)}\right] \\
&= \mathbb{P}_x^{\mathsf{S}}(A).
\end{aligned}$$

Moreover, by shrinking $\mathsf{S}' \subseteq \mathsf{S} \subseteq \mathbb{S}^{d-1}$ to a singleton $\theta \in \mathbb{S}^{d-1}$, one can similarly show that

$$\mathbb{P}_x^{\mathsf{S}}(A | X_{\zeta-} = \theta) = \mathbb{P}_x^{\{\theta\}}(A).$$

This has the flavour of a Williams' type decomposition that was shown for general Lévy processes conditioned to stay positive and subordinators conditioned to remain in an interval; see e.g [11] and [19].

3 Oscillatory Repulsion from S and Duality

Roughly speaking, we want to describe what we see when we time reverse the process $(X, \mathbb{P}^{\mathsf{S}})$ from its strike point on S, i.e. its so-called dual process. Such a process will necessarily avoid visiting S. Recalling that, for $\alpha \in (0, 1]$, the stable process hits spherical surfaces with probability zero (cf. [16, 25]), a heuristic guess for the aforesaid dual process is the stable process itself (see Fig. 1). This turns out to be precisely the case. In order to make this rigorous, we will use the language of Hunt-Nagasawa duality for Markov processes.

Suppose that $Y = (Y_t, t \le \zeta)$ with probabilities $\mathtt{P}_x$, $x \in E$, is a regular Markov process on an open domain $E \subseteq \mathbb{R}^d$ (or more generally, a locally compact Hausdorff space with countable base), with cemetery state Δ and killing time $\zeta = \inf\{t > 0 : Y_t = \Delta\}$. Let us additionally write $\mathtt{P}_\nu = \int_E \nu(\mathrm{d}a)\mathtt{P}_a$, for any probability measure ν on the state space of Y.

Suppose that $\mathcal{G}$ is the σ-algebra generated by Y and write $\mathcal{G}(\mathtt{P}_\nu)$ for its completion by the null sets of $\mathtt{P}_\nu$. Moreover, write $\overline{\mathcal{G}} = \bigcap_\nu \mathcal{G}(\mathtt{P}_\nu)$, where the intersection is taken over all probability measures on the state space of Y, excluding the cemetery state. A finite random time $\mathtt{k}$ is called an L-time (generalized last exit time) if, given a coordinate process $\omega = (\omega_t, t \ge 0)$ on $\mathbb{D}(E)$,

(i) $\mathtt{k}$ is measurable in $\overline{\mathcal{G}}$, and $\mathtt{k} \le \zeta$ almost surely with respect to $\mathtt{P}_\nu$, for all ν,
(ii) $\{s < \mathtt{k}(\omega) - t\} = \{s < \theta_t \circ \mathtt{k}\}$ for all $t, s \ge 0$,

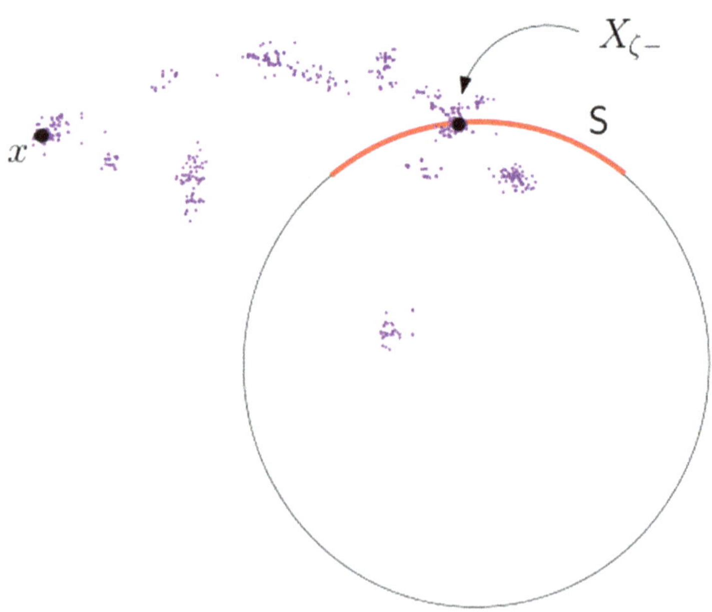

Fig. 1 The process $(X, \mathbb{P}^{\mathsf{S}})$ when time reversed is stochastically equal in law to $(X, \mathbb{P})$

where θ_t is the Markov shift of ω to time t. The most important examples of L-times are killing times and last exit times from closed sets.

Theorem 3 *Suppose that* $\alpha \in (0, 1]$. *For a given closed set* $\mathsf{S} \subset \mathbb{S}^{d-1}$ *with* $\sigma_1(\mathsf{S}) > 0$, *write*

$$\nu(\mathrm{d}a) := \frac{\sigma_1(\mathrm{d}a)}{\sigma_1(\mathsf{S})}, \qquad a \in \mathsf{S}. \tag{8}$$

For every L-time $\Bbbk$ *of* $(X, \mathbb{P})$, *the process* $(X_{(\Bbbk-t)-}, t < \Bbbk)$ *under* $\mathbb{P}_\nu$ *is a time-homogeneous Markov process whose transition probabilities agree with those of* $(X, \mathbb{P}^{\mathsf{S}})$.

4 The Setting of a Subset in an $\mathbb{R}^{d-1}$ Hyperplane

As alluded to in the introduction, the methods used in Sects. 2 and 3 are robust enough to deal with the setting of an arbitrary $(d-1)$-dimensional hyperplane in $\mathbb{R}^d$. Without loss of generality, we can describe such a hyperplane with unit orthogonal vector $v \in \mathbb{S}^{d-1}$ via

$$\mathbb{H}^{d-1} = \{x \in \mathbb{R}^d : (x, v) = 0\},$$

where $(\cdot, \cdot)$ is the usual Euclidean inner product. Henceforth, we will assume that $v \in \mathbb{S}^{d-1}$ is given, as it otherwise plays no role in the forthcoming. We are interested in defining the law of the stable process conditioned to hit $\mathsf{D} \subseteq \mathbb{H}^{d-1}$ in a similar spirit to the discussion in Sect. 2.

To this end, let us define

$$\kappa_\beta = \inf\{t > 0 : -\beta < (v, X_t) < \beta\}, \qquad \text{for } \beta > 0.$$

Whenever it is well defined, we will write, for $t \geq 0$, $\Lambda \in \mathcal{F}_t$ and $x \notin \mathsf{D}$,

$$\mathbb{P}^{\mathsf{D}}_x(\Lambda,\, t < \zeta) = \lim_{\beta\to 0}\lim_{\varepsilon\to 0} \mathbb{P}_x\left(\Lambda, t < \kappa_\beta \middle|\, \tau_{\mathsf{D}_\varepsilon} < \infty\right), \tag{9}$$

where

$$\tau_{\mathsf{D}_\varepsilon} = \inf\{t > 0 : X_t \in \mathsf{D}_\varepsilon\} \qquad \text{and} \qquad \mathsf{D}_\varepsilon := \{x \in \mathbb{R}^d : -\varepsilon \leq (v, x) \leq \varepsilon \text{ and } \hat{x} \in \mathsf{D}\}.$$

Here $\hat{x}$ denotes the orthogonal projection of x onto $\mathbb{H}^{d-1}$; in other words. $\hat{x} = x - v(v, x)$. We can gather the analogous conclusions of Theorems 1, 2, 3 and Corollary 1 into one theorem.

Theorem 4 *Suppose that $\alpha \in (0, 1]$ and the closed and bounded set $D \subseteq \mathbb{H}^{d-1}$ is such that $0 < \ell_{d-1}(D) < \infty$, where we recall that ℓ_{d-1} is $(d-1)$-dimensional Lebesgue measure.*

(i) *Suppose $\alpha \in (0, 1)$. For $x \notin D$,*

$$\lim_{\varepsilon \to 0} \varepsilon^{\alpha-1} \mathbb{P}_x(\tau_{D_\varepsilon} < \infty) = 2^{1-\alpha} \pi^{-(d-2)/2} \frac{\Gamma(\frac{d-2}{2})\Gamma(\frac{d-\alpha}{2})\Gamma(\frac{2-\alpha}{2})^2}{\Gamma(\frac{1-\alpha}{2})\Gamma(\frac{d-1}{2})\Gamma(2-\alpha)} M_D(x), \tag{10}$$

where

$$M_D(x) = \int_D |x-y|^{\alpha-d} \ell_{d-1}(\mathrm{d}y), \qquad x \notin D.$$

(ii) *Suppose $\alpha = 1$. For $x \notin D$,*

$$\lim_{\varepsilon \to 0} |\log \varepsilon| \, \mathbb{P}_x(\tau_{D_\varepsilon} < \infty) = \frac{\Gamma(\frac{d-2}{2})}{\pi^{(d-2)/2}} M_D(x). \tag{11}$$

(iii) *The limit (9) makes sense, therefore the process $(X, \mathbb{P}^D)$ is well defined and*

$$\left. \frac{\mathrm{d}\mathbb{P}_x^D}{\mathrm{d}\mathbb{P}_x} \right|_{\mathcal{F}_t} = \frac{M_D(X_t)}{M_D(x)}, \qquad t \ge 0, x \notin D. \tag{12}$$

(iv) *We have for all closed $D' \subseteq D$,*

$$\mathbb{P}_x^D(X_{\zeta-} \in D') = \frac{M_{D'}(x)}{M_D(x)}, \qquad x \notin D. \tag{13}$$

(v) *Write $\nu(\mathrm{d}a) := \ell_{d-1}(\mathrm{d}a)/\ell_{d-1}(D)$, $a \in D$. For every L-time $\Bbbk$ of $(X, \mathbb{P})$, the process $(X_{(\Bbbk-t)-}, t < \Bbbk)$ under $\mathbb{P}_\nu$ is a time-homogeneous Markov process whose transition probabilities agree with those of $(X, \mathbb{P}^D)$.*

Roughly speaking, Theorem 4 are to be expected as, following the ideas of [22] one may map $\mathbb{S}^{d-1}$ onto $\mathbb{H}^{d-1}$ via a standard sphere inversion transformation, which, thanks to the Riesz–Bogdan–Żak transform, also transforms the paths of the stable processes into that of a h-transformed stable processes; see [8]. The proofs we have given below, however, are direct nonetheless, following similar steps to those of Theorems 1, 2 and 3, as well as Corollary 1.

5 Heuristic for the Proof of Theorem 2

Let us begin with a sketch of the proof of Theorem 2. We start by recalling an identity that is known in quite a general setting from the potential analysis literature; see for example Section 13.11 of [13] and Section VI.2 of [7]. Suppose that A is a bounded closed set and let $\tau_A = \inf\{t > 0 : X_t \in A\}$. Let μ_A be a finite measure supported on A, which is absolutely continuous with respect to Lebesgue measure and define its potential by

$$U\mu_A(x) := \int_A |x-y|^{\alpha-d}\mu_A(\mathrm{d}y), \qquad x \in \mathbb{R}^d.$$

On account of the fact that μ_A is absolutely continuous, recalling that $|x|^{\alpha-d}$ is the potential of the stable process issued from the origin, stationary and independent increments allows us to identify

$$U\mu_A(x) = \int_A |x-y|^{\alpha-d} m_A(y)\ell_d(\mathrm{d}y) = \mathbb{E}_x\left[\int_0^\infty m_A(X_t)\mathrm{d}t\right], \qquad x \notin A,$$

where m_A is the density of μ_A with respect to Lebesgue measure, ℓ_d. As the support of μ_A is precisely A, we must have $m_A(y) = 0$ for all $y \notin A$. As such, the Strong Markov Property tells us that

$$U\mu_A(x) = \mathbb{E}_x\left[\mathbf{1}_{\{\tau_A<\infty\}}\int_{\tau_A}^\infty m_A(X_t)\mathrm{d}t\right] = \mathbb{E}_x\left[U\mu_A(X_{\tau_A})\mathbf{1}_{\{\tau_A<\infty\}}\right], \qquad x \notin A. \tag{14}$$

Note, the above equality is also true when $x \in A$ as, in that case, $\tau_A = 0$.

Replacing τ_A by a general stopping time τ in the above calculation changes the first equality in (14) to an inequality, thus giving the excessive property

$$U\mu_A(x) \geq \mathbb{E}_x\left[U\mu_A(X_\tau)\mathbf{1}_{\{\tau<\infty\}}\right], \qquad x \in \mathbb{R}^d. \tag{15}$$

This family of inequalities together with the Strong Markov Property easily gives us the classical result that $(U\mu_A(X_t), t \geq 0)$ is a supermartingale.

Let us now suppose that μ can be constructed in such a way that it is supported on A such that, for all $x \in A$, $U\mu(x) = 1$. We then recover from identity (14) the corollary to Theorem 1 in Chapter 5 of [13], see also equation (21) in the same chapter, which states that

$$\mathbb{P}_x(\tau_A < \infty) = U\mu(x), \qquad x \notin A.$$

Returning to the problem at hand, we can use the principals above to develop a 'guess and verify' approach to the proof, in particular, since we are not chasing an exact formula for $\mathbb{P}_x(\tau_{\mathbb{S}_\varepsilon} < \infty)$, but rather the asymptotic leading order behaviour.

Indeed, suppose we can 'guess' a measure, say $\mu_\varepsilon^{\mathsf{S}}$, supported on S_ε, such that

$$U\mu_\varepsilon^{\mathsf{S}}(x) = 1 + o(1), \qquad x \in \mathsf{S}_\varepsilon \text{ as } \varepsilon \to 0, \tag{16}$$

so that

$$(1 + o(1))\mathbb{P}_x(\tau_{\mathsf{S}_\varepsilon} < \infty) = U\mu_\varepsilon^{\mathsf{S}}(x), \qquad x \notin \mathsf{S}_\varepsilon. \tag{17}$$

Then, this would be a good basis from which to draw out the leading order decay in ε, especially if our guess of μ_ε is such that $U\mu_\varepsilon$ is tractable.

In one dimension, we know from Lemma 1 of [26], that for a one-dimensional symmetric stable process, the unique measure that satisfies (16) has density $(1 - y)^{-\alpha/2}(1 + y)^{-\alpha/2}$, i.e.

$$\int_{-1}^{1} |x - y|^{\alpha-1}(1 - y)^{-\alpha/2}(1 + y)^{-\alpha/2}\mathrm{d}y = 1, \qquad x \in [-1, 1]. \tag{18}$$

We can use this to build a reasonable choice of $\mu_\varepsilon^{\mathsf{S}}$. Indeed, writing $X = |X| \arg(X)$, when X begins in the neighbourhood of S, then $|X|$ begins in the neighbourhood of 1 and $\arg(X)$, essentially, from within S. On short-time scales and short-range, the time change $|X|$ behaves similarly to a one-dimensional stable process. Moreover, $\arg(X)$ is an isotropic process. A reasonable guess for $\mu_\varepsilon^{\mathsf{S}}$ would be to base it on the measure

$$\mu_\varepsilon(\mathrm{d}y) = c_{\alpha,d}(|y| - (1 - \varepsilon))^{-\alpha/2}(1 + \varepsilon - |y|)^{-\alpha/2}\ell_d(\mathrm{d}y), \tag{19}$$

restricted to S_ε, where we recall $c_{\alpha,d}$ is a constant to be determined so that (16) holds. As we will shortly see, when $\alpha \in (0, 1)$, the constant $c_{\alpha,d}$ does not depend on ε, however, when $\alpha = 1$, in order to respect (16) we need to make it depend on ε.

6 Proof of Theorem 2 (i)

As alluded to in the previous section, we will work with the guess $\mu_\varepsilon^{\mathsf{S}}$ given by the measure μ_ε defined in (19) restricted to S_ε. In order to show (16), we will take advantage of some of the symmetric features of μ_ε, when seen as a measure over $\mathbb{S}_\varepsilon^{d-1} = \{x \in \mathbb{R}^d : 1 - \varepsilon \le |x| \le 1 + \varepsilon\}$. For a subset $A \subset \mathbb{S}_\varepsilon^{d-1}$ we define μ_ε^A the restriction of μ_ε to A. In particular, writing $\mu_\varepsilon^{(1)}$ as μ_ε restricted to $\mathbb{S}_\varepsilon^{d-1}$ and $\mu_\varepsilon^{(2)}$ as μ_ε restricted to $\hat{\mathsf{S}}_\varepsilon := \mathbb{S}_\varepsilon^{d-1} \backslash \mathsf{S}_\varepsilon$, we have the obvious difference

$$U\mu_\varepsilon^{\mathsf{S}}(x) = U\mu_\varepsilon^{(1)}(x) - U\mu_\varepsilon^{(2)}(x), \qquad x \in \mathsf{S}_\varepsilon. \tag{20}$$

Moreover, we would like to introduce

$$\mu_{\varepsilon,\delta}^{(2)} := \mu_\varepsilon|_{\hat{\mathsf{S}}_\varepsilon^\delta}$$

where $\hat{\mathsf{S}}_\varepsilon^\delta = \mathbb{S}_\varepsilon^{d-1} \backslash \mathsf{S}_\varepsilon^\delta$ and

$$\mathsf{S}_\varepsilon^\delta := \{x \in \mathbb{R}^d : 1-\varepsilon < |x| < 1+\varepsilon \text{ and } \arg(x) \in \mathsf{S}^\delta\}, \text{ where } \mathsf{S}^\delta = \{x \in \mathbb{S}^{d-1} : \text{dist}(\arg(x), \mathsf{S}) < \delta\},$$

for some small $\delta > 0$, which, in due course, will depend on ε. Note that, since S is closed, S^δ (resp. $\mathsf{S}_\varepsilon^\delta$) shrinks to S (resp. S_ε) when $\delta \to 0$. Then, we also have that

$$U\mu_\varepsilon^{\mathsf{S}^\delta}(x) = U\mu_\varepsilon^{(1)}(x) - U\mu_{\varepsilon,\delta}^{(2)}(x), \qquad x \in \mathsf{S}_\varepsilon. \tag{21}$$

The estimate (21) will be useful for a certain lower bound that will give us what we need to prove Theorem 2. We need to prove two technical lemmas first. The first one deals with the term $U\mu_\varepsilon^{(1)}$.

Lemma 1 *Suppose that we choose*

$$c_{\alpha,d} = \frac{\Gamma((d+\alpha-2)/2)}{2^\alpha \pi^{d/2}\Gamma(1-\alpha)\Gamma((2-\alpha)/2)}.$$

Then,

$$\lim_{\varepsilon\to 0} \sup_{x\in\mathbb{S}_\varepsilon^{d-1}} |U\mu_\varepsilon^{(1)}(x) - 1| = 0.$$

Proof Appealing to (77), we have, for $x \in \mathbb{S}_\varepsilon^{d-1}$,

$$\begin{aligned}
&U\mu_\varepsilon^{(1)}(x)\\
&= c_{\alpha,d}\int_{\mathbb{S}_\varepsilon^{d-1}} |x-y|^{\alpha-d}(|y|-(1-\varepsilon))^{-\alpha/2}(1+\varepsilon-|y|)^{-\alpha/2}\ell_d(\mathrm{d}y)\\
&= \frac{2c_{\alpha,d}\pi^{(d-1)/2}}{\Gamma((d-1)/2)}\int_{1-\varepsilon}^{1+\varepsilon}\frac{r^{d-1}}{(r-(1-\varepsilon))^{\alpha/2}(1+\varepsilon-r)^{\alpha/2}}\mathrm{d}r\int_0^\pi \frac{\sin^{d-2}\theta\,\mathrm{d}\theta}{(|x|^2-2|x|r\cos\theta+r^2)^{(d-\alpha)/2}}\\
&= \frac{2c_{\alpha,d}\pi^{d/2}}{\Gamma(d/2)}|x|^{\alpha-d}\int_{1-\varepsilon}^{|x|}\frac{{}_2F_1\left(\frac{d-\alpha}{2},1-\frac{\alpha}{2};\frac{d}{2};(r/|x|)^2\right)r^{d-1}}{(r-(1-\varepsilon))^{\alpha/2}(1+\varepsilon-r)^{\alpha/2}}\mathrm{d}r\\
&\qquad + \frac{2c_{\alpha,d}\pi^{d/2}}{\Gamma(d/2)}\int_{|x|}^{1+\varepsilon}\frac{{}_2F_1\left(\frac{d-\alpha}{2},1-\frac{\alpha}{2};\frac{d}{2};(|x|/r)^2\right)r^{\alpha-1}}{(r-(1-\varepsilon))^{\alpha/2}(1+\varepsilon-r)^{\alpha/2}}\mathrm{d}r.
\end{aligned} \tag{22}$$

With a simple change of variables we can reduce this more simply to

$$U\mu_\varepsilon^{(1)}(x) = \frac{2c_{\alpha,d}\pi^{d/2}}{\Gamma(d/2)} \int_{\frac{1-\varepsilon}{|x|}}^{1} \frac{{}_2F_1\left(\frac{d-\alpha}{2}, 1-\frac{\alpha}{2}; \frac{d}{2}; r^2\right) r^{d-1}}{\left(r - \frac{1-\varepsilon}{|x|}\right)^{\alpha/2} \left(\frac{1+\varepsilon}{|x|} - r\right)^{\alpha/2}} \mathrm{d}r + \frac{2c_{\alpha,d}\pi^{d/2}}{\Gamma(d/2)} \int_{1}^{\frac{1+\varepsilon}{|x|}} \frac{{}_2F_1\left(\frac{d-\alpha}{2}, 1-\frac{\alpha}{2}; \frac{d}{2}; r^{-2}\right) r^{\alpha-1}}{\left(r - \frac{1-\varepsilon}{|x|}\right)^{\alpha/2} \left(\frac{1+\varepsilon}{|x|} - r\right)^{\alpha/2}} \mathrm{d}r. \tag{23}$$

For the first term on the right-hand side of (23), we can appeal to (71) and (72) to deduce that

$$\lim_{\varepsilon\to 0} \sup_{x\in\mathbb{S}_\varepsilon^{d-1}} \left| \frac{2c_{\alpha,d}\pi^{d/2}}{\Gamma(d/2)} \int_{\frac{1-\varepsilon}{|x|}}^{1} \frac{{}_2F_1\left(\frac{d-\alpha}{2}, 1-\frac{\alpha}{2}; \frac{d}{2}; r^2\right) r^{d-1}}{\left(r - \frac{1-\varepsilon}{|x|}\right)^{\alpha/2} \left(\frac{1+\varepsilon}{|x|} - r\right)^{\alpha/2}} \mathrm{d}r - \frac{2c_{\alpha,d}\pi^{d/2}\Gamma(1-\alpha)}{\Gamma((d-\alpha)/2)\Gamma((2-\alpha)/2)} \int_{\frac{1-\varepsilon}{|x|}}^{1} \frac{(1-r^2)^{\alpha-1} r^{d-1}}{\left(r - \frac{1-\varepsilon}{|x|}\right)^{\alpha/2} \left(\frac{1+\varepsilon}{|x|} - r\right)^{\alpha/2}} \mathrm{d}r - \frac{2c_{\alpha,d}\pi^{d/2}\Gamma(1-\alpha)}{\Gamma(\alpha/2)\Gamma((d+\alpha-2)/2)} \int_{\frac{1-\varepsilon}{|x|}}^{1} \frac{r^{d-1}}{\left(r - \frac{1-\varepsilon}{|x|}\right)^{\alpha/2} \left(\frac{1+\varepsilon}{|x|} - r\right)^{\alpha/2}} \mathrm{d}r \right| = 0. \tag{24}$$

Note that, by using the transformation $r = (1-\varepsilon+2\varepsilon u)/|x|$,

$$\begin{aligned} &\int_{\frac{1-\varepsilon}{|x|}}^{1} r^{d-1}\left(r - \frac{1-\varepsilon}{|x|}\right)^{-\alpha/2}\left(\frac{1+\varepsilon}{|x|} - r\right)^{-\alpha/2} \mathrm{d}r \\ &\quad = |x|^{\alpha-d}(2\varepsilon)^{1-\alpha} \int_0^{(|x|-1+\varepsilon)/2\varepsilon} (2\varepsilon u + 1 - \varepsilon)^{d-1} u^{-\alpha/2}(1-u)^{-\alpha/2} \mathrm{d}u \\ &\quad \le |x|^{\alpha-d}(2\varepsilon)^{1-\alpha} \frac{\Gamma((2-\alpha)/2)^2}{\Gamma(2-\alpha)}, \end{aligned} \tag{25}$$

which tends to zero uniformly in $x \in \mathbb{S}_\varepsilon^{d-1}$ as $\varepsilon \to 0$.

The asymptotic (25) also tells us that the approximating term of interest in (24) is the middle term. For that, we can use (78) to observe

$$\lim_{\varepsilon\to 0} \sup_{x\in\mathbb{S}_\varepsilon^{d-1}} \left| \int_{\frac{1-\varepsilon}{|x|}}^{1} (1-r^2)^{\alpha-1} r^{d-1}\left(r - \frac{1-\varepsilon}{|x|}\right)^{-\alpha/2}\left(\frac{1+\varepsilon}{|x|} - r\right)^{-\alpha/2} \mathrm{d}r - 2^{\alpha-1} \int_{\frac{1-\varepsilon}{|x|}}^{1} (1-r)^{\alpha-1}\left(r - \frac{1-\varepsilon}{|x|}\right)^{-\alpha/2}\left(\frac{1+\varepsilon}{|x|} - r\right)^{-\alpha/2} \mathrm{d}r \right| = 0 \tag{26}$$

and

$$
\begin{aligned}
&\frac{2^\alpha c_{\alpha,d}\pi^{d/2}\Gamma(1-\alpha)}{\Gamma(\alpha/2)\Gamma((d+\alpha-2)/2)}\int_{\frac{1-\varepsilon}{|x|}}^{1}(1-r)^{\alpha-1}\Big(r-\frac{1-\varepsilon}{|x|}\Big)^{-\alpha/2}\Big(\frac{1+\varepsilon}{|x|}-r\Big)^{-\alpha/2}\mathrm{d}r\\
&=\frac{2^\alpha c_{\alpha,d}\pi^{d/2}\Gamma(1-\alpha)}{\Gamma(\alpha/2)\Gamma((d+\alpha-2)/2)}\int_0^{1-\frac{1-\varepsilon}{|x|}}u^{\alpha-1}\Big(1-\frac{1-\varepsilon}{|x|}-u\Big)^{-\alpha/2}\Big(\frac{1+\varepsilon}{|x|}-1+u\Big)^{-\alpha/2}\mathrm{d}r\\
&=\frac{2^\alpha c_{\alpha,d}\pi^{d/2}\Gamma(1-\alpha)\Gamma((2-\alpha)/2)\Gamma(\alpha)}{\Gamma(\alpha/2)\Gamma((d+\alpha-2)/2)\Gamma((2+\alpha)/2)}\Big(\frac{|x|-1+\varepsilon}{1+\varepsilon-|x|}\Big)^{\alpha/2}\\
&\qquad\qquad {}_2F_1\Big(\alpha/2,\alpha;1+\alpha/2;-\frac{|x|-1+\varepsilon}{1+\varepsilon-|x|}\Big).
\end{aligned}
\tag{27}
$$

The second term on the right-hand side of (23) can be dealt with similarly. Indeed, using (72) we can produce an analogous statement to (24), from which, the leading order approximating term is the integral

$$
\begin{aligned}
&\frac{2c_{\alpha,d}\pi^{d/2}\Gamma(1-\alpha)}{\Gamma(\alpha/2)\Gamma((d+\alpha-2)/2)}\int_1^{\frac{1+\varepsilon}{|x|}}(1-r^{-2})^{\alpha-1}r^{d-1}\Big(r-\frac{1-\varepsilon}{|x|}\Big)^{-\alpha/2}\Big(\frac{1+\varepsilon}{|x|}-r\Big)^{-\alpha/2}\mathrm{d}r\\
&\sim\frac{2^\alpha c_{\alpha,d}\pi^{d/2}\Gamma(1-\alpha)}{\Gamma(\alpha/2)\Gamma((d+\alpha-2)/2)}\int_1^{\frac{1+\varepsilon}{|x|}}(r-1)^{\alpha-1}\Big(r-\frac{1-\varepsilon}{|x|}\Big)^{-\alpha/2}\Big(\frac{1+\varepsilon}{|x|}-r\Big)^{-\alpha/2}\mathrm{d}r\\
&=\frac{2^\alpha c_{\alpha,d}\pi^{d/2}\Gamma(1-\alpha)}{\Gamma(\alpha/2)\Gamma((d+\alpha-2)/2)}\int_0^{\frac{1+\varepsilon}{|x|}-1}u^{\alpha-1}\Big(u+1-\frac{1-\varepsilon}{|x|}\Big)^{-\alpha/2}\Big(\frac{1+\varepsilon}{|x|}-1-u\Big)^{-\alpha/2}\mathrm{d}u\\
&=\frac{2^\alpha c_{\alpha,d}\pi^{d/2}\Gamma(1-\alpha)\Gamma((2-\alpha)/2)\Gamma(\alpha)}{\Gamma(\alpha/2)\Gamma((d+\alpha-2)/2)\Gamma((2+\alpha)/2)}\Big(\frac{1+\varepsilon-|x|}{|x|-1+\varepsilon}\Big)^{\alpha/2}\\
&\qquad\qquad {}_2F_1\Big(\alpha/2,\alpha;1+\alpha/2;-\frac{1+\varepsilon-|x|}{|x|-1+\varepsilon}\Big),
\end{aligned}
\tag{28}
$$

uniformly for $x\in\mathbb{S}^{d-1}_\varepsilon$ as $\varepsilon\to 0$, where we have again used (78) to develop the right-hand side.

Somewhat remarkably, if we add together the right-hand side of (27) and (28), using the identity in (76), we see that the sum is equal to

$$
\frac{2^\alpha c_{\alpha,d}\pi^{d/2}\Gamma(1-\alpha)\Gamma((2-\alpha)/2)}{\Gamma((d+\alpha-2)/2)}=1,
\tag{29}
$$

where the equality with unity follows from the choice of $c_{\alpha,d}$ in the statement of the lemma. □

Piecing together then uniform estimates above as well as the simplification of the two integrals (27) and (28) as well as the decay of the term (25) in (24) and the analogous term when dealing with the second term on the right-hand side of (23), the statement of the lemma follows.

Next we deal with the term $U\mu^{(2)}_{\varepsilon,\delta}$.

Lemma 2 *Recalling that* $c_{\alpha,d}$ *is the constant given in Lemma 1, take* $\delta(\varepsilon) = \varepsilon^{(1-\alpha)/2(d-\alpha)}$, *then*

$$\limsup_{\varepsilon\to 0} \sup_{x\in \mathsf{S}_\varepsilon} \varepsilon^{(\alpha-1)/2} U\mu^{(2)}_{\varepsilon,\delta(\varepsilon)}(x) \le C_{\alpha,d},$$

where

$$C_{\alpha,d} = c_{\alpha,d}\frac{2^{2-\alpha}\pi^{(d-1)/2}\Gamma((2-\alpha)/2)^2}{\Gamma(2-\alpha)\Gamma((d-1)/2)}.$$

In particular,

$$\lim_{\varepsilon\to 0} \sup_{x\in \mathsf{S}_\varepsilon} U\mu^{(2)}_{\varepsilon,\delta(\varepsilon)}(x) = 0.$$

Proof Since $x \in \mathsf{S}_\varepsilon$ and $y \in \hat{\mathsf{S}}^\delta_\varepsilon$, i.e. $|x-y| > \delta$, we have,

$$\begin{aligned}\sup_{x\in \mathsf{S}_\varepsilon} U\mu^{(2)}_{\varepsilon,\delta}(x) &= \int_{\hat{\mathsf{S}}^\delta_\varepsilon} \frac{1}{|x-y|^{d-\alpha}}\mu_\varepsilon(\mathrm{d}y)\\ &\le \frac{1}{\delta^{d-\alpha}}\int_{\hat{\mathsf{S}}^\delta_\varepsilon}\mu_\varepsilon(\mathrm{d}y)\\ &\le \frac{1}{\delta^{d-\alpha}}\frac{2\pi^{(d-1)/2}}{\Gamma((d-1)/2)}\int_{1-\varepsilon}^{1+\varepsilon} r^{d-1}m_\varepsilon(r)\mathrm{d}r, \end{aligned} \tag{30}$$

where $m_\varepsilon(r) = c_{\alpha,d}(r-(1-\varepsilon))^{-\alpha/2}(1+\varepsilon-r)^{-\alpha/2}$. It is easy to see that

$$\begin{aligned}\int_{1-\varepsilon}^{1+\varepsilon} m_\varepsilon(r)\mathrm{d}r &= c_{\alpha,d}\int_{1-\varepsilon}^{1+\varepsilon}(r-(1-\varepsilon))^{-\alpha/2}(1+\varepsilon-r)^{-\alpha/2}\mathrm{d}r\\ &= c_{\alpha,d}\varepsilon^{1-\alpha}2^{1-\alpha}\frac{\Gamma((2-\alpha)/2)^2}{\Gamma(2-\alpha)}. \end{aligned} \tag{31}$$

Putting (30) and (31) together we have

$$\sup_{x\in \mathsf{S}_\varepsilon} U\mu^{(2)}_{\varepsilon,\delta}(x) \le c_{\alpha,d}\frac{2^{2-\alpha}\pi^{(d-1)/2}\Gamma((2-\alpha)/2)^2}{\Gamma(2-\alpha)\Gamma((d-1)/2)} \times \frac{\varepsilon^{1-\alpha}}{\delta^{d-\alpha}}(1+\varepsilon)^{d-1}. \tag{32}$$

By choosing $\delta = \delta(\varepsilon)$, the result follows. □

Let us now return to the proof of Theorem 2. We show that we can make careful sense of (16) and (17). Using (20) in (14) we see that for $x \notin \mathsf{S}$,

$$U\mu_\varepsilon^{\mathsf{S}}(x) = \mathbb{E}_x\left[(U\mu_\varepsilon^{(1)}(X_{\tau_{\mathsf{S}_\varepsilon}}) - 1);\, \tau_{\mathsf{S}_\varepsilon} < \infty\right] + \mathbb{P}_x(\tau_{\mathsf{S}_\varepsilon} < \infty) - \mathbb{E}_x\left[U\mu_\varepsilon^{(2)}(X_{\tau_{\mathsf{S}_\varepsilon}});\, \tau_{\mathsf{S}_\varepsilon} < \infty\right]$$

$$\leq \mathbb{E}_x\left[(U\mu_\varepsilon^{(1)}(X_{\tau_{\mathsf{S}_\varepsilon}}) - 1);\, \tau_{\mathsf{S}_\varepsilon} < \infty\right] + \mathbb{P}_x(\tau_{\mathsf{S}_\varepsilon} < \infty). \tag{33}$$

Then, due to Lemma 1, for each $x \notin \mathsf{S}$ and $\upsilon > 0$, we can choose ε sufficiently small such that

$$U\mu_\varepsilon^{\mathsf{S}}(x) \leq (1+\upsilon)\mathbb{P}_x(\tau_{\mathsf{S}_\varepsilon} < \infty). \tag{34}$$

Since we can take υ arbitrarily small, we have the lower bound on a liminf version of the statement of Theorem 2 given by

$$\liminf_{\varepsilon\to 0} \varepsilon^{\alpha-1} U\mu_\varepsilon^{\mathsf{S}}(x) \leq \liminf_{\varepsilon\to 0} \varepsilon^{\alpha-1}\mathbb{P}_x(\tau_{\mathsf{S}_\varepsilon} < \infty), \qquad x \notin \mathsf{S}. \tag{35}$$

On the other hand, suppose instead of S, we replace its role by $\mathsf{S}^{\delta(\varepsilon)}$, where $\delta(\varepsilon)$ was given in the statement of Lemma 2, we have from the excessive property (15) associated to $U\mu_\varepsilon^{\mathsf{S}^{\delta(\varepsilon)}}$ that

$$U\mu_\varepsilon^{\mathsf{S}^{\delta(\varepsilon)}}(x) \geq \mathbb{E}_x\left[U\mu_\varepsilon^{\mathsf{S}^{\delta(\varepsilon)}}(X_{\tau_{\mathsf{S}_\varepsilon}});\, \tau_{\mathsf{S}_\varepsilon} < \infty\right], \qquad x \notin \mathsf{S}. \tag{36}$$

Now appealing to (21), we get

$$U\mu_\varepsilon^{\mathsf{S}^{\delta(\varepsilon)}}(x) \geq \mathbb{E}_x\left[U\mu_\varepsilon^{(1)}(X_{\tau_{\mathsf{S}_\varepsilon}}) - 1;\, \tau_{\mathsf{S}_\varepsilon} < \infty\right] + \mathbb{P}_x(\tau_{\mathsf{S}_\varepsilon} < \infty) - \mathbb{E}_x\left[U\mu_{\varepsilon,\delta(\varepsilon)}^{(2)}(X_{\tau_{\mathsf{S}_\varepsilon}});\, \tau_{\mathsf{S}_\varepsilon} < \infty\right].$$

Appealing to Lemmas 1 and 2, for each $\upsilon > 0$, we can choose ε small enough such that, for each $x \notin \mathsf{S}$,

$$U\mu_\varepsilon^{\mathsf{S}^{\delta(\varepsilon)}}(x) \geq (1-\upsilon)\mathbb{P}_x(\tau_{\mathsf{S}_\varepsilon} < \infty). \tag{37}$$

Hence, since we can choose υ as small as we like, we have

$$\limsup_{\varepsilon\to 0} \varepsilon^{\alpha-1} U\mu_\varepsilon^{\mathsf{S}^{\delta(\varepsilon)}}(x) \geq \limsup_{\varepsilon\to 0} \varepsilon^{\alpha-1}\mathbb{P}_x(\tau_{\mathsf{S}_\varepsilon} < \infty), \qquad x \notin \mathsf{S}. \tag{38}$$

It follows from (35) and (38) that, as soon as

$$\limsup_{\varepsilon\to 0} \varepsilon^{\alpha-1} U\mu_\varepsilon^{\mathsf{S}^{\delta(\varepsilon)}}(x) = \liminf_{\varepsilon\to 0} \varepsilon^{\alpha-1} U\mu_\varepsilon^{\mathsf{S}}(x), \qquad x \notin \mathsf{S}, \tag{39}$$

and noting that $U\mu_\varepsilon^{\mathsf{S}} \le U\mu_\varepsilon^{\mathsf{S}^{\delta(\varepsilon)}}$, we have

$$\lim_{\varepsilon\to 0} \varepsilon^{\alpha-1} U\mu_\varepsilon^{\mathsf{S}}(x) = \lim_{\varepsilon\to 0} \varepsilon^{\alpha-1}\mathbb{P}_x(\tau_{\mathsf{S}_\varepsilon} < \infty), \qquad x \notin \mathsf{S}.$$

Let us thus complete the proof by verifying the limit on the equality (39) holds and by finding the left-hand side limit in the previous equation.

To this end, using that $|x-y|^{\alpha-d}$ is continuous on S_ε and, when $x \notin \mathsf{S}$, without loss of generality, we can take ε small enough so that $x \notin \mathsf{S}_\varepsilon$. For each $x \notin \mathsf{S}$, using the Mean Valued Theorem, there exists a $r_\varepsilon^* \in (1-\varepsilon, 1+\varepsilon)$ such that

$$\begin{aligned} U\mu_\varepsilon^{\mathsf{S}}(x) &= \int_{\mathsf{S}_\varepsilon} |x-y|^{\alpha-d} m_\varepsilon(|y|)\ell_d(\mathrm{d}y) \\ &= (r_\varepsilon^*)^{d-1} \int_{\mathsf{S}} |x - r_\varepsilon^*\theta|^{\alpha-d}\sigma_1(\mathrm{d}\theta) \int_{1-\varepsilon}^{1+\varepsilon} m_\varepsilon(r)\mathrm{d}r, \end{aligned} \tag{40}$$

where we recall that $m_\varepsilon(r) = c_{\alpha,d}(r-(1-\varepsilon))^{-\alpha/2}(1+\varepsilon-r)^{-\alpha/2}$. By using (31) we get

$$\varepsilon^{\alpha-1}U\mu_\varepsilon^{\mathsf{S}}(x) = (r_\varepsilon^*)^{d-1}2^{1-\alpha}c_{\alpha,d}\frac{\Gamma((2-\alpha)/2)^2}{\Gamma(2-\alpha)}\int_{\mathsf{S}}|x-r_\varepsilon^*\theta|^{\alpha-d}\sigma_1(\mathrm{d}\theta), \qquad x\notin\mathsf{S}. \tag{41}$$

Taking limits in (41) as $\varepsilon \to 0$ and recalling the value of $c_{\alpha,d}$ from the statement of Lemma 1, we have, for $x \notin \mathsf{S}$

$$\lim_{\varepsilon\to 0}\varepsilon^{\alpha-1}U\mu_\varepsilon^{\mathsf{S}}(x) = 2^{1-2\alpha}\frac{\Gamma((d+\alpha-2)/2)}{\pi^{d/2}\Gamma(1-\alpha)}\frac{\Gamma((2-\alpha)/2)}{\Gamma(2-\alpha)}\int_{\mathsf{S}}|x-\theta|^{\alpha-d}\sigma_1(\mathrm{d}\theta). \tag{42}$$

An application of the recursion formula for gamma functions allows us to identify the right-hand side as equal to that of the right-hand side of (5). Very little changes in the above calculation if we replace S by $\mathsf{S}^{\delta(\varepsilon)}$. As such, (42) allows us to conclude (39), and thus gives the statement of the Theorem 2. □

7 Proof of Theorem 2 (ii)

The proof needs some adaptation when we deal with the case $\alpha = 1$. Principally, we need to focus on Lemmas 1 and 2. What is different in these two lemmas is that the normalisation constant $c_{\alpha,d}$ must now depend on ε. The replacement for Lemma 1 and Lemma 2 (combined into one result) now takes the following form.

Lemma 3 *Suppose that we define, for* $0 < \varepsilon < 1$,

$$\mu_\varepsilon(\mathrm{d}y) = \frac{c_{1,d}}{|\log\varepsilon|}(|y| - (1-\varepsilon))^{-\alpha/2}(1+\varepsilon-|y|)^{-\alpha/2}\ell_d(\mathrm{d}y), \tag{43}$$

and

$$c_{1,d} = \frac{\Gamma((d-1)/2)}{\pi^{(d+1)/2}}.$$

(i) *We have*

$$\lim_{\varepsilon\to 0}\sup_{x\in\mathbb{S}_\varepsilon^{d-1}} |U\mu_\varepsilon^{(1)}(x) - 1| = 0.$$

(ii) *take* $\delta(\varepsilon) = |\log\varepsilon|^{-1/2(d-1)}$, *then*

$$\limsup_{\varepsilon\to 0}\sup_{x\in S_\varepsilon} \sqrt{|\log\varepsilon|}U\mu^{(2)}_{\varepsilon,\delta(\varepsilon)}(x) < \infty,$$

so that

$$\lim_{\varepsilon\to 0}\sup_{x\in S_\varepsilon} U\mu^{(2)}_{\varepsilon,\delta(\varepsilon)}(x) = 0.$$

Proof We give only a sketch proof of both parts for the interested reader to use as a guide to reproduce the finer details.

(i) The essence of the proof is an adaptation of the proof of Lemma 1. We pick up the proof of the latter at the analogue of (23), albeit $\alpha = 1$ and $c_{\alpha,d}$ is replaced by $c_{1,d}/|\log\varepsilon|$, i.e.

$$U\mu_\varepsilon^{(1)}(x) = \frac{2c_{1,d}\pi^{d/2}}{|\log\varepsilon|\Gamma(d/2)}\int_{\frac{1-\varepsilon}{|x|}}^{1} \frac{{}_2F_1\left(\frac{d-1}{2},\frac{1}{2};\frac{d}{2};r^2\right)r^{d-1}}{\left(r-\frac{1-\varepsilon}{|x|}\right)^{1/2}\left(\frac{1+\varepsilon}{|x|}-r\right)^{1/2}}\mathrm{d}r$$
$$+\frac{2c_{1,d}\pi^{d/2}}{|\log\varepsilon|\Gamma(d/2)}\int_1^{\frac{1+\varepsilon}{|x|}} \frac{{}_2F_1\left(\frac{d-1}{2},\frac{1}{2};\frac{d}{2};r^{-2}\right)r^{1-1}}{\left(r-\frac{1-\varepsilon}{|x|}\right)^{1/2}\left(\frac{1+\varepsilon}{|x|}-r\right)^{1/2}}\mathrm{d}r. \tag{44}$$

Appealing to (74), noting that $\log(1-r^2) \sim \log(1-r) + \log 2$, as $r \to 1$, we can deduce that there is an unimportant constant, say χ, such that

$$\lim_{\varepsilon\to 0}\sup_{x\in\mathbb{S}_\varepsilon^{d-1}}\left|\frac{2c_{1,d}\pi^{d/2}}{|\log\varepsilon|\Gamma(d/2)}\int_{\frac{1-\varepsilon}{|x|}}^{1}\frac{{}_2F_1\left(\frac{d-1}{2},\frac{1}{2};\frac{d}{2};r^2\right)r^{d-1}}{\left(r-\frac{1-\varepsilon}{|x|}\right)^{1/2}\left(\frac{1+\varepsilon}{|x|}-r\right)^{1/2}}dr\right.$$
$$+\frac{2c_{1,d}\pi^{d/2}}{|\log\varepsilon|\Gamma((d-1)/2)\Gamma(1/2)}\int_{\frac{1-\varepsilon}{|x|}}^{1}\frac{r^{d-1}\log(1-r)}{\left(r-\frac{1-\varepsilon}{|x|}\right)^{1/2}\left(\frac{1+\varepsilon}{|x|}-r\right)^{1/2}}dr$$
$$\left.-\frac{c_{1,d}\chi}{|\log\varepsilon|}\int_{\frac{1-\varepsilon}{|x|}}^{1}\frac{r^{d-1}}{\left(r-\frac{1-\varepsilon}{|x|}\right)^{1/2}\left(\frac{1+\varepsilon}{|x|}-r\right)^{1/2}}dr\right| = 0. \tag{45}$$

A similar uniform limiting control can be undertaken by subtracting off analogous terms from the second integral in (44), i.e. the integral

$$\frac{2c_{1,d}\pi^{d/2}}{|\log\varepsilon|\Gamma(d/2)}\int_{1}^{\frac{1+\varepsilon}{|x|}}\frac{{}_2F_1\left(\frac{d-1}{2},\frac{1}{2};\frac{d}{2};r^{-2}\right)}{\left(r-\frac{1-\varepsilon}{|x|}\right)^{1/2}\left(\frac{1+\varepsilon}{|x|}-r\right)^{1/2}}dr.$$

Using (25), again noting $\alpha = 1$, we can uniformly control the last term in (45) and note that it is $O(1/|\log\varepsilon|)$. Similarly to (26), the second term in (45) has the same behaviour as

$$-\frac{2c_{1,d}\pi^{d/2}}{|\log\varepsilon|\Gamma((d-1)/2)\Gamma(1/2)}\int_{0}^{1-\frac{1-\varepsilon}{|x|}}\frac{\log u}{(\frac{1+\varepsilon-|x|}{|x|}+u)^{1/2}(\frac{|x|-(1-\varepsilon)}{|x|}-u)^{1/2}}du. \tag{46}$$

To evaluate (46), using the change of variable $u = a - (a+b)/(t^2+1)$

$$\begin{aligned}\int_0^a \frac{\log u}{\sqrt{(b+u)(a-u)}}du &= 2\int_{\sqrt{\frac{b}{a}}}^{\infty}\log\left(a-\frac{a+b}{t^2+1}\right)\frac{dt}{t^2+1}\\ &= \int_0^{\arctan\sqrt{\frac{a}{b}}}\log(a-(a+b)\sin^2 w)dw\\ &= \int_0^{\arctan\sqrt{\frac{a}{b}}}\log a + \log\left(1-\frac{\sin^2 w}{\frac{a}{a+b}}\right)dw\\ &= \arctan\sqrt{\frac{a}{b}}\log(a+b) - L\left(\frac{\pi}{2}-2\arctan\sqrt{\frac{a}{b}}\right)-\frac{\pi}{2}\log 2,\end{aligned} \tag{47}$$

where we have used formula 4.226(5) of [15], which tells us that

$$\int_0^u \log\Big(1 - \frac{\sin^2 w}{\sin^2 v}\Big)\mathrm{d}w = -u \log \sin^2 v - L(\frac{\pi}{2} - v + u) - L(\frac{\pi}{2} - v - u), \tag{48}$$

for any $-\frac{\pi}{2} \le u \le \frac{\pi}{2}$ and $|\sin u| \le |\sin v|$ where $L(x)$ is the Lobachevsky function. Note that, Lobachevsky's function is defined and represented as

$$L(x) = -\int_0^x \log\cos\theta \,\mathrm{d}\theta = x\log 2 - \frac{1}{2}\sum_{k=1}^{\infty}(-1)^{k-1}\frac{\sin 2kx}{k^2}. \tag{49}$$

Using (47) to evaluate (46) as well to evaluate the partner integral to (46), which comes from the analogous control of the second integral in (44), we get a nice cancellation of terms (as happened at this stage of the argument for $\alpha \in (0, 1)$), to give us the controlled feature that

$$\lim_{\varepsilon\to 0}\sup_{x\in\mathbb{S}_\varepsilon^{d-1}}\left|U\mu_\varepsilon^{(1)}(x) + \frac{2c_{1,d}\pi^{d/2}}{|\log\varepsilon|\Gamma((d-1)/2)\Gamma(1/2)}\frac{\pi}{2}\log\varepsilon\right| = 0.$$

Noting that with the indicated choice of $c_{1,d}$, we have

$$\frac{2c_{1,d}\pi^{d/2}}{\Gamma((d-1)/2)\Gamma(1/2)}\frac{\pi}{2} = 1,$$

which concludes the proof of part (i).

(ii) For the second part, the proof is almost identical to the proof of Lemma 2. Indeed, following the calculations through to (32), recalling that we have replaced $c_{\alpha,d}$ by $c_{1,d}/|\log\varepsilon|$, we get, up to an unimportant constant χ',

$$\sup_{x\in\mathsf{S}_\varepsilon} U\mu_{\varepsilon,\delta}^{(2)}(x) \le \chi'\frac{1}{|\log\varepsilon|\delta^{d-1}}. \tag{50}$$

Hence, by taking $\delta = \delta(\varepsilon) = |\log\varepsilon|^{-1/2(d-1)}$ the statement of part (ii) follows. □

With Lemma 3 in hand, we can now complete the proof of Theorem 2 (ii). Inequalities (34) and (37) are still at our disposal for the same reasons as before. The proof thus boils down to the asymptotic treatment of the term $U\mu_\varepsilon^{\mathsf{S}}(x)$ as in (40) for $x \notin \mathsf{S}$. Recalling that we have replaced $c_{\alpha,d}$ by $c_{1,d}/|\log\varepsilon|$ we get from (31) and the constant $c_{1,d}$ given in the statement of Lemma 3,

$$\lim_{\varepsilon\to 0}|\log\varepsilon|\,\mathbb{P}_x(\tau_{\mathsf{S}_\varepsilon} < \infty) = \frac{\Gamma((d-1)/2)}{\pi^{(d+1)/2}}\Gamma(1/2)^2 H_{\mathsf{S}}(x) = \frac{\Gamma((d-1)/2)}{\pi^{(d-1)/2}}H_{\mathsf{S}}(x),$$

where we have used that $\Gamma(1/2) = \sqrt{\pi}$. □

8 Proof of Theorem 1

Recall the definition $\tau_\beta := \inf\{t > 0: 1/\beta < |X_t| < \beta\}$ for $\beta > 1$ and fix $\varepsilon_0 > 0$ such that, for all $0 < \varepsilon < \varepsilon_0$, $(1-\varepsilon, 1+\varepsilon) \subset (1/\beta, \beta)$. Then, by applying the Markov property at time t, we have, for any $\Lambda \in \mathcal{F}_t$,

$$\mathbb{P}_x(\Lambda, t < \tau_\beta | \tau_{\mathsf{S}_\varepsilon} < \infty) = \mathbb{E}_x\left[\mathbf{1}_{\{\Lambda, t<\tau_\beta\}} \frac{\mathbb{P}_{X_t}(\tau_{\mathsf{S}_\varepsilon} < \infty)}{\mathbb{P}_x(\tau_{\mathsf{S}_\varepsilon} < \infty)}\right]. \tag{51}$$

The event $\{t < \tau_\beta\}$ implies that either $|X_t| > \beta > 1$ or $|X_t| < 1/\beta < 1$. Hence, for all $0 < \varepsilon < \varepsilon_0$ and $y \in \mathbb{S}^{d-1}_\varepsilon$, on $\{t < \tau_\beta\}$,

$$|X_t - y|^{\alpha-d} < \max\{((1-\varepsilon_0) - 1/\beta)^{\alpha-d}, (\beta - (1+\varepsilon_0))^{\alpha-d}\}.$$

Hence, on $\{t < \tau_\beta\}$, we have from (37) and (41) that we can choose ε sufficiently small such that

$$\varepsilon^{\alpha-1}\mathbb{P}_{X_t}(\tau_{\mathsf{S}_\varepsilon} < \infty) < K_1,$$

for some constant $K_1 \in (0,\infty)$. In a similar spirit, using (34) and (41), since $x \notin \mathsf{S}$ and S is closed, it follows similarly that there is another constant $K_2 \in (0,\infty)$ such that, for x given in (51), we can choose ε sufficiently small such that

$$\varepsilon^{\alpha-1}\mathbb{P}_x(\tau_{\mathsf{S}_\varepsilon} < \infty) > K_2.$$

Theorem 2, dominated convergence and monotone convergence gives us, for all $\Lambda \in \mathcal{F}_t$, $t \geq 0$,

$$\lim_{\beta\to 1}\lim_{\varepsilon\to 0}\mathbb{P}_x(\Lambda, t < \tau_\beta|\tau_{\mathsf{S}_\varepsilon} < \infty) = \lim_{\beta\to 1}\mathbb{E}_x\left[\mathbf{1}_{\{\Lambda, t<\tau_\beta\}}\lim_{\varepsilon\to 0}\frac{\varepsilon^{\alpha-1}\mathbb{P}_{X_t}(\tau_{\mathsf{S}_\varepsilon} < \infty)}{\varepsilon^{\alpha-1}\mathbb{P}_x(\tau_{\mathsf{S}_\varepsilon} < \infty)}\right] = \mathbb{E}_x\left[\mathbf{1}_\Lambda\frac{H_{\mathsf{S}}(X_t)}{H_{\mathsf{S}}(x)}\right],$$

as required. □

9 Proof of Theorem 3

Recall the notation for a general Markov process (Y, P) on E preceding the statement of Theorem 3. We will additionally write $\mathcal{P} := (\mathcal{P}_t, t \geq 0)$ for the semigroup associated to (Y, P).

Theorem 3.5 of Nagasawa [23], shows that, under suitable assumptions on the Markov process, L-times form a natural family of random times at which the pathwise time-reversal

$$\overleftarrow{Y}_t := Y_{(\mathtt{k}-t)-}, \qquad t \in [0, \mathtt{k}],$$

is again a Markov process. Let us state Nagasawa's principle assumptions.

(A) The potential measure $U_Y(a, \cdot)$ associated to $\mathcal{P}$, defined by the relation

$$\int_E f(x)U_Y(a, \mathrm{d}x) = \int_0^\infty \mathcal{P}_t[f](a)\mathrm{d}t = \mathrm{E}_a\left[\int_0^\infty f(X_t)\,\mathrm{d}t\right], \qquad a \in E, \tag{52}$$

for bounded and measurable f on E, is σ-finite. Assume that there exists a probability measure, ν, such that, if we put

$$\mu(A) = \int U_Y(a, A)\,\nu(\mathrm{d}a) \quad \text{for } A \in \mathcal{B}(E), \tag{53}$$

then, there exists a Markov transition semigroup, say $\hat{\mathcal{P}} := (\hat{\mathcal{P}}_t, t \geq 0)$ such that

$$\int_E \mathcal{P}_t[f](x)g(x)\,\mu(\mathrm{d}x) = \int_E f(x)\hat{\mathcal{P}}_t[g](x)\,\mu(\mathrm{d}x), \quad t \geq 0, \tag{54}$$

for bounded, measurable and compactly supported test-functions f, g on E.

(B) For any continuous test-function $f \in C_0(E)$, the space of continuous and compactly supported functions, and $a \in E$, assume that $\mathcal{P}_t[f](a)$ is right-continuous in t for all $a \in E$ and, for $q > 0$, $U^{(q)}_{\hat{Y}}[f](\overleftarrow{Y}_t)$ is right-continuous in t, where, for bounded and measurable f on E,

$$U^{(q)}_{\hat{Y}}[f](a) = \int_0^\infty \mathrm{e}^{-qt}\hat{\mathcal{P}}_t[f](a)\mathrm{d}t, \qquad a \in E,$$

is the q-potential associated to $\hat{\mathcal{P}}$.

Nagasawa's duality theorem, Theorem 3.5. of [23], now reads as follows.

Theorem 5 (Nagasawa's duality theorem) *Suppose that assumptions* **(A)** *and* **(B)** *hold. For the given starting probability distribution ν in* **(A)** *and any L-time* $\mathtt{k}$, *the time-reversed process $\overleftarrow{Y}$ under P_ν is a time-homogeneous Markov process with transition probabilities*

$$\mathrm{P}_\nu(\overleftarrow{Y}_t \in A \mid \overleftarrow{Y}_r, 0 < r < s) = \mathrm{P}_\nu(\overleftarrow{Y}_t \in A \mid \overleftarrow{Y}_s) = p_{\hat{Y}}(t - s, \overleftarrow{Y}_s, A), \quad \mathrm{P}_\nu\text{-almost surely}, \tag{55}$$

for all $0 < s < t$ and closed A in E, where $p_{\hat{Y}}(u, x, A)$, $u \geq 0$, $x \in E$, is the transition measure associated to the semigroup $\hat{\mathcal{P}}$.

9.1 Completing the Proof of Theorem 3

We will make a direct application of Theorem 5, with Y taken to be the process $(X, \mathbb{P}_\nu)$ where ν satisfies (8). Recall that its potential is written U and we will denote its transition semigroup by $(\mathcal{P}_t, t \geq 0)$. Moreover, the dual process, formerly $\hat{Y}$, is taken to be $(X, \mathbb{P}^{\mathsf{S}})$ and we will, in the obvious way, work with the notation U^{S} in place of $U_{\hat{Y}}$, $\mathcal{P}^{\mathsf{S}}$ in place of $\hat{\mathcal{P}}$ and so on. We need only to verify the two assumptions **(A)** and **(B)**.

In order to verify **(A)**, writing

$$U(x, \mathrm{d}y) = \int_0^\infty \mathbb{P}_x(X_t \in \mathrm{d}y)\mathrm{d}t = \frac{\Gamma((d-\alpha)/2)}{2^\alpha \pi^{d/2}\Gamma(\alpha/2)}|x-y|^{\alpha-d}\ell_d(\mathrm{d}y), \qquad x, y \in \mathbb{R}^d,$$

we have, up to a multiplicative constant,

$$\eta(\mathrm{d}x) = \int_{\mathbb{R}^d} U(a, \mathrm{d}x)\nu(\mathrm{d}a) = \frac{1}{\sigma_1(\mathsf{S})}\int_{\mathsf{S}} |x-a|^{\alpha-d}\sigma_1(\mathrm{d}a) \propto H_{\mathsf{S}}(x)\mathrm{d}x. \tag{56}$$

Now, we need to verify that (54) holds. Hunt's switching identity (cf. Chapter II.1 of [4]) for $(X, \mathbb{P})$, states that

$$\mathcal{P}_t(y, \mathrm{d}x)\mathrm{d}y = \mathcal{P}_t(x, \mathrm{d}y)\mathrm{d}x, \qquad x, y \in \mathbb{R}^d.$$

Using Hunt's switching identity together with (56), we have for $x, y \in \mathbb{R}^d \setminus \mathsf{S}$

$$\mathcal{P}_t(y, \mathrm{d}x)\eta(\mathrm{d}y) = \mathcal{P}_t(y, \mathrm{d}x)H_{\mathsf{S}}(y)\mathrm{d}y = \mathcal{P}_t(x, \mathrm{d}y)\frac{H_{\mathsf{S}}(y)}{H_{\mathsf{S}}(x)}H_{\mathsf{S}}(x)\mathrm{d}x = \mathcal{P}_t^{\mathsf{S}}(x, \mathrm{d}y)\eta(\mathrm{d}x).$$

Let us now turn to the verification of assumption **(B)**. This assumption is immediately satisfied on account of the fact that $\mathcal{P}^{\mathsf{S}}$ is a right-continuous semigroup by virtue of its definition as a Doob h-transform with respect to the Feller semigroup $\mathcal{P}$ of the stable process.

With both **(A)** and **(B)** in hand, we are ready to apply Theorem 5 and the desired result thus follows. □

10 Proof of Theorem 4

For the proof of Theorem 4, we focus on just part (i) and (ii) as the proof of parts (iii)–(v) are essentially verbatim the same as for the case of $\mathsf{S} \in \mathbb{S}^{d-1}$. Moreover, for both parts (i) and (ii) we will provide only a sketch proof as the reader will quickly see that the proof is not hugely different form that of Theorem 2, albeit for a few technical details.

(i) The substance of the proof of part (i) is thus to follow a similar strategy as with Theorem 2 and build a measure $\rho_\varepsilon^{\mathsf{D}}$ such that the analogue of (16) holds, i.e. $U\rho_\varepsilon^{\mathsf{D}}(x) = 1 + o(1)$, for $x \in \mathsf{D}$ so that $(1 + o(1))\mathbb{P}_x(\tau_{\mathsf{D}\varepsilon} < \infty) = U\rho_\varepsilon^{\mathsf{D}}(x), x \notin \mathsf{D}\varepsilon$. More precisely, we develop analogues of Lemmas 1 and 2 to help make this precise.

Following what we have learned for $\mu_\varepsilon^{\mathsf{S}}$, our choice of $\rho_\varepsilon^{\mathsf{D}}$ is built from the base measure

$$\rho_\varepsilon(\mathrm{d}y) = k_{\alpha,d}((v, y) + \varepsilon)^{-\alpha/2}(\varepsilon - (v, y))^{-\alpha/2}\ell_d(\mathrm{d}y). \tag{57}$$

for an appropriate choice of $k_{\alpha,d}$. As in (20) we can work with the decomposition,

$$U\rho_\varepsilon^{\mathsf{D}}(x) = U\rho_\varepsilon^{(1)}(x) - U\rho_\varepsilon^{(2)}(x), \qquad x \in \mathsf{D}_\varepsilon, \tag{58}$$

where $\rho_\varepsilon^{(1)}$ (resp. $\rho_\varepsilon^{(2)}$) is the restriction of ρ_ε to $\mathbb{H}_\varepsilon^{d-1} := \{x \in \mathbb{R}^d : -\varepsilon < (v, x) < \varepsilon\}$ (resp. to $\hat{\mathsf{D}}_\varepsilon := \mathbb{H}_\varepsilon^{d-1} \setminus \mathsf{D}_\varepsilon$). This helps with lower bounding $\liminf_{\varepsilon\to 0} \varepsilon^{\alpha-1}\mathbb{P}_x(\tau_{\mathsf{D}_\varepsilon} < \infty)$ by following steps of (33)–(35) together with the last paragraph of the Proof of Theorem 2, for which an analogue of Lemma 1 is needed.

For each $|u| < \varepsilon$, define the following sets: $\mathsf{D}^\delta = \{x \in \mathbb{H}^{d-1} : \mathrm{dist}(x, \mathsf{D}) < \delta\}$, $\mathsf{D}_\varepsilon^\delta = \{y \in \mathbb{H}_\varepsilon^{d-1} : \hat{y} \in \mathsf{D}^\delta\}$ (recalling $\hat{y}$ is the orthogonal projection of y on to $\mathbb{H}^{d-1}$) and $\hat{\mathsf{D}}_\varepsilon^\delta = \mathbb{H}_\varepsilon^{d-1} \setminus \mathsf{D}_\varepsilon^\delta$. Moreover, for any $u \in \mathbb{R}$, we define $\mathbb{H}^{d-1}(u) = \{x \in \mathbb{R}^d : (v, x) = u\}$, $\mathsf{D}(u) := \{y \in \mathbb{H}^{d-1}(u) : \hat{y} \in \mathsf{D}\}$, $\mathsf{D}^\delta(u) := \{y \in \mathbb{H}^{d-1}(u) : \hat{y} \in \mathsf{D}^\delta\}$, and $\hat{\mathsf{D}}^\delta(u) = \mathbb{H}^{d-1}(u) \setminus \mathsf{D}^\delta(u)$. Similarly, in the spirit of (21) we can use the decomposition

$$U\rho_\varepsilon^{\mathsf{D}^\delta}(x) = U\rho_\varepsilon^{(1)}(x) - U\rho_{\varepsilon,\delta}^{(2)}(x), \qquad x \in \mathsf{D}_\varepsilon, \tag{59}$$

where $\rho_{\varepsilon,\delta}^{(2)}$ is the restriction of ρ_ε to $\hat{\mathsf{D}}_\varepsilon^\delta$. which helps with $\limsup_{\varepsilon\to 0} \varepsilon^{\alpha-1}\mathbb{P}_x(\tau_{\mathsf{D}_\varepsilon} < \infty)$ by following steps (36)–(39) together with the last paragraph of the Proof of Theorem 2, for which an analogue of Lemma 2 is needed.

Let us address the technical detail that differs from the proof of Theorem 2 that we alluded to above. For $x \in \mathsf{D}_\varepsilon$,

$$\begin{aligned}
&U\rho_\varepsilon^{(1)}(x)\\
&= k_{\alpha,d}\int_{\mathbb{H}_\varepsilon^{d-1}} |x - y|^{\alpha-d}((v, y) + \varepsilon)^{-\alpha/2}(\varepsilon - (v, y))^{-\alpha/2}\ell_d(\mathrm{d}y)\\
&= k_{\alpha,d}\int_{-\varepsilon}^{\varepsilon} (u + \varepsilon)^{-\alpha/2}(\varepsilon - u)^{-\alpha/2}\mathrm{d}u \int_{\mathbb{H}^{d-1}(u)} |x - y|^{\alpha-d}\ell_{d-1}(\mathrm{d}y)\\
&= k_{\alpha,d}\int_{-\varepsilon}^{\varepsilon} (u + \varepsilon)^{-\alpha/2}(\varepsilon - u)^{-\alpha/2}\mathrm{d}u \int_{\mathbb{H}^{d-1}((v,x))} (|x - \hat{z}|^2 + |(v, x) - u|^2)^{\frac{\alpha-d}{2}}\ell_{d-1}(\mathrm{d}\hat{z}),
\end{aligned}$$

where $\hat{z}$ is the orthogonal projection of $y \in \mathbb{H}^{d-1}(u)$ onto $\mathbb{H}^{d-1}((v,x))$, which satisfies $|\hat{z}-y| = |(v,x)-u|$ and $\ell_{d-1}(\mathrm{d}\hat{z}) = \ell_{d-1}(\mathrm{d}y)$. Note also that $(v, x-\hat{z}) = 0$, for $\hat{z} \in \mathbb{H}^{d-1}((v,x))$, and hence $x - \mathbb{H}^{d-1}((v,x))$ is equal to $\mathbb{H}^{d-1}(0)$, which, in turn, can otherwise be identified as $\mathbb{R}^{d-1}$. Therefore, if we used generalised polar coordinates to integrate over $\mathbb{H}^{d-1}((v,x))$ identified as $x - \mathbb{R}^{d-1}$, we have

$$
\begin{aligned}
&U\rho_\varepsilon^{(1)}(x)\\
&= k_{\alpha,d}\int_{-\varepsilon}^{\varepsilon}(u+\varepsilon)^{-\alpha/2}(\varepsilon-u)^{-\alpha/2}\mathrm{d}u\int_{\mathbb{H}^{d-1}((v,x))}\left(|x-\hat{z}|^2+|(v,x)-u|^2\right)^{\frac{\alpha-d}{2}}\ell_{d-1}(\mathrm{d}\hat{z})\\
&= \frac{2k_{\alpha,d}\pi^{(d-2)/2}}{\Gamma((d-2)/2)}\int_{-\varepsilon}^{\varepsilon}(u+\varepsilon)^{-\alpha/2}(\varepsilon-u)^{-\alpha/2}\mathrm{d}u\int_0^\infty\int_{\mathbb{S}^{d-2}}\left(r^2+|(v,x)-u|^2\right)^{\frac{\alpha-d}{2}}r^{d-2}\mathrm{d}r\sigma_1(d\theta)\\
&= \frac{2k_{\alpha,d}\pi^{(d-2)/2}}{\Gamma((d-2)/2)}\int_{-\varepsilon}^{\varepsilon}(u+\varepsilon)^{-\alpha/2}(\varepsilon-u)^{-\alpha/2}\mathrm{d}u\int_0^\infty\left(r^2+|(v,x)-u|^2\right)^{\frac{\alpha-d}{2}}r^{d-2}\mathrm{d}r\\
&= \frac{k_{\alpha,d}\pi^{(d-2)/2}}{\Gamma((d-2)/2)}\int_{-\varepsilon}^{\varepsilon}(u+\varepsilon)^{-\alpha/2}(\varepsilon-u)^{-\alpha/2}\mathrm{d}u\int_0^\infty\left(w+|(v,x)-u|^2\right)^{\frac{\alpha-d}{2}}w^{\frac{d-3}{2}}\mathrm{d}w \qquad (60)\\
&= \frac{k_{\alpha,d}\pi^{(d-2)/2}\Gamma(\frac{1-\alpha}{2})\Gamma(\frac{d-1}{2})}{\Gamma(\frac{d-2}{2})\Gamma(\frac{d-\alpha}{2})}\int_{-\varepsilon}^{\varepsilon}(u+\varepsilon)^{-\alpha/2}(\varepsilon-u)^{-\alpha/2}|(v,x)-u|^{\alpha-1}\mathrm{d}u\\
&= \frac{k_{\alpha,d}\pi^{(d-2)/2}\Gamma(\frac{1-\alpha}{2})\Gamma(\frac{d-1}{2})}{\Gamma(\frac{d-2}{2})\Gamma(\frac{d-\alpha}{2})}\int_{-1}^{1}(1+w)^{-\alpha/2}(1-w)^{-\alpha/2}|\varepsilon^{-1}(v,x)-w|^{\alpha-1}\mathrm{d}w, \qquad (61)
\end{aligned}
$$

where, in the penultimate equality, we used a classical representation of the Beta function (see formula 3.191.2 in [15]), which tells us that, for any $\mathrm{Re}(\nu) > \mathrm{Re}(\gamma) > 0$ and $z > 0$,

$$
\int_0^\infty (y+z)^{-\nu}y^{\gamma-1}\mathrm{d}y = z^{\gamma-\nu}\frac{\Gamma(\nu-\gamma)\Gamma(\gamma)}{\Gamma(\nu)},
$$

and in the final equality, we have changed variables using $w = \varepsilon u$. Next, we observe that $|\varepsilon^{-1}(v,x)| \le 1$ on account of the fact that $x \in \mathsf{D}_\varepsilon \subseteq \mathbb{H}_\varepsilon^{d-1}$. Now choose $k_{\alpha,d}$, so that

$$
\frac{k_{\alpha,d}\pi^{(d-2)/2}\Gamma(\frac{1-\alpha}{2})\Gamma(\frac{d-1}{2})}{\Gamma(\frac{d-2}{2})\Gamma(\frac{d-\alpha}{2})} = 1. \qquad (62)
$$

We can now appeal directly to (18) to deduce that, for $x \in \mathsf{D}_\varepsilon$

$$
U\rho_\varepsilon^{(1)}(x) = 1. \qquad (63)
$$

In the spirit of (33)–(35), it now follows that, for $x \notin \mathsf{D}$ and ε sufficiently small,

$$
U\rho_\varepsilon^{\mathsf{D}}(x) \le \mathbb{P}_x(\tau_{\mathsf{D}_\varepsilon} < \infty).
$$

So that

$$\liminf_{\varepsilon\to 0} U\rho_\varepsilon^{\mathsf{D}}(x) \le \liminf_{\varepsilon\to 0} \mathbb{P}_x(\tau_{\mathsf{D}_\varepsilon} < \infty), \qquad x \notin \mathsf{D}. \tag{64}$$

Now we turn our attention to (59). Noting that when $x \in \mathsf{D}_\varepsilon$, $|x - y| > \delta$ for $y \in \hat{\mathsf{D}}_\varepsilon^\delta$, we have, for all $x \in \mathsf{D}_\varepsilon$,

$$\begin{aligned}
U\rho_{\varepsilon,\delta}^{(2)}(x) &= k_{\alpha,d} \int_{\hat{\mathsf{D}}_\varepsilon^\delta} |x - y|^{\alpha-d}((v, y) + \varepsilon)^{-\alpha/2}(\varepsilon - (v, y))^{-\alpha/2}\ell_d(\mathrm{d}y) \\
&\le k_{\alpha,d}\delta^{\alpha-d} \int_{-\varepsilon}^{\varepsilon} (u + \varepsilon)^{-\alpha/2}(\varepsilon - u)^{-\alpha/2}\mathrm{d}u \int_{\mathsf{D}^\delta((v,x))} \ell_{d-1}(\mathrm{d}\hat{y}) \\
&\le \delta^{\alpha-d}k_{\alpha,d}\ell_{d-1}(\mathsf{D}^\delta) \int_{-\varepsilon}^{\varepsilon} (u + \varepsilon)^{-\alpha/2}(\varepsilon - u)^{-\alpha/2}\mathrm{d}u \\
&= \delta^{\alpha-d}\varepsilon^{1-\alpha}k_{\alpha,d}\ell_{d-1}(\mathsf{D}^\delta)2^{1-\alpha}\frac{\Gamma((2-\alpha)/2)^2}{\Gamma(2-\alpha)},
\end{aligned}$$

where we have used the calculation in (31) in the final equality. Choosing $\delta = \delta(\varepsilon) = \varepsilon^{(1-\alpha)/2(d-\alpha)}$, and noting that $\ell_{d-1}(\mathsf{D}^\delta)$ is uniformly bounded from above by an unimportant constant for e.g. all $\delta < 1$ (thanks to the assumption that $\ell_{d-1}(\mathsf{D}) < \infty$), we see that

$$\lim_{\varepsilon\to 0} \sup_{x\in\mathsf{D}_\varepsilon} U\rho_{\varepsilon,\delta(\varepsilon)}^{(2)}(x) = 0.$$

In a similar spirit to (36)–(38), we now have that

$$\limsup_{\varepsilon\to 0} \varepsilon^{\alpha-1}U\rho_\varepsilon^{\mathsf{D}^{\delta(\varepsilon)}}(x) \ge \limsup_{\varepsilon\to 0} \varepsilon^{\alpha-1}\mathbb{P}_x(\tau_{\mathsf{D}_\varepsilon} < \infty), \qquad x \notin \mathsf{D}. \tag{65}$$

Matching up the left-hand side of (64) with that of (65), we can proceed in a similar fashion to (41)–(42), leading to the statement of Theorem 4(i) as promised. The calculation is based around the fact that

$$\begin{aligned}
\lim_{\varepsilon\to 0} \varepsilon^{\alpha-1}U\rho_\varepsilon^{\mathsf{D}}(x) &= \lim_{\varepsilon\to 0} k_{\alpha,d}\varepsilon^{\alpha-1} \int_{\mathsf{D}_\varepsilon} |x - y|^{\alpha-d}((v, y) + \varepsilon)^{-\alpha/2}(\varepsilon - (v, y))^{-\alpha/2}\ell_d(\mathrm{d}y) \\
&= \lim_{\varepsilon\to 0} k_{\alpha,d}\varepsilon^{\alpha-1} \int_{-\varepsilon}^{\varepsilon} (u + \varepsilon)^{-\alpha/2}(\varepsilon - u)^{-\alpha/2}\mathrm{d}u \int_{\mathsf{D}(u)} |x - \hat{y}|^{\alpha-d}\ell_{d-1}(\mathrm{d}\hat{y}) \\
&= k_{\alpha,d}2^{1-\alpha}\frac{\Gamma((2-\alpha)/2)^2}{\Gamma(2-\alpha)} \int_{\mathsf{D}} |x - y|^{\alpha-d}\ell_{d-1}(\mathrm{d}y) \\
&= 2^{1-\alpha}\pi^{-(d-2)/2}\frac{\Gamma(\frac{d-2}{2})\Gamma(\frac{d-\alpha}{2})\Gamma(\frac{2-\alpha}{2})^2}{\Gamma(\frac{1-\alpha}{2})\Gamma(\frac{d-1}{2})\Gamma(2-\alpha)} \int_{\mathsf{D}} |x - y|^{\alpha-d}\ell_{d-1}(\mathrm{d}y),
\end{aligned} \tag{66}$$

where we have used the calculation in (31) and (62) in the third equality.

(ii) The setting $\alpha = 1$ requires yet another delicate handling of the associated potentials. Given that all the main ideas are now present in the paper, we simply lay out the key points of the proof, leaving the remaining detail for the reader.

Our calculations begin in the same way as in part (i), in particular, we work with the core measure ρ_ε as in (57), albeit (as with Theorem 2(ii)) replacing $k_{1,d}$ by $k_{1,d}/|\log\varepsilon|$, to be used in the constructions (58) and (59). An immediate complication we have is in evaluating $U\rho_\varepsilon^{(1)}(x)$, for $x \in \mathsf{D}_\varepsilon$, can be seen when we pick up the computations for part (i) at (60). Indeed, at that point, we are confronted with the integral

$$\int_0^\infty \left(w + |(v, x) - u|^2\right)^{\frac{1-d}{2}} w^{\frac{d-3}{2}} \mathrm{d}w = \infty.$$

The solution to this is to adjust the core measure ρ_ε as follows. Since D is bounded, we can choose an $R > 0$ sufficiently large that, $\mathsf{D} \subset \mathbb{S}^{d-2}(0, R) := \{y \in \mathbb{H}^{d-1} : |y| \le R\}$ strictly contains D. Denote $\mathbb{S}_\varepsilon^{d-2}(0, R) = \{x \in \mathbb{H}_\varepsilon^{d-1} : \hat{x} \in \mathbb{S}^{d-2}(0, R)\}$, where $\hat{x}$ is the orthogonal projection of x on to $\mathbb{H}^{d-1}$. Suppose we now make a slight adjustment and replace ρ_ε by

$$\rho_\varepsilon(\mathrm{d}y) = \frac{k_{1,d,R}}{|\log\varepsilon|}((v, y) + \varepsilon)^{-\alpha/2}(\varepsilon - (v, y))^{-\alpha/2}\mathbf{1}_{(y\in\mathbb{S}_\varepsilon^{d-2}(0,R))}\ell_d(\mathrm{d}y),$$

for an appropriate choice of $k_{1,d,R}$. We may now continue the argument from (60) with the calculation

$$|\log\varepsilon| U\rho_\varepsilon^{(1)}(x) = \frac{k_{1,d,R}\pi^{(d-2)/2}}{\Gamma((d-2)/2)} \int_{-\varepsilon}^{\varepsilon} (u + \varepsilon)^{-1/2}(\varepsilon - u)^{-1/2}\mathrm{d}u \int_0^R \left(w + |(v, x) - u|^2\right)^{\frac{1-d}{2}} w^{\frac{d-3}{2}} \mathrm{d}w. \tag{67}$$

Denote by $I(R, \varepsilon, x)$ the right-hand side of 67, ensuring that ε is small enough that $\varepsilon \ll R$.

Appealing to (78),

$$\begin{aligned} I(R, \varepsilon, x) &= \frac{k_{1,d,R}\pi^{(d-2)/2}}{\Gamma((d-2)/2)} \int_{-\varepsilon}^{\varepsilon} (u + \varepsilon)^{-1/2}(\varepsilon - u)^{-1/2}\mathrm{d}u \int_0^R \left(w + |(v, x) - u|^2\right)^{\frac{1-d}{2}} w^{\frac{d-3}{2}} \mathrm{d}w \\ &= \frac{k_{1,d,R}\pi^{(d-2)/2}}{\Gamma((d-2)/2)} \int_{-\varepsilon}^{\varepsilon} (\varepsilon^2 - u^2)^{-1/2}|(v, x) - u|^{1-d}\mathrm{d}u \int_0^R \left(\frac{w}{|(v, x) - u|^2} + 1\right)^{\frac{1-d}{2}} w^{\frac{d-3}{2}} \mathrm{d}w \\ &= \frac{k_{1,d,R}\pi^{(d-2)/2}}{\Gamma((d-2)/2)} \int_{-\varepsilon}^{\varepsilon} (\varepsilon^2 - u^2)^{-1/2}|(v, x) - u|^{1-d} \\ &\qquad \frac{R^{(d-1)/2}}{(d-1)/2} {}_2F_1\left(\frac{d-1}{2}, \frac{d-1}{2}; \frac{d+1}{2}; -\frac{R}{|(v, x) - u|^2}\right) \mathrm{d}u, \end{aligned}$$

where we have used the identity in (79). One of the many identities for hypergeometric functions, see [2], offers us the growth condition, for $c-a \in \mathbb{N}$, as $|z| \to \infty$,

$$ {}_2F_1(a,a;c;z) \sim \frac{\Gamma(c)(\log(-z) - \psi(c-a) - \psi(a) - 2\gamma)(-z)^{-a}}{\Gamma(a)(c-a-1)!} + \frac{\Gamma(c)2(-z)^{-c}}{\Gamma(a)^2((c-a)!)^2}, \tag{68} $$

where γ is an unimportant constant and $\psi(z) = \Gamma'(z)/\Gamma(z)$ is the di-gamma function. In the spirit of previous calculations, we can thus find to leading order, uniformly over $x \in \mathsf{D}_\varepsilon$,

$$ U\rho_\varepsilon^{(1)}(x) \sim 2\frac{\pi^{d/2}k_{1,d,R}}{\Gamma((d-2)/2)}, \tag{69} $$

which remarkably does not depend on R. This means we should choose the constant

$$ k_{1,d,R} = \frac{\Gamma((d-2)/2)}{2\pi^{d/2}} $$

for this asymptotic to serve our purpose.

At this point in the proof, recalling the fundamental decomposition (58), it is worth bringing in the term $U\mu_\varepsilon^{(2)}$ and noting that one can compute with relatively coarse estimates that

$$ \sup_{x \in \mathsf{D}_\varepsilon} \left| U\rho_\varepsilon^{(2)}(x) \right| \leq \frac{C}{|\log \varepsilon|}, $$

for some unimportant constant $C > 0$. Together with (69), in a calculation similar to (66) we can put the pieces together to get the asymptotic, for $x \notin \mathsf{D}$ and ε sufficiently small,

$$ \begin{aligned} \lim_{\varepsilon \to 0} |\log \varepsilon|\, \mathbb{P}_x(\tau_{\mathsf{D}_\varepsilon} < \infty) &= \lim_{\varepsilon \to 0} |\log \varepsilon|\, U\rho_\varepsilon^{\mathsf{D}}(x) \\ &= \lim_{\varepsilon \to 0} \frac{\Gamma((d-2)/2)}{2\pi^{d/2}} \int_{\mathsf{D}_\varepsilon} |x-y|^{1-d} (\varepsilon^2 - (v,y)^2)^{-1/2} \ell_d(\mathrm{d}y) \\ &= \lim_{\varepsilon \to 0} \frac{\Gamma((d-2)/2)}{\pi^{d/2}} \int_{-\varepsilon}^{\varepsilon} (\varepsilon^2 - u^2)^{-1/2} \mathrm{d}u \int_{\mathsf{D}(u)} |x - \hat{y}|^{1-d} \ell_{d-1}(\mathrm{d}\hat{y}) \\ &= \frac{\Gamma((d-2)/2)}{\pi^{(d-2)/2}} M_{\mathsf{D}}(x). \end{aligned} \tag{70} $$

The proof is complete. □

Appendix: Hypergeometric Identities

We work with the standard definition for the hypergeometric function,

$$ {}_2F_1(a,b,c;z) = \sum_{n=0}^{\infty} \frac{(a)_n(b)_n}{(c)_n}\frac{z^n}{n!}, \qquad |z|<1. $$

Of the many identities for hypergeometric functions, we need the following:

$$ {}_2F_1(a,b,c;z) = \frac{\Gamma(c)\Gamma(a+b-c)}{\Gamma(a)\Gamma(b)}(1-z)^{c-a-b}{}_2F_1(c-a,c-b,1+c-a-b;1-z) \\ + \frac{\Gamma(c)\Gamma(c-a-b)}{\Gamma(c-a)\Gamma(c-b)}{}_2F_1(a,b,a+b-c+1;1-z), \tag{71} $$

for $c-a-b\notin\mathbb{Z}$. Hence, thanks to continuity,

$$ \lim_{\varepsilon\to 0}\sup_{r\in[1-\varepsilon,1]}\Big|{}_2F_1\Big(\frac{d-\alpha}{2},1-\frac{\alpha}{2};\frac{d}{2};r^2\Big) \\ -\frac{\Gamma(d/2)\Gamma(1-\alpha)}{\Gamma((d-\alpha)/2)\Gamma((2-\alpha)/2)}(1-r^2)^{\alpha-1}-\frac{\Gamma(d/2)\Gamma(\alpha-1)}{\Gamma(\alpha/2)\Gamma((d+\alpha-2)/2)}\Big| = 0. \tag{72} $$

We will need to apply a similar identity to (71) but for the setting that $c-a-b=0$, which violates the assumption behind (71). In that case, we need to appeal to the formula

$$ {}_2F_1(a,b,a+b,z) = \frac{\Gamma(a+b)}{\Gamma(a)\Gamma(b)}\Big(\sum_{k=0}^{\infty}\frac{(a)_k(b)_k}{(k!)^2}(2\psi(k+1)-\psi(a+k)-\psi(b+k))(1-z)^k \\ -\log(1-z)\,{}_2F_1(a,b,1,1-z)\Big), \tag{73} $$

for $|1-z|<1$ where the di-gamma function $\psi(z)=\Gamma'(z)/\Gamma(z)$ is defined for all $z\neq -n, n\in\mathbb{N}$.

Again, thanks to continuity, we can write

$$ \lim_{\varepsilon\to 0}\sup_{r\in[1-\varepsilon,1]}\Big|{}_2F_1\Big(\frac{d-1}{2},\frac{1}{2};\frac{d}{2};r^2\Big)+\frac{\Gamma(d/2)}{\Gamma((d-1)/2)\Gamma(1/2)}\log(1-r^2) \\ -\frac{2\Gamma(d/2)(\psi(1)-\psi((d-1)/2)-\psi(1/2))}{\Gamma((d-1)/2)\Gamma(1/2)}\Big| = 0. \tag{74} $$

A second identity that is needed is the following combination formula, which states that for any $|z| < 1$, we have

$$
{}_2F_1(a, b; c; z) = \frac{\Gamma(b-a)\Gamma(c)}{\Gamma(c-a)\Gamma(b)}(-z)^{-a}{}_2F_1\left(a, a-c+1; a-b+1; \frac{1}{z}\right) + \frac{\Gamma(a-b)\Gamma(c)}{\Gamma(c-b)\Gamma(a)}(-z)^{-b}{}_2F_1\left(b-c+1, b; -a+b+1; \frac{1}{z}\right). \tag{75}
$$

Its proof can be found, for example at [1]. In the main body of the text, we use this identity for the setting that $a = \alpha/2$, $b = \alpha$ and $c = 1 + \alpha/2$. This gives us the identity

$$
\begin{aligned}
{}_2F_1\left(\frac{\alpha}{2}, \alpha; 1+\frac{\alpha}{2}; z\right) &= \frac{\Gamma(\alpha/2)\Gamma((2+\alpha)/2)}{\Gamma(\alpha)}(-z)^{-\alpha/2}{}_2F_1\left(\alpha/2, 0; 1-\alpha/2; \frac{1}{z}\right) \\
&\quad + \frac{\Gamma(-\alpha/2)\Gamma((2+\alpha)/2)}{\Gamma((2-\alpha)/2)\Gamma(\alpha/2)}(-z)^{-\alpha}{}_2F_1\left(\alpha/2, \alpha; 1+\alpha/2; \frac{1}{z}\right) \\
&= \frac{\Gamma(\alpha/2)\Gamma((2+\alpha)/2)}{\Gamma(\alpha)}(-z)^{-\alpha/2} \\
&\quad - (-z)^{-\alpha}{}_2F_1\left(\alpha/2, \alpha; 1+\alpha/2; \frac{1}{z}\right),
\end{aligned}
$$

where we have used the recursion formula for gamma functions twice in the final equality. This allows us to come to rest at the following useful identity

$$
(-z)^{-\alpha/2}{}_2F_1\left(\alpha/2, \alpha; 1+\alpha/2; \frac{1}{z}\right) + (-z)^{\alpha/2}{}_2F_1\left(\frac{\alpha}{2}, \alpha; 1+\frac{\alpha}{2}; z\right) = \frac{\Gamma(\alpha/2)\Gamma((2+\alpha)/2)}{\Gamma(\alpha)}. \tag{76}
$$

We are also interested in integral formulae, for which the hypergeometric function is used to evaluate an integral. The first is aversion of formula 3.665(2) in [15] which states that, for any $0 < |a| < r$ and $\nu > 0$, as

$$
\int_0^{\pi} \frac{\sin^{d-2}\phi}{(a^2 + 2ar\cos\phi + r^2)^{\nu}} \mathrm{d}\phi = \frac{1}{r^{2\nu}} B\Big(\frac{d-1}{2}, \frac{1}{2}\Big)\, {}_2F_1\Big(\nu, \nu - \frac{d}{2} + 1; \frac{d}{2}; \frac{a^2}{r^2}\Big), \tag{77}
$$

where $B(a, b) = \Gamma(a)\Gamma(b)/\Gamma(a+b)$ is the Beta function. The second is formula 3.197.8 in [15], which states that, for $Re(\mu) > 0$, $Re(\nu) > 0$ and $|\arg(u/\beta)| < \pi$,

$$
\int_0^u x^{\nu-1}(u-x)^{\mu-1}(x+\beta)^{\lambda}\mathrm{d}x = \beta^{\lambda}u^{\mu+\nu-1}B(\mu, \nu){}_2F_1\left(-\lambda, \nu; \mu+\nu; -\frac{u}{\beta}\right). \tag{78}
$$

The third is 3.194.1 of [15] and states that, for $|\arg(1+\beta u)| > \pi$ and $Re(\mu) > 0$, $Re(\nu) > 0$,

$$\int_0^u x^{\mu-1}(1+\beta x)^{-\nu}\mathrm{d}x = \frac{u^\mu}{\mu}{}_2F_1(\nu, \nu-\mu; 1+\mu; -\beta u), \tag{79}$$

where ${}_2F_1$ in the above identity is understood as its analytic extension in the event that $|\beta u| > 1$.

Acknowledgments TS acknowledges support from a Schlumberger Faculty of the Future award. SP acknowledges support from the Royal Society as a Newton International Fellow Alumnus (AL201023) and UNAM-DGAPA-PAPIIT grant no. IA103220.

References

1. https://functions.wolfram.com/HypergeometricFunctions/Hypergeometric2F1/17/02/09/
2. https://functions.wolfram.com/HypergeometricFunctions/Hypergeometric2F1/06/01/05/01/05/0001/
3. Alili, L., Chaumont, L., Graczyk, P., Zak, T.: Inversion, duality and Doob h-transforms for self-similar Markov processes. Electron. J. Probab. **22**(20), 18 (2017)
4. Bertoin, J.: Lévy Processes. Cambridge University Press (1996)
5. Bliedtner, J., Hansen, W.: Potential Theory. Universitext. Springer, Berlin (1986)
6. Blumenson, L.: A derivation of n-dimensional spherical coordinates. Am. Math. Mon. **67**, 63–66 (1960)
7. Blumenthal, R.M., Getoor, R.K.: Markov Processes and Potential Theory. Pure and Applied Mathematics, Vol. 29. Academic Press, New York/London (1968)
8. Bogdan, K., Żak, T.: On Kelvin transformation. J. Theoret. Probab. **19**, 89–120 (2006)
9. Caballero, M.E., Chaumont, L.: Conditioned stable L évy processes and Lamperti representation. J. Appl. Probab. **43**(4), 967–983 (2006)
10. Caballero, M.E., Pardo, J.C., Pérez, J.L.: Explicit identities for Lévy processes associated to symmetric stable processes. Bernoulli **17**, 1, 34–59 (2011)
11. Chaumont, L.: Conditionings and path decompositions for Lévy processes. Stoch. Proc. Appl. **64**, 39–54 (1996)
12. Chaumont, L., Pantí, H., Rivero, V.: The Lamperti representation of real-valued self-similar Markov processes. Bernoulli **19**(5B), 2494–2523 (2013)
13. Chung, K.L., Walsh, J.B.: Markov processes, Brownian motion, and time symmetry. Second edition. Grundlehren der Mathematischen Wissenschaften (Fundamental Principles of Mathematical Sciences), 249. Springer, New York (2005)
14. Döring, L., Weissman, P.: Stable processes conditioned to hit an interval continuously from the outside. Bernoulli **26**(2), 980–1015 (2020)
15. Gradshteyn, I.S., Ryzhik, I.M.: Table of Integrals, Series and Products. Academic Press, New York/London (1965)
16. Kyprianou, A.E.: Stable processes, self-similarity and the unit ball ALEA Lat. Am. J. Probab. Math. Stat. **15**, 617–690 (2018)
17. Kyprianou, A.E., Palau, S., Saizmaa, T.: Attraction to and repulsion from a subset of the unit sphere for isotropic stable Lévy processes arXiv:1911.05867 [math.PR] (2020)
18. Kyprianou, A.E., Pardo, J.C.: Stable Lévy Processes via Lamperti-Type Representations. Cambridge University Press (2021)

19. Kyprianou, A.E., Rivero V., Satitkanitkul, W.: Conditioned real self-similar Markov processes. Stoch. Proc. Appl. **129**, 954–977 (2019)
20. Kyprianou, A.E., Rivero V., Sengul, B.: Conditioning subordinators embedded in Markov processes. Stoch. Proc. Appl. **127**, 1234–1254 (2017)
21. Lebedev, N.N.: Special Functions and their Applications. Prentice-Hall (1965)
22. Luks, T.: Boundary behavior of α-harmonic functions on the complement of the sphere and hyperplane. Potential Anal. **39**, 29–67 (2013)
23. Nagasawa, M.: Time reversions of Markov processes. Nagoya. Math. J. **24**, 177–204 (1964)
24. Osypchuk, M.M., Portenko, M.I.: On the distribution of a rotationally invariant α-stable process at the hitting time of a given hyperplane. Dopov. Nac. akad. nauk Ukr. **12**, 14–20 (2018)
25. Port, S.C.: First hitting distribution of a sphere for symmetric stable processes. Trans. Am. Math. Soc. **135**, 115–125 (1969)
26. Profeta, C., Simon, T.: On the harmonic measure of stable processes. Séminaire de Probabilités **XLVIII**, 325–345, Lecture Notes in Mathematics, 2168. Springer (2016)

Angular Asymptotics for Random Walks

Alejandro López Hernández and Andrew R. Wade

Abstract We study the set of directions asymptotically explored by a spatially homogeneous random walk in d-dimensional Euclidean space. We survey some pertinent results of Kesten and Erickson, make some further observations, and present some examples. We also explore links to the asymptotics of one-dimensional projections, and to the growth of the convex hull of the random walk.

Keywords Random walk · Recurrent set · Spherical asymptotics · Asymptotic direction · Convex hull · Exceptional projections

1 Introduction

In this paper we examine some aspects of the way in which a random walk in d dimensions explores space, specifically through the limit points of the trajectory projected onto the sphere, and related questions concerning the growth of the convex hull of the walk. We ask, roughly speaking, in which directions does the walk grow without bound?

Let $d \in \mathbb{N} := \{1, 2, 3, \ldots\}$. Let $X, X_1, X_2, \ldots$ be i.i.d. random variables in $\mathbb{R}^d$, and define the associated random walk $(S_n; n \in \mathbb{Z}_+)$ by $S_0 := \mathbf{0}$ and $S_n = \sum_{k=1}^{n} X_k$ for $n \geq 1$; here and subsequently $\mathbf{0}$ is the origin in $\mathbb{R}^d$ and $\mathbb{Z}_+ := \{0, 1, 2, \ldots\}$. We suppose throughout that S_n is genuinely d-dimensional, i.e., supp X is not contained in a $(d-1)$-dimensional subspace of $\mathbb{R}^d$.

Denote by $\mathbf{x} \cdot \mathbf{y}$ the Euclidean inner product of vectors $\mathbf{x}, \mathbf{y} \in \mathbb{R}^d$, and by $\| \cdot \|$ the Euclidean norm on $\mathbb{R}^d$. Set $\mathbb{S}^{d-1} := \{\mathbf{x} \in \mathbb{R}^d : \|\mathbf{x}\| = 1\}$. For $\mathbf{x} \in \mathbb{R}^d \setminus \{\mathbf{0}\}$ define $\hat{\mathbf{x}} := \mathbf{x}/\|\mathbf{x}\|$; we also set $\hat{\mathbf{0}} := \mathbf{0}$. We view vectors in $\mathbb{R}^d$ as column vectors where necessary. Whenever the appropriate expectation exists, we write $\mu := \mathbb{E}X$, so $\mu \in \mathbb{R}^d$ is the mean drift vector of the random walk.

A. L. Hernández · A. R. Wade (✉)
Department of Mathematical Sciences, Durham University, Durham, UK
e-mail: alejandro.lopez-hernandez20@imperial.ac.uk; andrew.wade@durham.ac.uk

L. Chaumont, A. E. Kyprianou (eds.), *A Lifetime of Excursions Through Random Walks and Lévy Processes*, Progress in Probability 78,
https://doi.org/10.1007/978-3-030-83309-1_17

In Sect. 2 we look at the limit points in $\mathbb{S}^{d-1}$ of the sequence $\hat{S}_0, \hat{S}_1, \ldots$, drawing on closely related work of Kesten and Erickson [9–11, 19]. In particular, an adaptation of an idea of Kesten shows that the limit set is a.s. equal to a deterministic closed $\mathcal{D} \subseteq \mathbb{S}^{d-1}$ (see Theorem 2.1). In Sect. 3 we make more explicit the connection to the work of Kesten and Erickson [9–11, 19] on limit sets graded by particular speeds of growth. Section 4 considers the special case where $\mathcal{D}$ has a single element, in which the walk is transient with a limiting direction. In Sect. 5 we make some observations about the case where the walk has increments with mean zero (zero drift). Section 6 presents an argument due to Erickson which shows that an arbitrary closed $\mathcal{D} \subseteq \mathbb{S}^{d-1}$ can be achieved as the limit set by constructing a random walk with suitable heavy-tailed increments (Theorem 6.1). In Sect. 7 we introduce some relevant convexity ideas. Section 8 turns to considering the asymptotics of the one-dimensional projections $S_n \cdot \mathbf{u}$, $\mathbf{u} \in \mathbb{S}^{d-1}$. Section 9 studies the convex hull of the trajectory, and draws some connections to the preceding sections. In Sect. 10 we present some examples. These illustrate, for instance, that while walks whose increments are symmetric and have zero mean must have $\mathcal{D} = \mathbb{S}^{d-1}$ when $d = 2$ (Proposition 5.2), for $d \geq 4$ the set $\mathcal{D}$ can have measure zero in $\mathbb{S}^{d-1}$ (Example 10.3).

We make a few historical comments. As observed by Blackwell, and Chung and Derman (see [15, p. 493] and [2, p. 658]), it is a consequence of the Hewitt–Savage zero–one law that $\mathbb{P}(S_n \in A \text{ i.o.}) \in \{0, 1\}$ for any Borel set $A \subseteq \mathbb{R}^d$. Those authors raised the question of classifying sets A accordingly for a given random walk (see e.g. [5, p. 447]). For bounded sets A containing the origin in their interior, the question is that of recurrence vs. transience, and is answered by Chung and Fuchs [6].

Attention focused on determining infinite sets A visited infinitely often by (transient) random walks on $\mathbb{Z}^d$, $d \geq 3$, most notably for the case where the random walk converges to Brownian motion, where a classification of recurrent sets A is available in the form of 'Wiener's test': for the case of simple symmetric random walk, see [3, 4, 17], for bounded and symmetric increments, see [23, §6.5], and for increments with zero mean and finite second moments, see [18, 30, 31]. Wiener's test and its generalizations [4, 22, 27] give analytic criteria in terms of the capacity of A or Green's functions of the walk. An early paper of Doney [7] showed that Wiener's test can yield very useful information, but, according to Spitzer, "in general the computations are prohibitively difficult" [30, p. 320]. The present paper addresses questions related to the transience or recurrence of sets A that are cones or half-spaces.

2 Recurrent Directions

We say $\mathbf{u} \in \mathbb{S}^{d-1}$ is a *recurrent direction* for S_n if the sequence $\hat{S}_n$ has an accumulation point at $\mathbf{u}$, i.e., if $\hat{S}_n$ has $\mathbf{u}$ as a subsequential limit. Let L be the (random) set of all recurrent directions for S_n; equivalently,

$$L := \{\mathbf{u} \in \mathbb{S}^{d-1} : \liminf_{n\to\infty} \|\hat{S}_n - \mathbf{u}\| = 0\}.$$

Note that in L the possible accumulation point at $\mathbf{0}$ is excluded. Also define

$$\mathcal{D} := \{\mathbf{u} \in \mathbb{S}^{d-1} : \liminf_{n\to\infty} \|\hat{S}_n - \mathbf{u}\| = 0, \text{ a.s.}\},$$

i.e., the set of all a.s. recurrent directions for S_n.

For $d = 1$, ruling out the degenerate case where $\mathbb{P}(X = 0) = 1$, the well known trichotomy (see e.g. [8, Theorem 4.1.2]) states that either (i) $S_n \to +\infty$, a.s., (ii) $S_n \to -\infty$, a.s., or (iii) $\liminf_{n\to\infty} S_n = -\infty$ and $\limsup_{n\to\infty} S_n = +\infty$, a.s., corresponding to (i) $\mathcal{D} = \{+1\}$, (ii) $\mathcal{D} = \{-1\}$, and (iii) $\mathcal{D} = \{-1, +1\}$ (this latter case includes the case where S_n is recurrent). Our primary interest here is the case $d \geq 2$.

The following result is a consequence of a more general statement of Erickson [9] (see also §3 below), who pointed out that it can be obtained by adapting an argument of Kesten [19] (see also Lemma 1 of [21] for a generalization attributed to Neidhardt). An alternative proof of the fact that L is deterministic could be obtained by appealing to a general zero–one result for random closed sets such as Proposition 1.1.30 of [26], having first established that L is closed.

Theorem 2.1 *The set $\mathcal{D}$ is a non-empty, closed subset of $\mathbb{S}^{d-1}$, and $\mathbb{P}(L = \mathcal{D}) = 1$.*

We work towards the proof of Theorem 2.1. For $\mathbf{u} \in \mathbb{S}^{d-1}$ and $r > 0$, define the set

$$C(\mathbf{u}; r) := \{\mathbf{x} \in \mathbb{R}^d \setminus \{\mathbf{0}\} : \|\hat{\mathbf{x}} - \mathbf{u}\| < r\}$$

and the event

$$A(\mathbf{u}; r) := \{S_n \in C(\mathbf{u}; r) \text{ i.o.}\}.$$

By the Hewitt–Savage zero–one law (see e.g. [8, Theorem 4.1.1]), $\mathbb{P}(A(\mathbf{u}; r)) \in \{0, 1\}$.

Let $B(\mathbf{x}; r) := \{\mathbf{y} \in \mathbb{R}^d : \|\mathbf{x} - \mathbf{y}\| < r\}$ denote the open Euclidean ball centred at $\mathbf{x} \in \mathbb{R}^d$ with radius $r > 0$, and for $\mathbf{u} \in \mathbb{S}^{d-1}$ let $B_s(\mathbf{u}; r) := \mathbb{S}^{d-1} \cap B(\mathbf{u}; r)$. For $A \subseteq \mathbb{R}^d$, we write $\operatorname{cl} A$ for the closure of A in $\mathbb{R}^d$ in the usual topology.

Lemma 2.2 *For any $\mathbf{u} \in \mathbb{S}^{d-1}$ and any $r > 0$, we have*

$$\{L \cap B_s(\mathbf{u}; r) \neq \emptyset\} \subseteq A(\mathbf{u}; r) \subseteq \{L \cap \operatorname{cl} B_s(\mathbf{u}; r) \neq \emptyset\}.$$

Proof First note that

$$A(\mathbf{u}; r) = \{\hat{S}_n \in B_s(\mathbf{u}; r) \text{ i.o.}\}.$$

Hence $A(\mathbf{u}; r)$ implies that $\hat{S}_n \in \operatorname{cl} B_s(\mathbf{u}; r)$ i.o., and since $\operatorname{cl} B_s(\mathbf{u}; r)$ is compact, $\hat{S}_n$ must have an accumulation point in $\operatorname{cl} B_s(\mathbf{u}; r)$. On the other hand, if $\hat{S}_n$ has an accumulation point in $B_s(\mathbf{u}; r)$, then since $B_s(\mathbf{u}; r)$ is open in $\mathbb{S}^{d-1}$ we have $\hat{S}_n \in B_s(\mathbf{u}; r)$ i.o. □

The following continuity property is a key ingredient in the proof of Theorem 2.1.

Lemma 2.3 *Given any sequence* $\mathbf{x}_1, \mathbf{x}_2, \ldots \in \mathbb{S}^{d-1}$, *and any* $\mathbf{u}, \mathbf{v} \in \mathbb{R}^d$,

$$\left| \limsup_{n\to\infty} (\mathbf{x}_n \cdot \mathbf{u}) - \limsup_{n\to\infty} (\mathbf{x}_n \cdot \mathbf{v}) \right| \leq \| \mathbf{u} - \mathbf{v} \|.$$

Proof Suppose that $\mathbf{u}, \mathbf{v} \in \mathbb{R}^d$. Then

$$\begin{aligned} \limsup_{n\to\infty} (\mathbf{x}_n \cdot \mathbf{v}) &\leq \limsup_{n\to\infty} (\mathbf{x}_n \cdot \mathbf{u}) + \limsup_{n\to\infty} (\mathbf{x}_n \cdot (\mathbf{v} - \mathbf{u})) \\ &\leq \limsup_{n\to\infty} (\mathbf{x}_n \cdot \mathbf{u}) + \| \mathbf{v} - \mathbf{u} \|, \end{aligned}$$

since $\| \mathbf{x}_n \| = 1$. With a similar argument in the other direction, we get the result. □

Lemma 2.4 *The set $\mathcal{D}$ is closed in* $\mathbb{S}^{d-1}$.

Proof Note that for any $\mathbf{u} \in \mathbb{S}^{d-1}$,

$$\| \hat{S}_n - \mathbf{u} \|^2 = (\hat{S}_n - \mathbf{u}) \cdot (\hat{S}_n - \mathbf{u}) = 1 + \mathbf{1}\{ S_n \neq \mathbf{0} \} - 2 \hat{S}_n \cdot \mathbf{u},$$

so that

$$\liminf_{n\to\infty} \| \hat{S}_n - \mathbf{u} \| = 0 \text{ if and only if } \limsup_{n\to\infty} (\hat{S}_n \cdot \mathbf{u}) = 1.$$

Thus

$$\mathbb{S}^{d-1} \setminus \mathcal{D} = \left\{ \mathbf{u} \in \mathbb{S}^{d-1} : \limsup_{n\to\infty} (\hat{S}_n \cdot \mathbf{u}) < 1 \text{ a.s.} \right\}.$$

Consider $\mathbf{u} \in \mathbb{S}^{d-1} \setminus \mathcal{D}$. By the Hewitt–Savage theorem, $\limsup_{n\to\infty} (\hat{S}_n \cdot \mathbf{u}) = c$ a.s. for a constant $c < 1$. Lemma 2.3 shows that for any $\mathbf{v} \in \mathbb{S}^{d-1}$ with $\| \mathbf{u} - \mathbf{v} \| \leq \frac{1-c}{2}$, a.s.,

$$\limsup_{n\to\infty} (\hat{S}_n \cdot \mathbf{v}) \leq c + \frac{1-c}{2} = \frac{1+c}{2} < 1,$$

so that $\mathbf{v} \in \mathbb{S}^{d-1} \setminus \mathcal{D}$. Thus $\mathbb{S}^{d-1} \setminus \mathcal{D}$ is open in $\mathbb{S}^{d-1}$. □

Now we can complete the proof of Theorem 2.1.

Proof of Theorem 2.1 We adapt, in part, an argument from the proof of Theorem 1 of [19]. We call a ball $B_s(\mathbf{u}; r)$ *rational* if $\mathbf{u} \in \mathbb{S}^{d-1} \cap \mathbb{Q}^d$ and $r \in \mathbb{Q} \cap (0, \infty)$. Note that $\mathbb{S}^{d-1} \cap \mathbb{Q}^d$ is dense in $\mathbb{S}^{d-1}$, as follows from an argument based on stereographic projection (see e.g. [29]). Let $\mathcal{R}$ denote the (countable) set of all rational balls, and set

$$\mathcal{C} := \{B \in \mathcal{R} : \mathbb{P}(\hat{S}_n \in B \text{ i.o.}) = 1\}.$$

Then since $\mathcal{R}$ is countable, and, by the Hewitt–Savage theorem, $\mathbb{P}(\hat{S}_n \in B \text{ i.o.}) \in \{0, 1\}$ for any $B \in \mathcal{R}$, we have

$$\mathbb{P}(\hat{S}_n \in B \text{ i.o. for all } B \in \mathcal{C} \text{ but for no } B \in \mathcal{R} \setminus \mathcal{C}) = 1. \tag{2.1}$$

Observe that

$$\mathbf{u} \in L \text{ if and only if } \hat{S}_n \in B \text{ i.o. for every } B \in \mathcal{R} \text{ with } \mathbf{u} \in B, \tag{2.2}$$

and so $\mathbf{u} \in \mathcal{D}$ if and only if

$$\mathbb{P}(\hat{S}_n \in B \text{ i.o. for every } B \in \mathcal{R} \text{ with } \mathbf{u} \in B) = 1. \tag{2.3}$$

In particular, if $B \in \mathcal{R}$ contains some $\mathbf{u} \in \mathcal{D}$, then $B \in \mathcal{C}$. With (2.1), this means that

$$\mathbb{P}(\text{for all } \mathbf{u} \in \mathcal{D},\ \hat{S}_n \in B \text{ i.o. for every } B \in \mathcal{R} \text{ with } \mathbf{u} \in B) = 1.$$

Together with (2.2), it follows that $\mathbb{P}(\mathcal{D} \subseteq L) = 1$.

Let $\mathcal{C}_k$ be the set of $B \in \mathcal{C}$ with $\operatorname{diam} B < 1/k$. Let $W_k := \cup \mathcal{C}_k$ and $W := \cap_{k \in \mathbb{N}} W_k$. Then it follows from (2.3) that $\mathbf{u} \in \mathcal{D}$ if and only if for every $k \in \mathbb{N}$ there exists some $B \in \mathcal{C}_k$ with $\mathbf{u} \in B$. That is, $\mathbf{u} \in \mathcal{D}$ if and only if $\mathbf{u} \in W$, i.e., $\mathcal{D} = W$.

Let $\mathcal{R}_k$ be the set of $B \in \mathcal{R}$ with $\operatorname{diam} B < 1/k$. Now let $M_k := \cup\{B \in \mathcal{R}_k : L \cap B \neq \emptyset\}$. Let $B \in \mathcal{R}$. Since B is open in $\mathbb{S}^{d-1}$, we have that $B \cap L \neq \emptyset$ implies that $\hat{S}_n \in B$ i.o. So $M_k \subseteq \cup\{B \in \mathcal{R}_k : \hat{S}_n \in B \text{ i.o.}\}$. Hence by (2.1) we have that $\mathbb{P}(M_k \subseteq \cup \mathcal{C}_k) = 1$, i.e., $\mathbb{P}(M_k \subseteq W_k) = 1$. It follows that $\mathbb{P}(\cap_{k \in \mathbb{N}} M_k \subseteq \mathcal{D}) = 1$. Note that if $\mathbf{u} \in L$, then for all $k \in \mathbb{N}$ we have $B \cap L \neq \emptyset$ for some $B \in \mathcal{R}_k$, so $\mathbf{u} \in M_k$ for all k; hence $L \subseteq \cap_{k \in \mathbb{N}} M_k$ a.s. Hence we conclude that $\mathbb{P}(L \subseteq \mathcal{D}) = 1$.

To prove that $\mathcal{D}$ is non-empty, taking $r = 2$ in Lemma 2.2 shows that $\hat{S}_n$ has at least one accumulation point in L, since $C(\mathbf{u}; 2) = \mathbb{R}^d \setminus \{\mathbf{0}\}$ and, since S_n is genuinely d-dimensional, $S_n \neq \mathbf{0}$ i.o., a.s. □

Here is an alternative characterization of the set $\mathcal{D}$.

Proposition 2.5 *We have that*

$$\mathcal{D} = \{\mathbf{u} \in \mathbb{S}^{d-1} : \mathbb{P}(A(\mathbf{u}; r)) = 1 \text{ for all } r > 0\}.$$

Proof Define the set $\mathcal{D}' = \{\mathbf{u} \in \mathbb{S}^{d-1} : \mathbb{P}(A(\mathbf{u}; r)) = 1 \text{ for all } r > 0\}$. If $\mathbf{u} \in \mathcal{D}'$, then $\mathbb{P}(A(\mathbf{u}; 1/m)) = 1$ for all $m \in \mathbb{N}$, and so $\mathbb{P}(\cap_{m=1}^{\infty} A(\mathbf{u}; 1/m)) = 1$. In particular, $\mathbb{P}(\hat{S}_n \in B_s(\mathbf{u}; 1/m) \text{ i.o. for all } m \in \mathbb{N}) = 1$. In other words, a.s., $\liminf_{n\to\infty} \|\hat{S}_n - \mathbf{u}\| < 1/m$ for all $m \in \mathbb{N}$, and hence $\liminf_{n\to\infty} \|\hat{S}_n - \mathbf{u}\| = 0$, a.s., so $\mathbf{u} \in \mathcal{D}$. Thus $\mathcal{D}' \subseteq \mathcal{D}$.

On the other hand, suppose that $\mathbf{u} \in \mathbb{S}^{d-1} \setminus \mathcal{D}'$. Then there exists $r > 0$ such that $\mathbb{P}(A(\mathbf{u}; r)) < 1$, and, by the Hewitt–Savage theorem, in fact $\mathbb{P}(A(\mathbf{u}; r)) = 0$. Lemma 2.2 shows that $A(\mathbf{u}; r)^{\mathrm{c}} \subseteq \{L \cap B_s(\mathbf{u}; r) = \emptyset\}$ and hence $\mathbb{P}(L \cap B_s(\mathbf{u}; r) = \emptyset) = 1$. In particular, this means that $\mathbb{P}(\mathbf{u} \in L) = 0$ and so $\mathbf{u} \notin \mathcal{D}$. This shows that $\mathcal{D} \subseteq \mathcal{D}'$. □

We next show that the recurrent directions are determined solely by the behaviour of the walk at increasingly large distances from the origin. Define

$$L_\infty := \left\{ \mathbf{u} \in \mathbb{S}^{d-1} : \liminf_{n\to\infty} \left(\frac{1}{1 + \|S_n\|} + \|\hat{S}_n - \mathbf{u}\| \right) = 0 \right\}, \tag{2.4}$$

and

$$\mathcal{D}_\infty := \left\{ \mathbf{u} \in \mathbb{S}^{d-1} : \liminf_{n\to\infty} \left(\frac{1}{1 + \|S_n\|} + \|\hat{S}_n - \mathbf{u}\| \right) = 0, \text{ a.s.} \right\}.$$

In other words, $\mathbf{u} \in L_\infty$ if and only if there exists a (random) subsequence n_k of $\mathbb{Z}_+$ such that both $\lim_{k\to\infty} \|S_{n_k}\| = \infty$ and $\lim_{k\to\infty} \hat{S}_{n_k} = \mathbf{u}$. If $\mathbf{u} \in L_\infty$ we say that $\mathbf{u}$ is an *asymptotic direction* for the random walk. Clearly an asymptotic direction is a recurrent direction, so $\mathbb{P}(L_\infty \subseteq L) = 1$ and $\mathcal{D}_\infty \subseteq \mathcal{D}$.

Proposition 2.6 *If S_n is recurrent, then $\mathcal{D} = \mathcal{D}_\infty = \mathbb{S}^{d-1}$ and $\mathbb{P}(L = L_\infty = \mathbb{S}^{d-1}) = 1$.*

Proof Suppose that S_n is recurrent. Since $\mathcal{D}_\infty \subseteq \mathcal{D}$ and $L_\infty \subseteq L$, it suffices to show that $\mathcal{D}_\infty = \mathbb{S}^{d-1}$ and $\mathbb{P}(L_\infty = \mathbb{S}^{d-1}) = 1$. Proposition A.1 shows that there is some $h \in (0, \infty)$ such that, a.s., for every $\mathbf{x} \in \mathbb{R}^d$, $S_n \in B(\mathbf{x}; h)$ i.o. But for every $\mathbf{u} \in \mathbb{S}^{d-1}$, every $r > 0$, and every $R \in (h, \infty)$, $C(\mathbf{u}; r)$ contains some $B(\mathbf{x}; h)$ with $\|\mathbf{x}\| > 2R$, so that, a.s., for every $\mathbf{u} \in \mathbb{S}^{d-1}$, every $r > 0$, and every $R \in (h, \infty)$, there is a subsequence n_k along which $\|\hat{S}_{n_k} - \mathbf{u}\| < r$ and $\|S_{n_k}\| > R$. This shows that $\mathbb{P}(L_\infty = \mathbb{S}^{d-1}) = 1$, and essentially the same argument implies that $\mathcal{D}_\infty = \mathbb{S}^{d-1}$. □

Corollary 2.7 *If $\mathcal{D} \neq \mathbb{S}^{d-1}$, then S_n is transient.*

The next result says that, a.s., the sets of recurrent and asymptotic directions coincide.

Theorem 2.8 *We have $\mathcal{D}_\infty = \mathcal{D}$, and $\mathbb{P}(L_\infty = \mathcal{D}) = 1$.*

Proof The recurrent case is contained in Proposition 2.6; thus suppose that S_n is transient. Then since $\|S_n\| \to \infty$ a.s., we have that $\mathbb{P}(L = L_\infty) = 1$ and $\mathcal{D} = \mathcal{D}_\infty$. Combined with Theorem 2.1, this gives the result. □

Next we show how a distributional limit gives rise to recurrent directions. Here and elsewhere, '$\xrightarrow{\mathrm{d}}$' denotes convergence in distribution and 'supp' denotes the support of an $\mathbb{R}^d$-valued random variable.

Proposition 2.9

(i) *Suppose that there is a random vector* $\zeta \in \mathbb{S}^{d-1}$ *such that* $\hat{S}_n \xrightarrow{\mathrm{d}} \zeta$ *as* $n \to \infty$. *Then* $\operatorname{supp} \zeta \subseteq \mathcal{D}$.
(ii) *Suppose there is a sequence* a_n *of positive real numbers and a random vector* $\xi \in \mathbb{R}^d$ *with* $\mathbb{P}(\xi = \mathbf{0}) = 0$ *such that* $S_n/a_n \xrightarrow{\mathrm{d}} \xi$ *as* $n \to \infty$. *Then* $\operatorname{supp} \hat{\xi} \subseteq \mathcal{D}$.

Proof For part (i), suppose that $\hat{S}_n \xrightarrow{\mathrm{d}} \zeta$. Then, for a given $\mathbf{u} \in \mathbb{S}^{d-1}$, for all but countably many $\varepsilon > 0$,

$$\begin{aligned}
\mathbb{P}(\|\hat{S}_n - \mathbf{u}\| < \varepsilon \text{ i.o.}) &= \mathbb{P}\left(\bigcap_{n=1}^{\infty} \bigcup_{m=n}^{\infty} \{\|\hat{S}_m - \mathbf{u}\| < \varepsilon\}\right) \\
&= \lim_{n\to\infty} \mathbb{P}\left(\bigcup_{m=n}^{\infty} \{\|\hat{S}_m - \mathbf{u}\| < \varepsilon\}\right) \\
&\geq \lim_{n\to\infty} \mathbb{P}(\|\hat{S}_n - \mathbf{u}\| < \varepsilon) \\
&= \mathbb{P}(\|\zeta - \mathbf{u}\| < \varepsilon),
\end{aligned}$$

which is strictly positive provided $\mathbf{u} \in \operatorname{supp} \zeta$. It follows by the Hewitt–Savage theorem that if $\mathbf{u} \in \operatorname{supp} \zeta$, then $\mathbb{P}(\|\hat{S}_n - \mathbf{u}\| < \varepsilon \text{ i.o.}) = 1$ for all $\varepsilon > 0$, and hence $\mathbf{u} \in \mathcal{D}$.

For part (ii), we have that since $\mathbb{P}(\xi = \mathbf{0}) = 0$, and the function $\mathbf{x} \mapsto \hat{\mathbf{x}}$ is continuous on $\mathbb{R}^d \setminus \{\mathbf{0}\}$, the continuous mapping theorem implies that $\hat{S}_n \xrightarrow{\mathrm{d}} \hat{\xi}$, and then we may apply part (i). □

Here is a sufficient condition for $\mathcal{D} = \mathbb{S}^{d-1}$; if $d = 2$ the walk is recurrent and the result also follows from Proposition 2.6, while if $d \geq 3$ the walk is transient.

Corollary 2.10 *Suppose that* $\mathbb{E}(\|X\|^2) < \infty$ *and* $\mu = \mathbf{0}$. *Then* $\mathcal{D} = \mathbb{S}^{d-1}$.

Proof By assumption and the central limit theorem, $n^{-1/2} S_n$ converges in distribution to a non-degenerate normal distribution. Proposition 2.9 then shows that $\mathbb{S}^{d-1} \subseteq \mathcal{D}$. □

3 Compactification and Growth Rates

Let $\overline{\mathbb{R}^d}$ denote the compactification of $\mathbb{R}^d$ obtained by adjoining the "sphere at ∞". More formally, $\overline{\mathbb{R}^d}$ is the compact metric space obtained by the completion of $\mathbb{R}^d$ with respect to the metric

$$\rho(\mathbf{x}, \mathbf{y}) = \left\| \frac{\mathbf{x}}{1 + \|\mathbf{x}\|} - \frac{\mathbf{y}}{1 + \|\mathbf{y}\|} \right\|.$$

Then we can represent $\overline{\mathbb{R}^d}$ as $\overline{\mathbb{R}^d} = \mathbb{R}^d \cup \mathbb{R}^d_\infty$ where $\mathbb{R}^d_\infty$ is in bijection to $\mathbb{S}^{d-1}$. We write elements of $\mathbb{R}^d_\infty$ as $\infty \cdot \mathbf{u}$ for $\mathbf{u} \in \mathbb{S}^{d-1}$. The metric ρ on $\mathbb{R}^d$ is equivalent to the Euclidean metric, and extended to $\overline{\mathbb{R}^d}$ it is such that $\mathbf{x}_n \in \mathbb{R}^d$ has $\mathbf{x}_n \to \infty \cdot \mathbf{u}$ for $\mathbf{u} \in \mathbb{S}^{d-1}$ if $\|\mathbf{x}_n\| \to \infty$ and $\hat{\mathbf{x}}_n \to \mathbf{u}$.

The set of accumulation points of $S_0, S_1, S_2, \ldots$, taken in $\overline{\mathbb{R}^d}$, thus consists of any accumulation points in $\mathbb{R}^d$ (a.s. there are none if S_n is transient) and accumulation points in $\mathbb{R}^d_\infty$ represented by the set $\mathcal{L}_\infty$ of asymptotic directions, as defined at (2.4).

Erickson [9], generalizing one-dimensional work of Kesten and himself [11, 19], considers a finer graduation of asymptotic directions. For $\alpha \in \mathbb{R}_+$, set

$$\mathcal{L}_\infty^{>\alpha} := \left\{ \mathbf{u} \in \mathbb{S}^{d-1} : \liminf_{n\to\infty} \left(\frac{n^\alpha}{1 + \|S_n\|} + \|\hat{S}_n - \mathbf{u}\| \right) = 0 \right\}.$$

Then $\mathcal{L}_\infty^{>0} = \mathcal{L}_\infty$, while $\mathcal{L}_\infty^{>\alpha_2} \subseteq \mathcal{L}_\infty^{>\alpha_1}$ for any $0 \le \alpha_1 \le \alpha_2 < \infty$. Similarly, set

$$\mathcal{D}_\infty^{>\alpha} := \left\{ \mathbf{u} \in \mathbb{S}^{d-1} : \liminf_{n\to\infty} \left(\frac{n^\alpha}{1 + \|S_n\|} + \|\hat{S}_n - \mathbf{u}\| \right) = 0, \text{ a.s.} \right\}.$$

Roughly speaking, the set $\mathcal{L}_\infty^{>\alpha}$ consists of those directions in which the walk grows at rate faster than n^α. Also for $\alpha > 0$ set

$$\mathcal{A}^\alpha = \left\{ \mathbf{x} \in \mathbb{R}^d : \liminf_{n\to\infty} \left\| n^{-\alpha} S_n - \mathbf{x} \right\| = 0 \right\}, \tag{3.1}$$

and $\mathcal{L}_\infty^\alpha = \{\hat{\mathbf{x}} : \mathbf{x} \in \mathcal{A}^\alpha \setminus \{\mathbf{0}\}\}$. Then $\mathcal{L}_\infty^\alpha \subseteq \mathcal{L}_\infty$ are those asymptotic directions in which the walk grows at rate precisely n^α.

Erickson [9, 10] studies in detail $\mathcal{A}^\alpha$ and $\mathcal{L}_\infty^{>\alpha}$, with particular focus on the case $\alpha = 1$, which has some peculiar features. The version of Theorem 2.1 stated by Erickson [9, p. 802] is that $\mathbb{P}(\mathcal{L}_\infty^{>\alpha} = \mathcal{D}_\infty^{>\alpha}) = 1$, and $\mathcal{D}_\infty^{>\alpha}$ is a closed subset of $\mathbb{S}^{d-1}$.

For $d \ge 3$, the value $\alpha = 1/2$ is special, since a remarkable paper of Kesten [20] shows that $n^{-\alpha}\|S_n\| \to \infty$ for any $\alpha < 1/2$ and any genuinely d-dimensional random walk S_n in $\mathbb{R}^d$, $d \ge 3$. Thus for $d \ge 3$ we have $\mathcal{D}_\infty^{>\alpha} = \mathcal{D}_\infty$ for any $0 \le \alpha < 1/2$.

4 Limiting Direction

By the Hewitt–Savage theorem, $\mathbb{P}(\lim_{n\to\infty} \hat{S}_n \text{ exists}) \in \{0, 1\}$, and if the limit exists, then it is a.s. constant. If $\lim_{n\to\infty} \|S_n\| = \infty$ a.s. and $\lim_{n\to\infty} \hat{S}_n = \mathbf{u}$ a.s. for some $\mathbf{u} \in \mathbb{S}^{d-1}$, we say that S_n is *transient with limiting direction* $\mathbf{u}$.

Lemma 4.1 *Let* $\mathbf{u} \in \mathbb{S}^{d-1}$. *The following are equivalent.*

(i) $\mathcal{D} = \{\mathbf{u}\}$.
(ii) $\lim_{n\to\infty} \hat{S}_n = \mathbf{u}$, *a.s.*
(iii) S_n *is transient with limiting direction* $\mathbf{u}$.

Proof The result will follow from the sequence of implications (iii) $\Rightarrow$ (ii) $\Rightarrow$ (i) $\Rightarrow$ (iii). That (iii) implies (ii) is trivial. If (ii) holds, then clearly $\mathbf{u} \in \mathcal{D}$, and for any $r > 0$ we have $\hat{S}_n \in B_s(\mathbf{u}; r)$ for all but finitely many n. For any $\mathbf{v} \in \mathbb{S}^{d-1} \setminus \{\mathbf{u}\}$, we may choose $r > 0$ sufficiently small so that $B_s(\mathbf{u}; r)$ and $B_s(\mathbf{v}; r)$ are disjoint, so that $\mathbb{P}(\hat{S}_n \in B_s(\mathbf{v}; r) \text{ i.o.}) = 0$, and hence Proposition 2.5 shows that $\mathbf{v} \notin \mathcal{D}$. Thus (i) holds.

Finally, suppose that (i) holds. Then Corollary 2.7 shows that S_n is transient, and in particular $S_n = \mathbf{0}$ only finitely often. By the Hewitt–Savage theorem, $\limsup_{n\to\infty} \|\hat{S}_n - \mathbf{u}\|$ is a.s. constant. If $\mathbf{u}$ is not a limiting direction for the walk, then this constant is strictly positive, so that, for some $\varepsilon > 0$, $\|\hat{S}_n - \mathbf{u}\| \geq \varepsilon$ i.o., a.s. Since the set $\{\mathbf{v} \in \mathbb{S}^{d-1} : \|\mathbf{v} - \mathbf{u}\| \geq \varepsilon\}$ is compact, it follows that $\hat{S}_n$ has an accumulation point $\mathbf{v} \neq \mathbf{u}$, and hence $\mathbf{v} \in \mathcal{D}$, which gives a contradiction. Hence (i) implies (iii). □

The following result is contained in Theorem 1.6.1(i) of [25].

Proposition 4.2 *Suppose that* $\mathbb{E}\|X\| < \infty$. *If* $\mu \neq \mathbf{0}$, *then* $\mathcal{D} = \{\hat{\mu}\}$.

Remark 4.3 If $\mu = \mathbf{0}$ there is no limiting direction: see Proposition 5.1 below.

Proof of Proposition 4.2 The strong law of large numbers (SLLN) shows that $n^{-1}S_n \to \mu$, a.s., and $n^{-1}\|S_n\| \to \|\mu\|$, a.s. If $\mu \neq \mathbf{0}$, then $\|S_n\| \to \infty$, so $S_n \neq \mathbf{0}$ for all but finitely many n, and then

$$\lim_{n\to\infty} \hat{S}_n = \lim_{n\to\infty} \frac{n^{-1}S_n}{n^{-1}\|S_n\|} = \hat{\mu}, \text{ a.s.}$$

□

5 The Zero-Drift Case

In this section we turn to the case where the walk has zero drift, i.e., $\mu = \mathbf{0}$. If $d = 1$, then zero drift implies recurrence, and hence $\mathcal{D} = \{-1, +1\}$ (see e.g. [8, Theorem 4.2.7]). If $\mathbb{E}(\|X\|^2) < \infty$, then Corollary 2.10 shows that $\mathcal{D} = \mathbb{S}^{d-1}$.

Thus the most interesting cases are when $d \geq 2$ and $\mathbb{E}(\|X\|^2) = \infty$. The following result contrasts with Proposition 4.2, and improves on Theorem 1.6.1(ii) of [25].

Proposition 5.1 *Suppose that* $d \geq 2$, $\mathbb{E}\|X\| < \infty$, *and* $\mu = \mathbf{0}$. *Then* $\mathcal{D}$ *is uncountable.*

In the case where $d = 2$, we can say more. For measurable $A \subseteq \mathbb{S}^{d-1}$ we write $|A|$ for the Haar measure of A. Write '$\stackrel{d}{=}$' for equality in distribution; $X \stackrel{d}{=} -X$ means that random variable $X \in \mathbb{R}^d$ has a centrally symmetric distribution.

Proposition 5.2 *Suppose that* $d = 2$, $\mathbb{E}\|X\| < \infty$, *and* $\mu = \mathbf{0}$.

(i) *We have* $|\mathcal{D}| \geq \frac{1}{2}|\mathbb{S}^1|$.
(ii) *If* $X \stackrel{d}{=} -X$, *then* $\mathcal{D} = \mathbb{S}^1$.

Remarks 5.3

(a) Example 10.2 below gives a walk with $d = 2$, $X \stackrel{d}{=} -X$, and $\mathbb{E}\|X\| = \infty$, for which $\mathcal{D}$ has only two elements, so the condition $\mathbb{E}\|X\| < \infty$ in Proposition 5.2 cannot be removed.
(b) Example 10.3 below gives a family of random walks in $\mathbb{R}^d$, $d \geq 4$, for which $\mu = \mathbf{0}$ and $X \stackrel{d}{=} -X$, but $\mathcal{D}$ is a set of measure zero, so in higher dimensions the hypotheses of Proposition 5.2 do not guarantee that $\mathcal{D}$ occupies a positive fraction of the sphere.

For further results in the zero-drift case, see Corollary 9.4 below. In the rest of this section we prove Propositions 5.1 and 5.2.

Lemma 5.4 *Suppose that* $d \geq 2$, $\mathbb{E}\|X\| < \infty$, *and* $\mu = \mathbf{0}$. *Then for every* $\mathbf{u} \in \mathbb{S}^{d-1}$, *there exists* $\mathbf{v} \in \mathcal{D}$ *with* $\mathbf{u} \cdot \mathbf{v} = 0$.

Proof If S_n is recurrent, then the result follows from Proposition 2.6. So suppose that S_n is transient. Fix $\mathbf{u} \in \mathbb{S}^{d-1}$. For $\varepsilon > 0$, let $O_\varepsilon(\mathbf{u}) = \{\mathbf{v} \in \mathbb{S}^{d-1} : |\mathbf{v} \cdot \mathbf{u}| \leq \varepsilon\}$. Since $\mathbb{E}(X \cdot \mathbf{u}) = \mu \cdot \mathbf{u} = 0$, the random walk $S_n \cdot \mathbf{u}$ is recurrent, and $\liminf_{n\to\infty} |S_n \cdot \mathbf{u}| < \infty$. Since S_n is transient we have $\|S_n\| \to \infty$, so that $\liminf_{n\to\infty} |\hat{S}_n \cdot \mathbf{u}| = 0$. In other words, for every $\varepsilon > 0$ we have that for infinitely many $n \in \mathbb{N}$, $\hat{S}_n$ is in the compact set $O_\varepsilon(\mathbf{u})$. Hence $O_\varepsilon(\mathbf{u})$ must contain an element of $\mathcal{D}$. Thus there is a sequence $\mathbf{v}_1, \mathbf{v}_2, \ldots \in \mathcal{D}$ with $|\mathbf{v}_j \cdot \mathbf{u}| \to 0$, and (since $\mathcal{D}$ is compact) this sequence has a subsequence which converges to $\mathbf{v} \in \mathcal{D}$ with $\mathbf{v} \cdot \mathbf{u} = 0$. □

Proof of Proposition 5.1 Suppose, for the purpose of deriving a contradiction, that $\mathcal{D}$ is countable. Set $O(\mathbf{u}) = \{\mathbf{v} \in \mathbb{S}^{d-1} : \mathbf{v} \cdot \mathbf{u} = 0\}$. Then $O = \cup_{\mathbf{u} \in \mathcal{D}} O(\mathbf{u})$ is a countable union of subsets of $\mathbb{S}^{d-1}$ of measure zero (since each $O(\mathbf{u})$ is a copy of $\mathbb{S}^{d-2}$). Thus O is measure zero, and so there exists $\mathbf{v} \in \mathbb{S}^{d-1} \setminus O$. This $\mathbf{v}$ has $\mathbf{v} \cdot \mathbf{u} \neq 0$ for all $\mathbf{u} \in \mathcal{D}$, which contradicts Lemma 5.4. Hence $\mathcal{D}$ cannot be countable. □

To prove Proposition 5.2, we need some additional notation. Let

$$\mathcal{D}_1 := \{\mathbf{u} \in \mathbb{S}^{d-1} : \mathbf{u} \in \mathcal{D}, \ -\mathbf{u} \notin \mathcal{D}\},$$
$$\mathcal{D}_2 := \{\mathbf{u} \in \mathbb{S}^{d-1} : \mathbf{u} \in \mathcal{D}, \ -\mathbf{u} \in \mathcal{D}\},$$
$$\mathcal{C}_1 := \{\mathbf{u} \in \mathbb{S}^{d-1} : \mathbf{u} \notin \mathcal{D}, \ -\mathbf{u} \in \mathcal{D}\} = -\mathcal{D}_1,$$
$$\mathcal{C}_2 := \{\mathbf{u} \in \mathbb{S}^{d-1} : \mathbf{u} \notin \mathcal{D}, \ -\mathbf{u} \notin \mathcal{D}\}.$$

Then $\mathcal{D} = \mathcal{D}_1 \cup \mathcal{D}_2$ and $\mathbb{S}^{d-1} \setminus \mathcal{D} = \mathcal{C}_1 \cup \mathcal{C}_2$.

Lemma 5.5 *Suppose that $d = 2$, $\mathbb{E}\|X\| < \infty$, and $\mu = \mathbf{0}$. Then $\mathcal{C}_2 = \emptyset$.*

Proof Lemma 5.4 shows that for every $\mathbf{u} \in \mathbb{S}^1$, there exists $\mathbf{v} \in \mathbb{S}^1$ such that $\mathbf{u}\cdot\mathbf{v} = 0$ and $\mathbf{v} \in \mathcal{D}$. As $\mathbf{u}$ runs over $\mathbb{S}^1$, the set of $\pm\mathbf{v}$ such that $\mathbf{u} \cdot \mathbf{v} = 0$ runs over the whole of $\mathbb{S}^1$, and so in this case we conclude that for every $\mathbf{u} \in \mathbb{S}^{d-1}$, at least one of $\pm\mathbf{u}$ is in $\mathcal{D}$. Hence $\mathcal{C}_2 = \emptyset$. □

Proof of Proposition 5.2 Note that $|\mathcal{D}| = |\mathcal{D}_1| + |\mathcal{D}_2|$. If Lemma 5.5 applies, then we have $|\mathbb{S}^1 \setminus \mathcal{D}| = |\mathcal{C}_1| = |\mathcal{D}_1|$. Hence $|\mathbb{S}^1| = 2|\mathcal{D}_1| + |\mathcal{D}_2|$, and part (i) follows. If $X \overset{d}{=} -X$, then $\mathcal{D} = -\mathcal{D}$, so $\mathcal{D}_1 = \mathcal{C}_1 = \emptyset$. Thus $\mathbb{S}^{d-1} = \mathcal{D}_2 \cup \mathcal{C}_2$. If Lemma 5.5 applies, then $\mathcal{D} = \mathcal{D}_2 = \mathbb{S}^1$, giving part (ii). □

6 An Arbitrary Set of Recurrent Directions

We know from Theorem 2.1 that the set $\mathcal{D}$ is closed. The aim of this section is to show that there are, in general, no other restrictions on $\mathcal{D}$: it can be an arbitrary closed subset of the sphere. This result is essentially due to Erickson [10, pp. 508–510]; we reproduce the argument here.

Theorem 6.1 *Let A be a non-empty closed subset of $\mathbb{S}^{d-1}$. Suppose that the increment distribution of the random walk is given by $X = Q\xi$ where $Q \in \mathbb{S}^{d-1}$ and $\xi \in \mathbb{R}_+$ are independent, $\mathbb{P}(\xi > 0) > 0$, and* supp $Q = A$. *Let $\xi_1, \xi_2, \ldots$ be independent copies of ξ, and suppose that*

$$\lim_{n\to\infty} \frac{\max_{1\le i\le n} \xi_i}{\sum_{i=1}^n \xi_i} = 1, \ a.s. \tag{6.1}$$

Then the recurrent directions of the random walk $S_n = \sum_{i=1}^n X_i$ are $\mathcal{D} = A$.

Remarks 6.2

(a) Pruitt, in Theorem 2 of [28], shows that (6.1) holds if and only if $\sum_{k\ge1} u_k^2 < \infty$, where $u_k = \mathbb{P}(2^k < \xi \le 2^{k+1})/\mathbb{P}(2^k < \xi)$. Examples that work have very heavy tails, and include $\mathbb{P}(\xi > r) = 1/\log r$ for $r \ge \mathrm{e}$ (see [10, pp. 509–510]) and $\mathbb{P}(\xi > r) = \exp(-(\log r)^\beta)$ for $r \ge 1$ with $\beta \in (0, 1/2)$ (see [28, p. 895]).

(b) The intuition behind Theorem 6.1 is as follows. The condition (6.1) means that the biggest jump so far is a.s. on a bigger scale than all the other jumps combined, and so the projection on the sphere is determined by the Q corresponding to the current biggest jump. As times goes on, one sees an i.i.d. subsequence of the Qs associated with the biggest jumps, and so the walk explores the sphere over the set A.
(c) Theorem 6.1 can be compared to the construction of random walks with desired limit properties of [9–11, 19].

Proof of Theorem 6.1 Write $X_i = Q_i \xi_i$ where the Q_i are i.i.d. copies of Q and the ξ_i are i.i.d. copies of ξ. Let $T_n = \sum_{i=1}^n \xi_i$, $M_n = \max_{1 \le i \le n} \xi_i$, and $B_n = T_n - M_n$; then (6.1) is equivalent to $B_n / M_n \to 0$, a.s. Also set $k(1) := 1$ and, for $n \in \mathbb{N}$,

$$k(n+1) := \begin{cases} k(n) & \text{if } \xi_{n+1} \le M_n, \\ n+1 & \text{if } \xi_{n+1} > M_n. \end{cases}$$

Then $M_n = \xi_{k(n)}$. Define $R_n := S_n - M_n Q_{k(n)}$. Since $\| Q_{k(n)} \| = 1$, repeated application of the triangle inequality yields

$$\begin{aligned} \| \hat{S}_n - Q_{k(n)} \| &= \left\| \frac{M_n Q_{k(n)} + R_n - \| S_n \| Q_{k(n)}}{\| S_n \|} \right\| \\ &\le \frac{| M_n - \| S_n \| |}{\| S_n \|} + \frac{\| R_n \|}{\| S_n \|} \\ &\le \frac{2 \| R_n \|}{M_n - \| R_n \|}. \end{aligned}$$

But $\| R_n \| = \| \sum_{i \in \{1,\ldots,n\} \setminus \{k(n)\}} X_i \| \le B_n$ where $B_n = T_n - M_n$, so

$$\| \hat{S}_n - Q_{k(n)} \| \le \frac{2 (B_n / M_n)}{1 - (B_n / M_n)} \to 0, \text{ a.s.},$$

by (6.1).

Since M_n is a non-decreasing sequence in $\mathbb{R}_+$ with $M_n \to \infty$ a.s. (as easily follows from (6.1) and the fact that $\mathbb{P}(\xi > 0) > 0$) the sequence $k(1), k(2), \ldots$ is a non-decreasing subsequence of $\mathbb{Z}_+$ with $k(n) \to \infty$ a.s., and since the Q_i are independent of the ξ_i, the sequence $k(1), k(2), \ldots$ is independent of the sequence $Q_1, Q_2, \ldots$. Let $\ell_1 = 1$ and for $n \in \mathbb{N}$ define $\ell_{n+1} = \min\{m > \ell_n : k(m) > k(\ell_n)\}$, so that $1 = k(\ell_1) < k(\ell_2) < k(\ell_3) < \cdots$. Then the sequence $Q_{k(\ell_1)}, Q_{k(\ell_2)}, \ldots$ has the same law as a sequence of i.i.d. copies of Q. Hence if $\mathbf{u} \in A$ we have

$$\liminf_{n \to \infty} \| \hat{S}_n - \mathbf{u} \| \le \lim_{n \to \infty} \| \hat{S}_n - Q_{k(\ell_n)} \| + \liminf_{n \to \infty} \| Q_{k(\ell_n)} - \mathbf{u} \| = 0, \text{ a.s.}$$

Thus $\mathbf{u} \in \mathcal{D}$. This shows that $A \subseteq \mathcal{D}$.

On the other hand, if $\mathbf{u} \notin A$ we have that since $\mathbb{S}^{d-1} \setminus A$ is open in $\mathbb{S}^{d-1}$ there is some $r > 0$ such that $\mathbb{P}(Q \in B_s(\mathbf{u}; r)) = 0$, and

$$\liminf_{n\to\infty} \|\hat{S}_n - \mathbf{u}\| \geq \liminf_{n\to\infty} \|Q_{k(\ell_n)} - \mathbf{u}\| - \lim_{n\to\infty} \|\hat{S}_n - Q_{k(\ell_n)}\| \geq r, \text{ a.s.,}$$

so that $\mathbf{u} \notin \mathcal{D}$. Thus $\mathcal{D} \subseteq A$ and the proof is complete. □

7 Convexity and an Upper Bound

We start this section with a straightforward result (Theorem 7.1) that is sometimes useful for giving an upper bound on $\mathcal{D}$ in terms of the support of $\hat{S}_n$. We then present (in Proposition 7.3 below) a simpler description of the upper bound in terms of the distribution of X alone, rather than its convolutions. To do so, we need an appropriate notion of convexity, which will also be useful in Sects. 8 and 9 below when we look at one-dimensional projections and the convex hull of the walk.

Let $\mathcal{X}_n = (\operatorname{supp} \hat{S}_n) \setminus \{\mathbf{0}\}$, and let $\mathcal{X}^\star = \operatorname{cl}(\cup_{n\geq 1} \mathcal{X}_n)$. Here is the upper bound.

Theorem 7.1 *We have that* $\mathcal{D} \subseteq \mathcal{X}^\star$.

Proof Suppose that $\mathbf{u} \in \mathbb{S}^{d-1} \setminus \mathcal{X}^\star$. Since $\mathcal{X}^\star$ is closed, there exists $r > 0$ such that $B_s(\mathbf{u}; r) \cap \mathcal{X}_n = \emptyset$ for all $n \in \mathbb{N}$, and so $\mathbb{P}(\hat{S}_n \in B_s(\mathbf{u}; r)) = 0$ for all $n \in \mathbb{N}$. Then the Borel–Cantelli lemma shows that $\mathbb{P}(A(\mathbf{u}; r)) = \mathbb{P}(\hat{S}_n \in B_s(\mathbf{u}; r) \text{ i.o.}) = 0$. Hence, by Proposition 2.5, we have $\mathbf{u} \notin \mathcal{D}$. Hence $\mathcal{D} \subseteq \mathcal{X}^\star$. □

For $\mathbf{u}, \mathbf{v} \in \mathbb{S}^{d-1}$ and $\alpha \in [0, 1]$, let

$$I_\alpha(\mathbf{u}, \mathbf{v}) := \frac{\alpha\mathbf{u} + (1-\alpha)\mathbf{v}}{\|\alpha\mathbf{u} + (1-\alpha)\mathbf{v}\|},$$

unless $\mathbf{u} = -\mathbf{v}$ and $\alpha = 1/2$, in which case we set $I_{1/2}(\mathbf{u}, -\mathbf{u}) := \mathbf{0}$. If $\mathbf{u} \neq -\mathbf{v}$, set $I(\mathbf{u}, \mathbf{v}) := \{I_\alpha(\mathbf{u}, \mathbf{v}) : \alpha \in [0, 1]\}$, and set $I(\mathbf{u}, -\mathbf{u}) := \{\mathbf{u}, -\mathbf{u}\}$ (i.e., ignore $\alpha = 1/2$).

Definition 7.2 Say that $A \subseteq \mathbb{S}^{d-1}$ is *s-convex* if for every $\mathbf{u}, \mathbf{v} \in A$, one has $I(\mathbf{u}, \mathbf{v}) \subseteq A$.

Note that we only need to check the condition in Definition 7.2 for $\mathbf{v} \neq -\mathbf{u}$. In words, $A \subseteq \mathbb{S}^{d-1}$ is s-convex if for any $\mathbf{u}, \mathbf{v} \in A$, the radial projection onto $\mathbb{S}^{d-1}$ of the straight line segment from $\mathbf{u}$ to $\mathbf{v}$ in $\mathbb{R}^d$ lies in A. See also Lemma 7.5 below.

Denote by hull A the convex hull of $A \subseteq \mathbb{R}^d$. For $A \subseteq \mathbb{S}^{d-1}$, define

$$\text{s-hull}\, A := \{\hat{\mathbf{x}} : \mathbf{x} \in \text{hull}\, A,\ \mathbf{x} \neq \mathbf{0}\}.$$

We will show (see Lemma 7.7) that s-hull A is s-convex. Let $\mathcal{X} := (\operatorname{supp} \hat{X}) \setminus \{\mathbf{0}\}$.

Proposition 7.3 *We have that* $\mathcal{X}^\star = \operatorname{cl}$ *s-hull* $\mathcal{X}$*, and* $\mathcal{X}^\star$ *is s-convex.*

We work towards a proof of Proposition 7.3. Let $\mathcal{X}' := \{\hat{\mathbf{x}} : \mathbf{x} \in \operatorname{supp} X\}$.

Lemma 7.4 *For* $X \in \mathbb{R}^d$ *any random variable, we have that* $\mathcal{X} = (\operatorname{cl} \mathcal{X}') \setminus \{\mathbf{0}\}$.

Proof Recall that $\operatorname{supp} X$ is the smallest closed $A \subseteq \mathbb{R}^d$ such that $\mathbb{P}(X \in A) = 1$, or, equivalently, $\operatorname{supp} X = \{\mathbf{x} \in \mathbb{R}^d : \mathbb{P}(X \in B(\mathbf{x}; r)) > 0 \text{ for all } r > 0\}$. Since $\operatorname{supp} \hat{X}$ is a closed subset of $\mathbb{S}^{d-1} \cup \{\mathbf{0}\}$, it follows that $\mathcal{X}$ is a closed subset of $\mathbb{S}^{d-1}$.

Suppose that $\mathbf{u} \in \mathcal{X}'$ with $\mathbf{u} \neq \mathbf{0}$. Then $\mathbf{u}r \in \operatorname{supp} X$ for some $r > 0$. This means that $\mathbb{P}(X \in B(\mathbf{u}r; s)) > 0$ for all $s \in (0, r/2)$, say; but, for any $\mathbf{x} \in B(\mathbf{u}r; s)$,

$$\begin{aligned} \|\hat{\mathbf{x}} - \mathbf{u}\| &= \|\mathbf{x}\|^{-1} (\|\mathbf{x} - \|\mathbf{x}\|\mathbf{u}\|) \\ &\leq \|\mathbf{x}\|^{-1} (\|\mathbf{x} - r\mathbf{u}\| + |r - \|\mathbf{x}\||) \leq 4s/r, \end{aligned}$$

so $\mathbb{P}(\hat{X} \in B(\mathbf{u}; 4s/r)) \geq \mathbb{P}(X \in B(\mathbf{u}r; s)) > 0$ for all $s \in (0, r/2)$. Hence $\mathbf{u} \in \operatorname{supp} \hat{X}$. Thus $\mathcal{X}' \subseteq \mathcal{X} \cup \{\mathbf{0}\}$, and since $\mathcal{X} \cup \{\mathbf{0}\}$ is closed we get $\operatorname{cl} \mathcal{X}' \subseteq \mathcal{X} \cup \{\mathbf{0}\}$.

On the other hand suppose that $\mathbf{u} \in \mathcal{X}$. Let $r_n > 0$ be such that $r_n \to 0$. Then $\mathbb{P}(X \in C(\mathbf{u}; r_n)) = \mathbb{P}(\hat{X} \in B(\mathbf{u}; r_n)) > 0$ for all n, which means that $C(\mathbf{u}; r_n) \cap \operatorname{supp} X \neq \emptyset$, i.e., for every n there exists $\mathbf{x}_n \in \operatorname{supp} X$ with $\|\hat{\mathbf{x}}_n - \mathbf{u}\| \leq r_n$. Hence $\hat{\mathbf{x}}_n \in \mathcal{X}'$ with $\hat{\mathbf{x}}_n \to \mathbf{u}$, so $\mathbf{u} \in \operatorname{cl} \mathcal{X}'$, and we get $\mathcal{X} \subseteq \operatorname{cl} \mathcal{X}'$. □

The next result characterizes a set as s-convex if and only if all normalized conical combinations are contained within the set.

Lemma 7.5 *The set* $A \subseteq \mathbb{S}^{d-1}$ *is s-convex if and only if for all* $n \in \mathbb{N}$*, all* $\mathbf{u}_1, \ldots, \mathbf{u}_n \in A$*, and all* $\beta_1, \ldots, \beta_n \in (0, \infty)$*,*

$$\frac{\sum_{i=1}^n \beta_i \mathbf{u}_i}{\left\| \sum_{i=1}^n \beta_i \mathbf{u}_i \right\|} \in A, \text{ whenever } \sum_{i=1}^n \beta_i \mathbf{u}_i \neq \mathbf{0}. \tag{7.1}$$

Proof The 'if' half follows immediately (take $n = 2$ and $\beta_1 + \beta_2 = 1$). Suppose that A is s-convex. We proceed by an induction on n. Then (7.1) holds for $n = 2$, since

$$\frac{\beta_1 \mathbf{u}_1 + \beta_2 \mathbf{u}_2}{\|\beta_1 \mathbf{u}_1 + \beta_2 \mathbf{u}_2\|} = \frac{\frac{\beta_1}{\beta_1+\beta_2} \mathbf{u}_1 + \frac{\beta_2}{\beta_1+\beta_2} \mathbf{u}_2}{\left\| \frac{\beta_1}{\beta_1+\beta_2} \mathbf{u}_1 + \frac{\beta_2}{\beta_1+\beta_2} \mathbf{u}_2 \right\|}.$$

Suppose that (7.1) holds for all $n \in \{1, \ldots, m\}$ with $m \geq 2$, and consider $\mathbf{u}_1, \ldots, \mathbf{u}_{m+1} \in A$ and $\beta_1, \ldots, \beta_{m+1} \in (0, \infty)$ with $\sum_{i=1}^{m+1} \beta_i \mathbf{u}_i \neq \mathbf{0}$. We may also suppose that $\beta_m \mathbf{u}_m + \beta_{m+1} \mathbf{u}_{m+1} \neq \mathbf{0}$, or else the inductive hypothesis would apply directly. Set $\mathbf{u}_i' = \mathbf{u}_i$ for $1 \leq i \leq m - 1$ and

$$\mathbf{u}_m' = \frac{\frac{\beta_m}{\beta_m+\beta_{m+1}} \mathbf{u}_m + \frac{\beta_{m+1}}{\beta_m+\beta_{m+1}} \mathbf{u}_{m+1}}{\left\| \frac{\beta_m}{\beta_m+\beta_{m+1}} \mathbf{u}_m + \frac{\beta_{m+1}}{\beta_m+\beta_{m+1}} \mathbf{u}_{m+1} \right\|}.$$

Then since A is s-convex, $\mathbf{u}'_m \in A$, and

$$\frac{\sum_{i=1}^{m+1} \beta_i \mathbf{u}_i}{\left\| \sum_{i=1}^{m+1} \beta_i \mathbf{u}_i \right\|} = \frac{\sum_{i=1}^{m} \beta'_i \mathbf{u}'_i}{\left\| \sum_{i=1}^{m} \beta'_i \mathbf{u}'_i \right\|},$$

where $\beta'_i = \beta_i$ for $1 \leq i \leq m-1$ and $\beta'_m = \|\beta_m \mathbf{u}_m + \beta_{m+1} \mathbf{u}_{m+1}\|$. By inductive hypothesis, the expression in the last display is thus in A. This completes the inductive step. □

Corollary 7.6 *Suppose that $A \subseteq \mathbb{S}^{d-1}$ is s-convex. Then $A = \mathbb{S}^{d-1} \cap \operatorname{hull} A$.*

Proof It is clear that $A \subseteq \mathbb{S}^{d-1} \cap \operatorname{hull} A$. So suppose that $\mathbf{u} \in \mathbb{S}^{d-1} \cap \operatorname{hull} A$. Then (see e.g. Lemma 3.1 of [14, p. 42]) there exist $n \in \mathbb{N}$, $\mathbf{v}_1, \ldots, \mathbf{v}_n \in A$, and $\lambda_1, \ldots, \lambda_n \in [0, 1]$ with $\sum_{i=1}^n \lambda_i = 1$, for which $\mathbf{u} = \sum_{i=1}^n \lambda_i \mathbf{v}_i$. But, since A is s-convex and $\|\mathbf{u}\| = 1$, Lemma 7.5 shows that $\sum_{i=1}^n \lambda_i \mathbf{v}_i \in A$. So $\mathbb{S}^{d-1} \cap \operatorname{hull} A \subseteq A$. □

The next result shows that s-hull A has a similar characterization to the usual hull A.

Lemma 7.7 *For $A \subseteq \mathbb{S}^{d-1}$, s-hull A is the smallest s-convex $B \subseteq \mathbb{S}^{d-1}$ with $A \subseteq B$.*

Proof Let $\mathbf{u}, \mathbf{v} \in$ s-hull A with $\mathbf{v} \neq -\mathbf{u}$, and $\alpha \in (0, 1)$. Then $\mathbf{u} = \hat{\mathbf{x}}$ and $\mathbf{v} = \hat{\mathbf{y}}$ for some $\mathbf{x}, \mathbf{y} \in \operatorname{hull} A$ with $\mathbf{x}, \mathbf{y} \neq \mathbf{0}$. Choose $\beta \in (0, 1)$ given by

$$\beta = \frac{\alpha \|\mathbf{y}\|}{\alpha \|\mathbf{y}\| + (1 - \alpha)\|\mathbf{x}\|}.$$

Consider $\mathbf{w} = \beta \mathbf{x} + (1 - \beta)\mathbf{y}$. Then, since $\operatorname{hull} A$ is convex, $\mathbf{w} \in \operatorname{hull} A$, and $\mathbf{w} \neq \mathbf{0}$ since $\hat{\mathbf{x}} \neq -\hat{\mathbf{y}}$, so $\hat{\mathbf{w}} \in$ s-hull A. But

$$\frac{\mathbf{w}}{\|\mathbf{w}\|} = \frac{\alpha \hat{\mathbf{x}} + (1 - \alpha)\hat{\mathbf{y}}}{\|\alpha \hat{\mathbf{x}} + (1 - \alpha)\hat{\mathbf{y}}\|}$$

is thus in s-hull A for all $\alpha \in (0, 1)$, verifying that s-hull A is s-convex.

Next we claim that if $A \subseteq \mathbb{S}^{d-1}$ is s-convex, then s-hull $A = A$. Clearly $A \subseteq$ s-hull A. So suppose that A is s-convex, and consider $\mathbf{u} \in$ s-hull A. Then $\mathbf{u} = \hat{\mathbf{x}}$ for some $\mathbf{x} \in \operatorname{hull} A$, $\mathbf{x} \neq \mathbf{0}$, and thus (see e.g. Lemma 3.1 of [14, p. 42]) there exist $n \in \mathbb{N}$, $\mathbf{v}_1, \ldots, \mathbf{v}_n \in A$, and $\lambda_1, \ldots, \lambda_n \in [0, 1]$ with $\sum_{i=1}^n \lambda_i = 1$, for which $\mathbf{x} = \sum_{i=1}^n \lambda_i \mathbf{v}_i$. Then Lemma 7.5 shows that $\hat{\mathbf{x}} \in A$. In other words, s-hull $A \subseteq A$, as required.

Suppose B is s-convex with $A \subseteq B$; then the preceding paragraph shows that s-hull $A \subseteq$ s-hull $B = B$, which completes the proof of the lemma. □

Lemma 7.8 *Let $A \subseteq \mathbb{S}^{d-1}$ be s-convex. Then* cl A *is also s-convex.*

Proof It suffices to suppose $\mathbf{u}, \mathbf{v} \in \operatorname{cl} A$ with $\mathbf{u} \neq -\mathbf{v}$. Then there exist $\mathbf{u}_1, \mathbf{u}_2, \ldots \in A$ and $\mathbf{v}_1, \mathbf{v}_2, \ldots \in A$ with $\mathbf{u}_n \to \mathbf{u}$ and $\mathbf{v}_n \to \mathbf{v}$, and there exists $n_0 \in \mathbb{N}$ such that $\mathbf{u}_n \neq -\mathbf{v}_n$ for all $n \geq n_0$. Since A is s-convex, $I_\alpha(\mathbf{u}_n, \mathbf{v}_n) \in A$ for all $n \geq n_0$ and all $\alpha \in [0, 1]$. By continuity of the function $\mathbf{x} \mapsto \hat{\mathbf{x}}$ on $\mathbb{R}^d \setminus \{\mathbf{0}\}$, it follows that $I_\alpha(\mathbf{u}, \mathbf{v}) = \lim_{n\to\infty} I_\alpha(\mathbf{u}_n, \mathbf{v}_n) \in \operatorname{cl} A$ for all $\alpha \in [0, 1]$. Hence $\operatorname{cl} A$ is s-convex. □

Proof of Proposition 7.3 First we use induction to show that $\mathcal{X}_n \subseteq \operatorname{cl}\text{s-hull}\,\mathcal{X}$ for all $n \in \mathbb{N}$. Clearly this is true for $n = 1$. So suppose, for the inductive hypothesis, that $\mathcal{X}_m \subseteq \operatorname{cl}\text{s-hull}\,\mathcal{X}$ for all $m \in \{1, \ldots, n\}$. Now, provided that $S_{n+1} \neq \mathbf{0}$, we have

$$\hat{S}_{n+1} = \frac{\alpha_n \hat{S}_n + (1-\alpha_n)\hat{X}_{n+1}}{\|\alpha_n \hat{S}_n + (1-\alpha_n)\hat{X}_{n+1}\|}, \text{ where } \alpha_n = \frac{\|S_n\|}{\|S_n\| + \|X_{n+1}\|}.$$

In particular, since $\mathbb{P}(\hat{S}_n \in \mathcal{X}_n \cup \{\mathbf{0}\}) = 1$ and $\mathbb{P}(\hat{X}_{n+1} \in \mathcal{X} \cup \{\mathbf{0}\}) = 1$, we have

$$\mathbb{P}\left(\hat{S}_{n+1} \in (\cup\{I(\mathbf{u}, \mathbf{v}) : \mathbf{u}, \mathbf{v} \in \operatorname{cl}\text{s-hull}\,\mathcal{X}\}) \cup \{\mathbf{0}\}\right) = 1,$$

by the inductive hypothesis. But $\operatorname{cl}\text{s-hull}\,\mathcal{X}$ is s-convex, by Lemmas 7.7 and 7.8, so $\mathbb{P}(\hat{S}_{n+1} \in (\operatorname{cl}\text{s-hull}\,\mathcal{X}) \cup \{\mathbf{0}\}) = 1$, which means that $\mathcal{X}_{n+1} \subseteq \operatorname{cl}\text{s-hull}\,\mathcal{X}$, completing the induction. Thus we conclude that $\mathcal{X}^\star \subseteq \operatorname{cl}\text{s-hull}\,\mathcal{X}$.

Next we show that $\mathcal{X}^\star$ is s-convex. It suffices to suppose that $\mathbf{u}, \mathbf{v} \in \mathcal{X}^\star$ with $\mathbf{u} \neq -\mathbf{v}$. Then there exist sequences $\mathbf{u}_{n_k} \in \mathcal{X}_{n_k}$ and $\mathbf{v}_{m_k} \in \mathcal{X}_{m_k}$ with $\mathbf{u}_{n_k} \to \mathbf{u}$ and $\mathbf{v}_{m_k} \to \mathbf{v}$. Lemma 7.4 shows that, correspondingly, there exist sequences $\mathbf{x}_{n_{k,1}}, \mathbf{x}_{n_{k,2}}, \ldots \in \operatorname{supp} S_{n_k}$ and $\mathbf{y}_{m_{k,1}}, \mathbf{y}_{m_{k,2}}, \ldots \in \operatorname{supp} S_{m_k}$ with $\lim_{i\to\infty} \hat{\mathbf{x}}_{n_{k,i}} = \mathbf{u}_{n_k}$ and $\lim_{j\to\infty} \hat{\mathbf{y}}_{m_{k,j}} = \mathbf{v}_{m_k}$, and, for all k sufficiently large and all i, j sufficiently large, $\hat{\mathbf{x}}_{n_{k,i}} \neq -\hat{\mathbf{y}}_{m_{k,j}}$. Now for $s, t \in \mathbb{Z}_+$, $s\mathbf{x}_{n_{k,i}} + t\mathbf{y}_{m_{k,j}} \in \operatorname{supp} S_{sn_k+tm_k}$. Applying Lemma 7.4 with $X = S_{sn_k+tm_k}$ we see that $\mathbf{w} \in \mathcal{X}_{sn_k+tm_k} \subseteq \mathcal{X}^\star$, where

$$\mathbf{w} = \frac{s\mathbf{x}_{n_{k,i}} + t\mathbf{y}_{m_{k,j}}}{\|s\mathbf{x}_{n_{k,i}} + t\mathbf{y}_{m_{k,j}}\|} = I_{\alpha_{s,t,i,j}}(\hat{\mathbf{x}}_{n_{k,i}}, \hat{\mathbf{y}}_{m_{k,j}}),$$

with

$$\alpha_{s,t,i,j} = \frac{s\|\mathbf{x}_{n_{k,i}}\|}{s\|\mathbf{x}_{n_{k,i}}\| + t\|\mathbf{y}_{m_{k,j}}\|}.$$

For fixed k, i, j and $\alpha \in [0, 1]$, we may choose $s, t \to \infty$ such that $\alpha_{s,t,i,j} \to \alpha$, and since for $\mathbf{u} \neq -\mathbf{v}$, $\alpha \mapsto I_\alpha(\mathbf{u}, \mathbf{v})$ is continuous over $\alpha \in [0, 1]$, and $\mathcal{X}^\star$ is closed, we get

$$I_\alpha(\hat{\mathbf{x}}_{n_{k,i}}, \hat{\mathbf{y}}_{m_{k,j}}) = \lim_{s,t\to\infty} I_{\alpha_{s,t,i,j}}(\hat{\mathbf{x}}_{n_{k,i}}, \hat{\mathbf{y}}_{m_{k,j}}) \in \mathcal{X}^\star, \text{ for all } \alpha \in [0, 1].$$

Then by continuity of $(\mathbf{u}, \mathbf{v}) \mapsto I_\alpha(\mathbf{u}, \mathbf{v})$ away from $\mathbf{u} = -\mathbf{v}$ we get

$$I_\alpha(\mathbf{u}, \mathbf{v}) = \lim_{k\to\infty} I_\alpha(\mathbf{u}_{n_k}, \mathbf{v}_{m_k}) = \lim_{k\to\infty} \lim_{i,j\to\infty} I_\alpha(\hat{\mathbf{x}}_{n_{k,i}}, \hat{\mathbf{y}}_{m_{k,j}}) \in \mathcal{X}^\star,$$

for all $\alpha \in [0, 1]$. Hence $\mathcal{X}^\star$ is s-convex, and $\mathcal{X} \subseteq \mathcal{X}^\star$, so, by Lemma 7.7, we have s-hull $\mathcal{X} \subseteq \mathcal{X}^\star$, and since $\mathcal{X}^\star$ is closed, we get cl s-hull $\mathcal{X} \subseteq \mathcal{X}^\star$.

Thus we conclude that $\mathcal{X}^\star = \operatorname{cl} \text{s-hull}\, \mathcal{X}$, and the latter is s-convex by Lemmas 7.7 and 7.8. □

We finish this section with a result on the boundary of an s-convex set, which will be useful in Sect. 8 below. For $A \subseteq \mathbb{S}^{d-1}$, denote by s-int A the interior of A relative to $\mathbb{S}^{d-1}$, i.e., $\mathbf{u} \in \text{s-int}\, A$ if and only if $B_s(\mathbf{u}; \delta) \subseteq A$ for some $\delta > 0$. Also, for $A \subseteq \mathbb{S}^{d-1}$, we write $\partial_s A$ for the boundary of A relative to $\mathbb{S}^{d-1}$, i.e., $\partial_s A := (\operatorname{cl} A) \setminus (\text{s-int}\, A)$.

Lemma 7.9 *If $A \subseteq \mathbb{S}^{d-1}$ is s-convex, then (i)* $\text{s-int}\, A = \text{s-int} \operatorname{cl} A$*; and (ii)* $\partial_s A = \partial_s \operatorname{cl} A$.

Proof Suppose that $\mathbf{u} \in \text{s-int} \operatorname{cl} A$. Then there exist $m \in \mathbb{N}$ and $\mathbf{u}_1, \dots, \mathbf{u}_m \in \operatorname{cl} A$ such that $\mathbf{u} \in \text{s-int}\, P_s(\mathbf{u}_1, \dots, \mathbf{u}_m)$, where $P_s(\mathbf{u}_1, \dots, \mathbf{u}_m) := \text{s-hull}\, \{\mathbf{u}_1, \dots, \mathbf{u}_m\}$. Let

$$R_s(\mathbf{v}_1, \dots, \mathbf{v}_m; \mathbf{u}) = \inf\{\|\mathbf{v} - \mathbf{u}\| : \mathbf{v} \in \mathbb{S}^{d-1} \setminus P_s(\mathbf{v}_1, \dots, \mathbf{v}_m)\},$$

which is zero unless $\mathbf{u}$ lies in the interior of $P_s(\mathbf{v}_1, \dots, \mathbf{v}_m)$, when it is equal to the shortest distance from $\mathbf{u}$ to the boundary of $P_s(\mathbf{v}_1, \dots, \mathbf{v}_m)$. In particular, note that $R_s(\mathbf{u}_1, \dots, \mathbf{u}_m; \mathbf{u}) = \delta_0 > 0$. For $\mathbf{v}_1, \dots, \mathbf{v}_m \in \mathbb{S}^{d-1}$, the map $(\mathbf{v}_1, \dots, \mathbf{v}_m) \mapsto P_s(\mathbf{v}_1, \dots, \mathbf{v}_m)$, as a function from $(\mathbb{S}^{d-1})^m$ to compact subsets of $\mathbb{R}^d$ with the Hausdorff metric, is continuous. So the map from $(\mathbf{v}_1, \dots, \mathbf{v}_m)$ to $R_s(\mathbf{v}_1, \dots, \mathbf{v}_m; \mathbf{u})$ is also continuous. Hence for any $\delta \in (0, \delta_0)$, we can find $\varepsilon > 0$ sufficiently small such that $B_s(\mathbf{u}; \delta)$ is contained in $P_s(\mathbf{v}_1, \dots, \mathbf{v}_m)$ for all $\mathbf{v}_i \in \mathbb{S}^{d-1}$ with $\|\mathbf{v}_i - \mathbf{u}_i\| < \varepsilon$. Since $\mathbf{u}_i \in \operatorname{cl} A$, we can find $\mathbf{v}_i \in A$ with $\|\mathbf{v}_i - \mathbf{u}_i\| < \varepsilon$, which means that $B_s(\mathbf{u}; \delta) \subseteq P_s(\mathbf{v}_1, \dots, \mathbf{v}_m) \subseteq A$, since A is s-convex. Hence $\mathbf{u} \in \text{s-int}\, A$. This establishes (i). Then (ii) follows since $\partial_s \operatorname{cl} A = \operatorname{cl} A \setminus \text{s-int} \operatorname{cl} A = \operatorname{cl} A \setminus \text{s-int}\, A = \partial_s A$. □

8 Projection Asymptotics

In Sect. 9 we study the way in which the random walk fills space via the convex hull of the trajectory. Pertinent for this is the behaviour of one-dimensional projections of the walk, so we turn to this first. For fixed $\mathbf{u} \in \mathbb{S}^{d-1}$, the projection $S_n \cdot \mathbf{u}$ defines a random walk on $\mathbb{R}$, with increment distribution $X \cdot \mathbf{u}$, which either tends to $+\infty$, to $-\infty$, or oscillates (see Lemma 8.1 below). However, this, by itself, does not exclude that there might exist (random) $\mathbf{u} \in \mathbb{S}^{d-1}$ for which $S_n \cdot \mathbf{u}$ does something out of the ordinary, such as having a finite lim sup. While not central for what follows, we show that such *exceptional projections* do not exist, at least for $d \leq 2$.

Define the random sets

$$\mathcal{P}_+ := \left\{\mathbf{u} \in \mathbb{S}^{d-1} : \lim_{n\to\infty}(S_n \cdot \mathbf{u}) = +\infty\right\}, \quad \mathcal{P}_- := \left\{\mathbf{u} \in \mathbb{S}^{d-1} : \lim_{n\to\infty}(S_n \cdot \mathbf{u}) = -\infty\right\},$$
$$\mathcal{P}_\pm := \left\{\mathbf{u} \in \mathbb{S}^{d-1} : -\infty = \liminf_{n\to\infty}(S_n \cdot \mathbf{u}) < \limsup_{n\to\infty}(S_n \cdot \mathbf{u}) = +\infty\right\},$$

and their non-random counterparts

$$\mathcal{D}_+ := \left\{\mathbf{u} \in \mathbb{S}^{d-1} : \lim_{n\to\infty}(S_n \cdot \mathbf{u}) = +\infty, \text{ a.s.}\right\},$$
$$\mathcal{D}_- := \left\{\mathbf{u} \in \mathbb{S}^{d-1} : \lim_{n\to\infty}(S_n \cdot \mathbf{u}) = -\infty, \text{ a.s.}\right\},$$
$$\mathcal{D}_\pm := \left\{\mathbf{u} \in \mathbb{S}^{d-1} : -\infty = \liminf_{n\to\infty}(S_n \cdot \mathbf{u}) < \limsup_{n\to\infty}(S_n \cdot \mathbf{u}) = +\infty, \text{ a.s.}\right\},$$

Then $\mathcal{P}_+ = -\mathcal{P}_-$, $\mathcal{P}_\pm = -\mathcal{P}_\pm$, and similarly for the non-random versions.

Lemma 8.1 *The sets $\mathcal{D}_+$, $\mathcal{D}_-$, $\mathcal{D}_\pm$ partition $\mathbb{S}^{d-1}$.*

Proof Let $\mathbf{u} \in \mathbb{S}^{d-1}$. Then (see e.g. [8, Theorem 4.1.2]) exactly one of the following holds: (i) $\mathbf{u} \in \mathcal{D}_+$, (ii) $\mathbf{u} \in \mathcal{D}_-$, (iii) $\mathbf{u} \in \mathcal{D}_\pm$, or (iv) $\mathbb{P}(X \cdot \mathbf{u} = 0) = 1$. Case (iv) is ruled out by our assumption that the walk is genuinely d-dimensional. □

It is not immediately obvious that $\mathcal{P}_+$, $\mathcal{P}_-$, $\mathcal{P}_\pm$ also partition $\mathbb{S}^{d-1}$. We define

$$\mathcal{E}_+ := \left\{\mathbf{u} \in \mathbb{S}^{d-1} : \limsup_{n\to\infty}(S_n \cdot \mathbf{u}) \in \mathbb{R}\right\}, \quad \mathcal{E}_- := \left\{\mathbf{u} \in \mathbb{S}^{d-1} : \liminf_{n\to\infty}(S_n \cdot \mathbf{u}) \in \mathbb{R}\right\}.$$

We call $\mathbf{u} \in \mathcal{E} := \mathcal{E}_+ \cup \mathcal{E}_-$ an *exceptional projection* of the walk. Since $\mathcal{E}_- = -\mathcal{E}_+$, we have $\mathcal{E} = -\mathcal{E}$. Lemma 8.1 means that $\mathbb{P}(\mathbf{u} \in \mathcal{E}) = 0$ for all fixed $\mathbf{u} \in \mathbb{S}^{d-1}$. Recall the definition of s-convexity from Definition 7.2.

Lemma 8.2 *The sets $\mathcal{P}_+$, $\mathcal{P}_-$, $\mathcal{P}_+ \cup \mathcal{E}_-$, $\mathcal{P}_- \cup \mathcal{E}_+$, $\mathcal{D}_+$, and $\mathcal{D}_-$ are s-convex.*

Proof Suppose that $\mathbf{u}, \mathbf{v} \in \mathcal{P}_+$ with $\mathbf{v} \neq -\mathbf{u}$. Then

$$S_n \cdot (\alpha\mathbf{u} + (1-\alpha)\mathbf{v}) = \alpha S_n \cdot \mathbf{u} + (1-\alpha) S_n \cdot \mathbf{v},$$

and both $S_n \cdot \mathbf{u}$ and $S_n \cdot \mathbf{v}$ tend to infinity, so $I_\alpha(\mathbf{u}, \mathbf{v}) \in \mathcal{P}_+$ for all $\alpha \in [0, 1]$. Hence $\mathcal{P}_+$ is s-convex, and so is $\mathcal{P}_- = -\mathcal{P}_+$ as well. The argument for $\mathcal{D}_+$, $\mathcal{D}_-$ is essentially the same. Note that $\mathbf{u} \in \mathcal{P}_+ \cup \mathcal{E}_-$ if and only if $\liminf_{n\to\infty}(S_n \cdot \mathbf{u}) > -\infty$. Hence if $\mathbf{u}, \mathbf{v} \in \mathcal{P}_+ \cup \mathcal{E}_-$,

$$\liminf_{n\to\infty}(S_n \cdot (\alpha\mathbf{u} + (1-\alpha)\mathbf{v})) \geq \alpha \liminf_{n\to\infty}(S_n \cdot \mathbf{u}) + (1-\alpha) \liminf_{n\to\infty}(S_n \cdot \mathbf{v}) > -\infty,$$

so $\mathcal{P}_+ \cup \mathcal{E}_-$ is s-convex; similarly for $\mathcal{P}_- \cup \mathcal{E}_+$. □

The following result shows that random set $\mathcal{P}_+$ can differ from the non-random set $\mathcal{D}_+$ in a rather limited way. In particular, since $\mathcal{P}_+$ and $\mathcal{D}_+$ are s-convex (by Lemma 8.2), Proposition 8.3(i) with Lemma 7.9 shows that $\mathbb{P}(\partial_s \mathcal{P}_+ = \partial_s \mathcal{D}_+) = 1$. Similarly for $\mathcal{P}_-$ and $\mathcal{D}_-$.

Proposition 8.3

(i) *We have*

$$\mathbb{P}(\mathrm{cl}\,\mathcal{P}_+ = \mathrm{cl}\,\mathcal{P}_+ \cup \mathrm{cl}\,\mathcal{E}_- = \mathrm{cl}\,\mathcal{D}_+) = 1, \text{ and } \mathbb{P}(\mathrm{cl}\,\mathcal{P}_- = \mathrm{cl}\,\mathcal{P}_- \cup \mathrm{cl}\,\mathcal{E}_+ = \mathrm{cl}\,\mathcal{D}_-) = 1.$$

(ii) *Moreover,* $\mathbb{P}(\mathrm{cl}\,\mathcal{E}_+ \subseteq \partial_s \mathcal{D}_-) = \mathbb{P}(\mathrm{cl}\,\mathcal{E}_- \subseteq \partial_s \mathcal{D}_+) = 1$.

Proof For part (i), it suffices to prove the first statement. For ease of notation, write $\mathcal{P} = \mathcal{P}_+ \cup \mathcal{E}_-$. Since, by Lemma 8.2, $\mathcal{P}$ is s-convex, so is $\mathrm{cl}\,\mathcal{P}$, by Lemma 7.8. Thus, by Corollary 7.6, $\mathrm{cl}\,\mathcal{P} = \mathbb{S}^{d-1} \cap \mathrm{hull}\,\mathrm{cl}\,\mathcal{P}$. Since $\mathrm{cl}\,\mathcal{P}$ is bounded, $A = \mathrm{hull}\,\mathrm{cl}\,\mathcal{P} = \mathrm{cl}\,\mathrm{hull}\,\mathcal{P}$ [14, p. 45]. The set A is convex and compact, and so it is uniquely determined by its support function $h_A : \mathbb{R}^d \to \mathbb{R}$ given by $h_A(\mathbf{x}) = \sup\{\mathbf{x} \cdot \mathbf{y} : \mathbf{y} \in A\}$, which is continuous [14, p. 56]. Since $\mathbb{Q}^d$ is dense in $\mathbb{R}^d$, h_A is determined by $\{h_A(\mathbf{x}) : \mathbf{x} \in \mathbb{Q}^d\}$. By the Hewitt–Savage theorem, each member of this countable collection of random variables is a.s. constant, so h_A is a.s. constant. Thus the set A is non-random, and then $\mathbb{P}(\mathrm{cl}\,\mathcal{P} = S) = 1$ for the non-random closed, s-convex set $S = \mathbb{S}^{d-1} \cap A$. Note that

$$\mathbb{P}(\mathbf{u} \in \mathrm{cl}\,\mathcal{P}) = \begin{cases} 1 & \text{if } \mathbf{u} \in S, \\ 0 & \text{if } \mathbf{u} \notin S. \end{cases}$$

Since every $\mathbf{u} \in \mathcal{D}_+$ has $\mathbb{P}(\mathbf{u} \in \mathcal{P}_+ \subseteq \mathcal{P}) = 1$, we have $\mathcal{D}_+ \subseteq S$, and since S is closed, $\mathrm{cl}\,\mathcal{D}_+ \subseteq S$. On the other hand, if $S \setminus \mathrm{cl}\,\mathcal{D}_+ \neq \emptyset$, there is some $\mathbf{u} \in S \setminus \mathrm{cl}\,\mathcal{D}_+$ and some $\varepsilon > 0$ such that $S \cap B_s(\mathbf{u}; \varepsilon)$ does not intersect $\mathrm{cl}\,\mathcal{D}_+$. The compact set S contains a countable dense subset, Q, say, and every $\mathbf{v} \in Q \cap B_s(\mathbf{u}; \varepsilon)$ has $\mathbf{v} \notin \mathcal{D}_+$, so $\mathbb{P}(\mathbf{v} \in \mathcal{P}_+) = 0$. Also, $\mathbb{P}(\mathbf{v} \in \mathcal{E}_-) = 0$. Thus no member of $Q \cap B_s(\mathbf{u}; \varepsilon)$ is in $\mathcal{P}$. Since $\mathcal{P}$ is s-convex with closure S, this implies that there is a neighbourhood of $\mathbf{u}$ in S that does not intersect $\mathcal{P}$. Hence $\mathbf{u} \in S \setminus \mathrm{cl}\,\mathcal{P}$. But $\mathbb{P}(S \setminus \mathrm{cl}\,\mathcal{P} = \emptyset) = 1$. Thus $\mathbb{P}(\mathrm{cl}\,\mathcal{P}_+ \cup \mathrm{cl}\,\mathcal{E}_- = \mathrm{cl}\,\mathcal{D}_+) = 1$. Repeating the preceding argument, but taking $\mathcal{P} = \mathcal{P}_+$ throughout, gives $\mathbb{P}(\mathrm{cl}\,\mathcal{P}_+ = \mathrm{cl}\,\mathcal{D}_+) = 1$ too.

For part (ii), we have from (i) that $\mathbb{P}(\mathrm{cl}\,\mathcal{E}_- \subseteq \mathrm{cl}\,\mathcal{P}_+) = 1$. Moreover, we must have $\mathbb{P}(\mathrm{cl}\,\mathcal{E}_- \cap \text{s-int}\,\mathcal{P}_+ = \emptyset) = 1$, or else we would have $\mathcal{E}_- \cap \mathcal{P}_+ \neq \emptyset$. Thus $\mathbb{P}(\mathrm{cl}\,\mathcal{E}_- \subseteq \partial_s \mathcal{P}_+) = 1$. But since $\mathcal{P}_+$ and $\mathcal{D}_+$ are s-convex and a.s. have the same closure, Lemma 7.9 shows that $\mathbb{P}(\partial_s \mathcal{P}_+ = \partial_s \mathcal{D}_+) = 1$. This gives (ii). □

Corollary 8.4 *If* $\mathcal{D}_\pm = \mathbb{S}^{d-1}$, *then* $\mathbb{P}(\mathcal{P}_\pm = \mathbb{S}^{d-1}) = 1$.

Proof If $\mathcal{D}_\pm = \mathbb{S}^{d-1}$, then $\mathcal{D}_+ = \mathcal{D}_- = \emptyset$, by Lemma 8.1, and Proposition 8.3 shows that $\mathbb{P}(\mathrm{cl}\,\mathcal{P}_+ \cup \mathrm{cl}\,\mathcal{P}_- \cup \mathrm{cl}\,\mathcal{E} = \emptyset) = 1$. □

We turn briefly to the question of whether $\mathcal{E}$ is in fact empty.

Lemma 8.5 *With probability 1,* $\mathrm{cl}\,\mathcal{E}$ *is a perfect set.*

Proof For a measurable $B \subseteq \mathbb{S}^{d-1}$ let $N(B) = \#(B \cap \mathrm{cl}\,\mathcal{E})$, the number of points of $\mathrm{cl}\,\mathcal{E}$ in B. We claim that, for any B that is open in $\mathbb{S}^{d-1}$,

$$\mathbb{P}(N(B) = 0) = 1 \text{ or } \mathbb{P}(N(B) = \infty) = 1. \tag{8.1}$$

Indeed, the $\mathbb{Z}_+ \cup \{\infty\}$-valued random variable $N(B)$ is a.s. constant, by the Hewitt–Savage theorem: $\mathbb{P}(N(B) = K) = 1$ for some (non-random) K. If $1 \leq K < \infty$, we may label the elements of $B \cap \mathrm{cl}\,\mathcal{E} = B \cap \mathcal{E}$ in an arbitrary order as $\mathbf{u}_1, \ldots, \mathbf{u}_K$, and each is a.s. constant, by the Hewitt–Savage theorem again, so there exist constant $\mathbf{u}_1, \ldots, \mathbf{u}_K \in B$ with $\mathbb{P}(\mathbf{u}_j \in \mathcal{E}) = 1$ for each j. But $\mathbb{P}(\mathbf{u} \in \mathcal{E}) = 0$ for all $\mathbf{u}$. Hence $K \in \{0, \infty\}$. This establishes (8.1).

Recall that $\mathcal{R}$ denotes the (countable) set of all $B_s(\mathbf{u}; r)$ with $\mathbf{u} \in \mathbb{Q}^d \cap \mathbb{S}^{d-1}$ and $r \in \mathbb{Q} \cap (0, \infty)$. From (8.1) we have that $\mathbb{P}(N(B) \in \{0, \infty\}$ for all $B \in \mathcal{R}) = 1$, which means that $\mathrm{cl}\,\mathcal{E}$ contains no isolated points. □

Corollary 8.6 *Suppose that* $d \in \{1, 2\}$. *Then* $\mathbb{P}(\mathrm{cl}\,\mathcal{E} = \emptyset) = 1$.

Proof For $d = 1$ this is evident, so suppose that $d = 2$. By Proposition 8.3, $\mathbb{P}(\mathrm{cl}\,\mathcal{E}_- \subseteq \partial_s \mathcal{D}_+) = 1$, while Lemma 8.2 shows that $\mathcal{D}_+$ is s-convex, so $\partial_s \mathcal{D}_+$ contains at most two points. Similarly for $\mathrm{cl}\,\mathcal{E}_+$. Thus $\mathrm{cl}\,\mathcal{E}$ has at most four points. Lemma 8.5 then shows that $\mathbb{P}(\mathrm{cl}\,\mathcal{E} = \emptyset) = 1$. □

9 The Convex Hull

For $n \in \mathbb{Z}_+$ let $\mathcal{H}_n := \mathrm{hull}\{S_0, S_1, \ldots, S_n\}$ (a convex polytope). Set $\mathcal{H}_\infty := \cup_{n \geq 0} \mathcal{H}_n$. If $x, y \in \mathcal{H}_\infty$ then $x, y \in \mathcal{H}_n$ for some n, and since $\mathcal{H}_n$ is convex, $\theta x + (1 - \theta) y \in \mathcal{H}_n \subseteq \mathcal{H}_\infty$ for all $\theta \in [0, 1]$. Thus $\mathcal{H}_\infty$ is convex, and hence so is $\mathrm{cl}\,\mathcal{H}_\infty$ [14, p. 44]. Define

$$\mathcal{S}_\infty := \{S_0, S_1, \ldots\}. \tag{9.1}$$

If S_n is transient, then $\mathcal{S}_\infty$ has no finite limit points. Since $\mathcal{H}_n \subseteq \mathrm{hull}\,\mathcal{S}_\infty$, we have $\mathcal{H}_\infty \subseteq \mathrm{hull}\,\mathcal{S}_\infty$, while $\mathcal{H}_\infty$ is a convex set containing $\mathcal{S}_\infty$, so $\mathrm{hull}\,\mathcal{S}_\infty \subseteq \mathcal{H}_\infty$. That is,

$$\mathcal{H}_\infty = \mathrm{hull}\,\mathcal{S}_\infty = \mathrm{hull}\,\{S_0, S_1, S_2, \ldots\}.$$

Also define

$$r_n := \inf\{\|\mathbf{x}\| : \mathbf{x} \in \mathbb{R}^d \setminus \mathcal{H}_n\}.$$

Note that r_n is non-decreasing, so $r_\infty := \lim_{n\to\infty} r_n$ exists in $[0, \infty]$. In [24] it is shown that if $\mathbb{P}(r_\infty = \infty) = 1$, then there is a zero–one law for random variables that are tail-measurable for the sequence $\mathcal{H}_0, \mathcal{H}_1, \mathcal{H}_2, \ldots$: see [24, §3].

Lemma 9.1 *We have* $\mathbb{P}(r_\infty = \infty) = \mathbb{P}(\mathcal{H}_\infty = \mathbb{R}^d) \in \{0, 1\}$.

Proof By definition of r_n, we have $B(\mathbf{0}; r_n) \subseteq \mathcal{H}_n \subseteq \mathcal{H}_\infty$. Thus if $r_\infty = \infty$, we have $\mathcal{H}_\infty = \mathbb{R}^d$. On the other hand, if $\mathcal{H}_\infty = \mathbb{R}^d$, then for any $r \in (0, \infty)$ there exists some $n \in \mathbb{N}$ for which $B(\mathbf{0}; r) \subseteq \mathcal{H}_n$. (If not, there is some r and $\mathbf{x} \in B(\mathbf{0}; r)$ with $\mathbf{x} \notin \mathcal{H}_\infty$.) Then $r_n \geq r$, so $r_\infty \geq r$. Since r was arbitrary, we get $r_\infty = \infty$. Thus $\mathbb{P}(r_\infty = \infty) = \mathbb{P}(\mathcal{H}_\infty = \mathbb{R}^d)$, and the proof is completed by the Hewitt–Savage theorem. □

A consequence of a theorem of Carathéodory is that if $A \subseteq \mathbb{R}^d$ is compact, then hull A is also compact (see e.g. Corollary 3.1 of [14, p. 44]). Thus hull $\mathcal{D}$ is compact, by Theorem 2.1. The following result relates several concepts from earlier to the question of whether the convex hull eventually fills all of space. Here 'int' denotes interior.

Theorem 9.2 *Consider the following statements.*

(i) $\mathbf{0} \in \operatorname{int} \operatorname{hull} \mathcal{D}$.
(ii) $\mathbb{P}(r_\infty = \infty) = 1$.
(iii) $\mathbb{P}(\mathcal{H}_\infty = \mathbb{R}^d) = 1$.
(iv) $\mathcal{D}_\pm = \mathbb{S}^{d-1}$.
(v) $\mathbf{0} \in \operatorname{hull} \mathcal{D}$.

Then the following logical relationships apply: (i) ⇒ (ii) ⇔ (iii) ⇔ (iv) ⇒ (v).

Remarks 9.3

(a) If the random walk is recurrent, then $\mathcal{D} = \mathbb{S}^{d-1}$ (Proposition 2.6) and so (i) and hence (iv) hold, so that $\mathcal{D}_+ = \emptyset$. In other words, if $\mathcal{D}_+ \neq \emptyset$, then $\mathcal{D} \neq \mathbb{S}^{d-1}$, and the random walk is transient.
(b) Examples 10.1 and 10.2 below show that (i) is not necessary for (iii), and (v) is not sufficient for (iii).

In [24], it was shown that sufficient for $\mathbb{P}(\mathcal{H}_\infty = \mathbb{R}^d) = 1$ is that the random walk is recurrent; this follows from Theorem 9.2 and the fact that recurrence implies that $\mathcal{D} = \mathbb{S}^{d-1}$ (Proposition 2.6). Here are some further sufficient conditions.

Corollary 9.4 *Suppose that either (i)* $X \overset{d}{=} -X$, *or (ii)* $\mathbb{E}\|X\| < \infty$ *and* $\mu = \mathbf{0}$. *Then* $\mathbb{P}(\mathcal{H}_\infty = \mathbb{R}^d) = 1$.

Proof By Theorem 9.2, it suffices to show that $\mathcal{D}_\pm = \mathbb{S}^{d-1}$. But under either hypotheses (i) or (ii), the non-degenerate one-dimensional random walk with increment distribution $X \cdot \mathbf{u}$ oscillates. □

Proof of Theorem 9.2 First suppose that (i) holds. If $\mathbf{0} \in \operatorname{int} \operatorname{hull} \mathcal{D}$ then there exist $m \in \mathbb{N}$ and $\mathbf{u}_1, \ldots, \mathbf{u}_m \in \mathcal{D}$ such that $\mathbf{0}$ is also in the interior of the convex polytope $P(\mathbf{u}_1, \ldots, \mathbf{u}_m) := \operatorname{hull} \{\mathbf{u}_1, \ldots, \mathbf{u}_m\}$. Let

$$R(\mathbf{v}_1, \ldots, \mathbf{v}_m) = \inf\{\|\mathbf{x}\| : \mathbf{x} \in \mathbb{R}^d \setminus P(\mathbf{v}_1, \ldots, \mathbf{v}_m)\},$$

which is zero unless $\mathbf{0}$ lies in the interior of $P(\mathbf{v}_1, \ldots, \mathbf{v}_m)$, when it is equal to the shortest distance from $\mathbf{0}$ to the boundary of $P(\mathbf{v}_1, \ldots, \mathbf{v}_m)$. In particular, note that $R(\mathbf{u}_1, \ldots, \mathbf{u}_m) = \delta_0 > 0$.

For $\mathbf{v}_1, \ldots, \mathbf{v}_m \in \mathbb{R}^d$, the map $(\mathbf{v}_1, \ldots, \mathbf{v}_m) \mapsto P(\mathbf{v}_1, \ldots, \mathbf{v}_m)$, as a function from $\mathbb{R}^{md}$ to convex, compact subsets of $\mathbb{R}^d$ with the Hausdorff metric, is continuous. So the map from $(\mathbf{v}_1, \ldots, \mathbf{v}_m)$ to $R(\mathbf{v}_1, \ldots, \mathbf{v}_m)$ is also continuous. Hence for any $\delta \in (0, \delta_0)$, we can find $\varepsilon > 0$ sufficiently small such that $B(\mathbf{0}; \delta)$ is contained in $P(\mathbf{v}_1, \ldots, \mathbf{v}_m)$ for all $\mathbf{v}_i$ with $\|\mathbf{v}_i - \mathbf{u}_i\| < \varepsilon$. For such an $\varepsilon > 0$, let

$$C_i(r, \varepsilon) = \{\mathbf{x} \in \mathbb{R}^d : \|\hat{\mathbf{x}} - \mathbf{u}_i\| < \varepsilon, \ \|\mathbf{x}\| \geq r\}.$$

Then for any $\mathbf{x}_1, \ldots, \mathbf{x}_m$ with $\mathbf{x}_i \in C_i(r, \varepsilon)$, we have that $\operatorname{hull} \{\hat{\mathbf{x}}_1, \ldots, \hat{\mathbf{x}}_m\}$ contains the ball $B(\mathbf{0}; \delta)$. Thus, since $\|\mathbf{x}_i\| \geq r$,

$$B(\mathbf{0}; r\delta) \subseteq \operatorname{hull} \{r\hat{\mathbf{x}}_1, \ldots, r\hat{\mathbf{x}}_m\} \subseteq \operatorname{hull} \{\mathbf{x}_1, \ldots, \mathbf{x}_m\}.$$

Since $\mathbf{u}_i \in \mathcal{D} = \mathcal{D}_\infty$ (by Theorem 2.8), we have $S_n \in C_i(r, \varepsilon)$ i.o., a.s. Thus $B(\mathbf{0}; r\delta) \subseteq \mathcal{H}_n$ for all but finitely many n. That is $\liminf_{n\to\infty} r_n \geq r\delta$, a.s. Since $r > 0$ was arbitrary, we get $r_\infty = \infty$, a.s. Thus (i) implies (ii), and (ii) is equivalent to (iii) by Lemma 9.1.

Suppose that $\mathbf{u} \in \mathcal{D}_+$, so that $\mathbb{P}(\mathbf{u} \in \mathcal{L}_+) = 1$. Then $S_n \cdot \mathbf{u} \to \infty$, so that $\inf_{n\geq 0}(S_n \cdot \mathbf{u}) = c$ for some $c > -\infty$. It follows that $S_0, S_1, S_2, \ldots$ are contained in the half-space $H_+(\mathbf{u}) = \{\mathbf{x} \in \mathbb{R}^d : \mathbf{x} \cdot \mathbf{u} \geq c\}$. Thus $\mathcal{H}_n \subseteq H_+(\mathbf{u})$ for all n, and hence $\mathcal{H}_\infty \subseteq H_+(\mathbf{u})$. Thus $\mathcal{H}_\infty = \mathbb{R}^d$ implies $\mathcal{D}_+ = \mathcal{D}_- = \emptyset$, and so, by Lemma 8.1, (iii) implies (iv).

To show that (iv) implies (iii), we prove the contrapositive. By Lemma 9.1, it suffices to suppose that $\mathbb{P}(\mathcal{H}_\infty = \mathbb{R}^d) = 0$. Since $\operatorname{cl} \mathcal{H}_\infty$ is closed and convex, it can be written as an intersection of hyperplanes (see e.g. Corollary 4.1 of [14, p. 55]); in particular, if $\operatorname{cl} \mathcal{H}_\infty$ is not the whole of $\mathbb{R}^d$, it is contained in a half-space $H_-(\mathbf{u}) = \{\mathbf{x} \in \mathbb{R}^d : \mathbf{x} \cdot \mathbf{u} \leq c\}$ for some $\mathbf{u} \in \mathbb{S}^{d-1}$ and $c \in \mathbb{R}$. Thus $\sup_{n\geq 0}(S_n \cdot \mathbf{u}) < \infty$. In particular, $\mathcal{P}_\pm$ is not the whole of $\mathbb{S}^{d-1}$. By Corollary 8.4, this implies that $\mathcal{D}_\pm \neq \mathbb{S}^{d-1}$.

Finally, we show that (iv) implies (v). Suppose that $\mathbf{0} \notin \operatorname{hull} \mathcal{D}$. Since $\operatorname{hull} \mathcal{D}$ is closed, this means that there is a hyperplane that separates $\mathbf{0}$ from $\operatorname{hull} \mathcal{D}$, so there is a $\mathbf{u} \in \mathbb{S}^{d-1}$ and $c < 0$ such that $S(\mathbf{u}) = \{\mathbf{x} \in \mathbb{S}^{d-1} : \mathbf{x} \cdot \mathbf{u} \geq c\}$ contains no point of $\mathcal{D}$. Since $S(\mathbf{u})$ is compact, it must thus contain only finitely many of $\hat{S}_0, \hat{S}_1, \ldots$. That is $\limsup_{n\to\infty}(\hat{S}_n \cdot \mathbf{u}) \leq c$, and hence $\limsup_{n\to\infty}(S_n \cdot \mathbf{u}) \leq 0$. In particular, $\mathcal{P}_\pm$ is not the whole of $\mathbb{S}^{d-1}$, and Corollary 8.4 shows that $\mathcal{D}_\pm \neq \mathbb{S}^{d-1}$. □

10 Some Examples

Let $\mathbf{e}_1, \ldots, \mathbf{e}_d$ denote the standard orthonormal basis vectors of $\mathbb{R}^d$. For convenience we locate all our random walks on the integer lattice $\mathbb{Z}^d$, but this is not essential. We write $\xi \sim \text{Rad}$ to mean that $\mathbb{P}(\xi = +1) = \mathbb{P}(\xi = -1) = 1/2$ (a Rademacher distribution), and, for $\alpha > 0$, write $\zeta \sim S(\alpha)$ to mean that $\zeta \in \mathbb{Z}$ has $\mathbb{P}(\zeta \geq r) = \mathbb{P}(\zeta \leq -r) = \frac{1}{2} r^{-\alpha}$ for $r \in \mathbb{N}$. Our examples are constructed mostly from components that are copies of $\xi \sim \text{Rad}$ or $\zeta \sim S(\alpha)$.

If $\xi_1, \xi_2, \ldots$ are independent copies of $\xi \sim \text{Rad}$, then we write $W_n = \sum_{i=1}^n \xi_i$ for the associated simple symmetric random walk (SSRW) on $\mathbb{Z}$. If $\zeta_1, \zeta_2, \ldots$ are independent copies of $\zeta \sim S(\alpha)$, then we write $Y_n = \sum_{i=1}^n \zeta_i$.

We recall some well-known facts about W_n and Y_n. The local limit theorem for SSRW on $\mathbb{Z}$ (see e.g. [8, pp. 141–143]) says that, with ϕ the standard Gaussian density function,

$$\lim_{n\to\infty} \sup_{x\in\mathbb{Z}} \left| \frac{n^{1/2}}{2} \mathbb{P}(W_n = 2x - n) - \phi\left(\frac{2x - n}{\sqrt{n}}\right) \right| = 0. \tag{10.1}$$

If $\alpha \in (0, 1)$, then Y_n is transient and oscillates: $|Y_n| \to \infty$ and Y_n takes both signs i.o., and, moreover (see e.g. Theorem 3.5 of [13])

$$\text{if } \alpha \in (0, 1), \text{ then } \liminf_{n\to\infty} n^{-1} |Y_n| = \infty, \text{ a.s.} \tag{10.2}$$

If $\alpha \in (0, 2)$, $\alpha \neq 1$, then $n^{-1/\alpha} Y_n$ converges in distribution to (a constant multiple of) a symmetric α-stable random variable, since ζ is in the corresponding domain of normal attraction, with no centering (see e.g. Theorem 2.6.7 of [16] and [12, p. 580]). If g is the density of this limiting random variable, then Gnedenko's local limit theorem (see Theorem 4.2.1 of [16]) says that

$$\lim_{n\to\infty} \sup_{x\in\mathbb{Z}} \left| n^{1/\alpha} \mathbb{P}(Y_n = x) - g(n^{-1/\alpha} x) \right| = 0. \tag{10.3}$$

Note that g is uniformly bounded: this follows from the inversion formula for densities and the fact that the characteristic function of a symmetric stable random variable is of the form $\mathrm{e}^{-c|t|^\alpha}$, for some $c > 0$ (see e.g. [12, p. 570]).

Example 10.1 Suppose that $d = 2$. Let $X = \mathbf{e}_1 + \mathbf{e}_2 \zeta$ where $\zeta \sim S(\alpha)$ for $\alpha > 0$.

If $\alpha > 1$ then $\mathbb{E}\|X\| < \infty$ and $\mathbb{E}X = \mathbf{e}_1$, so the SLLN implies that S_n is transient with limiting direction $\mathbf{e}_1$, and Proposition 4.1 shows that $\mathcal{D} = \{\mathbf{e}_1\}$.

If $\alpha \in (0, 1)$, then $\|S_n\| \geq |S_n \cdot \mathbf{e}_1| = n$ so the walk is again transient. Write $X_i = \mathbf{e}_1 + \mathbf{e}_2 \zeta_i$ where the ζ_i are independent copies of ζ. Let $Y_n = \sum_{i=1}^n \zeta_i$. For $j = \pm 1$,

$$\|\hat{S}_n - j\mathbf{e}_2\| \leq \frac{n}{\|S_n\|} + \left| \frac{Y_n}{\|S_n\|} - j \right|.$$

By (10.2) we have that $n/\|S_n\| \le n/|Y_n| \to 0$, a.s., and so $\|S_n\|/|Y_n| \to 1$, a.s., and hence

$$\left| \frac{Y_n}{\|S_n\|} - \operatorname{sgn}(Y_n) \right| \le \left| \frac{|Y_n|}{\|S_n\|} - 1 \right| \to 0, \text{ a.s.}$$

It follows that, for $j = \pm 1$,

$$\liminf_{n\to\infty} \|\hat{S}_n - j\mathbf{e}_2\| = \liminf_{n\to\infty} |\operatorname{sgn}(Y_n) - j| = 0, \text{ a.s.}$$

Hence $\{\pm \mathbf{e}_2\} \subseteq \mathcal{D}$. On the other hand, if $\mathbf{u} \in \mathbb{S}^1 \setminus \{\pm \mathbf{e}_2\}$, we have $u_1 := \mathbf{u} \cdot \mathbf{e}_1 \neq 0$, and

$$\liminf_{n\to\infty} \|\hat{S}_n - \mathbf{u}\| \ge \liminf_{n\to\infty} \left| \frac{n}{\|S_n\|} - u_1 \right| = |u_1| > 0,$$

so $\mathbf{u} \notin \mathcal{D}$. Thus $\mathcal{D} = \{\pm \mathbf{e}_2\}$.

Finally, note that this example obviously has $\mathcal{H}_\infty \neq \mathbb{R}^2$ (since $S_n \ge 0$ for all n) while $\mathbf{0} \in \operatorname{hull} \mathcal{D}$, but $\mathbf{0} \notin \operatorname{int} \operatorname{hull} \mathcal{D}$. This shows that (iii) and (v) of Theorem 9.2 are not equivalent. △

Example 10.2 Suppose that $d = 2$. Let $X = \mathbf{e}_1\xi + \mathbf{e}_2\zeta$ where ξ and ζ are independent, $\xi \sim$ Rad, and $\zeta \sim S(\alpha)$ for $\alpha > 0$.

First suppose that $\alpha > 2$. Here $\mathbb{E}(\|X\|^2) < \infty$ and $\mathbb{E}X = \mathbf{0}$, so the central limit theorem applies, and Corollary 2.10 shows that $\mathcal{D} = \mathbb{S}^1$. Alternatively, note that the walk in this case is recurrent (see e.g. [8, Theorem 4.2.8]) and apply Proposition 2.6.

Next suppose that $\alpha \in (1, 2)$. In this case $\mathbb{E}X = \mathbf{0}$ but $\mathbb{E}(\|X\|^2) = \infty$. Here the walk is transient, as follows from the Borel–Cantelli lemma and the local limit theorems (10.1) and (10.3), which together show that $\mathbb{P}(S_n = \mathbf{0}) = \mathbb{P}(W_n = 0)\mathbb{P}(Y_n = 0) = O(n^{-(1/2)-(1/\alpha)})$. By construction, $X \overset{d}{=} -X$, so Proposition 5.2 shows that $\mathcal{D} = \mathbb{S}^1$.

Finally, suppose that $\alpha \in (0, 1)$. Since $|S_n \cdot \mathbf{e}_1| \le n$, a similar argument to that in Example 10.1 shows that $\mathcal{D} = \{\pm \mathbf{e}_2\}$. Note that this walk is transient, by Corollary 2.7, and, by Corollary 9.4, $\mathbb{P}(\mathcal{H}_\infty = \mathbb{R}^d) = 1$. This example has $\mathbf{0} \in \operatorname{hull} \mathcal{D}$, $\mathbf{0} \notin \operatorname{int} \operatorname{hull} \mathcal{D}$, and $\mathbb{P}(\mathcal{H}_\infty = \mathbb{R}^d) = 1$, showing that (i) and (iii) of Theorem 9.2 are not equivalent. △

Example 10.3 Suppose that $d \ge 4$. Let $X = \sum_{k=1}^{d-1} \mathbf{e}_k \zeta^{(k)} + \mathbf{e}_d \xi$ where $\xi, \zeta^{(1)}, \ldots, \zeta^{(d-1)}$ are independent, $\xi \sim$ Rad, and $\zeta^{(k)} \sim S(\alpha)$ for $\alpha \in (1, 2)$. This random walk has $X \overset{d}{=} -X$, $\mu = \mathbf{0}$, and is transient. Let $E_d := \{\mathbf{u} \in \mathbb{S}^{d-1} : \mathbf{u} \cdot \mathbf{e}_d = 0\}$, a copy of $\mathbb{S}^{d-2}$.

Recall that $C(\mathbf{u}; r) = \{\mathbf{x} \in \mathbb{R}^d \setminus \{\mathbf{0}\} : \|\hat{\mathbf{x}} - \mathbf{u}\| < r\}$. Fix $\varepsilon > 0$, and set

$$B_n := \{(x_1, x_2, \ldots, x_d) \in \mathbb{Z}^d : |x_d| \le n^{(1/2)+\varepsilon}\}.$$

Then we have the estimate

$$\mathbb{P}(S_n \in C(\mathbf{u}; r)) \le \mathbb{P}(|S_n \cdot \mathbf{e}_d| > n^{(1/2)+\varepsilon}) + \sum_{\mathbf{x} \in B_n \cap C(\mathbf{u}; r)} \mathbb{P}(S_n = \mathbf{x}).$$

Here we have from the local limit theorems (10.1) and (10.3) that, for some $C < \infty$,

$$\mathbb{P}(S_n = \mathbf{x}) = \mathbb{P}(W_n = x_d) \prod_{i=1}^{d-1} \mathbb{P}(Y_n = x_i) \le C n^{-(d-1)/\alpha} \cdot n^{-1/2},$$

for all $\mathbf{x} \in \mathbb{Z}^d$. Standard binomial tail bounds show that for SSRW $\mathbb{P}(|W_n| > n^{(1/2)+\varepsilon}) \le C \exp(-cn^{2\varepsilon})$ for constants $c > 0$ and $C < \infty$. Thus we get

$$\mathbb{P}(S_n \in C(\mathbf{u}; r)) \le C \exp(-cn^{2\varepsilon}) + C \sum_{\mathbf{x} \in B_n \cap C(\mathbf{u}; r)} n^{-(d-1)/\alpha} \cdot n^{-1/2}. \qquad (10.4)$$

Fix $\mathbf{u} \notin E_d$, and take $0 < r < |\mathbf{u} \cdot \mathbf{e}_d|$. Then any $\mathbf{x} = (x_1, x_2, \ldots, x_d) \in C(\mathbf{u}; r)$ has

$$\left| x_d - \|\mathbf{x}\| \mathbf{u} \cdot \mathbf{e}_d \right| \le \left\| \mathbf{x} - \|\mathbf{x}\| \mathbf{u} \right\| < r \|\mathbf{x}\|.$$

Thus $(|\mathbf{u} \cdot \mathbf{e}_d| - r)\|\mathbf{x}\| < |x_d| < (|\mathbf{u} \cdot \mathbf{e}_d| + r)\|\mathbf{x}\|$. It follows that there is a constant $C < \infty$ such that $|x_i| < C|x_d|$ for all $1 \le i \le d-1$ and all $\mathbf{x} \in C(\mathbf{u}; r)$. Hence the number of $\mathbf{x} \in B_n \cap C(\mathbf{u}; r)$ is at most $O(n^{(d/2)+d\varepsilon})$. Thus we obtain from (10.4) that

$$\mathbb{P}(S_n \in C(\mathbf{u}; r)) \le C \exp(-cn^{2\varepsilon}) + C n^{d\varepsilon} n^{-(d-1)(2-\alpha)/(2\alpha)},$$

where $C < \infty$ depends on $\mathbf{u}$ and r, but not ε. Thus for any α satisfying

$$1 < \alpha < \frac{2(d-1)}{1+d} \qquad (10.5)$$

we can choose $\varepsilon > 0$ small enough to ensure that $\sum_{n \ge 1} \mathbb{P}(S_n \in C(\mathbf{u}; r)) < \infty$. We can find α satisfying (10.5) provided $d > 3$.

Thus if we have $d \ge 4$ and α satisfying (10.5), the Borel–Cantelli lemma shows that $\mathbf{u} \notin \mathcal{D}$ for any $\mathbf{u} \notin E_d$, i.e., $\mathcal{D} \subseteq E_d$. On the other hand, we have $n^{-1/\alpha} S_n$ converges in distribution to $Z = (Z_1, \ldots, Z_{d-1}, 0)$, where the Z_i are independent α-stable random variables with supp $Z_i = \mathbb{R}$. It follows that supp $\hat{Z} = E_d$, and so, by Proposition 2.9, we conclude that $\mathcal{D} = E_d$. △

We write $\zeta \sim S_+(\alpha)$ to mean that $\zeta \in \mathbb{Z}_+$ has $\mathbb{P}(\zeta \ge r) = r^{-\alpha}$ for $r \in \mathbb{N}$.

Example 10.4 Let $d \in \mathbb{N}$ and $\alpha \in (0, 1)$. Let $X = \sum_{j=1}^{k} \mathbf{u}_j \zeta^{(j)}$ where $k \in \mathbb{N}$, the $\zeta^{(j)} \sim S_+(\alpha)$ are independent, and $\mathbf{u}_1, \ldots, \mathbf{u}_k$ are fixed vectors in $\mathbb{R}^d$. For $\mathbf{z} = (z_1, \ldots, z_k) \in \mathbb{R}^k$, set $\Lambda(\mathbf{z}) := \sum_{j=1}^{k} z_j \mathbf{u}_j$.

Write $X_i = \sum_{j=1}^{k} \mathbf{u}_j \zeta_i^{(j)}$, where the $\zeta_i^{(j)}$ are independent copies of $\zeta^{(j)}$, and let $Y_n^{(j)} = \sum_{i=1}^{n} \zeta_i^{(j)}$. Then $n^{-1/\alpha}(Y_n^{(1)}, \ldots, Y_n^{(k)})$ converges in distribution to $(Z_1, \ldots, Z_k)$, where $Z_1, \ldots, Z_k$ are independent, positive α-stable random variables supported on $\mathbb{R}_+$. By the continuous mapping theorem, $n^{-1/\alpha} S_n$ converges in distribution to $\sum_{j=1}^{k} \mathbf{u}_j Z_j =: V$. Since V is continuous, $\mathbb{P}(V = \mathbf{0}) = 0$, and so $\mathbb{P}(\hat{V} \in \mathbb{S}^{d-1}) = 1$. Thus

$$\operatorname{supp} \hat{V} = C := C(\mathbf{u}_1, \ldots, \mathbf{u}_k) := \operatorname{cl} \left\{ \frac{\Lambda(\mathbf{z})}{\|\Lambda(\mathbf{z})\|} : \mathbf{z} \in \mathbb{R}^k, \ z_1, \ldots, z_k > 0, \ \|\Lambda(\mathbf{z})\| > 0 \right\}.$$

Hence by Proposition 2.9(ii) we have that $C \subseteq \mathcal{D}$.

To get an inclusion in the other direction, we use the notation of Sect. 7. We have $\operatorname{supp} X = \operatorname{cl}\{\Lambda(\mathbf{z}) : \mathbf{z} \in \mathbb{N}^k\}$, and for any $\mathbf{x} \in \operatorname{supp} X$, either $\hat{\mathbf{x}} = \mathbf{0}$ (if $\mathbf{x} = \mathbf{0}$) or else $\hat{\mathbf{x}} = \lim_{n\to\infty} \hat{\mathbf{x}}_n \in \mathbb{S}^{d-1}$ with $\mathbf{x}_n = \Lambda(\mathbf{z}_n)$ and $\mathbf{z}_n \in \mathbb{N}^k$. It follows that

$$\left\{ \frac{\Lambda(\mathbf{z})}{\|\Lambda(\mathbf{z})\|} : \mathbf{z} \in \mathbb{N}^k, \ \|\Lambda(\mathbf{z})\| > 0 \right\} \subseteq \mathcal{X}' \subseteq \{\mathbf{0}\} \cup \operatorname{cl} \left\{ \frac{\Lambda(\mathbf{z})}{\|\Lambda(\mathbf{z})\|} : \mathbf{z} \in \mathbb{N}^k, \ \|\Lambda(\mathbf{z})\| > 0 \right\}.$$

Lemma 7.4 then shows that

$$\mathcal{X} = \operatorname{cl} \left\{ \frac{\Lambda(\mathbf{z})}{\|\Lambda(\mathbf{z})\|} : \mathbf{z} \in \mathbb{N}^k, \ \|\Lambda(\mathbf{z})\| > 0 \right\} = \operatorname{cl} \left\{ \frac{\Lambda(\mathbf{z})}{\|\Lambda(\mathbf{z})\|} : \mathbf{z} \in \lambda\mathbb{N}^k, \ \|\Lambda(\mathbf{z})\| > 0 \right\},$$

for any $\lambda > 0$, by scale invariance. It follows that

$$\mathcal{X} = \operatorname{cl} \left\{ \frac{\Lambda(\mathbf{z})}{\|\Lambda(\mathbf{z})\|} : \mathbf{z} \in \mathbb{Q}^k, \ z_1, \ldots, z_k > 0, \ \|\Lambda(\mathbf{z})\| > 0 \right\}.$$

Since $\mathbb{Q}^k$ is dense in $\mathbb{R}^k$, we get $\mathcal{X} = C$. Moreover, C is the closure of an s-convex set, and hence itself s-convex, by Lemma 7.8, and hence cl s-hull $\mathcal{X}$ = s-hull $\mathcal{X} = C$, by Lemma 7.7. Then Theorem 7.1 confirms that $\mathcal{D} = C$. △

11 Concluding Remarks

The Borel–Cantelli lemma shows that if for some $\varepsilon > 0$, $\sum_{n=1}^{\infty} \mathbb{P}(\|\hat{S}_n - \mathbf{u}\| < \varepsilon) < \infty$, then $\mathbb{P}(S_n \in C(\mathbf{u}; \varepsilon) \text{ i.o.}) = 0$, and so $\mathbf{u} \notin \mathcal{D}$, by Proposition 2.5. This is not sharp, however, as is already shown by the case of $d = 1$, when, for example, $+1 \in \mathcal{D}$ if and only if $\sum_{n=1}^{\infty} n^{-1}\mathbb{P}(S_n > 0) = \infty$ [12, p. 415].

Problem 11.1 *Is there a criterion for* $\mathbf{u} \in \mathcal{D}$ *in terms of* $\mathbb{P}(S_n \in \cdot)$*?*

We do not necessarily expect a simple answer to Problem 11.1: in $d = 1$, Kesten (Corollary 1 of [19, p. 1177]) gives a criterion for $\mathbf{x} \in \mathcal{A}^\alpha$ where $\mathcal{A}^\alpha$ is as defined at (3.1).

Proposition 5.2 leaves the following question.

Problem 11.2 *Suppose that $d = 2$, $\mathbb{E}\|X\| < \infty$, and $\mu = \mathbf{0}$. Is $\mathcal{D}$ always equal to $\mathbb{S}^1$?*

A The Recurrent Case

For most of the questions in the present paper, the main interest is the transient case, because, loosely speaking, any recurrent random walk explores all of space and hence all directions at all distances. Proposition A.1 is a precise version of this statement. Recall [8, p. 190] that S_n is *recurrent* if there is a non-empty set $\mathcal{R}$ of points $\mathbf{x} \in \mathbb{R}^d$ (the recurrent values) such that, for any $\varepsilon > 0$, $\|S_n - \mathbf{x}\| < \varepsilon$ i.o., a.s.

Proposition A.1 *If S_n is recurrent, then there exists $h > 0$ such that a.s., for any $\mathbf{x} \in \mathbb{R}^d$, $S_n \in B(\mathbf{x}; h)$ i.o.*

Proof Since S_n is recurrent, the set $\mathcal{R}$ of recurrent values is a closed subgroup of $\mathbb{R}^d$ and coincides with the set of *possible values* for the walk: see [8, p. 190]. Since S_n is genuinely d-dimensional, it follows from e.g. Theorem 21.2 of [1, p. 225] that $\mathcal{R}$ contains a further closed subgroup $\mathcal{R}'$ of the form $H\mathbb{Z}^d$ where H is a non-singular d by d matrix. Hence there exists $h > 0$ such that for every $\mathbf{x} \in \mathbb{R}^d$ there exists $\mathbf{y} \in \mathcal{R}'$ with $\|\mathbf{x} - \mathbf{y}\| < h/2$, and since $\mathcal{R}'$ is a countable set of recurrent values for the walk, we have that, a.s., for any $\mathbf{x} \in \mathbb{R}^d$, $S_n \in B(\mathbf{x}; h)$ i.o. □

Acknowledgments Some of this work was done while the first author was visiting Durham University in July–August 2018, supported by an international study scholarship awarded by the Secretariat of Public Education of Mexico and Universidad Nacional Autónoma de México (UNAM). The authors are grateful for the comments of an anonymous referee, and also to Nicholas Georgiou, James McRedmond, Mikhail Menshikov, and Vladislav Vysotskiy for discussions on the topic of this work.

References

1. Bhattacharya, R.N., Rao, R.R.: Normal Approximation and Asymptotic Expansions, updated reprint of the 1986 edition. SIAM, Philadelphia (2010)
2. Blackwell, D.: On transient Markov processes with a countable number of states and stationary transition probabilities. Ann. Math. Statist. **26**, 654–658 (1955)
3. Bucy, R.S.: Recurrent Events for Transient Markov Chains. PhD. Thesis, University of California, Berkeley (1963)
4. Bucy, R.S.: Recurrent sets. Ann. Math. Statist. **36**, 535–545 (1965)
5. Chung, K.L., Derman, C.: Non-recurrent random walks. Pacific J. Math. **6**, 441–447 (1956)
6. Chung, K.L., Fuchs, W.H.J.: On the distribution of values of sums of random variables. Mem. Am. Math. Soc. **6**, 12 (1951)
7. Doney, R.A.: Recurrent and transient sets for 3-dimensional random walks. Z. Wahrscheinlich. verw. Geb. **4**, 253–259 (1965)

8. Durrett, R.: Probability: Theory and Examples, 4th edn. Cambridge University Press, Cambridge (2010)
9. Erickson, K.B.: Recurrence sets of normed random walk in R^d. Ann. Probab. **4**, 802–828 (1976)
10. Erickson, K.B.: The limit points in $\overline{R^d}$ of averages of i.i.d. random variables. Ann. Probab. **28**, 498–510 (2000)
11. Erickson, K.B., Kesten, H.: Strong and weak limit points of a normalized random walk. Ann. Probab. **2**, 553–579 (1974)
12. Feller, W.: An Introduction to Probability Theory and its Applications, Volume II. 2nd edn. Wiley, New York (1971)
13. Griffin, P.S.: An integral test for the rate of escape of d-dimensional random walk. Ann. Probab. **11**, 953–961 (1983)
14. Gruber, P.M.L.: Convex and Discrete Geometry. Springer, New York (2007)
15. Hewitt, E., Savage, L.J.: Symmetric measures on Cartesian products. Trans. Am. Math. Soc. **80**, 470–501 (1955)
16. Ibragimov, I.A., Linnik, Y.V.: Independent and Stationary Sequences of Random Variables. Wolters-Noordhoff, Groningen (1971)
17. Itô, K., McKean, H.P. Jr.: Potentials and the random walk. Ill. J. Math. **4**, 119–132 (1960)
18. Jain, N.C., Orey, S.: Some properties of random walk paths. J. Math. Anal. Appl. **43**, 795–815 (1973)
19. Kesten, H.: The limit points of a normalized random walk. Ann. Math. Statist. **41**, 1173–1205 (1970)
20. Kesten, H.: Erickson's conjecture on the rate of escape of d-dimensional random walk. Trans. Am. Math. Soc. **240**, 65–113 (1978)
21. Kuelbs, J.: When is the cluster set of S_n/a_n empty? Ann. Probab. **9**, 377–394 (1981)
22. Lamperti, J.: Wiener's test and Markov chains. J. Math. Anal. Appl. **6**, 58–66 (1963)
23. Lawler, G.F., Limic, V.: Random Walk: A Modern Introduction. Cambridge University Press, Cambridge (2010)
24. McRedmond, J., Wade, A.R.: The convex hull of a planar random walk: perimeter, diameter, and shape. Electron. J. Probab. **23**, 131 (2018)
25. Menshikov, M., Popov, S., Wade, A.: Non-homogeneous Random Walks. Cambridge University Press, Cambridge (2016)
26. Molchanov, I.: Theory of Random Sets, 2nd edn. Springer, London (2017)
27. Murdoch, B.H.: Wiener's test for atomic Markov chains. Ill. J. Math. **12**, 35–56 (1968)
28. Pruitt, W.E.: The contribution to the sum of the summand of maximum modulus. Ann. Probab. **15**, 885–896 (1987)
29. Schmutz, E.: Rational points on the unit sphere. Cent. Eur. J. Math. **6**, 482–487 (2008)
30. Spitzer, F.L.: Principles of Random Walk, 2nd edn. Springer, New York (1970)
31. Uchiyama, K.: Wiener's test for random walks with mean zero and finite variance. Ann. Probab. **26**, 368–376 (1998)

First Passage Times of Subordinators and Urns

Philippe Marchal

Abstract It is well-known that the first time a stable subordinator reaches $[1, +\infty)$. is Mittag-Leffler distributed. These distributions also appear as limiting distributions in triangular Polya urns. We give a direct link between these two results, using a previous construction of the range of stable subordinators. Beyond the stable case, we show that for a subclass of complete subordinators in the domain of attraction of stable subordinators, the law of the first passage time is given by the limit of an urn with the same replacement rule but with a random initial composition.

Keywords Stable subordinator · First passage time · Polya urn

1 Introduction

Let $(S_t)_{t\geq 0}$ be a stable subordinator of index $\alpha \in (0, 1)$, started at 0, and let T be the first passage time in $[1, +\infty)$.

$$T = \inf\{t > 0, S_t > 1\}$$

Then it is well-known that the law of T is the Mittag-Leffler distribution with parameter α, which is characterized by its moments:

$$\mathbb{E}T^n = \frac{\Gamma(1/\alpha + n)}{\Gamma(1/\alpha)\Gamma(1 + n\alpha)}$$

See for instance [10], p.10. This same distribution also appears as the asymptotic number of white balls in a classical Polya urn scheme. Let us introduce some standard notation.

P. Marchal (✉)
LAGA, Université Paris Sorbonne Nord, Villetaneuse, France
e-mail: marchal@math.univ-paris13.fr

L. Chaumont, A. E. Kyprianou (eds.), *A Lifetime of Excursions Through Random Walks and Lévy Processes*, Progress in Probability 78,
https://doi.org/10.1007/978-3-030-83309-1_18

Definition We call an urn scheme with replacement matrix

$$\begin{pmatrix} a & b \\ c & d \end{pmatrix}$$

and initial condition (B_0, W_0) the following process. We initially have a black and a white ball with respective weights (B_0, W_0). Then sequentially, a ball is drawn at random with probability proportional to its weight. If this ball is black, it is replaced into the urn together with a black ball of weight a and a white ball of weight b. If the ball is white, it is put back into the urn together with a black ball with weight c and a black ball with weight d.

Consider the case with the replacement matrix

$$\begin{pmatrix} 1 & 0 \\ 1-\alpha & \alpha \end{pmatrix} \tag{1}$$

and initial condition (B_0, W_0). Let W_n be the total weight of white balls after n steps. Then $n^{-\alpha} W_n$ converges in law to a Mittag-Leffler random variable X which can be characterized by its moments, namely

$$\mathbb{E}X^n = \frac{\Gamma(B_0 + W_0)\Gamma(W_0/\alpha + n)}{\Gamma(W_0/\alpha)\Gamma(B_0 + W_0 + n\alpha)} \tag{2}$$

See e.g. [6]. In particular, X has the same law as the first passage time T defined above with the choice of parameters $(B_0, W_0) = (1-\alpha, \alpha)$. Note that (2) still holds when $\alpha = 1$.

We argue that these two results are directly related via a construction of stable subordinators that first appeared in [7] and that was then extended to complete subordinators in [8]. Complete subordinators can be indexed by all possible measurable functions $\beta : [0,1] \to [0,1]$ and have Lévy-Khintchine exponent given by

$$\phi^{(\beta)}(\lambda) = -\log \mathbb{E}[\exp(-\lambda S_1^{(\beta)})] = \exp \int_0^1 \frac{(\lambda-1)\beta(x)}{1+(\lambda-1)x} dx \tag{3}$$

For general references on subordinators, see e.g. [2] and [11]. Our result is that for a subclass of complete subordinators, the first passage time is also related to an urn process:

Theorem 1 *Let $\beta : [0,1] \to [0,1]$ be a measurable function which is constant, equal to $\alpha \in (0,1]$ on an interval $[0,h]$ for some $h \in (0,1]$. Let $(S_t^{(\beta)})$ be the subordinator with exponent given by* (3) *and let*

$$T^{(\beta)} = \inf\{t > 0, S_t^{(\beta)} > 1\}$$

be its first passage time to $[1, +\infty)$. *Then, up to a multiplicative constant,* $T^{(\beta)}$ *has the same law as the limit of* $n^{-\alpha} W_n$ *where* W_n *is the total weight of white balls in an urn scheme with replacement matrix* (1) *and random initial conditions as follows. Put* $\theta = (1/h) - 1$. *Then for all integers* $l, m \geq 0$,

$$\mathbb{P}((B_0, W_0) = (l+(m+1)(1-\alpha), (m+1)\alpha) = \frac{\theta^l e^{-\theta}}{l!} \frac{1}{2i\pi} \int_C \frac{dt}{t} \psi(t)^m (1-\psi(t)) \frac{1-(1/t)^{l+1}}{1-(1/t)}$$

where C *is the unit circle of the complex plane and the function* ψ *is given by*

$$\psi(t) = 1 - \exp\left(\int_0^1 \frac{t\gamma(x)}{1-tx} dt\right)$$

with

$$\gamma(x) = \beta\left(\frac{1}{\theta + 1 - \theta x}\right)$$

The presence of a multiplicative constant in Theorem 1 is not a real issue since this corresponds to replacing $(S_t^{(\beta)})$ with $(S_{ct}^{(\beta)})$ for some positive constant c. We stated our result for the entrance to $[1, +\infty)$ but of course, similar results hold with a staightforward adaptation for the entrance to $[a, +\infty)$ for any $a > 0$.

When $h = 1$, the subordinator is stable. When $h < 1$, the process is in the domain of attraction of an α-stable subordinator in small time: as $t \to 0$, $t^{-1/\alpha} S_t^{(\beta)}$ converges in law to the (unique) positive stable distribution with index α.

Conversely however, a complete subordinator $(S_t^{(\beta)})$ may belong to the domain of attraction of an α-stable subordinator in small time without the function β being constant near 0. Take for instance $\beta(x) = \alpha + (1 - \alpha)x$, then $(S_t^{(\beta)})$ belongs to this domain of attraction but Theorem 1 does not apply. It would be interesting to know how far Theorem 1 could be generalized for subordinators of this kind, that is, whether the first passage time can be related to an urn process.

Note that if the hypothesis of Theorem 1 on β is satisfied with $\alpha = 1$, the subordinator has positive drift whereas if it is satisfied with $\alpha = 0$, the subordinator is a compound Poisson process, see [8]. In the case $\alpha = 1$, Theorem 1 still holds. On the other hand, in the case $\alpha = 0$, the first passage time problem reduces to a problem on random walks which can be handled using the same tools as in Sect. 3.3. This last case is in fact very classical and we shall not review the corresponding literature here.

Using Theorem 1 and (2), one can compute the moments of the first passage time. Let us make these computations in two simple cases. First, suppose that

$$\beta(x) = \alpha \mathbf{1}_{\{x \in [0,h]}$$

Then almost surely, $W_0 = \alpha$ and the moments of T are given by (with c a positive constant and θ as in Theorem 1):

$$\mathbb{E}T^n = c^n \sum_{l \geq 0} \frac{\theta^l e^{-\theta}}{l!} \frac{\Gamma(1+n)}{\Gamma(1+l+n\alpha)}$$

Next, suppose that

$$\beta(x) = \alpha \mathbf{1}_{\{x \in [0,h)\}} + \mathbf{1}_{\{x \in [h,1]\}}$$

Then almost surely, $B_0 = 1 - \alpha$ and the moments of T are given by:

$$\mathbb{E}T^n = c^n \sum_{m \geq 0} \frac{\theta^m e^{-\theta}}{m!} \frac{\Gamma(1+m+n)}{\Gamma(1+m+n\alpha)}$$

In all other cases however, the computations are more intricate and there are no obvious simplifications.

Apart from exact computations, for which little is known, other results on first passage times for subordinators, regarding in particular the existence of a density or asymptotic estimates, can be found in [5] and references therein.

The remainder of this paper is organized as follows. We first recall the construction of regenerative sets from [8], both in the discrete and continuous case, in Sect. 2. We explain in Sect. 3 how urns are embedded in this construction and how the distributions described in Theorem 1 occur in that context. Finally, we show in Sect. 4 that the embedded urns described in Sect. 3 indeed correspond to first passage times for subordinators.

2 A Construction of Regenerative Sets

In the first two subsections, we recall the construction of regenerative sets given in [8], both in the discrete and in the continuous case. The proof of Theorems 2 and 3 can be found there. The class of regenerative sets obtained in Sect. 2.2 is exactly the class of ranges of complete subordinators, as noted in [1] and [4].

2.1 *The Lattice Case*

We begin by the construction of regenerative sets in $\mathbb{N}$.

Construction 1.

Fix a measurable function $\gamma : [0,1] \to [0,1]$. Let $(X_n, n \geq 1)$ be iid random variables, uniformly distributed on $[0,1]^2$. We denote $X_n = (h_n, U_n)$. One should

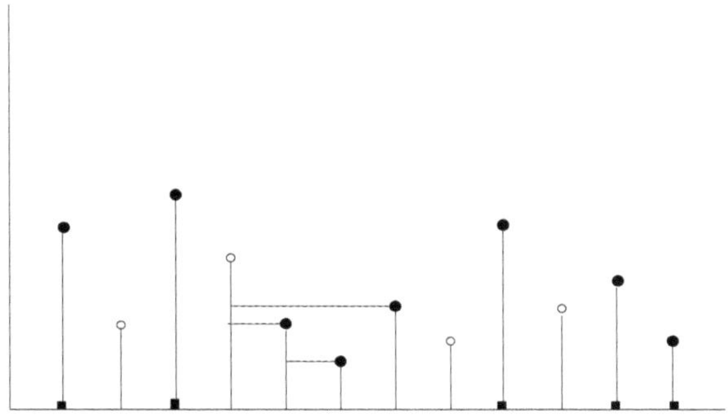

Fig. 1 Construction 1

view h as a height and U as a parameter. Say that X_n is green if $U_n \leq \gamma(h_n)$, and red otherwise. Say that an integer $k \in [1, n]$ is n-visible if $h_k \geq h_m$ for all integers $m \in [k, n]$. Finally, say that n percolates for γ if, for every $k \leq n$ such that k is n-visible, X_k is green. Let $R^{(\gamma)}$ be the set of integers that percolate for γ (by convention, 0 percolates for γ).

See Fig. 1. Green points are represented by black circles, red points by white circles and the black squares stand for the integers that percolate. The horizontal lines express the fact that the red point at 4 prevents 5, 6 and 7 from percolating.

Remark that if γ is a constant, then the X_n are green or red with probability γ (resp. $1 - \gamma$), independently of the height.

Theorem 2 *The set $R^{(\gamma)}$ defined by Construction 1 is a lattice regenerative set. It can be viewed as the image of a random walk $(S_n^{(\gamma)}, n \geq 0)$, where $S_n^{(\gamma)} = Y_1^{(\gamma)} + \ldots + Y_n^{(\gamma)}$, the $Y_i^{(\gamma)}$ being iid random variables taking values in $\mathbb{N} \cup \{\infty\}$, with generating function*

$$\psi^{(\gamma)}(t) = \mathbb{E}(t^{Y_1^{(\gamma)}}) = 1 - \exp\left(-\int_0^1 \frac{t\gamma(x)}{1-tx} dx\right)$$

2.2 The Continuous Case

Consider a Poisson Point process $\mathcal{N}$ on $\mathbb{R}_+ \times [0, 1] \times [0, 1]$ with intensity $dx \otimes y^{-2}dy \otimes dz$. Given a measurable function $\beta : [0, 1] \to [0, 1]$, we can define an analogue of Construction 1 as follows.

Construction 2.

Say that a point $X = (t, h, U)$ of $\mathcal{N}$ is green if $U \leq \beta(h)$, and red otherwise. Say that another point $X' = (t', h', U')$ of $\mathcal{N}$ is visible for X if $t' \leq t$ and if, for all points of $\mathcal{N}$ of the form $X'' = (t'', h'', u'')$ with $t' \leq t'' \leq t$, we have $h' \geq h''$. Finally, say that X percolates for β if, for every X' such that X' is visible for X,

X' is green. By convention, 0 percolates for β. We denote by $\mathcal{R}_1^{(\beta)}$ the set of first coordinates of percolating points, and we set

$$\mathcal{R}^{(\beta)} = \overline{\mathcal{R}_1^{(\beta)}}$$

For every point $X = (t, h, U)$ of $\mathcal{N}$, let $U(X)$ be the set of points of $\mathcal{N}$ of the form $X' = (t', h', u')$ with $t' \leq t$ and $h' \geq h$. Then almost surely, $U(X)$ is finite, since almost surely, every strip of the form $[0, t] \times [h, \infty] \times [0, 1]$ with $h > 0$ contains a finite number of points of $\mathcal{N}$. Moreover, determining whether X percolates only depends on $U(X)$, and therefore Construction 2 is well-defined.

Theorem 3 *The set $\mathcal{R}^{(\beta)}$ defined by Construction 2 is a regenerative set. It can be viewed as the image of a subordinator $(S_t^{(\beta)})_{t\geq 0}$ with Laplace exponent*

$$\phi^{(\beta)}(\lambda) = -\log \mathbb{E}[\exp(-\lambda S_1^{(\beta)})] = \exp \int_0^1 \frac{(\lambda - 1)\beta(x)}{1 + (\lambda - 1)x} dx$$

for $\lambda \geq 0$.

2.3 Relating the Discrete and the Continuous Case

Let $h > 0$. As noted above, if we only look at the points of $\mathcal{N}$ with y-coordinate $\geq h$ in Construction 2, we have a discrete set and we can determine whether these points percolate or not without taking into account the points whose y-coordinate is $< h$. Denote the points with y-coordinate $\geq h$ by

$$(x_1, y_1, U_1), (x_2, y_2, U_2), \ldots$$

with $x_1 < x_2 < \ldots$. From this discrete set, we can recover Construction 1 as follows.

Let $\theta = (1/h) - 1$ and consider the function $F : [h, 1] \to [0, 1]$ defined by

$$F(x) = 1 + \frac{1}{\theta} - \frac{1}{\theta x}$$

Its inverse is the function $F^{-1} : [0, 1] \to [h, 1]$ given by

$$F^{-1}(x) = \frac{1}{\theta + 1 - \theta x}$$

Put $h_n = F(y_n)$ for every $n \geq 1$. Then it is easily seen that the sequence $(h_n, n \geq 1)$ is a sequence of iid random variables, uniformly distributed on $[0, 1]$ and independent of the sequence $(x_n, n \geq 1)$. Therefore the sequence

$$((h_n, U_n), n \geq 1)$$

has the same law as in Sect. 2.1 and is independent of $(x_n, n \geq 1)$.

Consider the function

$$\gamma(x) = \beta(F^{-1}(x)) = \beta\left(\frac{1}{\theta + 1 - \theta x}\right)$$

so that $\beta(x) = \gamma(F(x))$. Then from the sequence (h_n, U_n) and the function γ, we can define a regenerative set R by Construction 1 and we check that $k \in R$ if and only if (x_k, y_k, U_k) percolates by Construction 2. Moreover, Theorem 2 tells us that R is the range of a random walk (S_n) with generating function

$$\psi(t) = 1 - \exp\left(\int_0^1 \frac{t\gamma(x)}{1 - tx}\right)$$

which is the same as in Theorem 1.

3 Embedded Urns

In this section, we use the construction of regenerative sets from Sect. 2.2. We shall always restrict ourselves to the subset of points of $\mathcal{N}$ with x-coordinate ≤ 1.

3.1 An Alternative Description of the Urn

Consider an urn scheme with replacement matrix

$$\begin{pmatrix} 1 & 0 \\ 1 - \alpha & \alpha \end{pmatrix}$$

and initial condition (B_0, W_0). This urn can be described by the following mechanism:

- At time 0, add a black ball with weight B_0 and a white ball with weight W_0.
- Recursively at time $N \geq 1$,

- choose, independently of the past, a random time $t_N \in \{0, \dots N-1\}$ with probability

$$\mathbb{P}(t_N = k) = \frac{1}{N - 1 + B_0 + W_0}$$

if $k \geq 1$ and

$$\mathbb{P}(t_N = 0) = \frac{B_0 + W_0}{N - 1 + B_0 + W_0}$$

- If $t_N = 0$, then at time N, with probability $B_0/(B_0+W_0)$, add a black ball with weight 1 and with probability $W_0/(B_0 + W_0)$, add a black ball with weight $1 - \alpha$ and a white ball with weight α.
- If, at time $t_N \geq 1$, a black ball with weight 1 had been added, then add at time N a black ball with weight 1.
- If, at time $t_N \geq 1$, a black ball with weight $1-\alpha$ and a white ball with weight α had been added, then at time N, with probability $1 - \alpha$, add a black ball with weight 1 and with probability α, add a black ball with weight $1 - \alpha$ and a white ball with weight α.

3.2 *The Stable Case*

We deal here with the case when β is constant and equal to $\alpha \in (0, 1)$. Let us denote the set of points of $\mathcal{N}$, re-arranged by decreasing y-coordinate, as

$$\{(x_1, y_1, U_1), (x_2, y_2, U_2), \dots\}$$

with $y_1 > y_2 \dots$. By convention, set $(x_0, y_0) = (0, \infty)$.

For two integers $N \geq 0$ and $k \in [0, N]$, put

$$z_k^{(N)} = \min(\{x \in \{1, x_0, \dots, x_N\}, x > x_k\})$$

and

$$I_k^{(N)} = (x_k, z_k^{(N)}]$$

In words, $x_0, \dots, x_N$ cut the interval $[0, 1]$ into $N + 1$ subintervals and $I_k^{(N)}$ is the subinterval with left extremity x_k. Denote the lengths of these subintervals

$$l_k^{(N)} = z_k^{(N)} - x_k$$

Let Q_{N+1} be the index of the interval where x_{N+1} lies, that is, put $Q_{N+1} = k$ if k is the (unique) integer $\in [0, N]$ such that $x_{N+1} \in I_k^{(N)}$. From the properties of Poisson point processes, the random variable $(l_0^{(N)}, \ldots, l_N^{(N)})$ is uniformly distributed on the N-dimensional simplex and is independent of the random variables Q_i, $1 \le i \le N$. Therefore, for every $k \in [0, N]$,

$$\mathbb{P}(Q_{N+1} = k | Q_1, \ldots Q_N) = 1/(N+1) \tag{4}$$

Say that $I_k^{(N)}$ percolates if the point (x_k, y_k, U_k) percolates. From Construction 2, we see that the point $(x_{N+1}, y_{N+1}, U_{N+1})$ percolates if and only if $I_{Q_N}^{(N)}$ percolates and $(x_{N+1}, y_{N+1}, U_{N+1})$ is green.

To put it formally, for every $k \in [1, N]$ let V_k be the indicator function that (x_k, y_k) is green and W_k be the indicator function that (x_k, y_k) percolates. Put also $W_0 = 1$. Then we have

$$W_{N+1} = V_{N+1} W_{Q_N} \tag{5}$$

Since the random variables V_n are independent of the random variables Q_n, we can extend (4) by further conditioning on the random variables V_n, W_n:

$$\mathbb{P}(Q_{N+1} = k | Q_1, \ldots Q_N, V_1, \ldots V_N, W_0, \ldots W_N,) = 1/(N+1) \tag{6}$$

Using (6) together with (5), we can describe the law of the family of random variables (W_n) as follows.

- First, $W_0 = 1$.
- Recursively at time $N \ge 1$,
  - choose Q_N uniformly at random on $[0, N]$, independently of the past.
  - If $W_{Q_N} = 0$, then $W_N = 0$.
  - If $W_{Q_N} = 1$, then independently of the past, choose either $W_N = 1$ or $W_N = 0$ with respective probabilities α, $1 - \alpha$.

Comparing with Sect. 3.1, we check that it is exactly the same mechanism as the urn scheme with initial condition $B_0 = 1 - \alpha$, $W_0 = \alpha$. So we can state

Proposition 1 *Let A_n be the number of percolating points in the set*

$$\{(x_1, y_1, U_1), (x_2, y_2, U_2), \ldots (x_n, y_n, U_n)\}$$

Then the sequence (A_n) has the same law as (W_n), where W_n is the total weight of white balls in an urn scheme with replacement matrix

$$\begin{pmatrix} 1 & 0 \\ 1-\alpha & \alpha \end{pmatrix}$$

and initial condition $(1 - \alpha, \alpha)$.

3.3 The General Case

We use here the same assumptions on the function β as in Theorem 1 and we keep the notation from Sect. 3.2.

Let M be the number of points of the process $\mathcal{N}$ with y-coordinate greater that h. Then conditionally on M, the interval $[0, 1]$ is cut into $M + 1$ subintervals. We define percolating and non-percolating subintervals as in the previous subsection and denote by W the number subintervals that percolate. Note that even if $M = 0$, $W = 1$ since by convention, we say that (x_0, y_0) percolates.

Using the same arguments as in Sect. 3.2, we see that the family of lengths of these subintervals, which we can denote by $(l_1, \dots l_{M+1})$, is uniformly distributed on the simplex and that $(x_{M+1}, y_{M+1}, U_{M+1})$ percolates if and only if it is green and x_{M+1} lies in a subinterval which percolates.

Then adding x_{M+1}, we cut $[0, 1]$ into $M + 2$ subintervals and then we can see in which subintervals x_{M+2} and whether the point (x_{M+2}, y_{M+2}) percolates or not. Reasoning this way by induction, as in Sect. 3.2, we see that conditionally on M and W, we get an urn scheme with the same replacement matrix (1) but now the initial condition is $(M + 1 - W\alpha, W\alpha)$.

Proposition 2 *Let A_n be the number of percolating points in the set*

$$\{(x_1, y_1), (x_2, y_2), \dots (x_{M+n}, y_{M+n})\}$$

Then the sequence (A_n) has the same law as (W_n), where W_n is the total weight of white balls in an urn scheme with replacement matrix

$$\begin{pmatrix} 1 & 0 \\ 1-\alpha & \alpha \end{pmatrix}$$

and random initial condition $(M + 1 - W\alpha, W\alpha)$.

It remains to study the joint law of (M, W). First, the law of M is Poisson with mean $\theta = (1/h) - 1$. Next, conditionally on M, using Sect. 2.3, we get that W has the same law as the number of points in $[0, M]$ in the regenerative set R obtained from Construction 1 in Sect. 2.3.

This regenerative set R is the trace of a random walk (S_n) and the generating function of S_1 is the function ψ given in Theorem 1. Conditionally on M, we have

$$\mathbb{P}(W = n+1|M) = \mathbb{P}(Y_n \leq M, Y_{n+1} > M) = \mathbb{P}(Y_n \leq M) - \mathbb{P}(Y_{n+1} \leq M) \quad (7)$$

For each k, we have

$$\mathbb{P}(Y_n = k) = [t^k]\mathbb{E}(t^{Y_n}) = [t^k]\psi(t)^n$$

where $[t^k]f(t)$ stands for the coefficient of the momomial t^k in the function $f(t)$ viewed as a power series. By the theorem of residues,

$$[t^k]\psi(t)^n = \frac{1}{2i\pi}\int_C \frac{\psi(t)^n}{t^{k+1}}$$

Summing over k in (7) yields

$$\mathbb{P}(W = n+1|M) = \frac{1}{2i\pi}\int_C \frac{dt}{t}\psi(t)^n(1-\psi(t))\frac{1-(1/t)^{M+1}}{1-(1/t)}$$

and so finally,

$$\mathbb{P}(M = m, W = n+1) = \frac{\theta^m e^{-\theta}}{m!}\frac{1}{2i\pi}\int_C \frac{dt}{t}\psi(t)^n(1-\psi(t))\frac{1-(1/t)^{m+1}}{1-(1/t)} \quad (8)$$

Comparing (8) with Theorem 1 and using Proposition 1, we can state:

Proposition 3 *The sequence (A_n) from Proposition* 1 *has the same law as the number of white balls in the urn process described in Theorem* 1.

4 Proof of Theorem 1

We assume in this section that the conditions of Theorem 1 are satisfied. We shall use the following property, see for instance [3] or [9] for a recent use of it:

Proposition 4 *Suppose that a sequence of subordinators $S^{(n)}$ which are not compound Poisson converges in law to S. Then the law of the first passage time for $S^{(n)}$ converges in distribution to the law of the first passage time for S.*

Let us go back to the construction of Sect. 2.2. If we only consider the points that percolate and that have a y-coordinate more than $1/n$, this yields a regenerative set $R^{(n)}$ associated with the function $\alpha_n(x) = \alpha(x)\mathbf{1}_{\{x\geq 1/n\}}$. According to Theorem 2, $R^{(n)}$is the range of a subordinator $S^{(n)}$ with exponent

$$\phi^{(n)}(\lambda) = \exp\int_0^1 \frac{(\lambda-1)\alpha(x)}{1+(\lambda-1)x}\mathbf{1}_{\{x\geq 1/n\}}dx$$

Thus $S^{(n)}$ converge to the subordinator S with exponent given in Theorem 1. Using Proposition 4, we get the convergence

$$T_n^{(\beta)} \overset{law}{\to} T^{(\beta)}$$

where $T_n^{(\beta)}$ stands for the first passage time for $S^{(n)}$

Since $S^{(n)}$ has a finite number of jumps inside a finite interval, it is a compound Poisson process. This means that the times between two consecutive jumps are iid, exponentially distributed random variables whose mean is given by

$$m_n = \frac{1}{\phi^{(n)}(\infty)} = \exp\left(-\int_{1/n}^{1} \frac{\alpha(x)}{x} dx\right) \sim cn^{-\alpha}$$

for some constant $c > 0$. Now let $K^{(n)}$ denote the number of jumps of $S^{(n)}$ before exiting from $[0, 1]$. Then conditionally on $K^{(n)} = k$, the first passage time $T_n^{(\beta)}$ has the same law as the sum of k iid, exponentially distributed random variables with mean m_n and variance m_n^2. Using the Chebyshev inequality, we get

$$P(|T_n^{(\beta)} - K^{(n)}m_n| > A|K^{(n)}) \leq \frac{K^{(n)}m_n^2}{A^2} = \frac{K^{(n)}c^2n^{-2\alpha}}{A^2} \tag{9}$$

Next, remark that $K^{(n)}$ is the cardinal of the set $R^{(n)} \cap [0, 1]$, that is, the number of percolating points with y-coordinate greater than $1/n$. Using Proposition 3, we get that $K^{(n)}$ has the same law as the total weight of white balls in the urn scheme described in Theorem 1:

$$\alpha K^{(n)} \stackrel{law}{=} W_{L^{(n)}}$$

where $L^{(n)}$ is the number of points of $\mathcal{N}$ with y-coordinate greater than $1/n$. Note that $L^{(n)}$ is Poisson distributed with mean $n - 1$ and therefore

$$P(|L^{(n)} - n| \geq n^{2/3}) \to 0 \tag{10}$$

as $n \to \infty$. It follows from (10) that $n^{-\alpha}W_{L^{(n)}}$ and $n^{-\alpha}W_n$ have the same limit law, which is also the limit law of $n^{-\alpha}\alpha K^{(n)}$.

Taking $\delta > 0$ and $A = n^{-\alpha/4}$ in (9) yields

$$P(|T_n^{(\beta)} - K^{(n)}m_n| > n^{-\alpha/4}) \leq \mathbb{P}(K^{(n)} > \delta n^{\alpha}) + \frac{\delta}{n^{\alpha/2}} \tag{11}$$

This is true for every $\delta > 0$ and we have seen that

$$\mathbb{P}(K^{(n)} > \delta n^{\alpha})$$

has the same limit as

$$\mathbb{P}(W_n > \delta\alpha n^{\alpha})$$

Since we know that the sequence $(n^{-\alpha}W_n)$ converges, it is tight and therefore, the upper bound in (11) goes to 0 as n goes to infinity. So $T_n^{(\beta)}$ has the same limit law as $K^{(n)}m_n$, that is, the limit law described in Theorem 1. This concludes the proof.

Acknowledgments This research was carried out while I was visiting Mexico during a thematic semester. I thank my mexican colleagues, in particular Geronimo Uribe, Maria Emila Caballero, Victor Rivero, Juan Carlos Pardo-Millan and Arno Siri-Jégousse for their warm welcome and interesting discussions on the topic. I thank the referee for a careful reading of the manuscript. And of course I would like to pay a tribute to Ron Doney, who has been interested in my work since I was a PhD student, invited me to Manchester, was a member of my *habilitation* jury and whose contributions to the study of Lévy processes are an inspiration for all probabilists working in the field.

References

1. Alili, L., Jedidi, W., Rivero, V.: On exponential functionals, harmonic potential measures and undershoots of subordinators. ALEA, Lat. Am. J. Probab. Math. Stat. **11**(2), 711–735 (2014)
2. Bertoin, J.: Subordinators: examples and applications. Lectures on probability theory and statistics (Saint-Flour, 1997), 1–91. Lecture Notes in Mathematics **1717**. Springer, Berlin (1999)
3. Chaumont, L., Doney, R.: Invariance principles for local times at the maximum of random walks and Lévy processes. Ann. Probab. **38**(4), 1368–1389
4. Deng, C., Schilling, R.: Complete Bernstein functions and subordinators with nested ranges. A note on a paper by P. Marchal. Electron. Commun. Probab. **21**(78), 5 (2016)
5. Doney, R., Rivero, V.: Asymptotic behaviour of first passage time distributions for subordinators. Electron. J. Probab. **20**(91), 28 (2015)
6. Janson, S.: Limit theorems for triangular urn schemes. Probab. Theory Relat. Fields **134**(3), 417–452 (2006)
7. Marchal, P.: Nested regenerative sets and their associated fragmentation process. Mathematics and computer science. III, 461–470, Trends Math., Birkhäuser, Basel (2004)
8. Marchal, P.: A class of special subordinators with nested ranges. Ann. Inst. Henri Poincar Probab. Stat. **51**(2), 533–544 (2015)
9. Mijatović, A., Uribe, G.: Invariance principles for local times in regenerative settings. (2019) Preprint
10. Pitman, J.: Combinatorial stochastic processes. Ecole d'Eté de Probabilités de Saint-Flour XXXII – 2002. Lecture Notes in Mathematics 1875. Springer, Berlin (1875)
11. Schilling, R., Song, R., Vondracek, Z.: Bernstein functions. Theory and applications. de Gruyter Studies in Mathematics, **37**. Walter de Gruyter, Berlin (2010)

GPSR Compliance
The European Union's (EU) General Product Safety Regulation (GPSR) is a set of rules that requires consumer products to be safe and our obligations to ensure this.

If you have any concerns about our products, you can contact us on

ProductSafety@springernature.com

In case Publisher is established outside the EU, the EU authorized representative is:

Springer Nature Customer Service Center GmbH
Europaplatz 3
69115 Heidelberg, Germany

www.ingramcontent.com/pod-product-compliance
Ingram Content Group UK Ltd.
Pitfield, Milton Keynes, MK11 3LW, UK
UKHW022226051026
15624UKWH00002B/373
9783030833084